Lecture Notes in Computer Science 1017

Edited by G. Goos, J. Hartmanis and J. van Leeuwen

D0293594

Springer
Berlin
Heidelberg
New York
Barcelona
Budapest
Hong Kong
London
Milan
Paris
Santa Clara
Singapore
Tokyo

Manfred Nagl (Ed.)

Graph-Theoretic Concepts in Computer Science

21st International Workshop, WG '95
Aachen, Germany, June 20-22, 1995
Proceedings

Springer

Series Editors

Gerhard Goos
Universität Karlsruhe
Vincenz-Priessnitz-Straße 3, D-76128 Karlsruhe, Germany

Juris Hartmanis
Department of Computer Science, Cornell University
4130 Upson Hall, Ithaca, NY 14853, USA

Jan van Leeuwen
Department of Computer Science, Utrecht University
Padualaan 14, 3584 CH Utrecht, The Netherlands

Volume Editor

Manfred Nagl
Lehrstuhl für Informatik III, RWTH Aachen
Ahornstr. 55, D-52074 Aachen, Germany

Cataloging-in-Publication data applied for

Die Deutsche Bibliothek - CIP-Einheitsaufnahme

Graph theoretic concepts in computer science : 21th
international workshop ; proceedings / WG '95, Aachen,
Germany, June 20 - 22, 1995 / Manfred Nagl (ed.). - Berlin ;
Heidelberg ; New York ; Barcelona ; Budapest ; Hong Kong ;
London ; Milan ; Paris ; Tokyo : Springer, 1995
  (Lecture notes in computer science ; Vol. 1017)
  ISBN 3-540-60618-1
NE: Nagl, Manfred [Hrsg.]; WG <21, 1995, Aachen>; GT

CR Subject Classification (1991): G.2.2, F.2, F.1.2-3, F.3-4, E.1, I.3.5

ISBN 3-540-60618-1 Springer-Verlag Berlin Heidelberg New York

Typesetting: Camera-ready by author
SPIN 10512279   06/3142 – 5 4 3 2 1 0   Printed on acid-free paper

Preface

The WG '95 21st "International Workshop on Graph-Theoretic Concepts in Computer Science" was held at *Haus Eich, Aachen, Germany,* from *June 20–22, 1995.* It was organized by Lehrstuhl für Informatik III of Aachen University of Technology. Haus Eich is an education institution of the Roman Catholic Church, managed by the Diocese of Aachen. The familiar atmosphere of Haus Eich considerably contributed to the success of this workshop. It facilitated intensive discussions on scientific problems as well as the exchange of personal opinions and interests.

The aim of this workshop series is to contribute to *integration in computer science*. This striving for integration is achieved by applying graph-theoretic concepts. Thereby integration appears in two ways. First, graph-theoretic concepts are applied to various fields of specialization in computer science and thus commonalities between the fields are detected. Second, the workshops aims to combine theoretical aspects with practice and applications; this is achieved either by applying theoretical concepts to practice or by taking up problems from practice and trying to solve them theoretically.

This workshop is thus a *rarity* in computer science, as it is neither purely theoretical nor practical or oriented to applications. It is also a vertical cut through the different fields of computer science in which graphs and graph-theoretic concepts can be applied. On the other hand the workshop has a *tradition:* this volume collects the papers of its twenty-first occurrence (see list below). That alone is remarkable in a field like computer science where new topics appear and disappear quite rapidly.

The *Program Committee* of the WG '95 consisted of

G. Engels, Leiden, Netherlands
L. Kucera, Prague, Czech Republic
J. van Leeuwen, Utrecht, Netherlands
A. Marchetti-Spaccamela, Rome, Italy
E. Mayr, Munich, Germany
R. Moehring, Berlin, Germany
M. Nagl, Aachen, Germany

H. Noltemeier, Wuerzburg, Germany
G. Schmidt, Munich, Germany
P. Spirakis, Patras, Greece
G. Tinhofer, Munich, Germany
P. Widmayer, Zurich, Switzerland
J. Winkowski, Warsaw, Poland

In the Call for Papers contributions were solicited describing original results in the study and application of graph-theoretic concepts in various fields of computer science. From 54 submissions the program committee selected 30 papers after a careful refereeing process. This *selection* reflects several *current research directions* and up-to-date snapshots that are representative for the topic of the workshop. They are a starting point for continued work in computer science oriented graph theory.

The selected papers are related to (1) different underlying *formalisms*, such as directed or undirected graphs, labeled or unlabeled graphs, graphs or hypergraphs. Furthermore,

(2) the following *topics* of graph-theoretic concepts are studied: subgraphs/ isomorphism, graph rewriting, routing, colouring, recognition, embedding/ placement, graph properties, special graph classes, structural graph theory, and graph decomposition. *Techniques* used are (3) getting exact or approximate solutions, static or incremental algorithms, sequential or parallel algorithms, defining them to work globally or to use local properties. Furthermore, in every paper complexity considerations were an important matter. The *applications* of graph-theoretic concepts of this volume are (4) data structures, data bases, geometry, communication, networks of processors, fault tolerance and detection, VLSI, optimization, and graph theory itself.

The papers of this volume are given in a reasonable total ordering according to the dimensions (1), (2), (3), and (4) of above. They are the accepted and revised versions of all papers presented at the workshop. Revisions have been based on comments and suggestions received by the authors from referees and/or participants at the workshop. Several papers are in the form of preliminary reports on ongoing research. More elaborate versions of the papers of this volume will eventually appear in scientific journals. The members of the program committee hope that this volume gives a *good impression* of *current activities* in the field of graph-theoretic concepts of computer science.

The workshop was attended by 57 participants from 14 countries (Australia, Austria, Canada, Czech Republic, France, Germany, Greece, Japan, Italy, Moldavia, the Netherlands, Slovakia, Switzerland, and USA). Their names and addresses are given in a list at the end of this volume. The *success* of the workshop is due to the activeness of *participants* contributing to presentations and discussions during the workshop. Furthermore, a remarkable number of *reviewers* (a list of their names is also given below) was engaged in the refereeing process to select the best papers among the submitted ones. Of course, the quality of this process is mainly due to the *program committee*, namely to the extra load of work they accepted and to the expertise of its members. The editor would like to thank all of them. Finally, the *organization committee* consisting of Roland Baumann, Angelika Fleck, Peter Heimann, Katja Keimer, Peter Klein, Carl-Arndt Krapp, Peter Möckel, Ansgar Radermacher, Anita Wagner, Andreas Winter, Andreas Schürr, and Margot Schürr has done a good job.

Aachen University of Technology (RWTH) in 1995 celebrates its *125th anniversary.* The workshop was part of the official program.

The organizer of this workshop would like to express his gratitude to the *sponsors* of this workshop, namely Deutsche Forschungsgemeinschaft, Ministry for Science and Research of the State Northrhine-Westphalia, and the Rector of Aachen University of Technology. Their support allowed us to finance parts of the travel expenses for some of the participants and a part of the organizational costs of this workshop.

Aachen, Germany, October 1995   M. Nagl

The 21 WGs and Their Organizers

Berlin	1975	U. Pape
Göttingen	1976	H. Noltemeier
Linz	1977	J. Mühlbacher
Castle Feuerstein near Erlangen	1978	M. Nagl, H. J. Schneider
Berlin	1979	U. Pape
Bad Honnef	1980	H. Noltemeier
Linz	1981	J. Mühlbacher
Neunkirchen near Erlangen	1982	H. J. Schneider, H. Göttler
Haus Ohrbeck near Osnabrück	1983	M. Nagl, J. Perl
Berlin	1984	U. Pape
Castle Schwanenberg near Würzburg	1985	H. Noltemeier
Monastery Bernried near München	1986	G. Tinhofer, G. Schmidt
Castle Banz near Bamberg	1987	H. Göttler, H. J. Schneider
Amsterdam	1988	J. van Leeuwen
Castle Rolduc near Aachen	1989	M. Nagl
Johannesstift Berlin	1990	R. H. Möhring
Judges' Home Fischbachau n. München	1991	G. Schmidt, R. Berghammer
W.–Kempf–Haus, Wiesbaden–Naurod	1992	E. W. Mayr
Sports Center Papendal near Utrecht	1993	J. van Leeuwen
Herrsching near München	1994	G.Tinhofer, E.W.Mayr, G. Schmidt
Haus Eich at Aachen	1995	M. Nagl

List of Reviewers

P. Altimonti, Rome
L. Babel, Munich
A. Bayer, Munich
St. Bylka, Warsaw
H. L. Bodlaender, Utrecht
A. Corradini, Pisa
F. d'Amore, Rome
R. De Nicola, Rome
N. Dendris, Patras
H. Ehrig, Berlin
G. Engels, Leiden
S. Fei, Zurich
M. Flammini, Rome
P.G. Franciosa, Rome
P. Giaccio, Rome
J. Gustedt, Berlin
A. Habel, Hildesheim
St. Hartmann, Berlin
C. Hattensperger, Munich
F. Huber–Wäschle, Zurich
E. Ihler, St. Augustin
W. Kahl, Munich
V. Kapoulas, Patras
P. Kempf, Munich
L. Kirousis, Patras
J. N. Kok, Leiden
B. Konikowska, Warsaw
P. Koopmann, Leiden
H.-J. Kreowski, Bremen
S.O. Krumke, Würzburg
L. Kucera, Prague
J. van Leeuwen, Leiden
M. Lenzerini, Rome
S. Leonardi, Rome
R. McConnell, Amherst
E. Malesinska, Berlin
A. Marchetti–Spaccamela, Rome
E. W. Mayr, Munich
A. Mazurkiewicz, Warsaw
R. Möhring, Berlin

M. Müller–Hannemann, Berlin
M. Nagl, Aachen
S. Nokoletseas, Patras
H. Noltemeier, Würzburg
E. Nuutila, Helsinki
W. Oberschelp, Aachen
R. Pajarola, Zurich
F. Parisi Presicce, Rome
A. Parra, Berlin
A. Radermacher, Aachen
J. Rekers, Leiden
T. Roos, Zurich
F. Rossi, Pisa
M. Schäffter, Berlin
P. Scheffler, Berlin
I. Schiermeyer, Cottbus
F. Schmalhofer, Munich
G. Schmidt, Munich
L. Schmitz, Munich
A. S. Schulz, Berlin
A. Schürr, Aachen
P. Spirakis, Patras
Y. Stamatiou, Patras
U. Stege, Zurich
B. Tampakas, Patras
B. Thalheim, Cottbus
D. Thilikos, Patras
G. Tinhofer, Munich
B. Triantafillou, Patras
P. Trunz, Zurich
K. Verbarg, Würzburg
L. Vismara, Rome
E. Wanke, Düsseldorf
P. Widmayer, Zurich
E. Welzl, Berlin
B. Westfechtel, Aachen
J. Winkowski, Warsaw
A. Winter, Aachen
M. Winter, Munich
A. Zündorf, Paderborn

Contents

VC-Dimensions for Graphs
(Extended Abstract)

Evangelos Kranakis[1*], Danny Krizanc[1] Berthold Ruf[3**] , Jorge Urrutia,[2] and Gerhard J. Woeginger[4***]

[1] Carleton University, School of Computer Science, Ottawa, ON, K1S 5B6, Canada, {kranakis,krizanc}@scs.carleton.ca
[2] University of Ottawa, Department of Computer Science, Ottawa, ON, K1N 9B4, Canada,jorge@csi.uottawa.ca
[3] TU Graz, Institute of Theoretical Computer Science, Klosterwiesgasse 32/2, A-8010 Graz, Austria, bruf@igi.tu-graz.ac.at
[4] TU Graz, Department of Mathematics, Kopernikusgasse 24, A-8010 Graz, Austria.woe@fmatbds01.tu-graz.ac.at

Abstract. We study set systems over the vertex set (or edge set) of some graph that are induced by special graph properties like clique, connectedness, path, star, tree, etc. We derive a variety of combinatorial and computational results on the VC (Vapnik-Chervonenkis) dimension of these set systems.

For most of these set systems (e.g. for the systems induced by trees, connected sets, or paths), computing the VC-dimension is an NP-hard problem. Moreover, determining the VC-dimension for set systems induced by neighborhoods of single vertices is complete for the class LOGNP. In contrast to these intractability results, we show that the VC-dimension for set systems induced by stars is computable in polynomial time. For set systems induced by paths, we determine the extremal graphs G with the minimum number of edges such that $VC_\mathcal{P}(G) \geq k$. Finally, we show a close relation between the VC-dimension of set systems induced by connected sets of vertices and the VC dimension of set systems induced by connected sets of edges; the argument is done via the line graph of the corresponding graph.

1 Introduction

The *Vapnik-Chervonenkis-dimension* of a set system dates back to a seminal paper by Vapnik and Chervonenkis [13] in 1971 on the uniform convergence of relative frequencies of events to their probabilities. It is defined as follows. For $\mathcal{F}$

* Research supported in part by NSERC (National Science and Engineering Research Council of Canada) grant.
** Research supported in part by a grant from the DAAD (German Academic Exchange Service).
*** Research supported by the Spezialforschungsbereich F 003 "Optimierung und Kontrolle", Projektbereich Diskrete Optimierung.

a family of subsets of a finite set X and $D \subseteq X$, set D is said to be *shattered* by $\mathcal{F}$ iff any subset of D is of the form $D \cap F$ for some $F \in \mathcal{F}$. The Vapnik-Chervonenkis (or VC, for short) *dimension* of $\mathcal{F}$ is the maximum size of a subset of X that is shattered by $\mathcal{F}$. In the meantime, the VC-dimension has proved useful in many areas as in probability theory, in learnability theory (PAC-learnable concept classes can be characterized via the VC-dimension, cf. Blumer, Ehrenfeucht, Haussler, and Warmuth [2]) and in computational geometry (geometric range spaces allow linear sized data structures with sublinear query time iff their VC dimension is finite, cf. Chazelle and Welzl [4]).

Papadimitriou and Yannakakis [10] investigated the computational complexity of computing the VC-dimension. Since $VC(\mathcal{F}) \leq \log(|\mathcal{F}|)$ holds, the VC-dimension can be computed in $O(|X|^{\log(|\mathcal{F}|)})$ time by simply checking all subsets of X of cardinality $\leq \log(|\mathcal{F}|)$. This indicates that the problem is not NP-complete. To provide stronger evidence against NP-completeness, Papadimitriou and Yannakakis introduced the complexity class LoGNP and proved that the following problem is LoGNP-complete: Given a family $\mathcal{C}$ of c sets over a set X (by explicit enumeration of all sets in the family) and an integer k, is the VC-dimension of $\mathcal{C}$ at least k ? The class LoGNP is sandwiched between P and NP, $P \subseteq$ LoGNP $\subseteq$ NP, and the general belief is that both inclusions are proper. Hence, with high probability LoGNP-complete problems are neither NP-complete nor solvable in polynomial time.

A special class of set systems arises in connection with graphs. Haussler and Welzl [6] introduced the VC-dimension of a graph as an example in their study of simplex-range queries with epsilon nets. Their definition is as follows. For $G = (V, E)$ a simple, loopless, undirected graph with vertex set V and edge set E, the closed neighborhood $N(v)$ of a vertex $v \in V$ is the set consisting of the vertex v together with all vertices adjacent to it. A set $D \subseteq V$ of vertices is called shattered if it is shattered by the family $\mathcal{F}_{nbd} = \{N(v) : v \in V\}$ of neighborhoods of G (in the sense of the above definition of shatteredness). Since a graph has as many neighborhoods as it has vertices, its VC-dimension clearly is at most $\log |V|$. Anthony et al [1] study this VC-dimension in more detail and show that the threshold probability for a random graph to have VC-dimension $\geq d$ is about $p = n^{-1/d}$, for d sufficiently large, where n is the number of vertices of the graph.

1.1 Results of the paper

The VC-dimension of a graph as defined by Haussler and Welzl is defined via subsets of V that are neighborhoods of single vertices. It is natural to investigate a more general concept where the VC dimension results from set systems induced by other properties on sets of vertices as e.g. cliques, connected sets, paths, stars, trees, etc. In this paper we will introduce and study the VC-dimensions for all these properties.

Connectedness. We will study in detail the VC-dimension for set systems induced by *connected sets* and show that for a given graph, the maximum size of a shattered set for the connectedness property differs by at most one from the number of leaves in a maximum leaf spanning tree. Hence, we can approximate this VC-dimension by applying the approximation algorithms for maximum leaf spanning trees derived by Lu and Ravi [11]. Moreover, it can be shown that computing the VC-dimension for set systems induced by *connected sets* is NP-complete.

The reader should note that the LogNP-completeness complexity result derived by Papadimitriou and Yannakakis [10] is *not* in contradiction to our NP-completeness result: [10] considered a problem where the input is given by explicit enumeration of all sets, whereas in our case the input is implicitly described via the graph and hence a potentially exponential number of sets (all connected subsets of the graph) is encoded by a structure of polynomial size (edge and vertex set of the graph).

Paths. Computing the VC-dimension of set systems induced by *paths* can also be proved to be NP-hard. We give a complete combinatorial characterization of the graphs for which this VC-dimension equals three, and we provide upper and lower bounds on the number of edges in terms of the number of vertices and the VC-dimension.

Neighborhoods and stars. In contrast to the two NP-hardness results above, computing the VC-dimension for set systems induced by *neighborhoods* is LogNP-complete. The computation of the VC-dimension for set systems induced by *stars* can be done even in polynomial time.

Connected sets of edges. Finally, we will study shattering principles for families of edge-sets and the corresponding *edge-VC (or EVC for short) dimension* of graphs. We will show that the EVC-dimension for set systems over edges induced by connected edge sets in some graph G is related in a specific way to the VC-dimension for set systems over vertices induced by connected vertex sets in the corresponding line graph of G. Moreover the problem of computing the EVC dimension for connected sets is shown to be NP-complete.

1.2   Organization of the paper

The paper is organized into sections as follows. Section 2 introduces several general concepts and gives all basic definitions. Section 3 deals with the VC-dimension resulting from neighborhoods and stars, and Section 4 with the VC dimension resulting from connected sets and trees. Section 5 treats the VC-dimension for paths and Section 6 states the results on the VC-dimension for edges, and Section 7 finishes the paper with the conclusion.

2   VC-dimensions for Vertices

In this section we give precise definitions for the notion of the VC-dimension of a graph G with respect to certain graph properties. Let $G = (V, E)$ be a graph with vertex set V and edge set E. Let $\mathcal{P}$ be a family of subgraphs of G. Typical choices for $\mathcal{P}$ include families of subgraphs which are cliques, connected sets, neighborhoods, paths, trees, etc.

Definition 2.1 *Let $\mathcal{P}$ be a family of subgraphs of G. We say that a subset $A \subseteq V$ is $\mathcal{P}$-shattered if and only if for all $B \subseteq A$ there exists a subgraph in $\mathcal{P}$ on a set of vertices $C \subseteq V$ such that $B = C \cap A$. Then the VC-dimension with respect to $\mathcal{P}$ of G is defined by*

$$VC_{\mathcal{P}}(G) = \max\{|A| : A \text{ is } \mathcal{P} - shattered\} \tag{1}$$

Thus, depending on the family $\mathcal{P}$, we get the following notions of VC-dimensions: VC_{con}, VC_{path}, VC_{star}, VC_{tree}, VC_{cycle}, VC_{nbd}, for the properties connected, path, star, tree, cycle, neighborhood, respectively.

Note that the VC-dimensions as defined in Anthony et al. [1] is the same as our VC_{nbd}. The following examples might be helpful for a better understanding of these definitions.

Example 2.1 *For the complete graph K_n on n vertices, $VC_{con}(K_n) = n$ holds since any subset of the vertices is connected. For a path P_n on $n \geq 2$ vertices, we get $VC_{con}(P_n) = 2$: A set of 3 vertices cannot be shattered since it is impossible to connect the outer two vertices without using the inner vertex. By the same argument we see that $VC_{tree}(P_n) = VC_{path}(P_n) = 2$. For a cycle C_n on $n \geq 3$ vertices, it can be checked that $VC_{con}(C_n) = 3$.* ∎

Lemma 2.1 *If $\mathcal{P} \subseteq \mathcal{P}'$ then $VC_{\mathcal{P}}(G) \leq VC_{\mathcal{P}'}(G)$.* ∎

The problem of computing the $VC_{\mathcal{P}}$-dimension of a graph for a given graph property $\mathcal{P}$ can be formulated as the following decision problem.

Problem $VC_{\mathcal{P}}$:
Instance: A graph $G = (V, E)$, a positive integer $k \leq |V|$.
Question: Is there an $A \subseteq V$ with $|A| \geq k$ such that for all $B \subseteq A$ there is a subgraph $G' = (V', E')$ of G having property $\mathcal{P}$ such that $B = V' \cap A$?

Computing the VC-dimension for some graph property $\mathcal{P}$ is sometimes equivalent to well-known problems studied in complexity theory. For example, if $\mathcal{P}$ is the family of *cliques*, or the family of *independent sets*, it is easy to verify that $VC_{clique}(G)$ (respectively, $VC_{independent}(G)$) equals the size of the largest clique (respectively, largest independent set) in the graph G. It is well-known that both of these problems are NP-complete (see e.g. Garey and Johnson [5]).

3 Neighborhoods and Stars

In this section, we investigate the VC-dimension for *neighborhoods* and for *stars*. The notion of VC-dimension for neighborhoods was introduced by Haussler and Welzl [6].

Theorem 3.1 *There is an $O(\min\{n^2 2^d, n^{\log n}\})$ algorithm for computing a maximum size set of vertices shattered by neighborhoods in a graph $G = (V, E)$, where $|V| = n$ and d is the maximum degree of G. Hence, this algorithm is polynomial for maximum degree $d = O(\log n)$.*

PROOF (OUTLINE) A maximum size shattered set must be a subset of a neighborhood. There are as many neighborhoods as vertices, i.e. n, and each neighborhood has at most 2^d subsets. We can test if a given set of size $\leq d$ is shattered by neighborhoods in time $O(n2^d)$. This gives the $O(n^2 2^d)$ upper bound. Since $\mathrm{VC}_{nbd}(G) \leq \log(n)$ the $n^{\log n}$ upper bound is obvious. ∎

For the complexity of computing VC_{nbd} in general graphs (without bounded maximum degree) we get the following result: (for definition of LOGNP-completeness see page 2 of this paper and [10]).

Theorem 3.2 *It is LOGNP-complete to decide for a given graph $G = (V, E)$ and an integer k, whether $\mathrm{VC}_{nbd}(G) \geq k$ holds.*

PROOF See [9]. ∎

Next, we deal with the VC-dimension of set systems induced by stars. We start with a precise definition of the term 'star'.

Definition 3.1 *Given a graph $G = (V, E)$. For any vertex $u \in V$, a star of u is a subset of $\{v \mid (u, v) \in E\} \cup \{u\}$. An open-star of u is a subset of $\{v \mid (u, v) \in E\}$.*

Theorem 3.3 *If G is a graph with maximum degree d then*

1. $d \leq \mathrm{VC}_{star}(G) \leq d + 1$,
2. $d = \mathrm{VC}_{open-star}(G)$.

PROOF Let $u \in V$ denote a vertex of degree d. Then the neighborhood of u, excluding u itself, is star-shattered. Depending on the graph, u might be added to the shattered set. This shows part 1 of the theorem. Part 2 is trivial. ∎

Theorem 3.4 *There is an $O(nd^2)$ algorithm for computing a maximum size set of vertices shattered by stars, for an arbitrary graph with n vertices and maximum degree d.*

PROOF (OUTLINE) Let G be an arbitrary graph with maximum degree d. For each vertex x, let $N'(x)$ (respectively, $N(x)$) be the open (respectively, closed) neighborhood of x, i.e. the set of vertices adjacent to x ($N'(x) \cup \{x\}$). Consider the set M of vertices of G of maximum degree. For each $x \in M$, the maximum size shattered set is either $N'(x)$ or $N(x)$. It is clear that every subset B of $N(x)$ such that $x \in B$ is shattered. It is therefore sufficient to find a "polynomial" condition guaranteeing that $N'(x)$ as well is shattered.

Let $x \in M$ be fixed. Define $X_0 := N'(x) := \{x_1, x_2, \ldots, x_d\}$ to be the set of neighbors of x. The idea of the algorithm is as follows. Check if there is a vertex $u \neq x$ such that $N'(x) = N'(u)$. If yes, the whole set $N(x)$ is shattered and

$VC_{star}(G) = d + 1$. If no, then compute the set $S_0 := \{v \in X_0 : X_0 \subseteq N(v)\}$. If S_0 is empty then $N(x)$ is not shattered. If $S_0 \neq \emptyset$ then any set $B \subseteq N'(x)$ such that $B \cap S_0 \neq \emptyset$ is shattered. Hence, it is enough to test whether or not all subsets of the set $X_1 = N'(x) \setminus S_0$ are shattered. Now replace X_0 by X_1 above and repeat.

Clearly, on any given input x the maximum number of iterations is d. Each step may take time $O(n)$. The above algorithm must be executed on all vertices of maximum degree d. It is easy to check that its complexity is $O(n^2 d)$.

We can improve this complexity to $O(nd^2)$ by using a more sophisticated data structure for testing "neighborhood equality", namely whether or not $N'(x) = N'(u)$, for $x \neq u$. The idea is to look at the adjacency matrix of the graph. Now neighborhood equality corresponds to equality of two rows of the adjacency matrix and can be tested in time linear in the number of edges of the graph, which is $O(nd)$. Since the above algorithm requires $O(d)$ iterations, the proof of the theorem is complete. ∎

4 Connected Sets and Trees

In this section, we show that VC_{tree} and VC_{con} are identical. We also investigate the close relationship between VC_{con} and the so-called maximum leaf spanning tree. Unless otherwise specified, in this section we will deal with VC dimensions for *connected* sets. Thus when speaking of a shattered set we always mean that it is shattered by connected sets.

Lemma 4.1 *For any graph* $G = (V, E)$, $VC_{tree}(G) = VC_{con}(G)$ *holds.*

PROOF In view of Lemma 2.1 it is sufficient to show that $VC_{tree}(G) \geq VC_{con}(G)$. Consider a set $A \subseteq V$ which is shattered by connected subsets of G. Then for every $B \subseteq A$ there exists a connected set C with $B = C \cap A$. Replace the edges that connect C by a subset of these edges forming a spanning tree for C. The claim follows. ∎

Lemma 4.2 *For any tree* T *with* l *leaves,* $VC_{con}(T) = l$.

PROOF Let L denote the set of leaves of T. For any $B \subseteq L$ we can find a subtree of T whose leaves are exactly the elements of B. Thus L is shattered and $VC_{con}(T) \geq l$. For proving $VC_{con}(T) \leq l$, we consider an arbitrary set A which is shattered by connected sets of G. For each vertex u of T let $u_1, \ldots, u_k$ denote the children of u and $T_1, \ldots, T_k$ the corresponding subtrees of T rooted at $u_1, u_2, \ldots, u_k$. If $u \in A$ then there can be at most one index i, $1 \leq i \leq k$, with $T_i \cap A \neq \emptyset$ (two vertices $x \in T_i \cap A$ and $y \in T_j \cap A$ with $i \neq j$ cannot be connected without using u). ∎

There is a nice characterization of the VC_{con} dimension by relating it to maximum leaf spanning trees (abbreviated MLST). A *maximum leaf spanning tree* is a spanning tree with a maximum number of leaves among all spanning trees.

Definition 4.1 *For any arbitrary graph G let*

$$l(G) := \max\{k \mid \text{there exists a spanning tree } T \text{ of } G \text{ with } k \text{ leaves}\}.$$

Theorem 4.1 *For any graph G*

$$l(G) \leq \text{VC}_{con}(G) \leq l(G) + 1. \tag{2}$$

PROOF The inequality in the lefthand side follows from Lemma 4.2. To prove the inequality in the righthand side, consider a shattered set A of maximum cardinality. We show that there exists a spanning tree T with at least $|A| - 1$ leaves. Choose any vertex $r \in A$ as the root of T. Since A is shattered, there exists a path in G between any two vertices in A avoiding all the other vertices in A. Connect all $v \in A$, $v \neq r$ by these paths to r. This yields a connected subgraph G' of G where all vertices in $A \setminus \{r\}$ are of degree one. Destroy all cycles in G' by removing appropriate edges while keeping the subgraph connected. This eventually results in a tree T with $|A| - 1$ leaves. So far T is not necessarily a spanning tree. While there exist vertices not connected to the subgraph, perform the following procedure: Find an edge between some vertex that belongs to the subgraph and another vertex that does not belong to the subgraph and add it to the subgraph. This procedure cannot decrease the number of leaves in the tree (the just connected vertex always is a leaf). Finally, we will end up with a spanning tree with $|A| - 1$ leaves. ∎

A possible characterization when $\text{VC}_{con}(G)$ and $l(G)$ differ by exactly one is given in [9].

Theorem 4.2 *It is NP-complete to decide for an input consisting of some graph $G = (V, E)$ and a number $k \geq 1$, whether $\text{VC}_{con}(G) \geq k$ holds.*

PROOF First we show that the VC-problem is in NP: Just guess a subset $A \subseteq V$ with $|A| \geq k$. It is easy to see that one can check in polynomial time whether A is shattered: Test for all $O(n^2)$ possible pairs $a, b \in A$ if they are connected by a path avoiding the other vertices in A.

NP-hardness is proved by a transformation from the MINIMUM SET COVER problem [5], which is defined as follows: Given a finite set $S = \{a_1, \ldots, a_n\}$, a collection of m subsets $S_1, \ldots, S_m \subseteq S$ and an integer $t \leq m$, one wants to know whether there exists an index set $I \subseteq \{1, \ldots, m\}$ such that $|I| \leq t$ and $\bigcup_{i \in I} S_i = S$.

Consider the following graph $G = (V, E)$ for a given instance of the MINIMUM SET COVER problem: The set of vertices is given by four pairwise disjoint sets A, B, C and D with $V = A \cup B \cup C \cup D$ (see figure 1 for an illustration). A has $n \cdot (m + 1)$ vertices, arranged in n columns of $m + 1$ vertices each. $B = \{v_1, \ldots, v_m\}$, where v_i corresponds to the set S_i. C consists of only one vertex and D of $m + 2$ vertices. The vertices are connected as follows: The vertex from C is directly connected to all vertices in B and D. There are edges between $v_i \in B$ and all vertices of the jth column of A iff $a_j \in S_i$. Note that $|V| = mn + 2m + n + 3$.

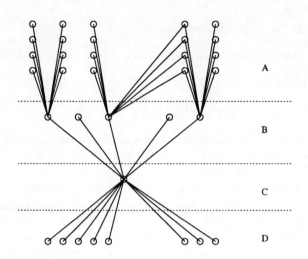

Fig. 1. Construction of the graph

We claim that $\mathrm{VC}_{con}(G) \geq |V| - (t+1)$ if and only if the instance of MINIMUM SET COVER has a solution. (If). Assume that there exists a solution $I \subseteq \{1, \ldots, m\}$ of the MINIMUM SET COVER instance. Then the set $V' = D \cup A \cup \{v_i \in B \mid i \notin I\}$ is shattered in G: Every vertex in V' is connected to C by a path that avoids V'. This trivially holds for the vertices in B and in D. For every vertex in A, there is an adjacent vertex in $v_i \in B \setminus V'$ (since the corresponding element in S is contained in some S_i with $i \in I$) and thus it is connected to C via this vertex v_i. Every subset of V' can be covered by the corresponding set of paths. This set of paths is connected (via the vertex in C) and avoids $V \setminus V'$.

(Only if). Consider a shattered set V' with $V' \geq |V| - t - 1$. How does V' look like? The single vertex in C cannot be in V' since otherwise none of the $m + 2 > t + 1$ vertices in D could be included in V'. Hence, we may assume w.l.o.g. that $D \subseteq V'$. Since all elements in any fixed column in A have identical neighborhoods, we may assume that either all or no elements from any column are in V'. Since every column contains $m + 1$ vertices, it follows that *all* elements in *all* columns have to be in V' (otherwise at least $m + 2 \geq t + 2$ vertices would be outside of V'). Now $A \subseteq V'$, $D \subseteq V'$ and $C \nsubseteq V'$, and consequently at least $m - t$ vertices $\in B$ have to be in V'. Since V' is shattered, for every vertex in A there must exist an adjacent $v_i \in B \setminus V'$. With this it is straightforward to see that $I = \{i | v_i \in B \setminus V'\}$ constitutes a solution for the given instance of the MINIMUM SET COVER problem. ∎

5 Paths

We derive three types of results in this section. First, we give a precise characterization of all graphs fulfilling $\mathrm{VC}_{path}(G) = 3$. Then we derive upper and lower bounds for the number $|E|$ of edges in terms of the number $|V|$ of vertices and the VC-dimension $VC_{path}(G)$. Finally, we deal with the computational complexity of computing VC_{path}. This problem is NP-hard in general, but it can be solved in polynomial time if the dimension is not part of the input.

First we characterize those connected graphs $G = (V, E)$ with $VC_{path}(G) = 3$ (observe that $VC_{path}(G) = 2$ iff G is a tree).

Theorem 5.1 *The graphs G having $VC_{path}(G) = 3$ are the graphs depicted in Figure 2, where from each of the vertices may emanate trees and the cycles depicted in the right hand side are adjacent on a single edge.*

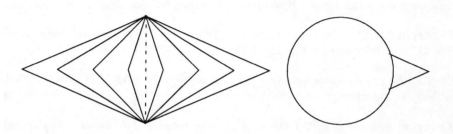

Fig. 2. Graphs G with $VC_{path}(G) = 3$.

PROOF (OUTLINE) See [9]. ■

VC_{path} of a graph can be easily related with the number of its edges. For each $k \leq n$ let e_k be the minimal number of edges of a connected graph G with $VC_{path}(G) \geq k$. It is clear that $e_2 = n - 1$ and $e_n = n(n-1)/2$.

Theorem 5.2 $e_{n-k} = \Theta(n^2/k)$ *holds.*

PROOF (OUTLINE) We proceed in two steps. First, we prove that $e_{n-k} \leq c_1 n^2/k$ for some appropriate positive constant c_1 and then we prove $e_{n-k} \geq c_2 n^2/k$ for another positive constant c_2. Both constants c_1 and c_2 can be made arbitrarily close to $\frac{1}{2}$.

For the upper bound, consider k complete graphs, where for all $i \in \{1, ..., k-1\}$ the ith and the $(i+1)$st graph have exactly one vertex in common. The first $k - 1$ graphs each consist of $\lfloor n/k \rfloor$ vertices; the kth graph has $n - (k-1)\lfloor n/k \rfloor$ vertices It is easy to see that for such a graph G, $VC_{path}(G) = n - k$, where the total number of edges in G is $O(n^2/k)$. To prove the lower bound, we first argue that a graph $G = (V, E)$ with $VC_{path}(G) = n - k$ cannot contain an independent set of size $k + 2$: Suppose otherwise. Then let $I \subseteq V$, $|I| = k + 2$, denote an independent set, let $A \subseteq V$, $|A| = n - k$, denote a set that is shattered by paths and define $A' = A \cap I$, $|A'| \geq 2$. Since A is shattered, there exists a path that contains A' but no other vertices of A. Since A' is independent, there must be at least $|A'| - 1$ vertices on this path that neither belong to I nor to A. Now G contains these $|A'| - 1$ vertices that are neither in I nor in A, $n - k$ vertices in A, and $k + 2 - |A'|$ vertices in $I \setminus A$. Alltogether, this yields that there are at least $n + 1$ vertices, a contradiction.

Next we apply a celebrated theorem of Turán [12] that essentially states that a sparse graph contains a large independent set (see pp. 292–295 [3] for a modern discussion with extensions): Every graph contains an independent set of size at least $n^2/(2|E| + n)$. Combining this with the above discussion yields $k + 2 > n^2/(2|E| + n)$, which in turn gives $|E| = \Omega(n^2/k)$. ∎

We do not know whether computing VC_{path} is in the class NP (since there is an exponential number of conditions that have to be checked in order to verify that some set is shattered). Hence, we only prove NP-hardness of the problem.

Theorem 5.3 *It is NP-hard to decide for an input consisting of some graph $G = (V, E)$ and a number $k \geq 1$, whether $VC_{path}(G) \geq k$ holds.*

PROOF The proof is done by a transformation from the HAMILTONIAN PATH problem in bipartite graphs (for details see [9]). ∎

Theorem 5.4 *For any fixed number k (that is not part of the input), the problem of deciding whether $VC_{path}(G) \geq k$ for some input graph $G = (V, E)$ is solvable in polynomial time.*

PROOF This problem is a special case of the FIXED-VERTEX SUBGRAPH HOMEOMORPHISM(H) problem, which is solvable in polynomial time (see [9]). ∎

6 VC Dimensions for Edges

So far we considered the VC-dimensions in a graph only for vertices. Next, we give an analogous definition for edges.

Definition 6.1 *Let $G = (V, E)$ be a graph and let $\mathcal{P}$ be a family of sets of edges of the graph. We say that a subset $A \subseteq E$ is $\mathcal{P}$-edge-shattered (or shattered by sets of edges) if and only if for all $B \subseteq A$ there exists a set $C \subseteq E$ satisfying property $\mathcal{P}$ such that $B = C \cap A$. Then the EVC-dimensions of G with respect to $\mathcal{P}$ are given by*

$$EVC_{\mathcal{P}}(G) := \max\{|A| : A \text{ is P-edge-shattered }\} \tag{3}$$

Defining the EVC-dimensions for connectedness, trees, paths and stars yields $EVC_{con}(G)$, $EVC_{tree}(G)$, $EVC_{path}(G)$ and $EVC_{star}(G)$. If clear from the context we will simply say "shattered" instead of "edge-shattered".

Lemma 2.1 also holds for the EVC-dimensions. However EVC_{tree} and EVC_{con} are not necessarily equal. For example $EVC_{tree}(C_3) = 2$, but $EVC_{con}(C_3) = 3$, where C_3 is the cycle on three vertices. For $n \geq 3$, $\text{EVC}_{tree}(C_n) = \text{EVC}_{con}(C_n)$. Also, as in Example 2.1, we get that $\text{EVC}_{con}(P_n) = 2$. However for complete graphs the situation is quite different as the following theorem indicates.

Theorem 6.1 *Let K_n be the complete graph on n vertices, $n \geq 5$. Then*

$$\text{EVC}_{con}(K_n) = \frac{n(n-1)}{2} - (n-2).$$

PROOF First we show that $\text{EVC}_{con}(K_n) \geq \frac{n(n-1)}{2} - (n-2)$. Consider a set $B = \{e_1, \ldots, e_{n-1}\}$ of all edges being adjacent to one arbitrary vertex u. Let A denote the set of all remaining edges of K_n. Clearly $|A| = \frac{n(n-1)}{2} - (n-1)$. For any subset $\{c_1, \ldots, c_k\} \subseteq A$ we can find for every c_i an edge $e_{j_i} \in B$ connecting c_i to u. Now we can choose $C = \{c_1, \ldots, c_k\} \cup \{e_{j_1}, \ldots, e_{j_k}\}$ and we get $C \cap A = \{c_1, \ldots, c_k\}$. Thus A is shattered. However we can add one more edge to A because in the construction above we have for every c_i two choices for e_{j_i}. Thus one element from B can be moved to A.

Next we show that a set A of cardinality greater than $\frac{n(n-1)}{2} - (n-2)$ cannot be shattered. Suppose $|A| \geq \frac{n(n-1)}{2} - (n-3)$ holds and consider the complement B of A, $B = E \setminus A$. Since $|B| \leq n-3$, the corresponding subgraph (V, B) of K_n has at least three connected components. We distinguish two cases: In case there are at least four connected components, select an edge e_1 that connects two distinct components and an edge e_2 that connects two other components. Clearly, the subset $A' = \{e_1, e_2\}$ of A is not shattered. In case there are exactly three connected components $C_1 = (V_1, E_1)$, $C_2 = (V_2, E_2)$ and $C_3 = (V_3, E_3)$, at least one of them, say C_1, spans three vertices. If C_1 is not a complete subgraph we may choose an edge e_1 over V_1 that is not in E_1 and an arbitrary edge e_2 that connects C_2 to C_3. Similarly as above one sees that $A' = \{e_1, e_2\}$ is not shattered. Finally, if C_1 is a complete subgraph, $|E_1| \geq |V_1|$. Trivially, $|E_2| \geq |V_2| - 1$ and $|E_3| \geq |V_3| - 1$ holds. Summarizing, this yields $|E_1| + |E_2| + |E_3| \geq n - 2$ which is in contradiction to $|B| \leq n - 3$. ∎

For any connected graph, the EVC-dimensions for trees and connected sets can lie only within a small interval. For example it can be shown that for any graph G with n vertices, $n - 1 \geq \text{EVC}_{tree}(G) \geq \text{VC}_{tree}(G) - 1$ and $|E| \geq \text{EVC}_{con}(G) \geq |E| - (n-1)$. For details see [9].

The precise relationship between the VC-dimensions and the EVC-dimensions can be characterized via line graphs. For a good overview on line graphs see e.g. [7]. We say $L(G) = (V^*, E^*)$ is the line graph of $G = (V, E)$, if $V^* = E$ and $E^* = \{\{e_1, e_2\} | e_1, e_2 \in E$ *and* e_1, e_2 *have a common vertex*$\}$

Now it is easy to show that the EVC_{con}-dimensions of a graph are equal to the VC_{con}-dimensions of its line graph.

Theorem 6.2 *If $L(G)$ is the line graph of a graph G then*

$$\text{VC}_{con}(L(G)) = \text{EVC}_{con}(G)$$

PROOF We only show that $\mathrm{VC}_{con}(L(G)) \leq \mathrm{EVC}_{con}(G)$. The $\geq$ case can be done in a very similar way.

Consider a set $A^* \subseteq V^*$, which is shattered in $L(G)$. This means by definition that for all $B^* \subseteq A^*$ there exists a connected C^* with $B^* = A^* \cap C^*$. Now consider B and C. Since a graph is connected iff its line graph is connected C is connected. It follows that also A is shattered by edges. ∎

Theorem 6.3 *It is* NP-*complete to decide for an input consisting of some graph* $G = (V, E)$ *and a number* $k \geq 1$, *whether* $\mathrm{EVC}_{con}(G) \geq k$ *holds.*

PROOF The proof is very similar to the NP-completeness proof for VC_{con} in Theorem 4.2 and can also be done by transformation from the MINIMUM SET COVER. ∎

7 Conclusion

We have investigated several set systems resulting from special graph properties of simple loopless graphs and the associated VC-dimensions for vertices (like VC_{con}, VC_{path}, VC_{star}, VC_{tree}, VC_{nbd} and VC_{cycle}) as well as for edges (like EVC_{con}, EVC_{path}, EVC_{star}, EVC_{tree}). We studied the computational complexity of $\mathrm{VC}_{\mathcal{P}}$ for several graph properties $\mathcal{P}$ and showed that they all are NP-hard with the exception of the neighborhood property (which is complete for the class LOGNP) and the star property (which is computable in polynomial time). We derived several combinatorial properties of these set systems and related them to special graph parameters (like the maximum number of leaves in any spanning tree). In addition, for the path and cycle properties we constructed graphs G with the minimum number of edges under the condition $\mathrm{VC}_{\mathcal{P}}(G) \geq k$.

This paper is just a first step towards a systematic investigation of the Vapnik-Chervonenkis dimension on graphs. Problems that deserve further studies are e.g. the investigation of set systems induced by other graph properties (like planarity, bounded genus, k-connectivity, bounded diameter, k-colorability, or forbidden subgraphs) or the problem of determining the complexity of computing VC_{con}, VC_{path}, etc. for specially structured graph classes (like interval graphs, cographs, partial k-trees, or planar graphs).

References

1. M. Anthony, G. Brightwell, and C. Cooper, "On the Vapnik-Chervonenkis Dimension of a Graph", Technical Report, London School of Economics, 1993.
2. A. Blumer, A. Ehrenfeucht, D. Haussler, and M. K. Warmuth, "Learnability and the Vapnik-Chervonenkis Dimension", Journal of the ACM, 36: 929–965, 1989.
3. B. Bollobás, "Extremal Graph Theory", Academic Press, London, 1978.
4. B. Chazelle and E. Welzl, "Quasi-Optimal Range Searching and VC-Dimensions", Discrete & Computational Geometry, 4: 467–490, 1989.
5. M. R. Garey and D. S. Johnson, "Computers and intractability" Freeman, San Francisco, 1979.

6. D. Haussler and E. Welzl, "Epsilon-nets and Simplex Range Queries", Discrete & Computational Geometry, 2: 127–151, 1987.

7. R. L. Hemminger and L. W. Beineke, "Line Graphs and Line Digraphs", in: L.W. Beineke and R.J. Wilson, editors, Selected topics in graph theory, pages 271–305, Academic Press, London, 1978.

8. D. S. Johnson, "The NP-completeness column: An ongoing guide", Journal of Algorithms, 8: 285–303 (1987).

9. E. Kranakis, D. Krizanc, B. Ruf, J. Urrutia and G. Wöginger, "VC-Dimensions for Graphs", Technical Report SCS-TR-255, Carleton University, School of Computer Science, Ottawa, 1994.

10. C. H. Papadimitriou and M. Yannakakis, "On limited nondeterminism and the complexity of VC-dimension", Proceedings of the eighth annual Conference on Structure in Complexity Theory, 12–18, IEEE, 1993.

11. H.-I. Lu and R. Ravi, "The Power of Local Optimization: Approximation Algorithms for Maximum-Leaf Spanning Tree", Proceedings of 1992 Allerton Conference.

12. P. Turán, "On an extremal problem in graph theory", Mat. Fiz. Lapok, 48: 436–452, 1941 (in Hungarian).

13. V. N. Vapnik and A. Ya. Chervonenkis, "On the Uniform Convergence of Relative Frequencies of Events to their Probabilities", Theory of Probability and its Applications, 16(2): 264–280, 1971.

14. M. Yannakakis, "Node- and Edge-deletion NP-complete Problems", Proc. 10th Annual ACM Symposium on Theory of Computing, Association for Computing Machinery, New York, 253–264, 1978.

Finding and Counting Small Induced Subgraphs Efficiently

T. Kloks[1]*, D. Kratsch[2] and H. Müller[2]

[1] Department of Mathematics and Computing Science
Eindhoven University of Technology
P.O.Box 513, 5600 MB Eindhoven, The Netherlands
[2] Fakultät für Mathematik und Informatik
Friedrich-Schiller-Universität Jena,
Universitätshochhaus, 07740 Jena, Germany

Abstract. We give two algorithms for listing all simplicial vertices of a graph. The first of these algorithms takes $O(n^\alpha)$ time, where n is the number of vertices in the graph and $O(n^\alpha)$ is the time needed to perform a fast matrix multiplication. The second algorithm can be implemented to run in $O(e^{\frac{2\alpha}{\alpha+1}}) = O(e^{1.41})$, where e is the number of edges in the graph.

We present a new algorithm for the recognition of diamond-free graphs that can be implemented to run in time $O(n^\alpha + e^{3/2})$.

We also present a new recognition algorithm for claw-free graphs. This algorithm can be implemented to run in time $O(e^{\frac{\alpha+1}{2}}) = O(e^{1.69})$.

It is a fairly easy observation that, within time $O(e^{\frac{\alpha+1}{2}}) = O(e^{1.69})$ it can be checked whether a graph has a K_4. This improves the $O(e^{\frac{3\alpha+3}{\alpha+3}}) = O(e^{1.89})$ algorithm mentioned by Alon, Yuster and Zwick.

Furthermore, we show that *counting* the number of K_4's in a graph can be done within the same time bound $O(e^{\frac{\alpha+1}{2}})$.

Using the result on the K_4's we can count the number of occurences as induced subgraph of any other fixed connected graph on four vertices within $O(n^\alpha + e^{1.69})$.

1 Introduction

The first problem we consider is the problem of finding all simplicial vertices of a graph. This is of interest, since the complexity of many problems (e.g., CHROMATIC NUMBER, MAXIMUM CLIQUE) can be reduced by first removing simplicial vertices from the graph.

We first show an algorithm with running time $O(n^\alpha)$, which is the time needed to compute the square of an $n \times n$ 0/1-matrix, where n is the number of vertices of the graph. (Currently $\alpha < 2.376$.) Then, using an idea of Alon, Yuster and Zwick in [1] we obtain an alternative algorithm with running time $O(e^{\frac{2\alpha}{\alpha+1}})$, where e is the number of edges in the graph.

---

* Email: ton@win.tue.nl

A very basic problem in theoretical computer science is the problem of finding a triangle in a graph. In 1978 Itai and Rodeh presented two solutions for this problem. The first algorithm has a running time of $O(n^\alpha)$. The second algorithm needs $O(e^{3/2})$.

In [2] this second result was refined to $O(ea(G))$, where $a(G)$ is the arboricity of the graph. Since $a(G) = O(\sqrt{e})$ in a connected graph (see [2]), this extends the result of Itai and Rodeh. These were for almost 15 years the best known algorithms.

A drastic improvement was made recently by Alon et. al. In [1] they showed the following surprisingly elegant and easy result. Deciding whether a directed or an undirected graph $G = (V, E)$ contains a triangle, and finding one if it does, can be done in $O(e^{\frac{2\alpha}{\alpha+1}}) = O(e^{1.41})$.

For finding a certain induced connected subgraph with four vertices there are not many non-obvious results known. Two notable exceptions are the recognition of paw-free graphs and the recognition of P_4-free graphs. A *paw* is the graph consisting of a triangle and one pendant vertex. Using a characterization of Olariu [9], the class of paw-free graphs can be recognized in $O(n^\alpha)$ time. P_4-free graphs (cographs) can even be recognized in linear time [3]. We present new efficient algorithms for all other induced connected subgraphs on four vertices.

First we adopt the idea of [1] to obtain an efficient algorithm that checks if a graph contains a diamond and finds one if it does. The running time of our algorithm is $O(n^\alpha + e^{3/2})$.

Using standard techniques, it is easy to see that connected claw-free graphs can be recognized in $O(n^{\alpha-1}e)$ time. We show that there is also a $O(e^{\frac{\alpha+1}{2}}) = O(e^{1.69})$ recognition algorithm for claw-free graphs.

In [1] an easy $O(n^\alpha)$ algorithm counting the number of triangles in a graph is given. Furthermore the authors ask for an efficient algorithm counting the K_4's. We give an $O(e^{\frac{\alpha+1}{2}}) = O(e^{1.69})$ algorithm counting the K_4's. Moreover we show that for any fixed graph H on four vertices the number of copies of H in a given graph G can be counted in time $O(n^\alpha + e^{\frac{\alpha+1}{2}})$.

2   Listing all Simplicial Vertices

Definition 1 *A vertex x in a graph $G = (V, E)$ is called a* simplicial vertex *if its neighborhood $N(x)$ is complete.*

For many problems (e.g., coloring) simplicial vertices can be safely removed, tackling the problem on the reduced graph. It is therefore of interest to find these simplicial vertices quickly.

For any graph $G = (V, E)$ and $X \subseteq V$ we denote by $G[X]$ the subgraph of G *induced* by X. We denote by $N[x]$ the *closed* neighborhood of x, i.e., $N[x] = \{x\} \cup N(x)$. Let $d(x)$ be the degree of x, i.e., $d(x) = |N(x)|$.

Lemma 1 *A vertex x is simplicial if and only if for all neighbors y of x, $N[x] \subseteq N[y]$.*

Proof. The 'only if' part is obvious. Assume x is a vertex such that for every neighbor y, $N[x] \subseteq N[y]$. Assume x is not simplicial. Let y and z be two non adjacent neighbors of x. Then $z \in N[x]$, but $z \notin N[y]$, contradicting $N[x] \subseteq N[y]$.

Corollary 1 *A vertex x is simplicial if and only if for all neighbors y of x,* $|N[x] \cap N[y]| = |N[x]|$.

Now let A be the 0/1-adjacency matrix of G with 1's on the diagonal, i.e., $A_{x,x} = 1$ for all x and for $x \neq y$, $A_{x,y} = 1$ if x and y are adjacent in G and $A_{x,y} = 0$ otherwise. Hence A is a symmetric $n \times n$ 0/1-matrix. Consider A^2. The following is a key observation. For all $x, y \in V$:

$$\left(A^2\right)_{x,y} = |N[x] \cap N[y]|$$

(In particular $\left(A^2\right)_{x,x} = d(x) + 1$).)

Theorem 1 *There exists an $O(n^\alpha)$ algorithm which finds a list of all simplicial vertices of a graph G with n vertices.*

Proof. We assume that we have for each vertex a list of its neighbors. Construct the 0/1-adjacency matrix A with 1's on the diagonal. Compute A^2 in time $O(n^\alpha)$. By Corollary 1 a vertex x is simplicial if and only if $\left(A^2\right)_{x,y} = \left(A^2\right)_{x,x}$ holds for all $y \in N(x)$. Hence for each vertex x this test can be performed in time $O(d(x))$. The result follows.

The next algorithm is based on a technique presented in [1]. Let D be some integer (to be determined later). We call a vertex x of *low degree* if $d(x) \leq D$. A vertex which is not of low degree, is said to be of *high degree*. Our algorithm consists of four phases.

Phase 1 Search all simplicial vertices that are of low degree.
Phase 2 Mark all vertices of high degree that have a low degree neighbor.
Phase 3 Remove all low degree vertices from the graph. Call the resulting graph G^*.
Phase 4 Perform a matrix multiplication for the 0/1-adjacency matrix of G^* with 1's on the diagonal. Make a list of all simplicials of G^* which are *not* marked in phase 2, using the algorithm of Theorem 1.

Correctness follows from the following observation.

Lemma 2 *If a vertex x of high degree is simplicial, then all its neighbors are of high degree.*

Proof. Assume a vertex x of high degree is simplicial. Then for every neighbor y of x, $d(y) + 1 = |N[y]| \geq |N[x]| = d(x) + 1 > D + 1$, by Lemma 1. Hence y is also of high degree.

Theorem 2 *There exists an $O(e^{\frac{2\alpha}{\alpha+1}})$ time algorithm to compute a list of all simplicial vertices of a graph.*

Proof. Let $D = e^{\frac{\alpha-1}{\alpha+1}}$. We assume we have for every vertex a list of its neighbors. Let L be the set of vertices of low degree and let H be the set of vertices of high degree.

The first phase of the algorithm can be implemented as follows. For each $x \in L$, we check if every pair of its neighbors is adjacent. Hence the first phase can be performed in time proportional to $\sum_{x \in L} d(x)^2 \leq 2De$.

Phase 2 and phase 3 can clearly be implemented in linear time.

Now notice that $2e \geq \sum_{x \in H} d(x) \geq |H|D$, hence the number of vertices of G^* is at most $2e/D$. Computing the square of the 0/1-adjacency matrix for G^* with 1's on the diagonal can be performed in time $O((2e/D)^\alpha)$. Hence the total time needed by the algorithm is $O(eD + (2e/D)^\alpha) = O(e^{\frac{2\alpha}{\alpha+1}})$ by our choice of D.

3 Recognizing Diamond-free Graphs

In this section we consider the recognition of diamonds in graphs.

Definition 2 *A* diamond *is a graph isomorphic to the graph depicted in Figure 1 on the left.*

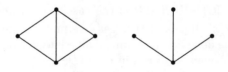

Fig. 1. diamond (left) and claw (right)

For the following result, see also [11].

Lemma 3 *A graph $G = (V, E)$ is diamond-free if and only if for every vertex x, the graph $G[N(x)]$ is a disjoint union of cliques.*

Proof. Consider a vertex x of degree 3 in a diamond. Then the neighborhood of x contains a P_3. Hence $G[N(x)]$ cannot be a disjoint union of cliques. Conversely, if for some vertex x, $G[N(x)]$ is not a disjoint union of cliques, then $G[N(x)]$ must contain a P_3. It follows that G contains a diamond.

Corollary 2 MAXIMUM CLIQUE *is solvable in time $O(n(n+e))$ for diamond-free graphs.*

Our algorithm for checking if a graph is diamond-free works as follows. Again, let D be some number. We partition the vertices into vertices of low degree, i.e., vertices of degree at most D and vertices of high degree. Let L be the set of low degree vertices, and H the set of high degree vertices.

Phase 1 Check if there is an induced diamond with a vertex of degree 3 which is of low degree in G.

Phase 2 Check if there is an induced diamond with a vertex of degree 2 which is of low degree in G.

Phase 3 If no diamond is found in the two previous steps, then remove all vertices of low degree from the graph (otherwise stop). Call the resulting graph G^*.

Phase 4 Check if G^* has a diamond.

We can implement this algorithm as follows. We assume we have the 0/1-adjacency matrix A with 0's on the diagonal. For each vertex x of low degree, construct adjacency lists for $G[N(x)]$. This can be accomplished in time $O(d(x)^2)$. Compute the connected components of $G[N(x)]$ (for example using depth first search). Check if each connected component is a clique. This can be done in $O(d(x)^2)$ time. Notice that if some connected component in $N(x)$ is *not* a clique, a P_3, and hence a diamond, can easily be computed in $O(d(x)^2)$ time (for example if y and z are non adjacent vertices in the same connected component, then start a breadth first search from y). It follows that Phase 1 can be implemented to run in $\sum_{x \in L} d(x)^2 \leq De$ time.

Next consider Phase 2. For each vertex x of low degree, we now have the cliques in the neighborhood (as a result of the previous phase). First compute A^2 in time $O(n^\alpha)$. Let C be a clique in $N(x)$. For each pair y, z in C, we check whether $\left(A^2\right)_{y,z} > |C| - 1$. If this is the case, then y and z have a common neighbor outside $N[x]$. Hence in that case we find a diamond. (Producing the diamond can then be done in linear time, if all adjacency lists are sorted.) Hence, Phase 2 can be implemented to run in time $(De + n^\alpha)$.

Finally, assume that neither Phase 1 nor Phase 2 produces a diamond. Compute G^*. The number of vertices in G^* is at most $2e/D$. Now, simply repeat the procedure described in Phase 1, for *all* vertices of G^*. This takes time proportional to $\sum_{x \in H} d_H(x)^2 = O(e|H|) = O(e^2/D)$.

Hence we obtain an overall time bound of $O(De + e^2/D + n^\alpha)$.

Theorem 3 *There exists an $O(e^{3/2} + n^\alpha)$ algorithm that checks if a graph has a diamond and produces one if it does.*

Proof. Take $D = \sqrt{e}$.

4 Recognition of Claw-free Graphs

The importance of claw-free graphs follows from matching properties, line graphs, hamiltonian properties and the polynomial time algorithm for computing the independence number [4, 7]. However, although many characterizations are known, there is no fast recognition algorithm known. Even the extensive survey on claw-free graphs in [4] mentiones only a $O(n^{3.5})$ recognition algorithm. We present in this section an $O(e^{\frac{\alpha+1}{2}}) = O(e^{1.69})$ recognition algorithm.

Definition 3 *A* claw *is a graph isomorphic to the graph depicted in Figure 1 on the right. A graph is* claw-free *if it does not have an induced subgraph isomorphic to a claw. We denote the vertex of degree 3 in a claw as the* central vertex.

Let G be a graph with n vertices and e edges. We start with an easy observation.

Lemma 4 *If G is claw-free then every vertex has at most $2\sqrt{e}$ neighbors.*

Proof. Consider a vertex x. If G is claw-free, then $\overline{G[N(x)]}$ is triangle-free. Let $p = |N(x)|$. By Turán's theorem, a graph with p vertices and without triangles can have at most $\frac{p^2}{4}$ edges (see, e.g., [10, 5]). Hence, there must be at least $\frac{1}{2}p(p-1) - \frac{1}{4}p^2 = \frac{1}{4}p^2 - \frac{1}{2}p$ edges in $G[N(x)]$. Then, (adding the edges incident with x), $G[N[x]]$ must contain at least $\frac{1}{4}p^2 - \frac{1}{2}p + p \geq \frac{1}{4}p^2$ edges. This can be at most the number of edges of G. Hence $p \leq 2\sqrt{e}$.

Our algorithm for recognizing claw-free graphs works as follows. First we check whether every vertex has at most $2\sqrt{e}$ neighbors. If there is a vertex with more than $2\sqrt{e}$ neighbors, there must be a claw with this vertex as the central vertex by Lemma 4.

If every vertex has at most $2\sqrt{e}$ neighbors, we perform a fast matrix multiplication for each neighborhood to check if the complement of such a neighborhood contains a triangle. This step of the algorithm can be performed in time proportional to $\sum_x d(x)^\alpha \leq (2\sqrt{e})^{\alpha-1} \sum_x d(x) \leq 2^\alpha e^{\frac{\alpha+1}{2}}$.

This proves the following theorem.

Theorem 4 *There exists an $O(e^{\frac{\alpha+1}{2}})$ algorithm to check whether a connected graph is claw-free.*

5   Counting the Number of K_4's

In this section we first describe an algorithm that decides whether a connected graph has a K_4 as induced subgraph and outputs one if there exists one. Moreover, we show how to extend it to an algorithm that counts the number of occurences of K_4 as an induced subgraph of the given connected graph. Both algorithms have running time $O(e^{\frac{\alpha+1}{2}}) = O(e^{1.69})$. The first one improves upon an $O(e^{1.89})$ algorithm mentioned in [1].

As in the preceeding sections the vertex set of the input graph G is partitioned into the sets L and H. The vertices of L are those with degree at most D. The recognition algorithm works as follows.

Phase 1 For each vertex $x \in H$ compute the square of the adjacency matrix of $G[N(x) \cap H]$ to decide if there is a triangle contained in $G[N(x) \cap H]$. This can be done in time $\sum_{x \in H} d_H(x)^\alpha = O(e(\frac{e}{D})^{\alpha-1})$.

Phase 2 For each vertex $x \in L$ compute the square of the adjacency matrix of $G[N(x)]$ to decide whether $G[N(x)]$ contains a triangle. This can be done in time $\sum_{x \in L} d(x)^\alpha = O(D^{\alpha-1}e)$.

Theorem 5 *There is an $O(e^{\frac{\alpha+1}{2}})$ algorithm that checks whether a connected graph has a K_4 and outputs one if it does.*

Proof. G has a K_4 if and only if the algorithm finds a triangle in $G[N(x) \cap H]$ in phase 1 or a triangle in $G[N(x)]$ in phase 2. The time bound follows by taking $D = \sqrt{e}$.

In [1] an easy $O(n^{\alpha+1})$ algorithm for counting the K_4's in a graph is given and the authors ask whether it can be improved to an $o(n^{\alpha+1})$ algorithm. We present an $O(e^{(\alpha+1)/2})$ algorithm that counts the number of K_4's in a given connected graph. Note that this is at least as good as the algorithm of [1] for all graphs. However it is an $o(n^{\alpha+1})$ algorithm only for certain sparse graphs.

We distinguish five different types of a K_4, depending on the number of low and high vertices, respectively, in the K_4. We denote by $L(i)$ the set of K_4's with exactly i vertices of low degree. Clearly, each K_4 is of exactly one of the types $L(0)$, $L(1)$, $L(2)$, $L(3)$ and $L(4)$. The counting algorithm determines the number of K_4's in G from each type, denoted by $\#(L(4))$, $\#(L(3))$, $\#(L(2))$, $\#(L(1))$ and $\#(L(0))$, respectively. The algorithm starts as the recognition algorithm, however in each phase it counts the corresponding number of triangles using the $O(n^\alpha)$ algorithm given in [1].

Phase 1 For each vertex $x \in H$ compute the square of the adjacency matrix of $G[N(x) \cap H]$. Compute the sum of the number of triangles contained in $G[N(x) \cap H]$ taken over all $x \in H$. This number is exactly $4\#(L(0))$. The running time is $\sum_{x \in H} d_H(x)^\alpha = O(e(\frac{e}{D})^{\alpha-1})$.

Phase 2 For each vertex $x \in L$ compute the square of the adjacency matrix of $G[N(x)]$. Compute the sum of the number of triangles contained in $G[N(x)]$ taken over all $x \in L$. This number is equal to $4\#(L(4)) + 3\#(L(3)) + 2\#(L(2)) + \#(L(1))$ since each K_4 is counted exactly i times, $i \in \{1, 2, 3, 4\}$, if and only if it contains i low vertices. Phase 2 can be done in time $\sum_{x \in L} d(x)^\alpha = O(D^{\alpha-1}e)$.

The next two phases compute $\#(L(4))$ and $\#(L(1))$, respectively, in a similar fashion.

Phase 3 For every $x \in L$ compute $G[N(x) \cap L]$. Compute the sum of the number of triangles in $G[N(x) \cap L]$. This number is equal to $4\#(L(4))$. This takes time $\sum_{x \in L} d(x)^\alpha = O(eD^{\alpha-1})$.

Phase 4 For every $x \in L$ compute $G[N(x) \cap H]$. Compute the sum of the number of triangles in $G[N(x) \cap H]$. This number is equal to $\#(L(1))$. This takes time $\sum_{x \in L} d(x)^\alpha = O(eD^{\alpha-1})$.

Finally we are going to count the triangles in the neighbourhood of a vertex in a sligthly different way.

Phase 5 For every low vertex x compute the adjacency matrix $A(x)$ of $G[N(x)]$ and then compute its square $A(x)^2$. Then $(A(x)^2)_{i,j}$ is exactly the number of common neighbours of $i, j \in N(x)$ belonging to $N(x)$. We compute the sum over all $(A(x)^2)_{i,j}$ for which $\{i,j\} \in E$, $i < j$ and $i, j \in H$. Moreover, we sum these values over all low vertices x. Hence, we only count those K_4 of type $L(2)$ or $L(1)$. The final value we get is exactly $2\#(L(2)) + 3\#(L(1))$. This can be done in time $\sum_{x \in L} d(x)^\alpha = O(eD^{\alpha-1})$.

Theorem 6 *There is an $O(e^{(\alpha+1)/2}) = O(e^{1.69})$ time algorithm counting the number of K_4's in a given graph.*

Proof. Using all the values we computed by the algorithm it is easy to compute $\#(L(0)) + \#(L(1)) + \#(L(2)) + \#(L(3)) + \#(L(4))$ i.e., the number of K_4's occuring as induced subgraph in G. The time bound follows by taking $D := \sqrt{e}$.

6 Recognizing K_ℓ's

Nešetřil and Poljak have given in [8] an $O(n^{\alpha\lfloor \ell/3 \rfloor + i})$ time algorithm that decides whether a given graph contains a K_ℓ. Here i is the remainder of ℓ by division by 3 which we will also denote by $i = mod(\ell, 3)$. As mentioned in [1] it is easy to see that the method of [8] can also be used for counting K_ℓ's within the same time bound. Alon, Yuster and Zwick mention in [1] that combining this K_ℓ counting algorithm of [8] with their methods leads to $O(n \, d(G)^{\alpha\lfloor (\ell-1)/3 \rfloor + i})$ and $O(e \, d(G)^{\alpha\lfloor (\ell-2)/3 \rfloor + i})$ time algorithms for counting the number of K_ℓ's in a given graph where $d(G)$ denotes the degeneracy of the input graph.

We show that the algorithm of [8] can also be used to design another type of algorithm that recognizes K_ℓ's by generalizing the $O(e^{\frac{2\alpha}{\alpha+1}})$ time triangle recognition algorithm of [1] and the $O(e^{\frac{\alpha+1}{2}})$ time K_4 recognition algorithm of section 5.

As in the preceeding sections the vertex set of the input graph G is partitioned into the sets L and H. The vertices of L are those with degree at most D. The algorithm recognizing K_ℓ's works as follows.

Phase 1 For each vertex $x \in L$ compute the adjacency matrix of $G[N(x)]$ and decide whether $G[N(x)]$ contains a $K_{\ell-1}$ by applying the $K_{\ell-1}$ recognition algorithm of [8].

Phase 2 Compute the adjacency matrix of $G[H]$ and decide whether there is a K_ℓ contained in $G[H]$ by applying the K_ℓ recognition algorithm of [8].

Theorem 7 *There is an $O(e^{\frac{1}{2}(\alpha\lfloor \ell/3 \rfloor + i)})$ time algorithm deciding whether a given graph contains a K_ℓ if $i = mod(\ell, 3) \in \{1, 2\}$.*
There is an $O(e^{\beta \frac{\beta-\alpha+2}{2\beta-\alpha+1}})$, where $\beta = (\alpha\ell)/3$, time algorithm that decides whether a given graph contains a K_ℓ if ℓ is a multiple of 3.

Proof. The adjacency matrices of $G[N(x)]$, $x \in L$, are $d(x) \times d(x)$ matrices with $d(x) \leq D$. Hence the running time of Phase 1 is

$$\sum_{x \in L} O(d(x)^{\alpha\lfloor (\ell-1)/3 \rfloor + mod(\ell-1,3)}) = O(e \, D^{\alpha\lfloor (\ell-1)/3 \rfloor + mod(\ell-1,3)-1}).$$

Since $|H| \le 2e/D$ the running time of Phase 2 is $O((e/D)^{\alpha\lfloor \ell/3 \rfloor + mod(\ell,3)})$.

Now we have $\alpha\lfloor (\ell-1)/3 \rfloor + mod(\ell-1,3) = \alpha\lfloor l/3 \rfloor + mod(\ell,3) - 1$ if ℓ is not a multiple of 3 and $\alpha\lfloor (\ell-1)/3 \rfloor + mod(\ell-1,3) = \beta - \alpha + 2$ if ℓ is a multiple of 3.

Suppose ℓ is not a multiple of 3, i.e., $i = mod(\ell,3) \in \{1,2\}$. Then we choose $D := \sqrt{e}$ implying that the running time of the K_ℓ recognition algorithm is $O(e^{\frac{1}{2}(\alpha\lfloor \ell/3 \rfloor + i)})$.

Suppose ℓ is a multiple of 3. Choosing $D := e^{\frac{\beta-1}{2\beta-\alpha+1}}$ we get the stated running time of the K_ℓ recognition algorithm.

It is worth to mention that the running time of our algorithms is at least as fast as that of the algorithms of Nešetřil and Poljak on all input graphs if ℓ is not a multiple of 3. If ℓ is a multiple of 3 then this is not the case in general. However, notice that the running time of the algorithm given in [8] is better only when the number of edges in the graph exceeds $\Omega\left(n^{2-\frac{3-\alpha}{\beta-\alpha+2}}\right) = \Omega\left(n^{2-\Theta\left(\frac{1}{\ell}\right)}\right)$. Hence, for large values of ℓ, the algorithm of [8] is faster only for very dense graphs.

It is an interesting question whether there is an $O(e^{\alpha \ell/6})$ algorithm for recognizing whether a graph contains a K_ℓ, ℓ is a multiple of 3, in particular whether there is a $O(e^{\alpha/2})$ algorithm for recognizing triangles.

7 Counting Other Small Subgraphs

We show that for any connected subgraph H on four vertices there is a $O(n^\alpha + e^{(\alpha+1)/2})$ algorithm counting the number of occurences of H in the given graph G. More precisely, on input $G = (V,E)$ we compute the cardinality of $\{W : W \subseteq V$ and $G[W]$ is isomorphic to $H\}$ for a fixed graph H on four vertices. The connected graphs on four vertices are K_4, $K_4 - e$ (the diamond), C_4, P_4, $K_{1,3}$ (the claw) and $\overline{P_3 + K_1}$ (the paw).

Theorem 8 *Let $\tilde{H}$ be a connected graph on four vertices such that there is an $O(t(n,e))$ time algorithm counting the number of $\tilde{H}$'s in a given graph. Then there is an $O(n^\alpha + t(n,e))$ time algorithm counting the number of H's for all connected graphs H on four vertices.*

Proof. The number of occurences of the connected subgraphs on four vertices in a graph G fullfill the following system of linear equations.

On left hand side, A denotes the adjacency matrix of the graph $G = (V,E)$, C the adjacency matrix of $\overline{G}$, both with zeros on the main diagonal. We are able to compute the values on left sides of these equations in time $O(n^\alpha)$.

$$\sum_{(x,y)\in E} \binom{(A^2)_{x,y}}{2} = 6\#K_4 + \#\text{diamond}$$

$$\sum_{(x,y)\notin E} \binom{(A^2)_{x,y}}{2} = \#\text{diamond} + 2\#C_4$$

$$\sum_{(x,y)\in E}(AC)_{x,y}(CA)_{x,y} = 4\#C_4 + \#P_4$$

$$\sum_{x\in V}(A^3C)_{x,x} = 4\#\text{diamond} + 2\#P_4 + 4\#\text{paw}$$

$$\sum_{(x,y)\in E}\left(\binom{(AC)_{x,y}}{2} + \binom{(AC)_{y,x}}{2}\right) = \#\text{paw} + 6\#\text{claw}$$

Let us explain how to see the first equation. We consider an edge $(x,y) \in E$ such that $\binom{(A^2)_{x,y}}{2} = k$. Then there exist exactly k pairs of different vertices p and q such that $p, q \in N(x) \cap N(y)$, i.e., p, q, x and y induce either a K_4 (if $(p,q) \in E$) or a diamond (if $(p,q) \notin E$). Summing up we count each K_4 exactly six times, since every edge of K_4 will play the role of (x,y), and every diamond exactly once, where x and y are the vertices of degree three in the diamond. The equation follows. Similar observations give the other equations.

On right side we need $\#\tilde{H}$ for one subgraph $\tilde{H}$ to be able to solve the system and find the values $\#H$ for all the other subgraphs H.

A similar theorem can be shown for all graphs on four vertices by extending the system of linear equations. This and Theorem 6 imply

Corollary 3 *For any graph H on four vertices there is a $O(n^\alpha + e^{\frac{\alpha+1}{2}})$ algorithm counting the H's in a given graph G.*

References

1. Alon, N., R. Yuster and U. Zwick, Finding and counting given length cycles, *Algorithms–ESA '94. Second Annual European Symposium*, Springer-Verlag, Lecture Notes in Computer Science 855, (1994), pp. 354–364.
2. Chiba, N. and T. Nishizeki, Arboricity and subgraph listing algorithms, *SIAM J. Comput.*, **14**, (1985), pp. 210–223.
3. Corneil, D. G., Y. Perl and L. K. Stewart, A linear recognition algorithm for cographs, *SIAM J. Comput.*, **4**, (1985), pp. 926–934.
4. Faudree, R., E. Flandrin and Z. Ryjáček, Claw-free graphs–A survey. Manuscript.
5. Harary, F., *Graph Theory*, Addison Wesley, Publ. Comp., Reading, Massachusetts, (1969).
6. Itai, A. and M. Rodeh, Finding a minimum circuit in a graph, *SIAM J. Comput.*, **7**, (1978), pp. 413–423.
7. Minty, G. J., On maximal independent sets of vertices in claw-free graphs, *J. Combin. Theory B*, **28**, (1980), pp. 284–304.
8. Nešetřil, J. and S. Poljak, On the complexity of the subgraph problem, *Commentationes Mathematicae Universitatis Carolinae*, **14** (1985), no. 2, pp. 415–419.
9. Olariu, S., Paw-free graphs, *Information Processing Letters*, **28**, (1988), pp. 53–54.
10. Turán, P., Eine Extremalaufgabe aus der Graphentheorie, *Mat. Fiz. Lapok*, **48**, (1941), pp. 436–452.
11. Tucker, A., Coloring perfect $K_4 - e$-free graphs, *Journal of Combinatorial Theory, series B*, **42**, (1987), pp. 313–318.

On the Isomorphism of Graphs with Few P_4s

Luitpold Babel[1] and Stephan Olariu[*2]

[1] Institut für Mathematik, Technische Universität München,
80290 München, Germany
[2] Department of Computer Science, Old Dominion University,
Norfolk, VA 23529, U.S.A.

Abstract. We present new classes of graphs for which the isomorphism problem can be solved in polynomial time. These graphs are characterized by containing – in some local sense – only a small number of induced paths of length three. As it turns out, every such graph has a unique tree representation: the internal nodes correspond to three types of graph operations and the leaves are basic graphs with a simple structure. The paper extends and generalizes results on cographs, P_4-reducible graphs, and P_4-sparse graphs.

1   Introduction

In recent years the study of the P_4-structure of graphs turned out to be of considerable importance. The starting point and original motivation for many investigations was the class of graphs where no induced P_4 is allowed to exist (hereinafter P_k denotes a chordless path on k vertices and $k-1$ edges). For these graphs, commonly termed *cographs*, some interesting structural results have been obtained which helped to solve efficiently many graph-theoretic problems which are hard in general (see [6] for a discussion). The study of cographs has been extended by B. Jamison and S. Olariu to graphs which contain a restricted number of paths of length three. Besides P_4-*extendible graphs* [13] and P_4-*lite graphs* [14] they studied P_4-*reducible graphs* [12], defined as those graphs where no vertex belongs to more than one P_4, and P_4-*sparse graphs* [10], which generalize both cographs and P_4-reducible graphs. A graph is P_4-sparse if no set of five vertices induces more than one P_4.

We propose to call a graph a (q, t) *graph* if no set of at most q vertices induces more than t distinct P_4s. In this sense, the cographs are precisely the $(4, 0)$ graphs, the P_4-sparse graphs coincide with the $(5, 1)$ graphs and P_4-lite graphs turn out to be special $(7, 3)$ graphs. The main contribution of this paper is to investigate the structure of $(q, q-4)$ graphs for any fixed $q \geq 4$.

Tree representations for special graphs are often the basis for fast solutions of algorithmic problems which are hard in general. One of the best known paradigms is the isomorphism problem whose complexity is still unknown for arbitrary graphs. Using tree representations, polynomial isomorphism tests have

[*] work supported by NSF grant CCR-9407180 and by ONR grant N00014-95-1-0779

been obtained among others for hook-up graphs [15], transitive series parallel digraphs [16], interval graphs [4], rooted directed path graphs [2], cographs [6], P_4-extendible graphs [13] and P_4-sparse graphs [10].

We consider the concept of encoding a graph into a rooted tree whose internal nodes represent certain graph operations and whose leaves correspond to certain basic graphs. If the encoding is unique and can be be obtained in polynomial time, and if the basic graphs can efficiently be tested for isomorphism then we are able to solve the isomorphism problem for two such graphs in polynomial time. We will prove that the $(q, q-4)$ graphs admit of such a tree representation.

The remainder of the paper is organized as follows. In Section 2 we review the concept of p-connectedness and recall some fundamental facts. Section 3 studies minimally p-connected graphs. The obtained results are used in Section 4 to classify all p-connected $(q, q - 4)$ graphs and, furthermore, to prove that $(q, q - 4)$ graphs are brittle graphs for $q \leq 8$. Thus, as a very interesting by-product, we are provided with new classes of brittle graphs, distinct from all the previously known brittle graphs. Section 5 discusses the tree representation and an efficient isomorphism test for $(q, q - 4)$ graphs. Finally, in the last section we summarize the results and pose some open problems.

2   Background and Terminology

Let $G = (V, E)$ be a simple graph with vertex-set V and edge-set E. For a vertex v of G define $N(v)$ to be the set of vertices adjacent to v. A vertex of G is said to be an *articulation point* if its removal disconnects G. Given a set A of vertices of G, we let $G(A)$ denote the subgraph of G induced by A. We shall use $G - \{v\}$ as a shorthand for $G(V - \{v\})$.

A chordless path P_4 with vertices u, v, w, x and edges uv, vw, wx is denoted by $uvwx$. The vertices u and x are termed the *endpoints*, while v and w are the *midpoints* of the P_4. A graph is a *clique* if its vertices are pairwise adjacent. A *stable set* denotes a set of pairwise non-adjacent vertices. For other graph-theoretic notations we refer to Golumbic [8].

In the following we shall adopt the terminology introduced by B. Jamison and S. Olariu [9]. A graph $G = (V, E)$ is *p-connected* if for every partition of V into nonempty disjoint sets A and B there exists a *crossing* P_4, that is, a P_4 containing vertices from both A and B. The *p-connected components* of a graph are the maximal induced subgraphs which are p-connected. Every p-connected component has at least four vertices. Vertices which are not contained in a p-connected component are called *weak*. It is easy to see that each graph has a unique partition into p-connected components and weak vertices. Furthermore, the p-connected components are closed under complementation and are connected subgraphs of G and $\overline{G}$.

A p-connected graph $G = (V, E)$ is called *separable* if there exists a partition of V into nonempty disjoint sets V_1, V_2 such that each P_4 which contains vertices from both sets has its endpoints in V_2 and its midpoints in V_1. We say that (V_1, V_2) is a *separation* of G. Obviously, the complement of a separable p-connected graph is also separable. If (V_1, V_2) is a separation of G then (V_2, V_1) is a separation of $\overline{G}$. We now recall some important facts that form the basis for the results derived in this paper.

Theorem 2.1 (B. Jamison and S. Olariu [9]). *Every separable p-connected component H has a unique separation (H_1, H_2). Furthermore, every vertex of H belongs to a crossing P_4 with respect to (H_1, H_2).* $\square$

Let $G = (V, E)$ be an arbitrary graph. A set Z of vertices of G is called *homogeneous* if $1 < |Z| < |V|$ and each vertex outside Z is either adjacent to all vertices of Z or to none of them. A homogeneous set Z is *maximal* if no other homogeneous set properly contains Z. Let H be a p-connected component. The graph obtained from H by replacing every maximal homogeneous set by one single vertex is called *characteristic* p-connected component of H. Recall that a graph is a *split graph* if its vertex-set can be partitioned into a clique and a stable set.

Theorem 2.2 (B. Jamison and S. Olariu [9]). *A p-connected component H is separable if and only if the characteristic p-connected component of H is a split graph.* $\square$

The introduction and study of separable p-connected graphs is justified by the following general structure theorem for arbitrary graphs.

Theorem 2.3 (B. Jamison and S. Olariu [9]). *Let $G = (V, E)$ be a graph. Exactly one of the following statements holds:*
(i) G is disconnected.
(ii) $\overline{G}$ is disconnected.
(iii) There exists a unique proper separable p-connected component H with separation (H_1, H_2) such that every vertex outside H is adjacent to all vertices in H_1 and to no vertex in H_2.
(iv) G is p-connected. $\square$

As already pointed out in [9], this structure theorem suggests, in a natural way, a tree representation for every graph G. The leaves of the tree correspond, essentially, to the p-connected components of G. If these subgraphs have a simple structure then we may hope to solve the isomorphism problem in polynomial time. This observation motivates a further study of p-connected graphs. As a first step in this direction, in the next section of this work, we shall look at graphs that are critical in the sense of p-connectedness.

3 Minimally p-connected Graphs

A graph $G = (V, E)$ is *minimally p-connected* if G is p-connected and, for every vertex v of G, $G - \{v\}$ is not p-connected. Following the notation in [10] a p-connected graph $G = (V, E)$ is called a *spider* if V admits a partition into disjoint sets S and K such that:

(i) $|S| = |K| \geq 2$, S is stable, K is a clique;

(ii) There exists a bijection $f : S \to K$ such that either

$$N(s) = \{f(s)\} \text{ for all vertices } s \text{ in } S,$$

or else

$$N(s) = K - \{f(s)\} \text{ for all vertices } s \text{ in } S.$$

If the first of the two alternatives of (ii) holds then G is said to be a spider with *thin* legs, otherwise the spider has *thick* legs. As a technicality, a P_4 is considered to be a spider with thin legs. Obviously, the complement of a spider with thin legs is a spider with thick legs and vice versa. The main goal of this section is to prove that each minimally p-connected graph is a spider. Our first result shows that no minimally p-connected graph contains a homogeneous set.

Lemma 3.1 *Let $G = (V, E)$ be a p-connected graph and let Z be a homogeneous set in G. Then, for every vertex v in Z, $G - \{v\}$ is p-connected.*

Proof. Since G is p-connected there is a P_4 containing vertices from both Z and $V - Z$. This P_4 contains exactly one vertex from Z, say u. If u is replaced by any other vertex w from Z then we again get a P_4.

Assume that $G^* = G - \{v\}$ is not p-connected. Then there is a partition A, B of the vertex set $V^* = V - \{v\}$ of G^* without a crossing P_4. Let $Z^* = Z - \{v\}$. Z^* is a subset of one of the sets A, B. This can be seen as follows. Let $Z^* \cap A \neq \emptyset$ and $Z^* \cap B \neq \emptyset$. Take a P_4 with vertices from both Z^* and $V^* - Z^*$ (the existence follows from the above observation). This P_4 is contained in one of the sets A or B, say A. Replace the vertex from $Z^* \cap A$ by a vertex from $Z^* \cap B$. Then we get a crossing P_4, a contradiction.

Therefore let without loss of generality $Z^* \subseteq A$. In G there exists a P_4 containing vertices from both $A \cup \{v\}$ and B. This P_4 contains v but no vertex from Z^*. If v is replaced by any vertex from Z^* then we obtain a new P_4 which is crossing between A and B, contrary to the assumption. $\qquad\square$

Let G be p-connected and $G^* = G - \{v\}$ not p-connected. By Theorem 2.3 exactly one of the following statements is true:

(i) G^* is disconnected, i.e. v is an articulation point in G.

(ii) $\overline{G^*}$ is disconnected, i.e. v is an articulation point in $\overline{G}$.

(iii) There is a unique proper separable p-connected component H of G^* with separation (H_1, H_2) such that every vertex outside H is adjacent to all vertices in H_1 and to no vertex in H_2.

According to the different cases we call the vertex v to be of *type* 1, 2 or 3.

Lemma 3.2 *Let $G = (V, E)$ be p-connected. If each vertex of G is of type 1 or 2 then G is a P_4.*

Proof. A connected graph has at most $|V| - 2$ articulation points. Therefore G contains vertices of both types. In particular, since $|V| \geq 4$ there exist at least two vertices which are articulation points in $\overline{G}$. Furthermore, since G is connected there are vertices of different type, say x of type 1 and y of type 2, with $xy \in E$.

Suppose first that $|N(y)| > 1$.

Denote $G(U_1), G(U_2), \ldots, G(U_r)$ the components of $G - \{x\}$ and let $y \in U_1$. Note that under the above assumption we have $U_1 - \{y\} \neq \emptyset$ and $r \geq 2$. Since there is no edge in G connecting vertices from different sets $U_1 - \{y\}, U_2, \ldots, U_r$ we conclude that $\overline{G} - \{x, y\}$ is connected. Now let $G(W_1), G(W_2), \ldots,$ be the components of $\overline{G} - \{y\}$. Then we get $W_1 = \{x\}$ and $W_2 = V - \{x, y\}$. This means that x is adjacent to all other vertices in G. However, then there is no P_4 containing x and this contradicts to the fact that G is p-connected. Therefore $|N(y)| = 1$.

Since there exist at least two articulation points in $\overline{G}$ and since G is connected, there is a second vertex y' of type 2 which is adjacent to a vertex x' of type 1. Analogously as above we conclude that $|N(y')| = 1$. Thus we have $N(y) = \{x\}$ and $N(y') = \{x'\}$. Again denote $G(W_1), G(W_2), \ldots$ the components of $\overline{G} - \{y\}$. Since $|N(y')| = 1$ we have $W_1 = \{x'\}$ and $W_2 = V - \{x', y\}$. If $x = x'$ then x would be adjacent to all other vertices in G. This is not possible since G is p-connected. Therefore $x \neq x'$. $x' \in W_1$ and $x \in W_2$ implies $xx' \in E$. Therefore the vertex set $\{y, x, x', y'\}$ induces a P_4. Each further vertex w is adjacent to x' and also to x (exchange the parts of y and y'), thus exactly to the midpoints of the P_4. As a consequence, there is no crossing P_4 between $\{y, x, x', y'\}$ and the remaining vertices. Therefore no such vertex w exists. This proves the lemma.

$\square$

Lemma 3.2 implies that each nontrivial minimally p-connected graph contains a vertex of type 3. If v is of type 3 then we write $H(v)$ for the separable p-connected component and $(H_1(v), H_2(v))$ for the separation. Further we denote $R(v)$ to be the vertices of G^* outside $H(v)$.

Lemma 3.3 *Let $G = (V, E)$ be minimally p-connected and let $x \in V$ be a vertex of type 3 with $|R(x)|$ minimal. Then $|R(x)| = 1$.*

Proof. Assume that $|R(x)| \geq 2$. By virtue of Lemma 3.1, G contains no homogeneous set. Therefore, x is adjacent to some but not to all vertices in $R(x)$. Consequently, we find vertices u and u' in $R(x)$ with $xu \in E$ and $xu' \notin E$.

We consider vertex u and examine the possible types for u:

(i) Assume that u is of type 1, i.e. u is an articulation point in G. Since $G - \{u, x\}$ is connected we conclude that $N(x) = \{u\}$. Obviously, u' is not an articulation point in G and not in $\overline{G}$. Thus, u' is of type 3. x can neither be in $R(u')$ nor in $H_1(u')$ since each vertex from this two sets is adjacent to at least two vertices. Thus $x \in H_2(u')$ and as an immediate consequence $u \in H_1(u')$. Since both $H(x)$ and $H(u')$ are p-connected, we easily see that $H(x) \subset H(u')$. However, now $|R(u')| < |R(x)|$, contradicting the choice of x.

(ii) Assume that u is of type 2, i.e. u is an articulation point in $\overline{G}$. Since $\overline{G} - \{u, x\}$ is connected this would imply $N(x) = V - \{x\}$. However, this is not possible since $xu' \notin E$.

(iii) Assume that u is of type 3. Since $H(x)$ and $H(u)$ are p-connected, either $H(x) \subseteq H(u)$ or $H(x) \subseteq R(u)$ holds. The second case is not possible since some edges between $R(u)$ and $H_1(u)$ would be missing (take vertices $v \in H_2(x) \cap R(u)$ and $w \in R(x) \cap H_1(u)$, then $vw \notin E$).

Therefore $H(x) \subseteq H(u)$. Since, due to the choice of x, $|R(u)| \geq |R(x)|$ must hold, we conclude that $H(u) = H(x)$ and, due to the uniqueness of the separation (Theorem 2.1) $(H_1(u), H_2(u)) = (H_1(x), H_2(x))$. However, since we know from above that u is adjacent to all vertices in $H_1(x)$ and to none in $H_2(x)$, this would imply a homogeneous set $R(u) \cup \{u\}$, a contradiction.

This shows that the assumption $|R(x)| \geq 2$ is not correct. $\quad\square$

Lemma 3.4 *Let $G = (V, E)$ be minimally p-connected and let $x \in V$ be a vertex of type 3 with $R(x) = \{v\}$. Then $N(x) = R(x)$ or $N(x) = H_1(x) \cup H_2(x)$.*

Proof. Assume first that $xv \in E$. We distinguish the possible types for v.

If v is of type 2, i.e. an articulation point in $\overline{G}$ then $N(x) = V - \{x\}$. This is not possible since no P_4 would exist containing x in contradiction to the p-connectedness of G. If v is of type 3 then obviously $R(v) = \{x\}$ and therefore $N(x) = \{v\} \cup H_1(x)$. Thus $\{v, x\}$ would be a homogeneous set. Therefore v is of type 1, i.e. articulation point in G and $N(x) = \{v\}$. This shows the first part of the statement.

For the second part assume that $xv \notin E$.

If v is of type 1 then $N(x) = \emptyset$ which is not possible since G is connected. If v is of type 3 then $R(v) = \{x\}$ and therefore $N(x) = H_1(x)$. Again $\{v, x\}$ would be a homogeneous set. Therefore v is of type 2 and $N(x) = H_1(x) \cup H_2(x)$. $\quad\square$

We are now ready to prove the main result of this section.

Theorem 3.5 *Let G be minimally p-connected. Then G is a spider.*

Proof. If G contains no vertex of type 3 then, by Lemma 3.2, G is a P_4 and therefore a spider. Let x be a vertex of type 3 with $|R(x)|$ as small as possible. By virtue of Lemmas 3.3 and 3.4, we have $R(x) = \{v\}$ and $N(x) = R(x)$ or $N(x) = H_1(x) \cup H_2(x)$. It suffices to consider the case $N(x) = R(x)$, the second case being handled similarly.

Note that, if Z is a homogeneous set in the subgraph $H(x)$ then $Z \subseteq H_1(x)$ or $Z \subseteq H_2(x)$. This can be seen as follows. Assume that $Z \cap H_i(x) \neq \emptyset$ for $i = 1, 2$. Take a P_4 with vertices from both Z and $H(x) - Z$. Since Z is homogeneous, this P_4 contains exactly one vertex from Z, say z. As we have already seen, z may be replaced by any other vertex from Z to form another P_4. If $z \in H_1(x)$ then replace z by a vertex $z' \in Z \cap H_2(x)$, if $z \in H_2(x)$ then by a vertex $z'' \in Z \cap H_1(x)$. It is immediately clear that a P_4 results which is crossing between $H_1(x)$ and $H_2(x)$ and whose midpoints or endpoints are not both in $H_1(x)$ or $H_2(x)$.

We can conclude that Z is also homogeneous in G. However, Lemma 3.1 implies that G contains no homogeneous set. Therefore, no such set Z exists. Using Theorem 2.2 we conclude that $G(H_1(x) \cup H_2(x))$ is a split graph. For convenience denote K the vertex set of the clique induced by $H_1(x)$ and S the stable set $H_2(x)$. Note that each vertex of G is contained in a P_4 $xvks$ with $k \in K$ and $s \in S$.

Let $s' \in S$ with $N(s') = \{k'\}$. If $|N(k') \cap S| \geq 2$ then each vertex of $G - \{s'\}$ is contained in a path $xvks$ with $s \neq s'$, thus $G - \{s'\}$ would be p-connected, contradicting the minimality of G. Therefore $|N(k') \cap S| = 1$. Analogously, let $k'' \in K$ with $N(k'') \cap S = \{s''\}$. Then $|N(s'')| = 1$, otherwise $G - \{k''\}$ would be p-connected. Clearly, the vertices $k' \in K$ and $s' \in S$ with $|N(k') \cap S| = 1$ and $|N(s')| = 1$ together with x and v induce a spider with thin legs.

For all further vertices $k''' \in K$ and $s''' \in S$ which are not in the spider $|N(k''') \cap S| \geq 2$ resp. $|N(s''')| \geq 2$ holds. Assume that any of this vertices, say s''', is deleted. For each $k''' \in K$ with $s''' \in N(k''')$ there is at least one additional vertex in S which is adjacent to k'''. Therefore each vertex of $G - \{s'''\}$ is contained in a P_4 $xvks$ with $s \neq s'''$ and $G - \{s'''\}$ remains p-connected. Consequently, no further vertices exist and the proof is complete. □

Theorem 3.5 implies the following very useful property of p-connected graphs.

Corollary 3.6 *Let G be p-connected. Then there is an ordering $(v_n, v_{n-1}, \ldots, v_1)$ of the vertices of G and an integer $k \in \{4, 5, \ldots, n\}$ such that the following holds:*

$G(\{v_i, v_{i-1}, \ldots, v_1\})$ is p-connected for $i = k, \ldots, n$ and a spider for $i = k$. □

4 On p-connected $(q, q - 4)$ Graphs

We start with some properties concerning minimally p-connected graphs.

Observation 4.1 *In a spider each P_4 has its midpoints in the clique K and its endpoints in the stable set S, i.e. a spider is separable. For each pair $s, s' \in S$ ($k, k' \in K$) there is exactly one P_4 containing both vertices.* □

Observation 4.2 *A spider with $|K| = |S| = r$ contains exactly $\frac{r(r-1)}{2}$ P_4s.* □

Observation 4.3 *If H and G are spiders with thin (thick) legs and H has fewer vertices than G, then H is isomorphic to an induced subgraph of G.* □

Fact 4.4 *If q is even and G is a spider with q vertices then G is not a $(q, q-4)$ graph. If q is odd, $q \geq 9$, and G is a spider with $q - 1$ vertices then G is not a $(q, q - 4)$ graph.*

Proof. Let q be even. By virtue of Observation 4.2, the spider G contains $\frac{r(r-1)}{2}$ P_4s with $r = \frac{q}{2}$. Since $\frac{1}{8}q(q-2) > q - 4$ holds, G does not satisfy the definition of a $(q, q-4)$ graph.

Let q be odd. Then $r = \frac{q-1}{2}$ and G contains $\frac{1}{8}(q-1)(q-3)$ P_4s. For $q \geq 9$ we get $\frac{1}{8}(q-1)(q-3) > q - 4$. Therefore G is not a $(q, q-4)$ graph. □

The following theorem characterizes p-connected $(q, q - 4)$ graphs. Part a) already implicitly appeared in [10]. For the sake of completeness we restate it, giving, however, a completely different proof.

Theorem 4.5 *Let $G = (V, E)$ be p-connected.*
a) If G is a $(5, 1)$ graph then G is a spider.
b) If G is a $(7, 3)$ graph then $|V| < 7$ or G is a spider.
c) If G is a $(q, q - 4)$ graph, $q = 6$ or $q \geq 8$, then $|V| < q$.

Proof. By Corollary 3.6 there is an ordering $(v_n, \ldots, v_1)$ of the vertices of G and an integer $k \in \{4, 5, \ldots, n\}$ such that $G_i := G(\{v_i, v_{i-1}, \ldots, v_1\})$ is p-connected for $i = k, \ldots, n$ and G_k is a spider.

a) Let G be a $(5, 1)$ graph. It can easily be verified that each spider is a $(5, 1)$ graph. Assume that $k < n$, i.e. there is a vertex v_{k+1} which is not in the spider G_k.
Let X be the vertex set of an arbitrary P_4 in G_k. There are no three vertices in X such that v_{k+1} together with these vertices induces a P_4. Otherwise $G(X \cup \{v_{k+1}\})$ would be a graph with five vertices and at least two P_4s, thus not a $(5, 1)$ graph. Therefore, v_{k+1} is either adjacent to all vertices in X, to no vertex in X, or exactly to the two midpoints.
Using Observation 4.1 we conclude that v_{k+1} is either adjacent to all vertices of G_k, to none of them, or exactly to the vertices of the clique of G_k. However, in all three cases G_{k+1} is not p-connected since there is no P_4 in G_{k+1} containing v_{k+1}. This is a contradiction. Therefore $k = n$ and G is a spider.

b) Let G be a $(7, 3)$ graph. Again, it can easily be verified that each spider is a $(7, 3)$ graph.
If $k = 4$ then the spider G_k is a P_4. Since G_i is p-connected for $i = k, \ldots, n$, adding v_{i+1} to G_i increases the number of P_4s by at least one. Since G is a $(7, 3)$ graph no more than two vertices can be added. Therefore we get $|V| < 7$.
Let $k > 4$ and assume that $k < n$, i.e. there is a vertex v_{k+1} which is not in the spider G_k.
Since G_{k+1} is p-connected there exists a P_4 in G_{k+1} containing v_{k+1}. Let $X = \{x, y, z, v_{k+1}\}$ be the vertex set of this P_4. Further let H be the spider with smallest number of vertices which is a subgraph of G_k and which contains x, y and z. Obviously, H has four or six vertices. In the first case extend H to a spider with six vertices. Now adding v_{k+1} to H results in a graph with seven vertices and at least four P_4s. This is a contradiction. Therefore we have $k = n$ and G is a spider.

c) Let G be a $(q, q - 4)$ graph with $q = 6$ or $q \geq 8$. We know from Observation 4.3 and Fact 4.4 that $k < q$, i.e. the spider G_k has less than q vertices. By Observation 4.2 G_k contains exactly $\frac{1}{8}k(k - 2)$ P_4s. Since G_i is p-connected for $i = k, \ldots, n$, adding v_{i+1} to G_i strictly increases the number of P_4s. Therefore, G_i contains at least $\frac{1}{8}k(k - 2) + (i - k)$ P_4s.
Assume that G has at least q vertices, i.e. $n \geq q$. This would imply that the

number of P_4s which are contained in the graph G_q is at least
$$\tfrac{1}{8}k(k-2) + (q-k) = q + \tfrac{1}{8}k(k-10) \geq q - 3 > q - 4 \quad .$$
As a consequence, G_q would not be a $(q, q-4)$ graph, a contradiction. Therefore we have $|V| < q$.
This completes the proof. $\qquad\qquad\qquad\qquad\qquad\qquad\qquad\qquad\qquad\qquad\qquad\qquad$ □

This characterization can be used to derive interesting properties of $(q, q-4)$ graphs. A graph G is called *brittle* if each induced subgraph H of G contains a vertex which is either not the endpoint or not the midpoint of any P_4 in H. It is well known that brittle graphs are perfectly orderable. A graph G is *perfectly orderable* in the sense of Chvatal [5] if there exists a linear order on the set of vertices of G such that no induced path with vertices u, v, w, x and edges uv, vw, wx has $u < v$ and $x < w$. The importance of perfectly orderable graphs stems from the fact that these are precisely the graphs for which the coloring heuristic "always use the first available color" based on the linear order yields a coloring using the minimum number of colors. Chvatal has shown that perfectly orderable graphs are perfect.

It is easy to see that $(q, q-4)$ graphs, $q \geq 9$, are not brittle and not even perfect since the induced cycle of length five belongs to this classes. On the other side the following holds.

Theorem 4.6 *Every $(q, q-4)$ graph, $4 \leq q \leq 8$, is brittle.*

Proof. If a vertex v is not endpoint (midpoint) of any P_4 in a p-connected component of G then v is not endpoint (midpoint) of any P_4 in G. Therefore it suffices to prove that p-connected $(q, q-4)$ graphs, $4 \leq q \leq 8$, are brittle.
Let $q = 8$ and $G = (V, E)$ be a p-connected $(8, 4)$ graph with maximal number of vertices, i.e. $|V| = 7$. Further let $(v_7, v_6, \ldots, v_1)$ be an ordering of the vertices of V defined by Corollary 3.6. It is easy to see that v_7 is contained in exactly one P_4. For that reason v_7 is either not the endpoint or not the midpoint of any P_4 in G.
If we have at most six vertices, the conclusion follows by an exhaustive search. For $q \leq 7$ use Observation 4.1 to see that spiders are brittle. Then, as above, an exhaustive search should convince the reader that $(q, q-4)$ graphs, $q \leq 7$, with no more than six vertices are brittle. $\qquad\qquad\qquad\qquad\qquad\qquad\qquad\qquad\qquad\qquad$ □

5 The Tree Structure of $(q, q - 4)$ Graphs

Theorem 2.3 enables us to give for any graph a tree representation. The tree associated with a graph G carries labels on the interior nodes and is constructed by the obvious recursive procedure. The labels correspond to the cases in the theorem. Thus, label (1) indicates that the graph associated with this node as a root is the disjoint union of the graphs defined by its children. Label (2) defines the operation which we will call disjoint sum. All pairs of vertices belonging to different children are linked by an edge. Operation (3) adjoins the midpoints of the leftmost son - which has to represent a separable p-connected component - to all vertices of its other children. The leaves of the tree represent the p-connected components of the graph G along with its weak vertices.

It is well known that each cograph arises from single vertices by a sequence of operations disjoint union and disjoint sum. Thus, in this special case the leaves of the tree represent vertices and the labels of the interior nodes are (1) and (2).

Let $\mathcal{G}(q,t)$ denote the set of all (q,t) graphs. In particular, $\mathcal{G}(4,0)$ corresponds to the set of cographs, $\mathcal{G}(5,1)$ to the set of P_4-sparse graphs. The following theorem reflects the containment relations between the different classes.

Theorem 5.1
a) $\mathcal{G}(4,0) \subset \mathcal{G}(5,1)$, $\mathcal{G}(6,2) \subset \mathcal{G}(7,3)$.
b) $\mathcal{G}(6,2) \subset \mathcal{G}(q,q-4) \subset \mathcal{G}(q+1,q-3)$ *for $q \geq 8$.*
All inclusions are strict.

Proof. It is clear from the tree representation that it suffices to consider the p-connected components of the graphs. With this in mind all inclusions can immediately be deduced from Theorem 4.5.
Examples to confirm the strict inclusions are in case a) the P_4 respectively the graph consisting of a P_4 $uvwx$ extended by two vertices y, z which are adjacent to w. In case b) take the path P_6 with 6 vertices for the first and the path P_q with q vertices for the second inclusion.
The classes $\mathcal{G}(5,1)$ and $\mathcal{G}(6,2)$ are not comparable (take the path P_5 respectively a spider with 6 vertices). $\qquad\qquad\square$

As already indicated in Section 1 it is known from [12] that P_4-*reducible graphs* belong to the class $\mathcal{G}(5,1)$. We would like to mention another interesting set of graphs. A graph G is called P_4-*lite* [14] if every induced subgraph of G with at most six vertices either contains at most two P_4s or is isomorphic to a spider with six vertices. It is an easy observation that P_4-lite graphs are a proper superclass of $\mathcal{G}(5,1)$ and $\mathcal{G}(6,2)$ and a proper subclass of $\mathcal{G}(7,3)$. Up to now no polynomial isomorphism test for P_4-lite graphs was known.

It follows immediately from Theorem 2.3 that for any graph G the tree representation given above is unique up to isomorphism. It is known from [9] that it can be obtained in time polynomial in the number of vertices in G. Note that in our special case of $(q, q-4)$ graphs the nontrivial leaves of the tree represent

- spiders   if $q = 5$
- graphs with less than seven vertices or spiders   if $q = 7$
- graphs with less than q vertices   if $q = 6$ or $q \geq 8$.

With this information we are able to give an efficient isomorphism test. Here is an informal description. The algorithm tests whether two $(q, q-4)$ graphs are isomorphic or not. In the positive case it stops in state 'true', otherwise in state 'false'.

Algorithm ISOMORPH(G_1, G_2, *Boole*)

Input: Two $(q, q - 4)$ graphs G_1, G_2.
Output: A boolean variable *Boole*, which is true or false
depending on whether G_1 and G_2 are isomorphic.

Step 1. Construct the representing trees T_1, T_2 for G_1 and G_2.

Step 2. Test all pairs of graphs corresponding to leaves in T_1 and T_2
for isomorphism and assign two leaves the same label
if and only if the corresponding graphs are isomorphic.
As a result we obtain two labeled trees T_1^*, T_2^* (with integer labels
on the internal nodes and on the leaves).

Step 3. Perform a labeled tree isomorphism test for T_1^* and T_2^*.
If T_1^* is isomorphic to T_2^* then set *Boole* := *true* else set *Boole* := *false*.

The correctness of the algorithm is evident. It is well known that labeled tree
isomorphism can be tested in time linear in the number of vertices of the tree
(see e.g. [1]). Therefore it remains to ensure that transforming the representing
trees of G_1, G_2 into labeled trees can be done in polynomial time.

The crucial point is that the subgraphs associated with the leaves are very
simple. If the number of vertices is restricted by the constant q then isomorphism
testing for each pair of subgraphs requires only constant time. If the subgraphs
are spiders then isomorphism testing can be done in time linear in the size of the
spiders (note that the stable set of the spider consists of all vertices with minimal
number of neighbors). This considerations imply the following statement.

Theorem 5.2 *Let q be fixed. Then isomorphism of* $(q, q-4)$ *graphs can be tested
in polynomial time.*

6   Conclusions and Open Problems

In this work we proved that, for any fixed $q \geq 4$, $(q, q - 4)$ graphs admit a tree
representation which enables a polynomial isomorphism test. This generalizes
known results on cographs, P_4-reducible graphs and P_4-sparse graphs.

It is an open question whether the tree representation for arbitrary graphs
can be found in time linear in the size of the graph. If this is true then it
would immediately imply a linear isomorphism test and also a linear recognition
algorithm for $(q, q - 4)$ graphs (essentially we have to check the leaves of the
representing tree for membership in the class $\mathcal{G}(q, q - 4)$). Note that the naive
method "examine all subsets $U \subseteq V$ of cardinality q and count the P_4s in $G(U)$"
shows that the recognition problem is polynomial. Both the isomorphism and

the recognition problem are known to be solvable in linear time for cographs (see [7]) and P_4-sparse graphs (see [11]). We conjecture that this is also possible for $(q, q - 4)$ graphs with $q \geq 6$, using similar techniques.

Each $(q, q-4)$ graph is also a $(q, q-3)$ graph, therefore $\mathcal{G}(q, q-4) \subseteq \mathcal{G}(q, q-3)$ holds. Obviously $\mathcal{G}(4, 1)$ is the set of all graphs. It is easy to see that $\mathcal{G}(5, 2)$ coincides with the class of graphs which contain no induced cycle of length five. We conclude with an isomorphism completeness result (a problem is *isomorphism complete* if it is polynomial time equivalent to graph isomorphism).

Lemma 6.1 *Isomorphism testing of $(q, q - 3)$ graphs, $q \in \{4, 5, 6\}$, is isomorphism complete.*

Proof. The statement is trivial for $q = 4$. For $q = 5$ it follows from the fact that $\mathcal{G}(5, 2)$ contains all bipartite graphs, where the isomorphism problem is known to be isomorphism complete (see [3]).

Let $q = 6$. We give a polynomial reduction from the set of all graphs to the class $\mathcal{G}(6, 3)$ such that two graphs are isomorphic if and only if the corresponding $(6, 3)$ graphs are isomorphic.

Let $G = (V, E)$ be an arbitrary graph and $v \in V$. Assume that $N(v) = \{u_1, u_2, \ldots, u_r\}$. Replace each nonisolated vertex $v \in V$ by a clique with $|N(v)| = r$ vertices, say $w_1, \ldots, w_r$, and join all r pairs u_i, w_i by an edge. Furthermore, replace each edge which connects vertices from two different such cliques by a path of length two. It is an easy task to verify that the resulting graph is a $(6, 3)$ graph. $\qquad\qquad\square$

The complexity of the isomorphism problem remains unknown for the classes $\mathcal{G}(q, q - 3)$, $q \geq 7$.

Acknowledgment. The authors would like to thank Gottfried Tinhofer for stimulating discussions and valuable remarks.

References

1. A.V. Aho, J.E. Hopcroft and J.D. Ullman, *The Design and Analysis of Computer Algorithms*, Addison-Wesley, Reading, MA, 1974.
2. L. Babel, I. Ponomarenko and G. Tinhofer, Directed path graph isomorphism, in: E.W. Mayr, G. Schmidt, G. Tinhofer, eds., *Graph-Theoretic Concepts in Computer Science, 20th International Workshop, WG'94*, Lecture Notes in Computer Science, Springer-Verlag, Berlin, 1995.
3. K.S. Booth, C.J. Colbourn, Problems polynomially equivalent to graph isomorphism, *Report No. CS-77-04*, Computer Science Department, University of Waterloo (1979).
4. K.S. Booth, G.S. Lueker, A linear time algorithm for deciding interval graph isomorphism, *Journal of the ACM*, 26 (1979), 183–195.
5. V. Chvatal, Perfectly ordered graphs, in: C. Berge and V. Chvatal, eds., *Topics on Perfect Graphs*, North-Holland, Amsterdam, 1984, 63–65.

6. D.G. Corneil, H. Lerchs and L. Stewart Burlingham, Complement reducible graphs, *Discrete Applied Mathematics*, 3 (1981), 163–174.

7. D.G. Corneil, Y. Perl and L.K. Stewart, A linear recognition algorithm for cographs, *SIAM Journal on Computing*, 14 (1985), 926–934.

8. M.C. Golumbic, *Algorithmic Graph Theory and Perfect Graphs*, Academic Press, New York, 1980.

9. B. Jamison and S. Olariu, On the homogeneous decomposition of graphs, in: E.W. Mayr, ed., *Graph-Theoretic Concepts in Computer Science, 18th International Workshop, WG'92*, Lecture Notes in Computer Science, Springer-Verlag, Berlin 1993.

10. B. Jamison and S. Olariu, A Unique Tree Representation for P_4-sparse Graphs, *Discrete Applied Mathematics*, 35 (1992), 115–129.

11. B. Jamison and S. Olariu, Recognizing P_4-sparse graphs in linear time, *SIAM Journal on Computing*, 21 (1992), 381–406.

12. B. Jamison and S. Olariu, P_4-reducible graphs, a class of uniquely tree representable graphs, *Studies in Applied Mathematics*, 81 (1989), 79–87.

13. B. Jamison and S. Olariu, On a unique tree representation for P_4-extendible graphs, *Discrete Applied Mathematics*, 34 (1991), 151–164.

14. B. Jamison and S. Olariu, A new class of brittle graphs, *Studies in Applied Mathematics*, 81 (1989), 89–92.

15. M.M. Klawe, M.M. Corneil and A. Proskurowski, Isomorphism testing in hook-up graphs, *SIAM Journal on Algebraic and Discrete Methods*, 3 (1982), 260–274.

16. E.L. Lawler, Graphical algorithms and their complexity, *Math. Center Tracts*, 81 (1976), 3–32.

A Dynamic Algorithm for Line Graph Recognition

Daniele Giorgio Degiorgi[1] and Klaus Simon[2]

[1] via Maraini 4, CH-6900 Massagno, Switzerland[***]
[2] Institute for Theoretical Computer Science, Swiss Federal Institute of Technology
Zurich, CH-8092 Zurich, Switzerland[†]

Abstract. For a graph $G = (V, E)$ its line graph $L(G)$ has the node set
E and two nodes of $L(G)$ are adjacent if the corresponding edges of G
have a common endpoint. The problem of finding G for a given L was
already optimally solved by LEHOT[7] and ROUSSOPOULOS[11]. Here we
present a new dynamic solution to this problem, where we can add or
delete a node v in $L(G)$ in time proportional to the size of its adjacency
list.

1 Introduction

1.1 Motivation

In recent years, the dynamization of graph algorithms has become a current
research field. In particular if we think of testing predicates on graphs, see
for example EPPSTEIN ET AL.[1] or RAUCH[10]. Accordingly, we consider line
graphs which are a classical topic in the theory of special graphs, see for example GOLUMBIC[3] or SIMON[12].

Line graphs have some interesting algorithmic aspects. In order to illustrate
this, let L be a line graph and G its root graph. Then a matching in G is an
independent set in L and vice versa; the edges incident to a node of degree k in
G are the k nodes of a complete subgraph in L and the converse is true for all
$k > 3$. The maximum matching and the maximum degree problems can be solved
in polynomial time in G and it is easy to show that the maximum independent
set and maximum clique problems in L can also be solved in polynomial time.

A symmetric binary relation between n elements can be represented by an
undirected graph with n nodes. To achieve an $O(1)$ time by checking if two
elements are in relation one normally needs $O(n^2)$ space. Using intersection
graphs[5] the space can be reduced to $O(kn)$, where k is the cardinality of the
bigger of the sets representing the nodes, and the time is then $O(k)$. For line
graphs $k = 2$, as the sets are the edges of the root graph, thus achieving an $O(1)$
time for check and needing $O(n)$ space.

[***] Email: degiorgi@inf.ethz.ch. This author was partially supported by the Swiss
National Science Foundation
[†] Email: simon@inf.ethz.ch
[5] See for example HARARY[4]

1.2 Previous solutions

The first linear algorithm for line graph recognition is due to LEHOT[7]. He uses a characterization of line graphs due to VAN ROOIJ and WILF[14] who says that a graph G is a line graph if G does not have the complete bipartite $K_{1,3}$ as induced subgraph, and whenever two odd triangles[6] have a common edge, the subgraph induced by their nodes is the complete graph K_4.

The second solution was given by ROUSSOPOULOS[11], mainly using another characterization due to KRAUSZ[6], who states that G is a line graph if the edges of G can be partitioned into complete subgraphs in such a way that no node lies in more than two subgraphs.

1.3 Our solution

We will use here a new approach, based on a modification of the constructive proof due to ORE[9] of a theorem of WHITNEY[15] which says that if two connected graphs with more than four nodes are edge isomorphic, then there exists exactly one node isomorphism that induces the edge isomorphism.

This approach is a simplification of the problem: the cited algorithms uses global properties of line graphs while our algorithm uses only local ones and thus allowes an incremental recognition.

Further, we will show that this method allows to check in linear time whether a local modification in the original graph (i.e. adding or deleting an edge or a node) preserves the line graph property or not.

2 Notation

A *graph* G is a finite set of *nodes (vertices)* V together with a set E of unordered pairs of distinct nodes of V called *edges*. We represent the edge with the nodes v and w as vw (or wv which is the same edge) but we mean the set $\{v, w\}$. The *adjacency list* $\Gamma_G(v)$ of a node v in G is given by $\Gamma_G(x) = \{\, y \mid xy \in E \,\}$. A node not contained in any edge is said to be *isolated*. A graph is *complete* if every pair of distinct nodes is adjacent. The complete graph on n nodes is usually denoted by K_n. A subset $\{v_0, ..., v_k\}$ of V is a *path* of length k if all v_i, $0 \leq i \leq k$, are different and E contains all edges $v_{i-1}v_i$, $0 < i \leq k$. A graph is *connected* if each pair of nodes is contained in at least one path. We say that two graphs $G = (V, E)$ and $G' = (V', E')$ are *(node) isomorphic* (written $G \cong G'$) if there exists a bijective function $\phi: V \to V'$ such that

$$\forall v, w \in V : \quad vw \in E \iff \phi(v)\phi(w) \in E'.$$

Two graphs $G = (V, E)$ and $G' = (V', E')$ are *edge isomorphic* if there exists a bijective function $\psi: E \to E'$ such that

$$\forall e_1, e_2 \in E : \quad |e_1 \cap e_2| = |\psi(e_1) \cap \psi(e_2)|.$$

---

[6] An odd triangle is a triangle which is not even, and an even triangle is such that every node is adjacent to two or zero nodes of the triangle.

For any subset A of V the subgraph G_A of $G = (V, E)$ is defined by

$$G_A \stackrel{\text{def}}{=} (A, \{\, e \in E \mid e \subseteq A \,\}),$$

and is called the *induced* subgraph of A. For $B \subseteq E$ we designate $G_B = (V(B), B)$ the subgraph of G *spanned* by B, where $V(B)$ is the set of nodes which are incident with at least one edge in B (formally $V(B) = \bigcup B$). In this context $G_V = G$ for all graphs $G = (V, E)$ while $G_E = G$ if and only if G does not contain isolated nodes. We will also use the following abbreviations: Let $G = (V, E)$ be a graph, $v \in V$ a node and $e \in E$ an edge then $G - v = G_{V - \{v\}}$ and $G - e = G_{E - \{e\}}$. A set of nodes $T \subseteq V$ is a *stable set* if the subgraph of G induced by T contains no edges. For an integer r we say that a graph $G = (V, E)$ is r-colourable (*bipartite* for $r = 2$) if there exists a partition $V_1, \ldots, V_r$ of the nodes of G such that V_i, $1 \leq i \leq r$, is a stable set. The partition is also called an r-colouring and we say that all nodes in V_i have the same colour. With $K_{n,m}$ we denote a bipartite graph such that $|V_1| = n$, $|V_2| = m$, and xy is an edge for all $x \in V_1$, $y \in V_2$.

3   Line Graphs

The *line graph* $L(G) = (V_L, E_L)$ of G has E as its set of nodes and two nodes of $L(G)$ are adjacent whenever the respective edges in G have a common node. Formally:

$$V_L = E \quad \text{and} \quad E_L = \{e_1 e_2 \mid e_1 e_2 \subseteq E \wedge |e_1 \cup e_2| = 3\}.$$

On the other hand, we will say that G is the *root graph* of $L(G)$ or simply the *root*. Clearly if two line graphs are isomorphic, then the respective root graphs will be edge isomorphic. A graph $H = (V, E)$ is a line graph, if there exists a graph G for which $H \cong L(G)$. Our algorithm for recognizing line graphs is based on the following well-known theorem.

Theorem 1. WHITNEY[15] *Let G and G' be connected graphs with isomorphic line graphs. Then G and G' are isomorphic unless one is K_3 and the other is $K_{1,3}$.*

An elegant proof[7] of this theorem was given by JUNG[5]. His work includes

Theorem 2. *Let $G = (V, E)$ and $G' = (V', E')$ be connected edge isomorphic graphs with more than four nodes. Then G and G' are node isomorphic and there exists exactly one node isomorphism which generates the given edge isomorphism.*

The foregoing theorem is not true for small graphs. In Fig. 1 we have two edge isomorphic graphs which are not node isomorphic. For each of the three graphs in Fig. 2, the given edge isomorphism, by which each edge is mapped to the edge with the same name, is not induced by any node isomorphism. Fortunately the graphs in Figs. 1 and 2 are the only graphs for which this happens.

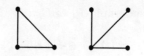

Fig. 1. The K_3 and $K_{1,3}$ are edge isomorphic but not node isomorphic.

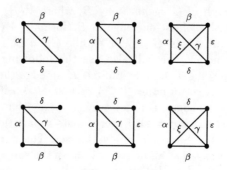

Fig. 2. Edge isomorphisms not induced by any node isomorphism

Our aim is to develop an incremental algorithm to recognize a line graph. We start with

Definition 3. Let $L = (E, F)$, $|E| \geq 4$, be a connected line graph with root $G = (V, E)$. Further, let e be a new node of L, $e \notin E$, and $\Gamma_L(e) \neq \emptyset$ its adjacency list in L. Then

$$X(e) = (V_X, E_X)$$

is defined as the subgraph of G spanned by $\Gamma_L(e)$.

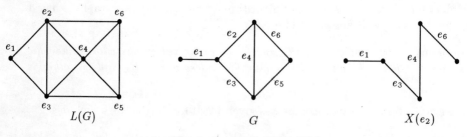

Fig. 3. An example for $X(e)$

Note that by definition $X(e)$ does not contain the edge e.

Let us now consider the typical situation in which we will apply our method:

– The root G of L is constructed,

---

[7] For a textbook description see *Harary*[4, p.72f]

– we know the graph $X(e)$ and
– we are looking for a possible placement of the edge e in $X(e)$.

We will see that the achievement of this goal is characterized by a special set of nodes called an anchor.

Definition 4. A subset T of V_X is an *anchor* if it fulfills:

A1. T is a stable set with $1 \leq |T| \leq 2$.
A2. Every edge in E_X has one endpoint in T, i.e.

$$\forall vw \in E_X : \quad |vw \cap T| = 1. \tag{1}$$

A3. Every edge not in E_X has no endpoint in T, i.e.

$$\forall vw \in E - E_X : \quad vw \cap T = \emptyset. \tag{2}$$

Definition 5. Let T be an anchor of $X(e)$. If $|T| = 2$ then let $T = \{x, y\}$ and otherwise $T = \{x\}$. In the second case y should be a new node of $G = (V, E)$. On these conditions the graph $G(e) = G + u$ with $u = xy$ is called the *extension* of G with respect to e.

Lemma 6. *In the situation above, let* $L' = (E \cup \{u\}, F')$ *be the line graph of* $G + u$, *where* $G + u, u = xy$, *is the extension of* G *with respect to* e. *Then* L' *and* $L + e$ *are isomorphic.*

Proof. At first, we claim that L' and $L + e$ are edge isomorphic. By Definition 4 part A2 we find for an edge $z \in \Gamma_L(e)$ of G

$$z \cap T \neq \emptyset.$$

Since A1 implies

$$T \subseteq u$$

we observe

$$z \cap u \neq \emptyset.$$

From this we infer for L' and $L + e$

$$\forall z \in \Gamma_L(e) : \quad uz \in F'. \tag{3}$$

Accordingly we can define a map $f : F_1 \to F'$ with

$$F_1 = F \cup \{ ez \mid z \in \Gamma_L(e) \}$$

by setting

$$\forall c \in F : \quad f(c) = c \tag{4}$$

and

$$\forall ez \in F_1 - F : \quad f(ez) = uz. \tag{5}$$

This function is well-defined by (3) and injective by definition. From the injectivity of f restricted to $\{\, ez \mid z \in \Gamma_L(e) \,\}$ follows its surjectivity, since with A3 we get

$$\Gamma_{L'}(u) \subseteq E_X = \Gamma_L(e)$$

and further

$$|\{\, uh \mid h \in \Gamma_{L'}(u) \,\}| \leq |\{\, ez \mid z \in \Gamma_L(e) \,\}|.$$

Therefore f is bijective and it remains to show that f is an edge isomorphism. By (4) we find immediately

$$\forall c_1, c_2 \in F : \quad |c_1 \cap c_2| = |f(c_1) \cap f(c_2)|.$$

Consider now $c_1 \in F$ and $c_2 = ez$ for any $z \in \Gamma_L(e)$. Then we get

$$c_1 \cap ez = \underbrace{(c_1 \cap e)}_{=\emptyset} \cup (c_1 \cap z) = \underbrace{(c_1 \cap u)}_{=\emptyset} \cup (c_1 \cap z) = c_1 \cap uz = f(c_1) \cap f(ez). \quad (6)$$

For $c_1 = ez_1$ and $c_2 = ez_2$, $\{z_1, z_2\} \subseteq \Gamma_L(e)$, we find in the same way

$$|ez_1 \cap ez_2| = |\{e\} \cup (z_1 \cap z_2)| = |\{u\} \cup (z_1 \cap z_2)| = |uz_1 \cap uz_2| = |f(ez_1) \cap f(ez_2)|. \quad (7)$$

This shows that L' and $L + e$ are edge isomorphic. From this and Theorem 2 follows that L' and $L + e$ are node isomorphic. □

The foregoing lemma gives a characterization of the right placement for the e in the subgraph $X(e)$. Moreover, it implies that if T exists, then $L + e$ is a line graph with root $G + u$. On the other hand, if $L + e$ is a line graph then the endpoints of e in G fulfill the definition of T. Furthermore, with Theorem 1 the root of $L + e$ is unique. Therefore we can summarize in

Corollary 7. *Let e be a new node of a line graph L such that $L+e$ is connected and has five or more nodes. Then the subgraph $X(e)$ of the root of L contains an anchor T if and only if $L + e$ is a line graph.*

The next step is the computation of an anchor T. Our central observation is

Lemma 8. *The subgraph $X(e) = (V_X, E_X)$ is bipartite and T together with $V_X - T$ build a 2-colouring of $X(e)$.*

Proof. With Definition 4 part A1, the anchor T is a stable set. By A2 no edge of E_X has both end points in $(V_X - T)$ which implies that $(V_X - T)$ is a stable set too. □

Next we find

Lemma 9. *If an anchor T exists then the subgraph $X(e)$ contains at most two connected components.*

Proof. Since $X(e)$ is spanned by the edge set $\Gamma_L(e)$ it contains no isolated nodes. Therefore every component of $X(e)$ contains at least one edge, which implies that every component has at least one node in each of the two colour classes of $X(e)$. For this reason if $X(e)$ would have three or more components, every colour class would contain more than two elements, against the definition of T. □

Now the situation becomes clear. In order to find T we determine the components of $X(e)$ and compute a 2–colouring for every component. The colouring classes of the components can be combined into the colouring classes of $X(e)$. One of them must be T. By Lemma 9 there are at most 4 possible candidates. Hence from a theoretical point of view the problem is solved, since we have only 4 possible cases and each of them can be checked in time $O(|\Gamma_L(e)|)$. But the right colouring class of $X(e)$ can be determined in a more efficient way.

Definition 10. We call a colouring class C of a component of $X(e)$ *impossible* if it fulfills one of the following conditions:

B1. There is an edge h of $E - E_X$ which has an endpoint in C.
B2. $|C| \geq 3$ and $X(e)$ is connected or $|C| \geq 2$ and $X(e)$ is not connected.

By Definition 4: If C is impossible it cannot be part of T. Now we are able to formulate our main theorem.

Theorem 11. *Let e be a new node of the graph $L = (E, F)$, $e \notin E$, with adjacency list $\Gamma_L(e)$. Furthermore let $G = (V, E)$ be the root graph of L.*

- *It can be checked in time $O(|\Gamma_L(e)|)$ wether or not $L + e$ has a root and –if it exists– the root of L can be modified in that complexity to a root of $L + e$.*
- *Every node v of L can be deleted in time $O(|\Gamma_L(v)|)$.*

Proof. We assume that L and G are represented in the usual way by an adjacency list. In addition we maintain a one-to-one map for the nodes of L to the edges of G and the degree of every node of G. More exactly, the nodes correspond to indices of an array of pointer to edges. Every edge embodies its endpoints and for each endpoint x a pointer to the previous and the next edge in the adjacency list of x. Double linking of our list allows inserting a new element before a given one or deleting any element in time $O(1)$. Therefore the deletion of a node v in L takes time $O(|\Gamma_L(v)|)$ and in G time $O(1)$. The correctness is obviously, since the line graph property is inhereted by definition.

The difficult part is the insertion. Our foregoing analysis was made under the precondition that G is connected and has 5 or more nodes. Now G could contain different components and some of them may have less than 5 nodes. Therefore we use an initialization procedure `small_root` which determines the root of a small connected graph $L_1 = (E_1, F_1)$, $|E_1| \leq 6$[8], by exhaustive search. Due to the bounded number of graphs with six or less nodes a call of `small_root` takes time $O(1)$.

---
[8] A graph with less than 5 nodes can not have more than 6 edges.

In order to extend G to the root of $L + e$, we first build its subgraph $X(e) = (V_X, E_X)$ spanned by $\Gamma_L(e)$. This can be done in time $O(|\Gamma_L(e)|)$. Now we have the following cases.

Case 1: $X(e)$ *is connected.* Let $H = (V_H, E_H)$ be the component of G containing $X(e)$.

    Case 1.1 $|V_H| \geq 5$. This is the normal case described above. Hence we compute the unique 2–colouring of $X(e)$. Next we check which colour class is impossible. One of them must be impossible, since for $|V_H| \geq 5$ either B1 is satisfied and or B2 must hold. If exactly one colouring class is impossible then the remaining class is the anchor T and we extend G like described in Definition 5. If both colour classes are impossible then T does not exist and $L + e$ is not a line graph.

    Case 1.2 $|V_H| \leq 4$. Then we replace H by the result of $\mathtt{small\_root}(L_{E_H} + e)$ if $L_{E_H} + e$ has a root; if not then L is not a line graph.

In order to decide whether $|V_H| \leq 4$ or $|V_H| \geq 5$ we start a breadth-first search from a node in $X(e)$ and stop at the latest when five nodes have been examined. This takes time $O(1)$. In Case 1.1 we compute a 2–colouring in the graph $X(e)$ in time $O(|V_X| + |E_X|)$ by breadth-first search. If this colouring is not found, then we have finished since by Lemma 8 the graph $L + e$ has no root. During this exploration the colour classes can be tested for "impossibility" in time $O(|V_X|)$, by computing their sizes and comparing the degrees of their elements in $X(e)$ and G. Therefore, the Case 1.1 can be handled in time $O(|V_X| + |E_X|)$. By definition of $\mathtt{small\_root}$ the Case 1.2 needs time $O(1)$.

Case 2. $X(e)$ *is not connected.* Then by Lemma 9 there are two connected components X_1 and X_2 of $X(e)$. Let $H_1 = (V_1, E_1)$ and $H_2 = (V_2, E_2)$ be the corresponding components of G which contain X_1 and X_2 ($H_1 = H_2$ is possible).

    Case 2.1 $|V_1| > 5 \wedge |V_2| > 5$. Here we proceed as in the foregoing discussion. We compute a unique 2–colouring for X_1 and X_2, respectively. Then we test which colour classes are impossible. By our preconditions if at least one class in each component remains, then the anchor T is their combination and we can extend G like in Definition 5. In all other cases T does not exist which implies by Corollary 7 that L is not a line graph.

    Case 2.2 $|V_1 \cup V_2| \leq 5$. Then remove $H_1 \cup H_2$ and recompute the root of $L_{E_1 \cup E_2 \cup \{e\}}$ with $\mathtt{small\_root}$.

    Case 2.3 $(|V_1 \cup V_2| > 5) \wedge (|V_1| < 5)$. First we remove the edges E_1 from G and the nodes E_1 from L, respectively. If $|V_2| < 5$ then we proceed as follows:

        1. We add e and some edges of E_1 to E_2 such that E_2 reaches seven edges and its induced subgraph in L remains connected. Now we are sure that the root will contain at least 5 nodes and Theorem 2 can be applied. If there are not enough edges in E_1 to expand sufficiently E_2 then we can apply Case 2.2.

        2. We build a set S which contains the edges of E_1 not added to E_2.

3. We replace H_2 by the result of a new call $\mathtt{small\_root}(L_{E_2})$.

If $|V_2| \geq 5$ then let S be the set $E_1 \cup \{e\}$.

Now we (re)insert the elements of S into L, if this is possible. We repeat the following step until $S = \emptyset$ or a failure results.

> Select an edge h of S such that the subgraph of L induced by $E_2 \cup \{h\}$ is connected. Extend H_2 to the root graph of $L_{E_2} + h$. Note that by the connectivity of $L_{E_2} + h$ this can be done with Case 1.1 or 2.1. Delete h from S and add it to E_2.

Case 2.4 $(|V_1 \cup V_2| > 5) \wedge (|V_2| < 5)$. Analogously to Case 2.3 with H_1 instead of H_2.

We come to the running time of Case 2. Analogously to Case 1 we first start breadth-first search from a node in $X_1(X_2)$ to determine a subgraph of $H_1(H_2)$. If BFS find five edges spanning a connected subgraph of $H_1(H_2)$ then the procedure breaks off.

Clearly, by this preparation the differentiation in Case 2.1–2.4 takes time $O(1)$. With a similar argumentation as for Case 1.1 we observe time $O(|\Gamma_L(e)|)$ also for the Case 2.1. The execution of the Case 2.2 needs time $O(1)$, since here $H_1 \cup H_2$ is of bounded size. In Case 2.3, by the same argument the removal of H_1 and the extension of H_2 to five nodes (if this is necessary) can be done in constant time. Let now S be the set in its initial state. Since $|S - \{e\}| \leq 4$ and every edge of $S - \{e\}$ is adjacent to at most three other edges in H_1, the reinsertion into L with Case 1.1 or 2.1 needs time $O(1)$. It probably remains the insertion of e with Case 1.1 which then gives costs of $O(|\Gamma_L(e)|)$. Altogether, we get $O(|\Gamma_L(e)|)$ as execution time of Case 2.3 and in the same way of Case 2.4.

Therefore, every part of our method is executable in time $O(|\Gamma_L(e)|)$. Because of this and the fact that we have only a fix number of parts, we reach as claimed the total running time of $O(|\Gamma_L(e)|)$. $\qquad\qquad\square$

4 Remarks

The insertion or deletion of an edge in a line graph can be easily reduced to the deletion of one of its endpoints with successive insertion (with the new adjacency list). The complexity is thus the same as that of node insertion.

The parallelisation of line graphs recognition was already studied. See for example NAOR AND NOVICK[8].

For a given connected graph L our algorithm can be simplified, since if we insert new nodes in the BFS scanning order, after the initialization we only need the cases 1.1 and 2.1 of Theorem 11.

This version of the algorithm, together with those of Lehot and Roussopoulos, are been implemented in our institute by SCHMOCKER[13]. The results of the comparison of the run times for the differents algorithms and different graph classes are shown in Figs. 4 thru 8.

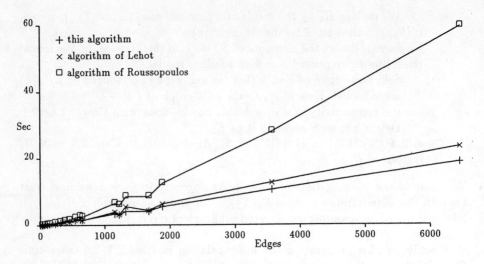

Fig. 4. Running time for sparse line-graphs

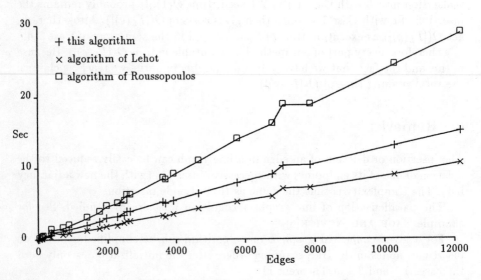

Fig. 5. Running time for dense line-graphs

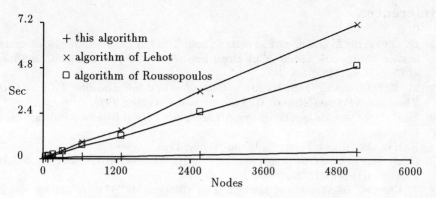

Fig. 6. Running time for planar non-line-graphs

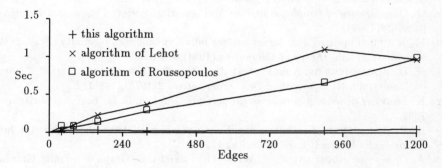

Fig. 7. Running time for sparse non-line-graphs

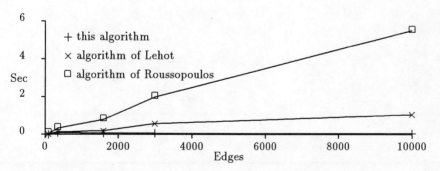

Fig. 8. Running time for dense non-line-graphs

References

1. D. EPPSTEIN, Z. GALIL, G.F. ITALIANO AND T.SPENCER. Separator based sparsi-fication for dynamic planar graph algorithms. *Proc. 25th Annual Symp. on Theory of Computing* (1993),208–217.
2. M. R. GAREY AND D. S. JOHNSON. *Computers and Intractability. A Guide to the Theory of NP-Completeness.* W. Freeman and Company, 1979.
3. M. C. GOLUMBIC. *Algorithmic Graph Theory and Perfect Graphs*, Academic Press, 1980.
4. F. HARARY. *Graph Theory*, Addison-Wesley, 1969.
5. H. A. JUNG. Zu einem Isomorphiesatz von H. Whitney für Graphen. *Math. Annalen* **164**(1966), 270–271.
6. J. KRAUSZ. Démonstration nouvelle d'un théorème de Whitney sur les résaux. *Mat. Fiz. Lapok* **50**(1943), 75–85.
7. P. G. H. LEHOT. An Optimal Algorithm to Detect a Line Graph and Output Its Root Graph. *Journal of the ACM* **21**(1974),569–575.
8. J. NAOR AND M.B.NOVICK. An Efficient Reconstruction of a Graph from Its Line Graph in Parallel. *Journal of Algorithms* **11**(1990),132–143.
9. O. ORE. *Theory of Graphs*, American Mathematical Society Colloquium Publications, **38**(1962).
10. M. RAUCH. Improved data structures for fully dynamic biconnectivity. *Proc. 26th Annual Symp. on Theory of Computing* (1994),686–695.
11. N. D. ROUSSOPOULOS. A max$\{m, n\}$ Algorithm for Detecting the Graph H from its Line Graph G. *Information Processing Letters* **2**(1973), 108–112.
12. K. SIMON. *Effiziente Algorithmen für perfekte Graphen.* B. G. Teubner, Stuttgart, 1992.
13. H.-P. SCHMOCKER. *Erkennung von Kantengraphen.* Diploma thesis, Institut für Theoretische Informatik, ETH Zürich, 1991.
14. A. C. M. VAN ROOIJ AND H. S. WILF. The Interchange Graph of a Finite Graph. *Acta Mathematica Academiae Scientiarum Hungaricae* **16**(1965), 263–269.
15. H. WHITNEY. Congruent Graphs and the Connectivity of Graphs. *American Journal of Mathematics* **54**(1932), 150–168.

Incremental hive graph*

Fabrizio d'Amore and Roberto Giaccio

Università di Roma "La Sapienza,"
Dipartimento di Informatica e Sistemistica,
Via Salaria 113, I-00198 Roma, Italy.

Abstract. The hive graph is a rectangular graph satisfying some additional condition widely used in computational geometry for solving several kinds of fundamental queries. It has been introduced by Chazelle [1] as a static structure that, after a preprocessing time of $O(n \log n)$, and coupled with one of the several data structure for optimally answering point location queries, allows to solve orthogonal segment intersection queries on a set of parallel line segments in optimal time and space.

In this paper we present an optimal algorithm for incrementally building a hive graph structure: while it retains the same performance in query answering, it also allows to incrementally insert new line segments with $O(\log n)$ worst case time per update. Our technique exploits a novel "eager" approach.

Some other dynamic operations performable on our structure in optimal time, such as *Purge* and *Backtrack,* are described. Also, we discuss some applications of our results.

1   Introduction

The hive graph [1] is a rectangular graph [5] in which any internal face is adjacent to at most two faces above and two faces below it. Hive graphs are generally built starting from a set of horizontal line segments which are enclosed in a larger isothetic rectangle so that the union of these segments and of those lying on the boundary of the enclosing rectangle is equal to the union of all the edge drawings.

The most attracting property of the hive graph is that it allows to efficiently navigate on the graph moving from any horizontal edge to that immediately above or below it in constant time. Such property is important in many application: in computational geometry, for example to solve orthogonal segment intersection queries [1], dominance and range searching problems [2]; in databases handling multidimensional and spatial data, for solving several kinds of queries (see, for instance, [9]).

In [1] the hive graph is introduced in a static setting: given a set of n horizontal line segments an algorithm is presented for building in $O(n \log n)$ time and

---

* Work partially supported by the Italian National Projects MURST 40% "Efficienza di Algoritmi e Progetto di Strutture Informative" and CNR "Ambienti e strumenti per la gestione di informazioni temporali."

$O(n)$ space a hive graph. The dynamic case, where line segments can be inserted and/or deleted, appears to be significantly more involved: in fact, in spite of the importance of the dynamic settings, we know only of one result in the literature, consisting of a fully dynamic algorithm capable of updating the structure in time $O(\log n + k)$, where k is the size of the output to the query defined by using the inserting (or deleting) line segment [7]. This leads to an overall worst case time $O(n)$. Hence, in the non-static setting no satisfactory solutions exist.

In this paper we provide an incremental algorithm for building a hive graph: while it allows to retain the same (optimal) performance of the static structure for the query time, it makes possible to update the structure to take account of the insertion of one or more line segments at the left and/or right of the graph. We show in this paper how to accomplish this in $O(\log n)$ worst case time per update. This bound can be achieved by virtue of our "eager" approach: it consists of anticipating some changes in the structure that could be required later because of updates. Our approach contrasts with the so-called "lazy" approach, where updates are deferred to the purpose of spanning their cost over successive operations.

We discuss also how to perform other update operations in the dynamic setting, among which the *Purge* of the first k updates and the *Backtrack* of the last k updates in worst case time $O(k)$. Moreover we present a database application where the incremental setting of the hive graph results particularly useful.

The paper is organized as follows. In Sect. 2 we give a formal definition of the hive graph and some related concepts; then, in Sect. 3 we show how to incrementally build a hive graph with logarithmic update time; in Sect. 4 we outline some applications of our results and finally, in Sect. 5 we point out further research directions.

2   Hive graphs

A *rectangular graph* [5] is a plane graph where all faces are four-sided and all edges are oriented in either the vertical or the horizontal direction. In addition, the graph enclosure must also be rectangular.

Given a set of horizontal line segments, a *hive graph* is a rectangular graph whose horizontal edges not on the boundary lie on and completely cover the segments of the given set, and whose internal faces satisfy the following condition: for each face there are at most two faces above and at most two faces below adjacent to it.

The hive graph was introduced by Chazelle in the framework of the filtering search technique [1] to the purpose of solving problems of the following form: preprocess a set of objects so that those satisfying a given property with respect to a query object can be listed very effectively.

Given a set of horizontal line segments,[2] we can obtain a rectangular graph

---

[2] In this paper for simplicity of notation we consider line segments in general positions, but our techniques also work in the more general case with no need of being adapted.

by enclosing them in a larger rectangle and drawing two vertical edges starting from each endpoint, the former to the nearest upper line segment and the latter to the nearest lower line segment. For sake of clarity in the following we use *line segments* for the original line segments and, while referring to any rectangular graph R, *rectangles* for the internal faces of R and *sides* for the sides of the faces of R. The graph defined above has a number of edges at most 9 times the number of line segments, since each line segment contributes with 4 vertical edges and each vertical edge causes at most a new split that creates a new edge.

In [1] a two-phases line sweep algorithm is presented to transform the rectangular graph defined above into a hive graph at most doubling the number of edges; the algorithm sweeps a horizontal line starting at the line segment with the lowest vertical coordinate, moving upward and splitting each rectangle whose lower side has more than one incident vertical edge. The process is then repeated moving downward from the upmost line segment.

According to the given definitions, in a hive graph the upper and lower sides of each rectangle can consist of one or two edges; in [1] the vertices that in the second case separate the two edges are called *upper* and *lower anomalies*, respectively. Hence, by definition, a rectangular graph is a hive graph if and only if all its rectangles have at most one upper and one lower anomaly. It is worth observing that the hive graph representing a given set of line segments is not unique.

On a hive graph H with edge set E we can efficiently solve the following query: given a vertical line segment q and an edge $e \in E$ intersecting q, report all horizontal edges of E above and below e intersecting q, ordered by ordinates. Algorithm *ReportAbove* illustrated in Fig. 1 solves the query for the edges above e; algorithm *ReportBelow* for the edges below e is similar and for simplicity is not described.

Since on a hive graph the *NextAbove* procedure runs in constant time, the *ReportAbove* algorithm has time complexity $O(k)$, where k is the number of reported edges. Furthermore, since each horizontal edge corresponds to a line segment, and horizontal edges with different ordinates to different line segments, the query defined above can also be used to solve the *orthogonal segment intersection query* at no extra asymptotic cost; this consists of reporting all the horizontal line segments intersecting a given vertical segment, assuming that a suitable entry point (any edge belonging to the solution) is input to algorithm *ReportAbove*. In practice, this means the capability of answering orthogonal segment intersection queries in $O(k + \log n)$ time [1] by coupling the hive graph with one of the structures capable of optimally answering *planar point location* queries: given a partition of the space and a point q locate the member of the partition containing q (see, for instance, [10, 11, 8, 3, 4]).

3 Incremental hive graph construction

In this section we discuss how to incrementally build a hive graph starting from an empty structure, and proceeding by inserting new line segments: this is carried

```
algorithm ReportAbove(e, q)
x := abscissa of q
e' := NextAbove(e, x)
while e' intersects q do
      report e'
      e' := NextAbove(e', x)
enddo

procedure NextAbove(e, x)
let R be the rectangle above e
let e' be one upper horizontal edge of R
if e' covers x
    then
           return e'
      else
           let e'' be the other upper horizontal edge of R
           return e''
endif
```

Fig. 1. The *ReportAbove* algorithm.

out by orderly inserting their endpoints. Even if we are going to examine only the case where the structure increases rightward, the same ideas apply for letting the structure increase leftward or bilaterally. The empty structure only consists of the enclosing rectangle, whose vertical sides have (by convention) abscissae $-\infty$ and $+\infty$ and whose horizontal sides have ordinate $-\infty$ and $+\infty$.

The main idea consists of sweeping a vertical line on the scenario, from left to right, performing an *OpenSegment* operation for any encountered left endpoint, and a *CloseSegment* operation for any encountered right endpoint. This sweeping algorithm only consists of one phase, versus the two-phases algorithm presented in [1]. The construction is incremental by virtue of the properties that at any time:

1. the rectangular graph built so far is a hive graph;
2. we can insert new line segments as long as they lie beyond the sweeping line.

Moreover, if a line segment has been opened but not yet closed, in the hive graph there is a horizontal edge corresponding to it (from its left endpoint up to a point on the right boundary); such edges are called *open horizontal edges*. This gives the possibility of closing at any time the line segment, which on the current structure becomes into moving the right endpoint of the corresponding open edge to the correct coordinate; this last operation *closes* the open horizontal edge at that coordinate.

If a left endpoint (x_1, y_1) is encountered by the sweeping line then three new edges are added to the hive graph: two vertical ones, the first upward and the second downward, starting at (x_1, y_1), and an open horizontal one starting at (x_1, y_1). As a side effect, the vertical edges split each of the two horizontal edges bounding the face containing (x_1, y_1) into two edges, and the open horizontal edge splits those on the right side of the face into two edges. This is accomplished by the *OpenSegment* operation.

If a right endpoint (x_2, y_2) is encountered by the sweeping line then the open horizontal edge at ordinate y_2 is closed at (x_2, y_2) and two new vertical edges, the first upward and the second downward, starting at (x_2, y_2) are added to the hive graph. Also in this case, these vertical edges split into two horizontal edges each of the two ones bounding the faces separated by the open edge and the vertical edges on the right side of those faces are joined together. This is accomplished by the *CloseSegment* operation.

What described actually only allows the incremental construction of a rectangular graph. In fact, whenever a new endpoint is encountered two new vertical edges are inserted: as a consequence, excepting for splits on the boundary, there will always be two faces that will incur in an increase of the number of their own upper and respectively lower anomalies. Therefore the condition on the maximum number of anomalies (see Sect. 2) is not guaranteed to be fulfilled.

A trivial approach to prevent such possibility would be to split a rectangle each time its number of upper or lower anomalies becomes two; the problem with this approach is that it could cause a new anomaly in another rectangle and so on, giving rise to a possible long splitting chain. It is easy to see that in this case the number of all splits is linear in the total number of line segments. Even if this insures that update operations have constant amortized time, single updates will require in the worst case a time linear in the number of the already swept line segments.

In order to prevent such behavior, we use a more clever (eager) approach. Instead of propagating splits when the number of anomalies becomes two, we insert "special" vertical edges beyond the sweeping line when the number of anomalies becomes one. We denote such edges as *dummy edges.* Their abscissae are not yet defined, but are treated as if they are greater than all already swept abscissae. In contrast, a non-dummy edge is called *regular edge*. Moreover let a *stripe* be the rectangular region corresponding to a rightmost rectangle on the graph obtained from the hive graph by removing the dummy edges; it can contain at most one dummy edge.

By this approach we obtain that any set of adjacent stripes, each of them having one anomaly, is delimited at right by a chain of dummy edges. If the number of anomalies of a stripe passes from one to two then we *fix* the abscissa of its dummy edge and of the involved chain of dummy edges to that of the endpoint which caused the new anomaly. Now observe that such fixes create the same chain of vertical edges we would have created in the trivial approach, with the difference that no new edge needs to be explicitly introduced at this time since the cost of creating the vertical edges of the splitting chain has been spanned over several update operations, according to the spirit of the eager approach. As a consequence, the entire cost of the splitting chain reduces to that of the fixes. In Subsect. 3.2 we show that this can be accomplished in constant time.

In the rest of this section we describe the algorithms in detail and study their complexity.

3.1 Update algorithms

We premise a few notations. Given a vertical edge α, its abscissa is denoted by $x(\alpha)$; this notation also holds for the vertices. Given a stripe A, we denote by up(A) and down(A) its adjacent upper and lower stripes, respectively. If α is a dummy edge we denote by high(α) and low(α) its upper and lower vertices, respectively, and by chain(α) the maximal path of dummy edges containing α. In Fig. 2 and 3 algorithms *OpenSegment* and *CloseSegment* are presented.

---

algorithm *OpenSegment*(x, y)
find the stripe A containing the point (x, y)
if *A has dummy edge* α
   then *fix* chain(α) *to x*
   else *insert an edge* α *with abscissa x*
      if up(A) *has dummy edge* β
         then *fix* chain(β) *to x*
         else *insert a dummy edge* β *in* up(A)
             join β *to the possible upper chain*
      endif
      if down(A) *has dummy edge* γ
         then *fix* chain(γ) *to x*
         else *insert a dummy edge* γ *in* down(A)
             join γ *to the possible lower chain*
      endif
endif
insert an open horizontal edge starting at (x, y)

Fig. 2. The *OpenSegment* algorithm.

These algorithms use some basic operations:

Find: find the stripe containing a given point;
InsertVerticalEdge: insert a dummy or a regular vertical edge in a stripe;
Join: join a newly created dummy edge to an adjacent chain;
Fix: fix a chain of dummy edges to a given abscissa.

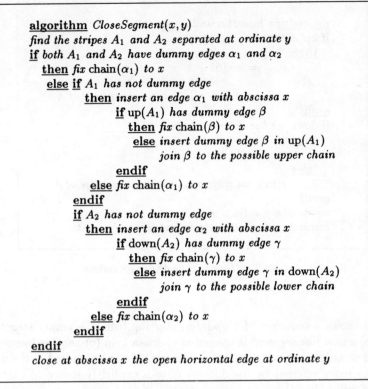

Fig. 3. The *CloseSegment* algorithm.

The *Find* operation is basically a search a dictionary whose items are the current open edges ordered by ordinates; as horizontal edges are opened or closed, the corresponding entries are inserted or deleted from the dictionary. The *InsertVerticalEdge* operation is described in detail in Fig. 4; note that it always concatenates the new edge to possible existing adjacent chains of dummy edges.

The main idea behind the *Join* and *Fix* operations is that if two dummy edges belong to the same chain they will be fixed to the same abscissa value; moreover, for each edge, we represent its abscissa as a reference to an entry holding its value; entries referred by dummy edges contain a special marker. When a new dummy edge is created we check if there exists an adjacent dummy edge; it is easy to convince that at most one adjacent dummy edge can exist; in this case we simply join it to the adjacent one, i.e., discard the new entry and let the new dummy edge refer to the same entry as the adjacent one. If a chain of dummy edges has to be fixed to a given abscissa, we simply store its value in the (unique) entry referred by all dummy edges in the chain.

---

procedure *InsertVerticalEdge*(A, x)
if up(A) *has dummy edge* β
   then
       $n_{low} := \text{low}(\beta)$
   else
       create the node n_{low} *on the upper edge of* A
endif
if down(A) *has dummy edge* γ
   then
       $n_{high} := \text{high}(\gamma)$
   else
       create the node n_{high} *on the lower edge of* A
endif
create edge $e = (n_{low}, n_{high})$
create an entry with value x *and let* e *refer to it*

Fig. 4. The *InsertVerticalEdge* procedure.

Fig. 5 shows a sequence of 5 updates, starting from the empty structure in (a). In (b) a new line segment is opened at abscissa 1; in (c) another line segment is opened at abscissa 2, which causes a dummy edge to be created in the upmost stripe; the entry referred by the dummy edge is explicitly shown to contain an undefined value. In (d) a new line segment is opened at abscissa 3, causing the previous dummy edge to be fixed at this value, and creating a new dummy edge in the lowest stripe. In (e) the line segment opened in (b) is closed at abscissa 4; that causes the creation of a new dummy edge above the existing one, whose abscissa entry is joined to that of the adjacent one. Finally, in (f) a last line segment is opened at abscissa 5, causing the chain formed by the two dummy edges to be fixed at 5; note that the fixing is accomplished by updating the unique entry referred by both the dummy segments in the chain. At the end of the sequence there are 3 currently open edges.

3.2 Analysis

Now we can analyze in detail the complexity of our algorithms.

Lemma 1. *If a stripe has one anomaly after an update operation then it has a dummy edge.*

Proof. It suffices to see that when *OpenSegment* or *CloseSegment* create a new vertical edge incident to the horizontal sides of a stripe A, either a dummy edge is created in A or such an edge already exists. $\qquad\square$

Lemma 2. *No rectangle has more than one anomaly.*

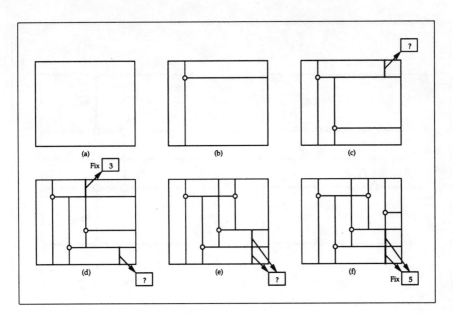

Fig. 5. A sequence of 5 updates.

Proof. Suppose that a rectangle A exists with more than one anomaly and let a_1 and a_2 be the two leftmost anomalies in A, with $x(a_1) < x(a_2)$; after a_1 was necessarily created (because of an *OpenSegment* or *CloseSegment* operation), A was a stripe and had to have a dummy edge α by Lemma 1; hence when a_2 was created chain(α) would have been fixed thus closing A and creating at right a new stripe with no anomalies. So A would have had one anomaly which contrasts the assumptions. $\square$

Theorem 3. *A sequence of n update operations builds a hive graph of size $O(n)$.*

Proof. On the ground of Lemma 2, starting from an empty graph any sequence of update operations builds a hive graph. Each *OpenSegment* operation can create at most 3 new regular edges, 2 dummy edges and cause 7 splits, which also create 7 more edges, giving a total of 12 new edges. Each *CloseSegment* operation can create 2 new regular edges, 2 dummy edges, cause 6 splits, which also create 6 more edges and 1 join, which deletes an existing edge, giving a total of 9 more edges. Hence the total number of edges is at most $12 \cdot n$. $\square$

In Fig. 6 the new edges and splits of the proof of Theorem 3 are shown, with part (a) and (b) referring to *OpenSegment* and *CloseSegment* operations respectively. In (a) the new edges e_1 to e_5 are created and circled vertices are the split ones; in (b) the new edges e_1 to e_4 are created and the vertex crossed out is a point whose incident edges are joined, thus reducing the total number of segments.

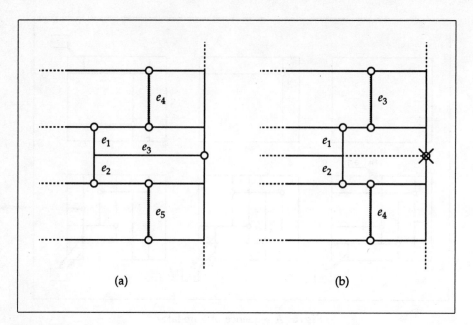

Fig. 6. New edges and splits caused by update operations. (a) Opening a new segment, (b) Closing a segment.

Theorem 4. *The OpenSegment and CloseSegment operations have time complexity $O(C(Find))$, where $C(Find)$ is the cost of the Find operation.*

Proof. The *Fix* operation is performed in constant time, since all the dummy edges in the same chain refer the same entry and fixing a chain to x is basically storing x in that entry. All the other operations have constant time complexity, since they are basically operations of edge-creation, edge-splitting and edge-joining on a graph with bounded vertex degree.   □

It is worth noting that the update cost is bounded by the cost of the *Find* operation. A straightforward approach should be storing all the current open edges in a dynamic balanced search tree ordered by ordinates; this allows a *Find* time $O(\log n)$, so giving the same time bound for the update operations. If we restrict the set of allowed ordinate values to the set $\{1, \ldots, U\}$ we can use van Emde Boas trees [13, 14, 6] to obtain a worst case update time $O(\log \log U)$.

3.3   Other operations

Some other operations can be performed on the incremental hive graph; it has been already mentioned that the structure can increase leftward or bilaterally with no changes on the technique. It is more interesting to point out that an incremental *Purge* operation can be performed that removes from the hive graph its leftmost endpoint. It is easy to convince that such an operation, which basically performs the same tasks of the previously introduced update operations in the reverse order, together with *OpenSegment* and *CloseSegment* allows to maintain a "moving window" on a larger hive graph. Moreover we do not need to carry out any *Find* operation, since it suffices to store a list of all endpoints already swept and to scan that list moving forward; hence the following theorem holds:

Theorem 5. *The Purge operation can be carried out in constant time.*

Another interesting operation that can be easily accomplished is the *Backtrack*, which undoes the last update; since each update modifies a constant fraction of the hive graph, we can associate with each endpoint already swept the information needed to restore the hive graph to the previous state. As in the case of the *Purge* operation we can locate the last endpoint in constant time by scanning backward the list of all endpoints already swept, so the following theorem also holds:

Theorem 6. *The Backtrack operation can be carried out in constant time.*

Of course the *Backtrack* operation can be iterated for undoing the last k updates in time $O(k)$.

4 Applications

For a description of the applications of the hive graph the reader is referred to [1, 2].

Here we wish to mention a database application in which the incremental setting of the hive graph results particularly useful. A *temporal database* is a database where information is handled by taking explicit account of its temporal components: *logical time,* referring to a time interval in which information is to be considered valid, and *physical time,* which is the time interval, possible semi-infinite, defined by the inserting and deleting (if any) time. Temporal databases are currently object of an extensive research activity (see, for instance, [12]). There are several different settings that can be of interest but all of them have in common the characteristic that information is never physically deleted from the database, even if when the user explicitly asks for the deletion. This guarantees at any instant the possibility of asking queries on present or past data.

A simple setting is that where the only temporal information of interest is the physical time (*rollback* databases). In these databases it is always possible to rollback any number of updates. In the case of mono-dimensional keys, information can be interpreted as a set of horizontal intervals whose ordinates are the stored keys and whose abscissae are the insertion and deletion times; inserted keys that have not been deleted yet correspond to semi-infinite horizontal intervals. Building an incremental hive graph for these intervals allows to efficiently solve range queries at any given instant and to update the structure in presence of insertions and/or deletions, thus giving the power of on-line handling updates. This is achieved without the need of adapting the presented approach to a different setting.

5 Conclusions and further research directions

In this paper we have presented an incremental algorithm for building a hive graph given a set of line segments. Our technique allows the insertion of new line segments at the two opposite sides of the graph at a time cost of $O(\log n)$: since the required space is $O(n)$, this is the first optimal non-static setting in which the hive graph has been studied. Although the hive graph is presented in this paper as a graph having rectangular enclosure, this is only for simplifying the presentation: in a real implementation there is no need to actually store in the data structure the large enclosing rectangle, and, as a consequence, all the vertices and edges lying on it.

Our future related research will be mainly concerned with two arguments. First, extending the use of the hive graph for solving more general queries, such as dominance queries in a planar graph. Second, trying to dynamize other intrinsically static data structures with the same eager approach.

References

1. B. Chazelle. Filtering search: a new approach to query-answering. *SIAM J. Comput.*, 15:703–724, 1986.
2. B. Chazelle and H. Edelsbrunner. Linear space data structures for two types of range search. *Discrete Comput. Geom.*, 2:113–126, 1987.
3. S. W. Cheng and R. Janardan. New results on dynamic planar point location. *SIAM J. Comput.*, 21:972–999, 1992.
4. Y.-J. Chiang and R. Tamassia. Dynamization of the trapezoid method for planar point location. In *Proc. 7th Annual ACM Symp. on Computational Geometry*, pages 61–70, 1991.
5. Y.-T. Lai and S. M. Leinwand. A theory of rectangular dual graphs. *Algorithmica*, 5:467–483, 1990.
6. K. Mehlhorn and S. Näher. Bounded ordered dictionaries in $O(\log \log n)$ time and $O(n)$ space. *Inform. Process. Lett.*, 35:183–189, 1990.
7. D. Mullis. Reporting overlaps in a dynamic interval set by filtering search. In *Proc. 4th Canad. Conf. Comput. Geom.*, pages 160–169, 1992.
8. F. P. Preparata and R. Tamassia. Dynamic planar point location with optimal query time. *Theoret. Comput. Sci.*, 74:95–114, 1990.
9. H. Samet. *The Design and Analysis of Spatial Data Structures*. Addison-Wesley, Reading, MA, 1990.
10. N. Sarnak and E. Tarjan. Planar point location using persistent search trees. *CACM*, 29(7):669–679, July 1986.
11. R. Tamassia. An incremental reconstruction method for dynamic planar point location. *Inform. Process. Lett.*, 37:79–83, 1991.
12. A. Tansel, J. Clifford, S. Gadia, S. Jajodia, A. Segev, and R. Snodgrass, editors. *Temporal Databases: Theory, Design, and Implementation*. Database Systems and Applications Series. Benjamin/Cummings, Redwood City, CA, 1993.

13. P. van Emde Boas. Preserving order in a forest in less than logarithmic time and linear space. *Inform. Process. Lett.*, 6:80–82, 1977.

14. D. E. Willard. Log-logarithmic worst case range queries are possible in space $O(n)$. *Inform. Process. Lett.*, 17:81–89, 1983.

Planarization of graphs embedded on surfaces [*]

Hristo N. Djidjev[1] and Shankar M. Venkatesan[2]

[1] Department of Computer Science, Rice University
P.O. Box 1892, Houston, TX 77251, USA
email: hristo@cs.rice.edu
[2] Camden Campus, Rutgers University, Camden, NJ 08102

Abstract. A planarizing set of a graph is a set of edges or vertices whose removal leaves a planar graph. It is shown that, if G is an n-vertex graph of maximum degree d and orientable genus g, then there exists a planarizing set of $O(\sqrt{dgn})$ edges. This result is tight within a constant factor. Similar results are obtained for planarizing vertex sets and for graphs embedded on nonorientable surfaces. Planarizing edge and vertex sets can be found in $O(n+g)$ time, if an embedding of G on a surface of genus g is given. We also construct an approximation algorithm that finds an $O(\sqrt{gn \log g})$ planarizing vertex set of G in $O(n \log g)$ time if no genus-g embedding is given as an input.

1 Introduction

A graph G is *planar* if G can be drawn in the plane so that no two edges intersect. Planar graphs arise naturally in many applications of graph theory, e.g. in VLSI and circuit design, in network design and analysis, in computer graphics, and is one of the most intensively studied class of graphs [23]. Many problems that are computationally hard for arbitrary graphs have efficient solutions for the case of planar graphs. Examples include minimum spanning trees, network flow problems, shortest path problems, and many others.

If the graph is not planar, then often a problem arises of how to find a planar subgraph that is as close to the given graph, as possible. A problem of this type is called a *graph planarization problem*. This problem has been intensively investigated in relation to its applications to circuit layout [21, 3, 24, 18]. One approach to the graph planarization problem is to construct a maximal planar subgraph of the input graph G. By solving the maximal planar subgraph problem one finds a minimal set of edges whose removal leaves a planar graph (called a *planarizing set* of edges). Several fast algorithms have been recently proposed for the maximal planar subgraph problem, including the $O(m \log n)$ algorithm of Cai et al. [2], the $O(m + n\alpha(m,n))$ algorithm of La Poutré [25] and the $O(m + n)$ algorithm of Djidjev [9] (n and m are the number of vertices and edges respectively). Although a solution of the maximal planar subgraph problem

---

[*] This work is partially supported by National Scientific Foundation grant CCR–9409191.

defines a *minimal* planarizing set of edges, there is no guarantee that that edge set is of small size. On the other hand the problem of finding a minimum planarizing set of edges is known to be NP-hard [11].

In this paper we show that for any n vertex graph of bounded genus g and maximum degree d there exists a planarizing edge set of size $O(\sqrt{dgn})$. This result is tight within a constant factor and improves the best previous bound of $O(d\sqrt{gn})$ [5]. We also consider the related problem of finding a small planarizing set of *vertices*. Although an asymptotically optimal $O(\sqrt{gn})$ bound for this problem is known [5], the previous proofs are quite complex and the constants are large. We present here very simple proof that also gives a leading constant 4, improving the previous leading constants $44/\sqrt{3}$ of [5, 6] and 26 of [17].

We consider also graphs embedded on nonorientable surfaces, showing that similar bounds hold for the sizes of the smallest edge and vertex planarizing sets.

Our proof technique combines a careful examination of the topology of the graph with a use of a radius reduction device [28]. Our proofs are constructive, giving linear algorithms that find the planarizing sets, if an embedding of the graph on its genus surface is given.

We also investigate the problem of finding a planarizing set in the case where an embedding of the graph on its genus surface is **not** given. This is motivated by the fact that the problem of embedding a graph on a surface of minimum genus is known to be NP-hard [27]. The best known algorithm for the latter problem is polynomial on the number of vertices n, but doubly exponential on the genus g [10]. In this paper we describe an approximation algorithm that finds an $O(\sqrt{dgn \log g})$ planarizing edge set that does not require a genus-g embedding to be given as an input. No comparable algorithm for this problem has been previously constructed.

The paper is organized as follows. Section 2 contains preliminaries. In Section 3 we prove our main result about planarizing edge sets and in Section 4 we prove a similar result about planarizing vertex sets. The nonorientable case is discussed in Section 5. In the last section we describe an efficient algorithm that constructs a planarizing set without knowing the embedding of the graph on its genus surface.

2 Preliminaries

By a *surface*, we mean a closed connected 2-manifold [14, 12]. There are two major types of surfaces: orientable and nonorientable surfaces. Informally, if an intelligent bug starts from some point on a closed curve drawn on a surface, traverses the curve, and returns to the initial point with its initial orientation reversed, then the curve is *orientation reversing*. Otherwise, the curve is *orientation preserving* [14]. A surface is *nonorientable* if there is at least one orientation reversing closed curve on the surface. Otherwise, it is *orientable*. Excluding the sphere, the simplest orientable surface is the torus, and the simplest nonorientable surface is the projective plane. Every orientable (nonorientable) surface is the sum of a certain number g of tori (projective planes), and this number

$g \geq 0$ is the *genus* of the surface [14]. By *orientable genus surface* of a graph, we mean the orientable surface of minimum genus on which the graph can be embedded. By *orientable genus* of a graph, we mean the genus of its orientable genus surface. The terms *nonorientable genus surface* and *nonorientable genus* are similarly defined. A *2-cell* is a region that is homeomorphic to the open unit disc in R^2.

An important relation between the number of vertices, edges, and faces of any 2-cell embedding of a graph on a surface is given by the Euler's formula [12, 14].

Theorem 1. *: Let an embedding of an n-vertex, m-edge f-face connected graph G on a surface S of genus g be given such that every one of the faces of the embedding is a 2-cell. Then the equality $n - m + f = X(S)$ holds, where $X(S) = 2 - 2g$, if S is orientable, or $X(S) = 2 - g$, if S is nonorientable.*

The number $X(S)$ from Theorem 1 is called the *Euler characteristic* of S.

A simple closed curve c is called *contractible* if c divides S into two disjoint regions one of which is a 2-cell. Otherwise c is *noncontractible*. If c is noncontractible curve, then c is either *noncontractible separating*, if c divides S into two disjoint regions (none of which is a 2-cell), or *noncontractible nonseparating* otherwise. Similarly, if G is a graph embedded on S, then any simple cycle of G will be called respectively contractible, noncontractible separating and noncontractible nonseparating depending on the corresponding curve on S.

Let G be a connected graph with spanning tree T of radius r and root t. Each nontree edge of G forms with the edges of T a unique simple cycle. Call each such cycle a *fundamental* cycle. The length of any fundamental cycle is at most $2r + 1$, if it contains t, or at most $2r - 1$, otherwise.

We are going to use the following known topological facts (for simple combinatorial proofs see, e.g. [7, 13].)

Lemma 2. *Let c be a noncontractible nonseparating curve on a surface S. Then $S - c$ can be embedded on a surface of characteristic $X(S) + 2$, if c is orientation preserving, or $X(S) + 1$, if c is orientation reversing.*

Lemma 3. *Consider a 2-cell embedding of a graph G with a spanning tree of radius r and a root t on an arbitrary surface S. Then there exists a set of $2 - X(S)$ fundamental cycles with a total of no more than $2r(2 - X(S)) + 1$ vertices, one of which is t, whose deletion from G leaves behind a planar graph.*

3 Planarizing sets for graphs embedded on orientable surfaces

Let G be an n-vertex connected graph of orientable genus $g > 0$. (If G is not connected we apply the same argument on any non-planar connected component of G.)

In order to apply Lemma 3 we need to reduce (in an appropriate sense) the radius of a spanning tree of the graph (to $O(\sqrt{n/g})$ for planarizing vertex sets and to $O(\sqrt{m/(4dg)})$ for edge sets). Since the construction for vertex sets is simpler, we will consider that case first. Then we will prove our main result for planarizing edge sets. We begin this section by proving two key lemmas.

3.1 The radius reducing device

Embed G on an orientable surface S of genus g and add new edges in order to obtain a triangulation. Divide the vertices of G into levels according to their distance from some fixed vertex of G.

The idea of our procedures constructing vertex and edge planarizing sets is the following. Firstly, we find a small set of vertices (resp. edges) whose removal divides G into components each containing $O(\sqrt{n/g})$ (resp. $O(\sqrt{m/(4dg)})$) levels. Secondly, we add a small number of new edges resulting in a graph of genus $O(g)$ and radius $O(\sqrt{n/g})$ (resp. $O(\sqrt{m/(4dg)})$). Finally, we apply Lemma 3. For implementing the second step we need the following two lemmas.

For some constant r to be defined below, denote by L_j, for $1 \leq j \leq r$, the set of vertices on all levels i such that $i \bmod r = j$. If the graph has at least r levels, then there exists a $i' \leq r$ such that $|L_{i'}| \leq \lfloor n/r \rfloor$. Let $M = L_{i'}$, $l_1 < \cdots < l_s$ be the set of vertices in levels in M, and E_M be the set of all edges (v, w) of G such that $w \in M$ and the level of v is lower than the level of w.

Lemma 4. *Let d be the number of components induced by the vertices of M and let the deletion of E_M from G results in a graph of k' components and genus g'. Then $g' \leq g - d + k' - 1$.*

Proof. Let $i \leq s$ and L be any component induced by the set of vertices at level l_i. Consider the subgraph L' of L induced by the set of the vertices of L adjacent to vertices at level $l_i + 1$. Then L' is a union of edge disjoint simple cycles [22] implying that L contains at least one simple cycle. Denote by c one such simple cycle. Delete all edges of E_M incident to vertices of c. Let G' be the resulting graph and k_c and g_c be the number of components and the genus of G'. If c is a contractible or a noncontractible separating cycle, then $k_c \geq k + 1$, $g_c \leq g$. If c is a noncontractible nonseparating cycle, then $k_c \geq k$, $g_c \leq g - 1$, where k is the number of components of G. In both cases $k_c - g_c \geq k - g + 1$. Delete all remaining edges of E_M incident to vertices of L. Let k_L and g_L be the number of components and the genus of the resulting graph. Then $k_L \geq k_c$ and $g_L \leq g_c$. From the previous inequality

$$k_L - g_L \geq k - g + 1. \tag{1}$$

Applying this construction on all connected components L of the subgraph of G induced by M, we obtain a k'-component graph of genus g'. By induction from (1) $k' - g' \geq k - g + d$ and since by assumption G is connected and $k = 1$ then $g' \leq g - d + k' - 1$. $\qquad \square$

Notice that in the proof of Lemma 4 we constructed an embedding of $G - E_M$ on an orientable surface S' such that

$$X = X(S') \geq X(S) + 2d. \tag{2}$$

For each level $l > 0$ from $L_{i'}$, delete all edges connecting a vertex on level $l - 1$ with a vertex on level l. Let k' and g' denote the number of components and the genus of the resulting graph G' and let K be any component of G'. Let l^* be the lowest level of K and K^* be the subgraph of K induced by the set of vertices on level l^*. Add to K a minimum number, e_K, of edges to make K^* connected. This increases the genus by no more than e_K. Let d_K denote the number of the components of K^*. Then

$$e_K \leq d_K - 1. \tag{3}$$

Obviously, if $l^* > 0$ then $l^* \in L_{i'}$. Shrink the subgraph induced by the vertices on level l^* in K (which after the addition of the new edges is connected) to a point t^*. The resulting component has a spanning tree rooted at t^* and radius not exceeding $r - 1$. Repeat the same procedure with all other components of G' and denote by G'' the resulting graph.

Lemma 5. *Each component of G'' has a spanning tree of radius not exceeding $r - 1$ and the genus of G'' is less or equal than g.*

Proof. Denote by d the sums of d_K over all components K of G' and by g'' the genus of G''. By Lemma 3, for each nonplanar component K'' of G'' a planarizing vertex set $P_{K''}$ of no more than $4g_{K''}(r-1)+1$ vertices of G'' exists, where $g_{K''}$ is the genus of K''. However, we will show that at most $4g_{K''}(r-1)$ of these vertices belong to G. If $P_{K''}$ contains a vertex corresponding to a shrunk subgraph induced by $L_{i'}$, then $|P_{K''}| \leq 4g_{K''}(r - 1)$. Otherwise $P_{K''}$ will not contain the root of the spanning tree of the corresponding component of G'' and thus $|P_{K''}| \leq 4g_{K''}(r - 2) + 1$. Since the genus of any graph is equal to sum of the genera of its connected components [1], then g'' is equal to the sum of $g_{K''}$ over all components K'' of G''. From this fact and (3) applied to all components K of G' we have $g'' - g' \leq d - k' + 1$, and by Lemma 4 $g' + d - k' + 1 \leq g$.

By combining the last two inequalities $g'' \leq g$ follows. □

3.2 Finding a planarizing vertex set

The following theorem improves the best previous bound by a factor of 6.5.

Theorem 6. *Let G be an n-vertex graph of orientable genus g. Then there exists a set of no more than $4\sqrt{gn}$ vertices of G whose removal leaves a planar graph.*

Proof. Suppose that G is connected. Embed G on a surface of genus g and add an appropriate number of edges to triangulate the embedding. Divide the vertices of G into levels according to their distance from some vertex t. Let $r = \lceil \sqrt{n/(4g)} \rceil$. Denote by L_j for $1 \leq j \leq r$ the set of vertices on all levels i such that $i \bmod r = j$. If there are no more than $r - 1$ levels in G, then Lemma 3 directly implies the existence of a $4\sqrt{gn}$ planarizing vertex set. Assume then that the number of levels is at least r. Find a level $i' \leq r$ as in the previous subsection such that $|L_{i'}| \leq \lfloor n/r \rfloor$ and for each level $l > 0$ from $L_{i'}$ delete all edges connecting a vertex on level $l-1$ with a vertex on level l. Modify the resulting graph by adding new edges and shrinking the vertices on the lowest level of each component so that (by Lemma 5) the resulting graph has genus at most g and each component has radius at most $r-1$. Then a planarizing vertex set for G exists that includes $L_{i'}$ and contains no more than $\lfloor n/r \rfloor + 4g(r-1) \leq n/\lceil \sqrt{n/(4g)} \rceil + 4g\lfloor \sqrt{n/(4g)} \rfloor \leq 4\sqrt{gn}$ vertices.

If G is not connected, then for each component K of G there exists a planarizing vertex set of no more than $4\sqrt{n_K g_K}$ vertices, where n_K and g_K are the number of vertices and genus of K, respectively. Since $\sum_K n_K = n$ and $\sum_K g_K = g$, then $\sum_K 4\sqrt{n_K g_K} \leq 4\sqrt{ng}$, which proves the theorem in the general case. □

3.3 Finding a planarizing edge set

The *degree* of a graph denote the maximum degree of its vertices.

Theorem 7. *For any m-edge graph of degree d and orientable genus g, there exists a planarizing set of no more than $4\sqrt{dgm}$ edges.*

Proof. Embed the given graph on some orientable surface S of genus g. Make each face a triangle by adding a suitable number of edges and call the resulting graph G.

To obtain a small planarizing edge set we shall define now $L_{i'}$ to be an edge set, instead of a vertex set as in Theorem 6. This is done as follows. In the graph G, assign a weight of 0 to all new triangulating edges and a weight of 1 to all m edges of the original graph. Divide the vertices into levels according to their distance from some vertex t. Choose $r = \lceil \sqrt{m'/(4dg)} \rceil$ where m' is the total weight of all edges connecting vertices at different levels. Denote by L_j for $0 \leq j < r$ the set of all edges between level i and level $i+1$ for all i such that $i \bmod r = j$. Assume again that the number of levels is at least r. Then there exists an $i' < r$ such that the weight of $L_{i'}$ is no more than $\lfloor m'/r \rfloor$. Therefore, $L_{i'}$ contains no more than $\lfloor m'/r \rfloor$ original edges.

Delete $L_{i'}$ from G, and add a number of edges to obtain a graph G' of genus not exceeding g and radius of each component not exceeding $r-1$ as in Lemma 5. As shown in the proof of Theorem 6, a planarizing vertex set of G' of no more than $4g(r-1)$ vertices can be found. Choosing the edges incident with these vertices, we obtain a planarizing edge set of G' of size at most $4gd(r-1)$. Then a planarizing edge set for G exists of size not exceeding

$$\lfloor m'/r \rfloor + 4dg(r-1) \leq m'/\sqrt{m'/(4dg)} + 4dg\sqrt{m'/(4dg)} = 4\sqrt{dgm'} \leq 4\sqrt{dgm}. \quad \square$$

Next we will give an estimation of the maximum size of a planarizing edge set of G, given the number of vertices of G (instead of the number of edges as in Theorem 7).

Lemma 8. *Consider a triangulation embedding of G with f faces on an arbitrary orientable surface. Let the vertices of G be divided into levels according to their* distance from some vertex t. Let U_t^G be the set of all the edges that connect two vertices at the same level in G. Then $|U_t^G| \geq f/2$.

Proof. Consider any face (x_1, x_2, x_3) of the triangulation. It is not possible that $level(x_i) - level(x_j) > 1$ for any $0 \leq i, j \leq 2$. Therefore, at least one of the three edges of each face must belong to U_t^G. Since each edge is adjacent to only two faces, the result follows. □

Theorem 9. *Let G be a connected n-vertex graph of degree d and orientable genus g. There exists a set of no more than $4\sqrt{2dg(n + 2g - 2)}$ edges whose removal makes G planar.*

Proof. Consider without a loss of generality a triangulation embedding of G on its genus surface. Then, by Lemma 8, $|U_t^G| \geq f/2 = m/3 = n + 2g - 2$. Furthermore, by the proof of Theorem 7, there exists a planarizing edge set of size not exceeding $4\sqrt{dgm'}$ for G. Then

$$4\sqrt{dgm'} = 4\sqrt{dg(m - |U_t^G|)} \leq 4\sqrt{2dg(n + 2g - 2)}. \ \square$$

The *skewness* of a graph is defined in [15] as the smallest number of edges whose removal leads to a planar graph. Then by Theorem 9 the skewness of the class of n-vertex g-genus d-degree graphs is not exceeding $4\sqrt{2dg(n + 2g - 2)}$.

4 Extension to the Nonorientable Case

We show here that results from the previous sections can be extended to graphs embedded on nonorientable surfaces. As we will show, we can use similar arguments as in the orientable case, with a few modifications. We will only sketch the main steps of the proof.

Let G be a connected graph. Consider a triangulation embedding of G on a nonorientable surface S of genus g. Divide the vertices of G into levels according to their distance to some vertex t. Denote by r the maximum level and let M be the set of all vertices on certain nonadjacent levels $0 < l_1 < ... < l_j \leq r$. Let E_M be the set of all edges (v, w) of G such that $w \in M$ and the level of v is lower than the level of w. Then we shall prove the following analogue of Lemma 4.

Lemma 10. *If d is the number of components induced by the vertices of M and if the deletion of E_M turns G into a k'-component graph G', then the components of G' can be suitably distributed to form a graph of orientable genus g^* and a graph of nonorientable genus g^{**} such that $2g^* + g^{**} \leq g - 2d + 2k' - 2$.*

Proof. The construction from the proof of Lemma 4 can be applied to G and S in order to obtain a graph G' embedded in k surfaces whose sum X of characteristics satisfies (2). Moreover, by construction each of these k' surfaces embeds a component of $G - E_M$. Thus $X \leq 2k - 2g^* - g^{**}$ and from (2) $2 - g + 2d \leq 2k' - 2g^* - g^{**}$, and thus $2g^* + g^{**} \leq g - 2d + 2k' - 2$. □

Theorem 11. *If G is an n-vertex graph of degree d and nonorientable genus g, then there exists an edge set of size $2\sqrt{2dgm}$ whose removal from G leaves a planar graph.*

Proof. Consider a triangulation embedding of G on its nonorientable genus surface S. Define m' as in the proof of Theorem 7. Choose $r = \lceil \sqrt{m'/(2dg)} \rceil$ and find the set $L_{i'}$ as in Theorem 7. Assume that the number of levels is at least $2r$ (otherwise the theorem follows directly from Lemma 3). As in the proof of Theorem 7, delete all edges connecting a vertex on level $l-1$ with a vertex on level l for each level $l > 0$ from $L_{i'}$. Let k denote the number of components of the resulting graph G'. Add a number of edges so that each component of the resulting graph G'' has a radius not exceeding $r-1$. Furthermore, by Lemma 10, the components of G'' form two subgraphs of G', one of orientable genus g^* and the other of nonorientable genus g^{**}, such that $2g^* + g^{**} \leq g$. Then, applying Lemma 3 on each nonplanar component of G'', we find a planarizing edge set of size no more than

$$\lfloor m'/r \rfloor + 4dg^*(r-1) + 2dg^{**}(r-1) \leq \lfloor m'/r \rfloor + 2dg(r-1)$$

$$\leq m'/\sqrt{m'/(2dg)} + 2dg\sqrt{m'/(2dg)} = 2\sqrt{2dgn}. \qquad \square$$

For planarizing vertex sets we have the following theorem.

Theorem 12. *If G is an n-vertex graph of nonorientable genus g, then there exists a vertex set of size $2\sqrt{2gn}$ whose removal from G leaves a planar graph.*

5 Tightness of the results

The tightness in the orientable case follows from the fact that $\Omega(\sqrt{gn})$ vertices or $\Omega(\sqrt{dgm})$ edges are needed in the worst case to separate an n-vertex m-edge graph of degree d and genus g [7, 13, 26]. If one can planarize a graph by removing $o(\sqrt{gn})$ vertices or $o(\sqrt{dgm})$ edges, then one can then find in the resulting planar graph a separator of $O(\sqrt{n})$ vertices [20] or $O(\sqrt{dn})$ edges [4], which will be a contradiction to the above lower bounds.

For graphs embedded on nonorientable surfaces the tightness of our results follows from the tightness in the orientable case. Let G be any n-vertex graph of orientable genus g. Consider a 2-cell embedding of G on an orientable surface of genus g. By adding a projective plane to S we obtain an embedding of G on a nonorientable surface of Euler characteristic $X(S) - 1 = 2 - 2g - 1$. Thus, by Theorem 1, the nonorientable genus of G is at most $2g + 1$.

6 Algorithmic aspects

The constructive proofs of the theorems from the previous sections can be transformed into linear algorithms that find planarizing sets satisfying the theorems, if an embedding of the graph on its orientable (resp. nonorientable) genus surfaces is known. Finding such an embedding, however, is a very hard computational problem itself. In this section we show that almost optimal planarizing vertex set (upto a factor of $O(\sqrt{\log g})$) can be found for any n-vertex g-genus graph in $O(n \log g)$ time. We will use an algorithm described in [8] which does not need a minimum genus embedding of the input graph. The idea of that algorithm is that if a non-planar graph G has no suitable simple cycle separator, then G will contain a short subgraph homeomorphic to $K_{3,3}$ whose contraction reduces the genus of G. Specifically, we are going to use the following result from [8].

Theorem 13. *For any n vertex graph G a partitioning A, B, C of the vertices of G can be found in $O(n)$ time such that no edge joins a vertex in A with a vertex in B, $|A|, |B| \leq n/2$, $|C| \leq c\sqrt{(g'+1)n}$, and the graph induced by the set of vertices $A \cup B$ has genus not exceeding $g - g'$, where g is the genus of G, $0 \leq g' \leq g$, and c is a global constant.*

The following algorithm finds a planarizing vertex set without knowing an embedding of the graph on its genus surface.

Algorithm PLANARIZE

Input: An arbitrary graph G.
Output: A planarizing vertex set for G.

1. Use the algorithm from [16] to test G for planarity. If G is planar then output the empty set and stop.
2. Initialize: $C = \emptyset$, $M = \{G\}$, where M denotes a set of nonplanar edge-disjoint subgraphs of $G - C$.
3. Perform Steps $3 \div 7$ until $M = \emptyset$.
4. Pick any graph G_i from M and and update $M := M - \{G_i\}$.
5. Divide G_i as in Theorem 13 into two disconnected subgraphs with no more than $n_i/2$ vertices each, by removing a set C_i of $O(\sqrt{(g_i+1)n_i})$ vertices of G_i, where g_i and n_i are the genus and the number of vertices of G_i.
6. Test each subgraph for planarity.
7. Let $C := C \cup C_i$. Add the nonplanar subgraphs found in Step 6 to M.

Let us estimate the maximum time $T(n, g)$ and the maximum size $S(n, g)$ of the set C found by Algorithm PLANARIZE for any n-vertex g-genus graph G. We have the relations

$$T(n, g) \leq kn + max\{T(n/2, g_1) + T(n/2, g_2)|g_1 + g_2 \leq g\}, \text{ if } n > 0, g > 0,$$
$$T(n, 0) = O(1), \quad T(0, g) = O(1)$$

and

$$S(n, g) \leq \max\{c\sqrt{n(g_3 + 1)} + S(n/2, g_1) + S(n/2, g_2)|g_1 + g_2 + g_3 = g,$$
$$g_3 \geq 0, \ g_1, g_2 > 0\}.$$
$$S(n, 0) = O(1), \quad S(0, g) = O(1).$$

Solving the recurrences we find $T(n, g) = O(n \log(2g))$ and $S(n, g) = O(\sqrt{gn \log g})$. Thus we have the following theorem.

Theorem 14. *For any n-vertex graph G a planarizing vertex set of G of size $O(\sqrt{gn\log g})$ can be found in $O(n\log(2g))$ time, where $g > 0$ is the orientable genus of G, without being given an embedding of G on its genus surface.*

For planarizing sets of edges we have the following result.

Theorem 15. *For any m-edge graph G of degree d a planarizing edge set of G of size $O(\sqrt{dgm\log g})$ can be found in $O(m\log(2g))$ time, where $g > 0$ is the orientable genus of G, without being given an embedding of G on its genus surface.*

References

1. J. Battle, F. Harary, Y. Kodama, and J.W.T. Youngs, Additivity of the Genus of a Graph, *Bulletin Amer. Math. Soc.* 68 (1962), 565-568.
2. J. Cai, X. Han, and R.E. Tarjan, An O(m log n)-Time Algorithm for the Maximal Planar Subgraph, *SIAM Journal on Computing*, 22 (1993), 1142-1162.
3. T. Chiba, I. Nishioka, and I. Shirakawa, An Algorithm of Maximal Planarization of Graphs, *Proc. IEEE Int. Symp. on Circuits and Systems*, 1978, 649-652.
4. K. Diks, H. N. Djidjev, O. Sykora, I. Vrto, Edge Separators for Planar Graphs and Their Applications, *Journal of Algorithms*, vol. 14, (1993), 258-279.
5. H. N. Djidjev, On Some Properties of Nonplanar Graphs, *Compt. rend. Acad. bulg. Sci.*, vol. 37 (1984), 9, 1183-1185.
6. H.N. Djidjev, Genus Reduction in Nonplanar Graphs, manuscript.
7. H.N. Djidjev, A Separator Theorem for Graphs of Fixed Genus, *SERDICA Bulgaricae Mathematicae Publicationes* 11 (1985), 319-329.
8. H.N. Djidjev, A Linear Algorithm for Partitioning Graphs of Fixed Genus, *SERDICA Bulgaricae Mathematicae Publicationes* 11 (1985), 369-387.
9. H. N. Djidjev, A Linear Algorithm for the Maximal Planar Subgraph Problem, Technical Report TR95-247, Department of Computer Science, Rice University, 1995.
10. H.N. Djidjev, J. Reif, An Efficient Algorithm for the Genus Problem with Explicit Construction of Forbidden Subgraphs, *Proc. Annual ACM Symposium on Theory of Computing* (1991), pp.337-347.
11. M.R. Garey and D.S. Johnson, *Algorithms and Intractability: A Guide to the Theory of NP Completeness.* San Francisco, Freeman, 1979.
12. P. J. Giblin, Graphs, Surfaces, and Homology : an Introduction to Algebraic Topology. Chapman and Hall, London, New York, 1981.
13. J.R. Gilbert, J.P. Hutchinson and R.E. Tarjan, A Separator Theorem for Graphs of Bounded Genus, *J. Algorithms* 5 (1984), 391-407.
14. J. L. Gross, T. W.Tucker. Topological Graph Theory, Wiley, New York, 1987.
15. F. Harary, Graph Theory, Addison-Wesley (1969).
16. J.E. Hopcroft, and R.E. Tarjan, Efficient Planarity Testing, *J. ACM 21* (1974), 549-568.
17. J.P.Hutchinson, and G.L. Miller, On Deleting Vertices to Make a Graph of Positive Genus Planar, *Discrete Algorithms and Complexity Theory*, Academic Press, Boston, 1986, 81-98.
18. R. Jayakumar, K. Thulasiraman, and M.N.S. Swamy, $O(n^2)$ Algorithms for Graph Planarization, *IEEE Trans. on Comp.-Aided Design* 8 (1989), 257-267.

19. M.S. Krishnamoorthy, and N. Deo, Node-Deletion NP- Complete Problems, *SIAM J. Comput.* 8 (1979), 619-625.

20. R.J. Lipton, and R.E. Tarjan, A Separator Theorem for Planar Graphs, *SIAM J. Appl. Math.* 36 (1979), 177-189.

21. M. Marek-Sadowska, Planarization Algorithm for Integrated Circuits Engineering, *Proc. IEEE Int. Symp. on Circuits and Systems,* 1979, 919-923.

22. G.L. Miller, Finding Small Simple Cycle Separators for Planar Graphs, *J. Computer System Sci.* 32 (1986), 265- 279.

23. T. Nishizeki, N. Chiba, *Planar Graphs: Theory and Algorithms*, North Holland, 1988.

24. T. Ozawa and H. Takahashi, A Graph-Planarization Algorithm and its Applications to Random Graphs, in *Graph Theory and Algorithms*, Lecture Notes in Computer Science, vol. 108, 95-107, Springer-Verlag, 1981.

25. J.A. La Poutré, Alpha-Algorithms for Incremental Planarity Testing, *Proc. of the Ann. ACM Symp. on Theory of Comput.*, 1994, 706-715.

26. O. Sykora and I Vrto, Edge Separators for Graphs of Bounded Genus with Applications, *Theoretical Computer Science*, 112 (1993).

27. C. Thomassen, The Graph Genus Problem is NP-Complete, *Journal of Algorithms*, 10, 1989.

28. S.M. Venkatesan, Improved Constants for Some Separator Theorems, *J. of Algorithms* 8 (1987), 572-578.

Complexity and Approximability of Certain Bicriteria Location Problems

S. O. Krumke[1], H. Noltemeier[1], S. S. Ravi[2] and M. V. Marathe[3,*]

[1] University of Würzburg, Am Hubland, 97074 Würzburg, Germany.
Email: {krumke, noltemei}@informatik.uni-wuerzburg.de.
[2] University at Albany - SUNY, Albany, NY 12222, USA.
Email: ravi@cs.albany.edu.
[3] Los Alamos Nat. Lab. P.O. Box 1663, MS M986, Los Alamos, NM 87545, USA.
Email: madhav@c3.lanl.gov.

Abstract. We investigate the complexity and approximability of some location problems when two distance values are specified for each pair of potential sites. These problems involve the selection of a specified number of facilities (i.e. a placement of a specified size) to minimize a function of one distance metric subject to a budget constraint on the other distance metric. Such problems arise in several application areas including statistical clustering, pattern recognition and load–balancing in distributed systems. We show that, in general, obtaining placements that are near-optimal with respect to the first distance metric is $\mathcal{NP}$-hard even when we allow the budget constraint on the second distance metric to be violated by a constant factor. However, when both the distance metrics satisfy the triangle inequality, we present approximation algorithms that produce placements which are near-optimal with respect to the first distance metric while violating the budget constraint only by a small constant factor. We also present polynomial algorithms for these problems when the underlying graph is a tree.

1 Introduction and Motivation

In this paper, we study some location problems with multiple constraints. The problems considered in this paper can be termed as *compact location* problems, since we will typically be interested in finding a "compact" placement of facilities, i.e. a placement minimizing some measure of the distances between the selected nodes. Compact location problems without multiple constraints have been studied extensively in the past (see [RKM+93, AI+91] and the references cited there in).

To illustrate the types of problems considered in this paper, we present the following example. Suppose we are given *two* weight–functions c, d on the edges of the network. Let the first weight function c represent the cost of constructing an

---

* Research supported by Department of Energy under contract W-7405-ENG-36.

edge, and let the second weight function d represent the actual transportation– or communication–cost over an edge (once it has been constructed). Given such a graph, we can define a general bicriteria problem $(\mathcal{A}, \mathcal{B})$ by identifying two minimization objectives of interest from a set of possible objectives. A budget value is specified on the second objective $\mathcal{B}$ and the goal is to find a placement of facilities having minimum possible value for the first objective $\mathcal{A}$ such that this solution obeys the budget constraint on the second objective. For example, consider the *Diameter–Constrained Minimum Diameter Problem* denoted by DC–MDP: Given an undirected complete graph $G = (V, E)$ with two non-negative integral edge weight functions c (modeling the building cost) and d (modeling the delay or the communication cost), an integer p denoting the number of facilities to be placed, and an integral bound B (on the total delay), find a placement of p facilities with minimum diameter under the c–cost such that the diameter of the placement under the d–costs (the maximum delay between any pair of nodes) is at most B. We term such problems as *bicriteria compact location problems*.

Here, we study the complexity and approximability of bicriteria compact location problems such as the ones mentioned above. Our study of these problems is motivated by practical problems arising in diverse areas such as statistical clustering, pattern recognition, processor allocation and load–balancing (see [HM79, MF90, KN+95a] and the references cited therein).

2 Preliminaries and Problem Formulation

We consider a complete undirected n–vertex graph $G = (V, E)$. Given an integer p, a *placement* P is a subset of V with $|P| = p$. The *set of neighbors* of a vertex v in G, denoted by $N(v, G)$, is defined by $N(v, G) := \{w : (v, w) \in E\}$. The *degree* $\deg(v, G)$ of v in G is the number of vertices in $N(v, G)$. For a subset $V' \subseteq V$ of nodes, we denote by $G[V']$ the subgraph of G induced by V'. Given a graph $G = (V, E)$, the graph $G^2 = (V, E^2)$ is defined by $(u, v) \in E^2$ if and only if there is a path in G between u and v consisting of at most two edges.

If the edge distances are allowed to be zero, then the optimal solution value may be zero. In a such case, obtaining a solution whose value is within some factor of the optimal solution value is trivially equivalent to finding an optimal solution itself. Therefore, we assume that the values of both the distance functions for any edge are strictly *positive*.

With $\delta \in \{c, d\}$ denoting one of the two edge–weight functions, we use $\mathcal{D}_\delta(P)$ to denote the *diameter* and $\mathcal{S}_\delta(P)$ to denote the *sum of the distances* between the nodes in the placement P; that is

$$\mathcal{D}_\delta(P) = \max_{\substack{u,v \in P \\ u \neq v}} \delta(u, v) \qquad \text{and} \qquad \mathcal{S}_\delta(P) = \sum_{\substack{u,v \in P \\ u \neq v}} \delta(u, v).$$

We note that the average length of an edge in a placement P equals $\frac{2}{p(p-1)}\mathcal{S}_\delta(P)$. Since the average length of an edge in a placement differs from the total length of all the edges in the placement by only the scaling factor $\frac{2}{p(p-1)}$, finding a placement of minimum average length is equivalent to finding a placement of minimum total length. We use this fact throughout this paper.

As usual, we say that $\delta \in \{c, d\}$ satisfies the *triangle inequality* if we have $\delta(v, w) \leq \delta(v, u) + \delta(u, w)$ for all $v, w, u \in V$. Following [HS86], the *bottleneck graph* bottleneck(G, δ, Δ) of $G = (V, E)$ with respect to δ and a bound Δ is defined by

$$\text{bottleneck}(G, \delta, \Delta) := (V, E'), \text{ where } E' := \{e \in E : \delta(e) \leq \Delta\}.$$

We now define the problems studied in this paper.

Definition 1 *Diameter Constrained Minimum Average Placement Problem* (DC–MAP).
Input: An undirected complete graph $G = (V, E)$ with two positive edge weight functions $c, d : E \to \mathbb{Q}^+$, an integer $2 \leq p \leq n$ and a number $\Omega \in \mathbb{Q}^+$.
Output: A set $P \subseteq V$, with $|P| = p$, minimizing the objective

$$\mathcal{S}_c(P) = \sum_{\substack{v, w \in P \\ v \neq w}} c(v, w)$$

subject to the constraint

$$\mathcal{D}_d(P) = \max_{\substack{v, w \in P \\ v \neq w}} d(v, w) \leq \Omega.$$

Definition 2 *Sum Constrained Minimum Average Placement Problem* (SC–MAP).
Input: Same as in DC–MAP above.
Output: A set $P \subseteq V$, with $|P| = p$, minimizing the objective

$$\mathcal{S}_c(P) = \sum_{\substack{v, w \in P \\ v \neq w}} c(v, w)$$

subject to the constraint

$$\mathcal{S}_d(P) = \sum_{\substack{v_i, v_j \in P \\ v_i \neq v_j}} d(v_i, v_j) \leq \Omega.$$

The *Sum Constrained Minimum Diameter Placement Problem* (SC–MDP) and the *Diameter Constrained Minimum Diameter Placement Problem* (DC–MDP) can be defined similarly. Given a problem Π, we use TI-Π to denote the problem Π restricted to graphs in which both the edge weight functions satisfy the triangle inequality.

We also investigate the existence of "good" solutions for bicriteria compact location problems when input graphs are restricted to be trees. In such a case, the distance between any two vertices u and v is the length of the path in the tree between u and v. Given a problem Π, we use TREE-Π to denote the problem Π restricted to trees.

Let $\Pi \in \{$SC–MAP, DC–MAP, TI–DC–MAP, TI–SC–MAP$\}$. Define an (α, β)–*approximation algorithm* for Π to be a polynomial–time algorithm, which for any instance I of Π does one of the following:

(a) It produces a solution within α times the optimal value with respect to the first distance function c, violating the constraint with respect to the second distance function d by a factor of at most β.

(b) It returns the information that no feasible placement exists at all.

Notice that if there is no feasible placement but there is a placement violating the constraint by a factor of at most β, an (α, β)–approximation algorithm has the choice of performing either action (a) or (b).

3   Summary of Results

In this paper, we present both $\mathcal{NP}$–hardness results and approximation algorithms with provable performance guarantees for several bicriteria compact location problems. For additional results on these types of problems, we refer the reader to a companion paper [KN$^+$95a]. Our results are based on two basic techniques. The first is an application of a *parametric search technique* discussed in [MR$^+$95] for network design problems. The second is the *power of graphs* approach introduced by Hochbaum and Shmoys [HS86]. Our results for complete graphs are summarized in Table 1. The table contains hardness results and performance ratios for finding compact placements for different pairs of minimization objectives. The horizontal entries denote the objective function. For example the entry in row i, column j denotes the performance guarantee for the problem of minimizing objective j with a budget on the objective i.

$\rightarrow$ Object. $\downarrow$ Budget	Diameter	Sum
Diameter	approximable within $(2, 2)$ [KN$^+$95a] not approximable within $(2-\varepsilon, 2)$ or $(2, 2-\varepsilon)$	approximable within $(2 - \frac{2}{p}, 2)^*$ not approximable within $(\alpha, 2 - \varepsilon)^*$
Sum	approximable within $(2, 2 - \frac{2}{p})$ [KN$^+$95a] not approximable within $(2-\varepsilon, \alpha)$	approximable within $((1 + \gamma)(2 - \frac{2}{p}), (1 + \frac{1}{\gamma})(2 - \frac{2}{p}))^*$

Table 1. Performance guarantee results for constrained compact location in a complete graph with edge weights obeying the triangle inequality. Asterisks indicate results obtained in this paper. $\gamma > 0$ is a fixed accuracy parameter. The non-approximability results stated assume that $\mathcal{P} \neq \mathcal{NP}$.

→ Object. ↓ Budget	Diameter	Sum
Diameter	polynomial time solvable	polynomial time solvable
Sum	polynomial time solvable	$\mathcal{NP}$–hard approximable within $(1+\gamma, 1+\frac{1}{\gamma})$

Table 2. Results for constrained compact location in tree networks.

4 Related Work

As mentioned earlier, problems involving the placement of p facilities so as to minimize suitable cost measures have been studied extensively in the literature. These problems can roughly be divided into two main categories. The first category of problems involves selecting a set of p facilities so as to minimize (or maximize) the distance (cost) from the unselected sites to the selected sites. Problems that can be cast in this framework include the p–center problem [HS86, DF85], the p–cluster problem [HS86, FG88, Go85] and the p–median problem [LV92, MF90]. The second category consists of problems where the goal is to select p facilities so as to optimize a certain cost measure defined on the set of selected facilities. Problems that can be cast in this framework include the p–dispersion problem [RRT91, EN89], the p–minimum spanning tree problem [RR⁺94, GH94, AA⁺94, BCV95] and the p–compact location problem [RKM⁺93, AI⁺91, KN⁺95a].

In contrast, not much work has been done in finding optimal location of facilities when there is more than one constraint. A notable work in this direction is by Bar-Ilan and Peleg [BP91] who considered the problem of assigning network centers, with a bound imposed on the number of nodes that any center can service. We refer the reader to [MR⁺95, RMR⁺93] for a survey of the work done in the area of algorithms for bicriteria network design and location theory problems. In [KN⁺95a], we studied the minimum diameter problems under sum and diameter constraints. There we gave efficient approximation algorithms with constant performance guarantees for these problems when both the edge weight functions obey the triangle inequality.

Due to lack of space, the rest of the paper consists of statements of results and selected proof sketches.

5 Problems for General Graphs

5.1 Diameter Constrained Problems

We begin with a non-approximability result for DC–MAP and TI–DC–MAP. The proof this result uses a reduction from the Clique problem [GJ79].

Theorem 3. *If the distance functions c, d are not required to satisfy the triangle inequality, there can be no polynomial time (α, β)-approximation algorithm for DC–MAP for any fixed $\alpha, \beta \geq 1$, unless $\mathcal{P} = \mathcal{NP}$. Moreover, if there is a polynomial time $(\alpha, 2 - \varepsilon)$-approximation algorithm for TI–DC–MAP for any fixed $\alpha \geq 1$ and $\varepsilon > 0$, then $\mathcal{P} = \mathcal{NP}$.*

Proof Sketch: We first consider the DC–MAP problem. Suppose there is a polynomial approximation algorithm $\mathcal{A}$ with a performance guarantee of (α, β) for some $\alpha, \beta \geq 1$. We will show that $\mathcal{A}$ can be used to solve an arbitrary instance of the Clique problem in polynomial time, contradicting the assumption that $\mathcal{P} \neq \mathcal{NP}$.

Let the graph $G = (V, E)$ and the integer J form an arbitrary instance of Clique. Construct the following instance I of DC–MAP. The vertex set for I is V itself. For all $u, v \in V$ ($u \neq v$), let $c(u, v) = 1$; also, let $d(u, v) = 1$ if $(u, v) \in E$ and $d(u, v) = \beta + 1$ otherwise. Finally set $p = J$ to complete the construction. In the remainder of this proof sketch, we will refer to any edge in the instance I with d value equal to $\beta + 1$ as a *long* edge; other edges are referred to as *short* edges.

If G has a clique of size J, then the nodes which form this clique constitute an optimal solution to the DC–MAP instance I with sum (under c–distance) equal to $J(J - 1)/2$ and diameter (under d–distance) equal to 1. Since $\mathcal{A}$ provides a performance guarantee of (α, β), the solution returned by $\mathcal{A}$ cannot include any long edges. If G does not have a clique of size J, then every subset of J nodes must include at least one long edge. Therefore, by merely examining the solution produced by $\mathcal{A}$, we can solve the Clique problem.

We use the same construction for TI–DC–MAP except that for every long edge, the d value is chosen as 2. This ensures that both the distance functions satisfy the triangle inequality. □

Using recent hardness results from [BS94] about the non-approximability of Max Clique, we obtain the following non-approximability result.

Theorem 4. *Let $\varepsilon > 0$ and $\varepsilon' > 0$ be arbitrary. Suppose that A is a polynomial time algorithm that, given any instance of TI–DC–MAP, either returns a subset $S \subseteq V$ of at least $\frac{2p}{|V|^{1/6 - \varepsilon'}}$ nodes satisfying $\mathcal{D}_d(S) \leq (2 - \varepsilon)\Omega$, or provides the information that no placement of p nodes having d–diameter of at most Ω does exist. Then $\mathcal{P} = \mathcal{NP}$.* □

Procedure HEUR–FOR–DIA–CONSTRAINT
1 $G' := \text{bottleneck}(G, d, \Omega)$
2 $V_{cand} := \{v \in G' : \deg(v, G') \geq p - 1\}$
3 **if** $V_{cand} = \emptyset$ **then return** "certificate of failure"
4 Let $best := +\infty$
5 Let $P_{best} := \emptyset$
6 **for each** $v \in V_{cand}$ **do**
7   Sort the neighbors $N(v, G')$ of v according to their c–distance from v
8   Assume now that $N(v, G') = \{w_1, \ldots, w_r\}$ with $c(v, w_1) \leq \cdots \leq c(v, w_r)$
9   Let $P(v) := \{v, w_1, \ldots, w_{p-1}\}$
10   **if** $\mathcal{S}_c(P(v)) < best$ **then** $P_{best} := P(v)$
11                                    $best := \mathcal{S}_c(P(v))$
12 **output** P_{best}

Fig. 1. Details of the heuristic for TI–DC–MAP.

We now consider the TI–DC–MAP problem where the distance functions satisfy the triangle inequality. For this problem, we present an approximation algorithm that provides a performance guarantee of $(2 - 2/p, 2)$. The algorithm is shown in Figure 1. The performance guarantee is established below.

Theorem 5. *Let I be any instance of* TI–DC–MAP *such that an optimal solution P^* of total c-cost $OPT(I) = \mathcal{S}_c(P^*)$ exists. Then the algorithm* HEUR–FOR–DIA–CONSTRAINT *returns a placement P satisfying $\mathcal{D}_d(P) \leq 2\Omega$ and $\mathcal{S}_c(P)/OPT(I) \leq 2 - 2/p$.*

Proof: Consider an optimal solution P^* such that $\mathcal{D}_d(P^*) \leq \Omega$. By definition, this placement forms a clique of size p in $G' := \text{bottleneck}(G, d, \Omega)$. Consequently, for any node $v \in P^*$ the set $N(v, G')$ has size at least p and V_{cand} is non–empty. Thus the heuristic will not output a "certificate of failure".

Moreover, any placement $P(v)$ considered by the heuristic will form a clique in $(G')^2$. By the definition of G' as a bottleneck graph with respect to d, the bound Ω and the assumption that the edge weights obey the triangle inequality, it follows that no edge e in $(G')^2$ has d–weight more than 2Ω. Thus, for *every* placement $P(v)$ considered by the heuristic, the value of $\mathcal{D}_d(P(v))$ is no more than 2Ω.

Now we are going to establish the performance guarantee with respect to the objective function value. To this end, define for a node $v \in P^*$: $S_v := \sum_{\substack{w \in P^* \\ w \neq v}} c(v, w)$. Then we have $\mathcal{S}_c(P^*) = \sum_{v \in P^*} S_v$. Now let $v \in P^*$ be so that S_v is a minimum among all nodes in P^*. Then clearly

$$OPT(I) = \mathcal{S}_c(P^*) \geq pS_v. \tag{1}$$

As mentioned earlier, $v \in V_{cand}$. Consider the step of the algorithm HEUR–FOR–DIA–CONSTRAINT in which it examines v. Let $N(v) := P(v) \setminus \{v\}$ denote the set of $p-1$ nearest neighbors of v in G' with respect to c. Then we have

$$\sum_{\substack{w \in N(v) \\ w \neq v}} c(v,w) \leq S_v, \tag{2}$$

by definition of $N(v)$ as the set of nearest neighbors. Let $w \in N(v)$ be arbitrary. Then

$$\sum_{u \in N(v) \cup \{v\} \setminus \{w\}} c(w,u) = c(w,v) + \sum_{u \in N(v) \setminus \{w\}} c(w,u)$$

$$\leq c(w,v) + \sum_{u \in N(v) \setminus \{w\}} (c(w,v) + c(v,u))$$

$$= (p-1)c(w,v) + \sum_{u \in N(v) \setminus \{w\}} c(v,u)$$

$$= (p-2)c(v,w) + \sum_{u \in N(v)} c(v,u)$$

$$\overset{(2)}{\leq} (p-2)c(v,w) + S_v. \tag{3}$$

Now using (3) and again (2), we obtain

$$S_c(P(v)) = S_c(N(v) \cup \{v\})$$

$$= \sum_{u \in N(v)} c(v,u) + \sum_{w \in N(v)} \sum_{u \in N(v) \cup \{v\} \setminus \{w\}} c(w,u)$$

$$\overset{(2)}{\leq} S_v + \sum_{w \in N(v)} \sum_{u \in N(v) \cup \{v\} \setminus \{w\}} c(w,u)$$

$$\overset{(3)}{\leq} S_v + \sum_{w \in N(v)} ((p-2)c(v,w) + S_v)$$

$$= S_v + (p-2)S_v + (p-1)S_v$$

$$= (2p-2)S_v$$

$$\overset{(1)}{\leq} (2 - 2/p)OPT(I).$$

As the algorithm chooses the placement P_{best} with the least S_c, the claimed performance guarantee follows. $\square$

5.2 Sum Constrained Problems

Next, we study bicriteria compact location problems where the objective is to minimize the sum of the distances S_c subject to a budget-constraint on S_d.

Again, it is not an easy task to find a placement P satisfying the budget-constraint or to determine that no such placement exists. Using a reduction from Clique [GJ79] similar to that used in the proof of Theorem 3, we get the following result.

Proposition 6. *If the distance functions* c, d *are not required to satisfy the triangle inequality, there can be no polynomial time* (α, β)*–approximation algorithm for* SC–MAP *for any fixed* $\alpha, \beta \geq 1$, *unless* $\mathcal{P} = \mathcal{NP}$. □

We proceed to present a heuristic for TI–SC–MAP. The main procedure shown in Figure 2 uses the test procedure from Figure 3. We note that γ is a fixed quantity that specifies the accuracy requirement.

Procedure HEUR–FOR–SUM–CONSTRAINT
1 Use a binary search to find the smallest integer $T \in [0, p^2 \max\{ c(e) : e \in E \}]$
  such that Sum–Test(T)=Yes.
2 **output** the placement generated by Sum–Test(T).

Fig. 2. Main procedure for TI–SC–MAP.

Procedure Sum–Test(T)
1 Let $\mu := \frac{T}{\Omega}$.
2 **for** each pair (v, w) of nodes define the distance function $h(v, w)$ by
  $h(v, w) := c(v, w) + \mu d(v, w)$.
3 Compute a $(2 - 2/p)$-approximation for the problem of finding a set of p nodes minimizing $\mathcal{S}_h$.
4 Let P_T be a set of p nodes with $\mathcal{S}_h(P_T) \leq (2 - 2/p) \cdot \min_{\substack{P \subseteq V \\ |P|=p}} \mathcal{S}_h(P)$.
5 **if** $\mathcal{S}_h(P_T) \leq (2 - 2/p)(1 + \gamma)T$ **then output Yes else output No.**

Fig. 3. Test procedure used for TI–SC–MAP.

For a value of T let $OPT_h(T)$ denote the sum of the distances of an optimal placement of p nodes with respect to the distance function $h(v, w) := c(v, w) + \frac{T}{\Omega}d(v, w) = c(v, w) + \mu d(v, w)$; i.e.,

$$OPT_h(T) = \min_{\substack{P \subseteq V \\ |P|=p}} \mathcal{S}_h(P).$$

Then we have the following lemma:

Lemma 7. *The function* $\mathcal{R}(T) = \frac{OPT_h(T)}{T}$ *is monotonically nonincreasing on* $\mathbb{Q} \setminus \{0\}$. □

Proof: Suppose for the sake of contradiction that for two values of T, say T_1 and T_2 with $T_1 < T_2$, we have that $\mathcal{R}(T_1) < \mathcal{R}(T_2)$. Let P_1 and P_2 denote optimal placements of p nodes under h when $T = T_1$ and $T = T_2$ respectively. For $i \in \{1, 2\}$, let C_i and D_i denote the costs of placement P_i under c and d respectively. Thus, we have that $\mathcal{R}(T_i) = \frac{C_i}{T_i} + \frac{D_i}{\Omega}$ for $i \in \{1, 2\}$.

Consider the cost under h of the placement P_1 when $T = T_2$. By the definition of C_1 and D_1, it follows that the cost of P_1 is $C_1 + \frac{D_1 \cdot T_2}{\Omega}$. Thus the value of $\mathcal{R}(T_2)$ is at most this cost divided by T_2 which is $\frac{C_1}{T_2} + \frac{D_1}{\Omega}$. This in turn is less than $\frac{C_1}{T_1} + \frac{D_1}{\Omega}$, since $T_1 < T_2$. But $\frac{C_1}{T_1} + \frac{D_1}{\Omega}$ is exactly $\mathcal{R}(T_1)$, and this contradicts the assumption that $\mathcal{R}(T_1) < \mathcal{R}(T_2)$. □

Now we can establish the result about the performance guarantee of the heuristic. Let $OPT(I) = \mathcal{S}_c(P^*)$ denote the function value of an optimal placement P^* of p nodes. To simplify the analysis, we assume that $OPT(I)/\gamma$ is an integer. This can be enforced by first scaling the cost function c so that all values are integers and then scaling again by γ.

Theorem 8. *Let I denote any instance of* TI–SC–MAP *and assume that there is an optimal placement P^* with $OPT(I) = \mathcal{S}_c(P^*)$. Then* HEUR–FOR–SUM–CONSTRAINT *with the test procedure* Sum–Test *returns a placement P with $\mathcal{S}_d(P) \leq (1 + \gamma)(2 - 2/p)\Omega$ and $\mathcal{S}_c(I)/OPT(I) \leq (2 - 2/p)(1 + 1/\gamma)$.*

Proof: Consider the call to the procedure Sum–Test when $T = T^* = OPT(I)/\gamma$. Notice that T^* is an integer by our assumption. The h–cost of the placement P^* is then $OPT(I) + \frac{T^*}{\Omega}\Omega = OPT(I) + T^* = (1 + \gamma)T^*$. Thus we have $OPT_h(T^*) \leq (1 + \gamma)T^*$ and the $(2 - 2/p)$–approximation P_T that is computed in step 3 will satisfy $\mathcal{S}_h(P_T) \leq (2 - 2/p)OPT_h(T^*) \leq (2 - 2/p)(1 + \gamma)T^*$.

Thus, we observe that the procedure will return **Yes** and that $\mathcal{R}(T^*) \leq 1 + \gamma$. Further, the value T found by the binary search in the main procedure satisfies $T \leq T^*$, since T is the minimum value such that Sum–Test(C') returns **Yes**. Let P_T be the corresponding placement that is returned by Sum–Test. Then we have

$$\mathcal{S}_c(P_T) \leq \mathcal{S}_h(P_T) \leq (2 - \frac{2}{p})(OPT(I) + \frac{T}{\Omega}\Omega) \leq (2 - \frac{2}{p})(1 + \frac{1}{\gamma}) \cdot OPT(I).$$

Moreover, we see that

$$\frac{T}{\Omega}\mathcal{S}_d(P_T) \leq \mathcal{S}_h(P_T) \leq (2 - \frac{2}{p})(1 + \gamma)T,$$

and multiplying the last chain of inequalities by Ω/T yields

$$\mathcal{S}_d(P_T) \leq (2 - 2/p)(1 + \gamma)\Omega$$

and this completes the proof. □

6 Problems for Tree Networks

In this section we study the constrained compact location problems for tree networks. In this case the distances between two vertices correspond to the path lengths along the trees.

Definition 9. A tree based distance structure τ is a set $V = \{v_1, v_2, \cdots v_n\}$ of n vertices, a spanning tree T on these vertices, and two non–negative lengths $c(e), d(e)$ assigned to each edge of the tree. For each pair v_i, v_j of vertices, the distances $c(v_i, v_j)$ and $d(v_i, v_j)$ implied by τ are the sum of the corresponding edge lengths along the unique path in T connecting v_i and v_j.

Versions of compact location problems can be defined for trees, in the same manner as we defined for arbitrary graphs but the distances are now specified by a tree-based distance structure. We denote these problems by **TREE–DC–MAP** and **TREE–SC–MAP** respectively. For instance, for the **TREE–DC–MAP** problem the input is a tree based distance structure, an integer p and a bound Ω. The requirement is to find a subset consisting of p nodes, such that the sum of the c–distances between the nodes is minimized and the diameter with respect to the d–distance does not exceed the bound Ω.

It has been shown in [RKM+93] that the *unconstrained* problems, **TREE–MAP** and **TREE–MDP**, which involve finding a subset of p nodes minimizing the sum of the c–distances and the c–diameter respectively (and ignoring the d–weights on the edges), can be solved in polynomial time.

6.1 The Complexity of TREE-MAP

The following result points out that obtaining an optimal solution to the SC–MAP problem is difficult even for trees.

Proposition 10. SC–MAP *is $\mathcal{NP}$–hard even when the underlying graph is a tree.*

Proof: We use a reduction from **Partition**: Given a multiset of (not necessarily distinct) positive integers $\{a_1, \ldots, a_n\}$ the question is whether there exists a subset $I \subseteq \{1, \ldots, n\}$ such that $\sum_{i \in I} a_i = \sum_{i \notin I} a_i$. Partition is known to be $\mathcal{NP}$–complete (cf. [GJ79]).

Given any instance of **Partition** we construct a star–shaped graph G having $n + 1$ nodes $\{x, y_1, \ldots, y_n\}$ and n edges (x, y_i), $i = 1, \ldots, n$. We then define $c(x, y_i) := a_i$ and $d(x, y_i) := D - a_i$, where $D := \sum_{i=1}^{n} a_i$.

We then run the hypothetical polynomial time algorithm A for SC–MAP for the instance I_j consisting of the graph defined as above and the parameters $p_j, \Omega_j, j = 1, \ldots, n$ where $p_j := j + 1$ and $\Omega_j := 2(j-1)jD - (j-1)D$. Observe that this will still result in an overall polynomial time.

Let $j \in \{1, n\}$ be fixed and assume that P is any placement of $p_j = j + 1$ nodes that includes the node x. It then follows that

$$\mathcal{S}_c(P) = 2(j-1) \sum_{x_j \in P} a_j \qquad \text{and} \qquad \mathcal{S}_d(P) = 2(j-1)jD - 2(j-1) \sum_{x_j \in P} a_j. \quad (4)$$

Moreover, if P is any *feasible* placement for I_j (i.e., $\mathcal{S}_d(P) \leq \Omega_j$) that includes the node x, then using the feasibility, equation (4) and the definition of Ω_j, we obtain

$$S_d(P) = 2(j-1)jD - 2(j-1)\sum_{x_j \in P} a_j \leq 2(j-1)jD - (j-1)D.$$

Thus for such a placement we get

$$\sum_{x_j \in P} a_j \geq D/2. \tag{5}$$

So far we have considered only placements that include the node x. The striking point now is that any optimal feasible placement for I_j must indeed include x. This follows from the fact that replacing any node x_j in the placement by the node x will decrease *both* S_c and S_d.

Hence using (4) and (5) we see that for any optimal placement P for I_j we have

$$\mathcal{S}_c(P) = 2(j-1)\sum_{x_j \in P} a_j \geq (j-1)D. \tag{6}$$

Assume that there is a partition I with $|I| = j$ elements. Then, if we choose the placement $P_j := \{x\} \cup \{y_j : a_j \in I\}$ for the instance I_j, we get

$$\mathcal{S}_d(P_j) = 2(j-1)jD - 2(j-1)\sum_{x_j \in P} a_j = 2(j-1)j - (j-1)D = \Omega_j.$$

Thus the placement is feasible. Moreover,

$$\mathcal{S}_c(P) = 2(j-1)\sum_{x_j \in P} a_j = (j-1)D. \tag{7}$$

Hence, by (6) this placement is optimal and the bound from equation (6) is satisfied as an equality.

Assume conversely that there is an optimal placement for some I_j where the bound from (6) is satisfied as an equality, i.e., equation (7) holds. If we let $I := \{j : x_j \in P\}$, we then have $\sum_{i \in I} a_i = \sum_{x_i \in P} a_i = D/2$.

Thus by running the hypothetical algorithm A on all the instances I_j, $j = 1, \ldots, n$ and inspecting the optimum function value $\mathcal{S}_c$ we can decide whether or not the given instance of **Partition** has a solution. $\square$

Given that **TREE–SC–MAP** is $\mathcal{NP}$–hard we investigate the existence of efficient approximation algorithms for it. By combining the parametric search technique from section 5.2 with the polynomial time algorithm in [RKM+93], for solving **TREE–MAP** (unconstrained version) optimally, we can obtain approximation algorithm for **TREE–SC–MAP** with performance guarantee $(1+\gamma, 1+1/\gamma)$. Thus we have the following theorem.

Theorem 11. *For any fixed $\gamma > 0$ there is a polynomial time algorithm which, given any instance of* TREE–SC–MAP *such that there exists an optimal solution P^* of total c–cost $OPT(I) = S_c(P^*)$ exists, finds a placement P of total d–cost $S_d(P)$ no more than $(1 + \gamma)\Omega$ and satisfying $S_c(P) \leq (1 + 1/\gamma)OPT(I)$.* □

6.2 Polynomial Time Solvable Subcases

While TREE–SC–MAP is $\mathcal{NP}$–hard, it turns out that the other three constrained compact location problems for trees (namely TREE–DC–MAP, TREE–DC–MDP and TREE–SC–MDP) are polynomial time solvable.

Here, we outline our idea for the TREE–DC–MAP problem. Polynomial time solvability for the other problems follows the same outline and is omitted in this version of the paper.

Theorem 12. TREE–DC–MAP *can be solved in polynomial time.*

Proof Sketch: It is easy to see that if two vertices a and b are in a solution, then each vertex on the unique path between a and b can also be added to the solution without violating the diameter constraint and also without increasing the value of the sum cost. Thus, there always exists an optimal solution which is connected; that is, there is an optimal solution which is a subtree of the original tree.

Consider an optimal solution T (i.e., a subtree of the original tree with p nodes) for an instance I of TREE–DC–MAP. Let L be the diameter of the tree with respect to distance function d and let a and b be the vertices in T which are at a distance of L from each other. For this proof sketch, let us assume that the cost with respect to the distance function d is integral and also that it is polynomially bounded. (The general case can be handled in a manner similar to the algorithm for the minimum diameter p–spanning tree problem discussed in [RR$^+$94].) Let us subdivide the edge by placing a dummy node r on it in such a way that $d(a, r) = d(b, r) = L/2$. Next, we prune the tree given by the instance I to obtain T_1 as follows. We delete all vertices in I which are at a distance more than $L/2$ from the point r. Then the pruned tree T_1 has the following desirable property. Every pair of vertices in T_1 is within a d–distance of L from each other. Now we solve the TREE–MAP problem on T_1 using the procedure outlined in [RKM$^+$93]. By repeating this procedure for each pair of vertices a and b such that the d–distance between a and b is at most Ω and choosing a placement with the minimum sum cost with respect to the c–distance, we obtain an optimal solution to the TREE–DC–MAP instance I. □

The algorithm for TREE–DC–MAP resulting from the above discussion is outlined in Figure 4.

Procedure TREE–DC–MAP

1 **for** each $v, w \in V$ **do**

2   Let L be the d–distance between u and v.

   if $L > \Omega$ (the diameter constraint) **then** go to the next iteration.

   Prune the tree I to obtain a new tree $T_{u,v}(V_1, E_1)$ such that

   every pair of nodes in $T_{u,v}$ is within a d–distance of L.

   if $|V_1| < p$, then start the next iteration of the for loop.

3   Solve the unconstrained compact location problem with distances given by c

   on the tree $T_{u,v}$ *optimally* in polynomial time using the algorithm in [RKM$^+$93].

   Let $P(u, v)$ the placement obtained this way.

4 **output** the best placement $P(u, v)$.

Fig. 4. Details of the heuristic for TREE–DC–MAP

References

[AI$^+$91]   A. Aggarwal, H. Imai, N. Katoh, and S. Suri. Finding k points with Minimum Diameter and Related Problems. *J. Algorithms*, 12(1):38–56, March 1991.

[BP91]   J. Bar-Ilan and D. Peleg. Approximation Algorithms for Selecting Network Centers. In *Proc. 2nd Workshop on Algorithms and Data Structures (WADS)*, pages 343–354, Ottawa, Canada, August 1991. Springer Verlag, LNCS vol. 519.

[AA$^+$94]   B. Awerbuch, Y. Azar, A. Blum, and S. Vempala, Improved approximation guarantees for minimum-weight k-trees and prize-collecting salesmen. *Proceedings of the 27th Annual ACM Symposium on the Theory of Computing (STOC'95)*, pp. 277–376.

[BCV95]   A. Blum, P. Chalasani and S. Vempala, A Constant-Factor Approximation for the k-MST Problem in the Plane. *Proceedings of the 27th Annual ACM Symposium on the Theory of Computing* (STOC'95), pp. 294–302.

[BS94]   M. Bellare and M. Sudan. Improved Non–Approximability Results. in *Proceedings of the 26th annual ACM Symposium on the Theory of Computing (STOC)*, May 1994.

[DF85]   M.E. Dyer and A.M. Frieze. A Simple Heuristic for the p–Center Problem. *Operations Research Letters*, 3(6):285–288, Feb. 1985.

[EN89]   E. Erkut and S. Neuman. Analytical Models for Locating Undesirable Facilities. *European J. Operations Research*, 40:275–291, 1989.

[FG88]   T. Feder and D. Greene. Optimal Algorithms for Approximate Clustering. In *ACM Symposium on Theory of Computing (STOC)*, pages 434–444, 1988.

[GJ79]   M.R. Garey and D.S. Johnson. *Computers and Intractability*. W.H. Freeman, 1979.

[Go85]   T.F. Gonzalez. Clustering to Minimize the Maximum Intercluster Distance. *Theoretical Computer Science*, 38:293–306, 1985.

[GH94]   N. Garg, D. Hochbaum, An $O(\log n)$ Approximation for the k-minimum spanning tree problem in the plane. *Proceedings of the 26th Annual ACM Symposium on the Theory of Computing* (STOC'95), pp. 294–302.

[HM79]   G.Y. Handler and P.B. Mirchandani. *Location on Networks: Theory and Algorithms.* MIT Press, Cambridge, MA, 1979.

[HS86]   D. S. Hochbaum and D. B. Shmoys. A Unified Approach to Approximation Algorithms for Bottleneck Problems. *Journal of the ACM*, 33(3):533–550, July 1986.

[KN$^+$95a] S.O. Krumke, H. Noltemeier, S.S. Ravi and M.V. Marathe, Compact Location Problems with Budget and Communication Constraints. In *1st International Conference on Computing and Combinatorics*, X'ian, China, August 1995.

[Lee82]   D.T. Lee. On k-nearest neighbor Voronoi diagrams in the plane. *IEEE Trans. Comput.*, C–31:478–487, 1982.

[LV92]   J.H. Lin and J. S. Vitter. ε-Approximations with Minimum Packing Constraint Violation. In *ACM Symposium on Theory of Computing (STOC)*, pages 771–781, May 1992.

[MR+95]   M.V. Marathe, R. Ravi, R. Sundaram, S.S. Ravi, D.J. Rosenkrantz, and H.B. Hunt III. Bicriteria Network Design Problems. To appear in *Proceedings of the 22nd International Colloquium on Automata Languages and Programming (ICALP)*, 1995.

[MF90]   P.B. Mirchandani and R.L. Francis. *Discrete Location Theory.* Wiley–Interscience, New York, NY, 1990.

[PS85]   F.P. Preparata and M.I. Shamos. *Computational Geometry: An Introduction.* Springer–Verlag Inc., New York, NY, 1985.

[RKM$^+$93] V. Radhakrishnan, S.O. Krumke, M.V. Marathe, D.J. Rosenkrantz, and S.S. Ravi. Compact Location Problems. In *13th Conference on the Foundations of Software Technology and Theoretical Computer Science (FST-TCS)*, volume 761 of *LNCS*, pages 238–247, December 1993.

[RMR$^+$93] R. Ravi, M.V. Marathe, S.S. Ravi, D.J. Rosenkrantz, and H.B. Hunt III. Many birds with one stone: Multi-objective approximation algorithms. In *Proceedings of the 25th Annual ACM Symposium on the Theory of Computing (STOC)*, pages 438–447, 1993.

[RRT91]   S.S. Ravi, D.J. Rosenkrantz, and G.K. Tayi. Facility Dispersion Problems: Heuristics and Special Cases. In *Proc. 2nd Workshop on Algorithms and Data Structures (WADS)*, pages 355–366, Ottawa, Canada, August 1991. Springer Verlag, LNCS vol. 519. (Journal version: *Operations Research*, 42(2):299–310, March-April 1994.)

[RR$^+$94] R. Ravi, R. Sundaram, M. V. Marathe, D. J. Rosenkrantz, and S. S. Ravi, Spanning trees short or small. *Proceedings, Fifth Annual ACM-SIAM Symposium on Discrete Algorithms*, (1994), pp 546–555. (Journal version to appear in *SIAM Journal on Discrete Mathematics.*)

[MR$^+$95] M. V. Marathe, R. Ravi, R. Sundaram, S. S. Ravi, D. J. Rosenkrantz, and H.B. Hunt III. Bicriteria Network Design problems, In *Proceedings of the 22nd International Colloquium on Automata, Languages and Programming (ICALP)*, pages 438–447, 1993.

[ST93]   D. B. Shmoys and E. Tardos. Scheduling unrelated parallel machines with costs. In *Proceedings of the 4th Annual ACM–SIAM Symposium on Discrete Algorithms (SODA)*, pages 438–447, 1993.

On Termination of Graph Rewriting

Detlef Plump*

Universität Bremen, Fachbereich 3 — Informatik
Postfach 33 04 40, D-28334 Bremen, Germany
e-mail: det@informatik.uni-bremen.de

Abstract

A necessary and sufficient condition for termination of graph rewriting
systems is established. Termination is equivalent to the finiteness of all
forward closures, being certain minimal derivations in which each step depends on previous steps. This characterization differs from corresponding
results for term rewriting in that the latter hold only for subclasses of
term rewriting systems. When applied to term graph rewriting, the result characterizes termination of arbitrary term rewriting systems under
graph rewriting. In particular, it captures non-terminating term rewriting
systems that are terminating under graph rewriting.

1 Introduction

Proving that a program terminates for all inputs is an important but in general difficult task (in fact, the problem is unsolvable in general). In the area of
term rewriting systems, proof methods for termination have been investigated
for a long time, leading to a comprehensive theory of termination (see Dershowitz's survey [Der87]). For graph rewriting systems, however, termination
criteria have not yet been developed to the author's knowledge. (Robertson and
Seymour [RS85] have proved a fundamental well-ordering result for graphs, but
it seems that this has not been applied to graph rewriting up to now.)

The objective of this paper is to show that termination of graph rewriting
is equivalent to termination of certain restricted derivations. These so-called
forward closures are derivations in which each step depends on previous steps,
and that are minimal in the sense that all nodes and edges are used or generated
by the rules. Forward closures can be inductively generated from the rules,
providing a proof method for termination which attempts to show that only
finite forward closures are possible.

For term rewriting systems, Dershowitz [Der81] and Geupel [Geu89] have
shown that right-linear respectively non-overlapping rewrite rules terminate if
their forward closures do. But, different from the present result, there are coun-
terexamples demonstrating that terminating forward closures need not guaran-

*Research partially supported by ESPRIT Basic Research Working Group 6112, *COMPASS*

tee termination in general. In contrast, the present result characterizes termination of arbitrary term rewriting systems in the computational model *term graph rewriting*. In particular, the result enables termination proofs for non-terminating term rewriting systems that are terminating under graph rewriting (see Section 4).

2 Graph rewriting

In this section a hypergraph variant of the algebraic approach to graph rewriting is briefly reviewed. For more information, the reader is referred to Ehrig's introduction [Ehr79] and Habel's recent survey [Hab92].

2.1 Hypergraphs and hypergraph morphisms

Let Σ be a *signature*, that is, a set such that each $l \in \Sigma$ comes with a natural number $type(l) \geq 0$. A *hypergraph* over Σ is a system $G = \langle V_G, E_G, lab_G, att_G \rangle$ consisting of two finite sets V_G and E_G of *nodes* and *hyperedges*, a labelling function $lab_G: E_G \rightarrow \Sigma$, and an attachment function $att_G: E_G \rightarrow V_G^*$ which assigns a sequence of nodes to each hyperedge e such that $type(lab_G(e))$ is the length of $att_G(e)$.

In pictures, a hyperedge e is represented as a box labelled with $lab_G(e)$. The box is connected with the attachment nodes in $att_G(e)$ by lines. For example, Figure 1 shows a hypergraph with three hyperedges, where $type(\mathbf{f}) = 4$ and $type(\mathbf{a}) = type(\mathbf{b}) = 1$ (for simplicity, the order among the attachment nodes of the $\mathbf{f}$-hyperedge is not indicated).

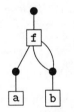

Figure 1: A hypergraph

Let G, H be hypergraphs. Then G is a *subhypergraph* of H, denoted by $G \subseteq H$, if $V_G \subseteq V_H$, $E_G \subseteq E_H$, and if lab_G and att_G are restrictions of lab_H and att_H.

A *hypergraph morphism* $f: G \rightarrow H$ consists of two functions $f_V: V_G \rightarrow V_H$ and $f_E: E_G \rightarrow E_H$ that preserve labels and attachment to nodes, that is, $lab_H \circ f_V = lab_G$ and $att_H \circ f_E = f_V^* \circ att_G$. (The extension $f^*: A^* \rightarrow B^*$ of a function $f: A \rightarrow B$ maps the empty string to itself and $a_1 \dots a_n$ to $f(a_1) \dots f(a_n)$.) The morphism f is *injective* (*surjective*) if f_V and f_E are injective (surjective). A morphism that is both injective and surjective is an *isomorphism*.

2.2 Rules and derivations

A *rule* $r = (L \leftarrow K \rightarrow R)$ consists of two hypergraph morphisms, where $K \rightarrow L$ is injective. L and R are called the *left-hand side* and the *right-hand side* of r.

Given two hypergraphs G and H, a *direct derivation* from G to H by r consists of two hypergraph pushouts[1] of the form shown in Figure 2. Intuitively,

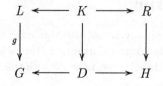

Figure 2: A direct derivation

D is obtained from G by removing the nodes and hyperedges in $g(L - K')$, where K' is the image of K in L. H is constructed from D by identifying items in D according to the morphism $K \rightarrow R$ and by adding all items of R that are not in the image of K. Such a direct derivation is denoted by $G \Rightarrow_{r,g} H$ or $G \Rightarrow_r H$ or simply by $G \Rightarrow H$. The subhypergraph $g(L)$ of G is the *redex* of this direct derivation and is denoted by $Redex(G \Rightarrow H)$.

A *derivation* from G to H, denoted by $G \Rightarrow^* H$, is either an isomorphism $G \rightarrow H$ (which is a derivation of length 0) or a non-empty sequence of direct derivations of the form $G = G_0 \Rightarrow G_1 \Rightarrow \ldots \Rightarrow G_n = H$. A derivation of the latter kind is also denoted by $G \Rightarrow^+ H$. An *infinite derivation* is an infinite sequence of the form $G_0 \Rightarrow G_1 \Rightarrow G_2 \Rightarrow \ldots$

A *graph rewriting system* $\mathcal{G} = (\Sigma, \mathcal{R})$ consists of a signature Σ and a set $\mathcal{R}$ of rules with hypergraphs over Σ.

3 Termination

In the first part of this section, the concept of independent derivation steps is recalled and generalized. Based on this concept, in the second part forward closures are introduced as special derivations and the main result of this paper is stated. The third part of this section is devoted to the proof of the main theorem.

Definition 3.1 (termination) A graph rewriting system $\mathcal{G}$ is *terminating* if it does not admit an infinite derivation.

---

[1] See [Ehr79] for the definition and construction of graph pushouts. The generalization to the hypergraph case is straightforward.

Example 3.2 Suppose that $\mathcal{G}$ contains only the following rule[2]:

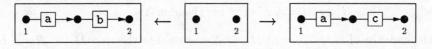

$\mathcal{G}$ is terminating since every rule application reduces the number of b-labels in a hypergraph by one.

3.1 Independence of derivations

Definition 3.3 (independence) Two direct derivations

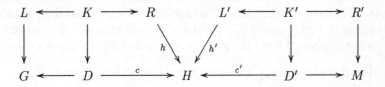

are *independent* if there are hypergraph morphisms $f\colon R \to D'$ and $f'\colon L' \to D$ such that $c' \circ f = h$ and $c \circ f' = h'$.

A necessary condition for independence is that $G \Rightarrow H$ does not generate items that are used by $H \Rightarrow M$ and that $H \Rightarrow M$ does not remove items that are used by $G \Rightarrow H$. More precisely, $h(R) \cap h'(L') \subseteq h(b(K)) \cap h'(b'(K'))$ must hold, where b and b' are the morphisms $K \to R$ and $K' \to L'$. This condition is sufficient for independence if the two rules are non-identifying, that is, if $K \to R$ and $K' \to R'$ are injective.

Independence is an important property as it allows to interchange rule applications. The following theorem was established by Ehrig and Kreowski for the graph case. Its proof can be easily adapted to the present setting.

Theorem 3.4 ([EK76]) *Let $G \Rightarrow_q H \Rightarrow_r M$ be independent direct derivations. Then there are direct derivations $G \Rightarrow_r H' \Rightarrow_q M$.*

The steps $G \Rightarrow_r H' \Rightarrow_q M$ need not be unique since identifying rules are allowed. But for the following considerations only the existence of such steps matters, not their uniqueness.

Definition 3.5 (generalized independence) A derivation $G \Rightarrow^* M$ and a direct derivation $M \Rightarrow_r N$ are *independent* if either

(1) $G \Rightarrow^* M$ has length 0, or

(2) $(G \Rightarrow^* M) = (G \Rightarrow^* H \Rightarrow_q M)$ and

---

[2]Node numbers indicate the hypergraph morphisms, and arrows are used to distinguish the second from the first attachment node of a hyperedge.

- $H \Rightarrow_q M \Rightarrow_r N$ are independent (in the sense of Definition 3.3),

- $G \Rightarrow^* H$ and $H \Rightarrow_r M'$ are independent for some $H \Rightarrow_r M'$ such that $H \Rightarrow_r M' \Rightarrow_q N$ exist by Theorem 3.4.

Independence of $G \Rightarrow^* M$ and $M \Rightarrow_r N$ means that the step $M \Rightarrow_r N$ can be shifted to the beginning, yielding a derivation $G \Rightarrow_r G' \Rightarrow^* N$. Two derivations that are not independent are said to be *dependent*.

3.2 Forward closures and termination

Forward closures are derivations in which each step depends on previous steps, and that are minimal in the sense that all nodes and hyperedges occur in some redex or are generated by the rules. To make this precise, instances of derivations are considered. An instance of a (possibly infinite) derivation is obtained by extending it with "context". Formally, an instance is a derivation whose pushouts are composed from the given pushouts and suitable new pushouts.

Definition 3.6 (instance) Consider a finite or infinite derivation[3]

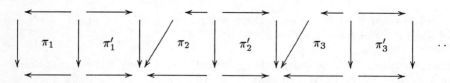

A derivation is an *instance* (or *embedding*) of this derivation if it can be written

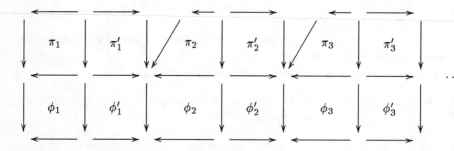

where for each $i \geq 1$, ϕ_i and ϕ_i' are pushouts with injective vertical morphisms. (So the pushouts constituting the derivation are composed of the pushouts π_i and ϕ_i, respectively π_i' and ϕ_i'.)

Given an instance $G \Rightarrow^+ H$ of a derivation $G^\ominus \Rightarrow^+ H^\ominus$ with associated injective morphism $f \colon H^\ominus \to H$, the difference $H - f(H^\ominus)$ (consisting of two sets of nodes and hyperedges) is denoted by $New(H)$. By pushout properties, hyperedges in $New(H)$ can have attachment nodes in $H^\ominus$ only if these nodes exist already in $G^\ominus$ and are not removed by any step in $G^\ominus \Rightarrow^+ H^\ominus$.

---

[3]The hypergraphs in the pushout diagrams are omitted to simplify the presentation.

Definition 3.7 (forward closure) *Forward closures* are inductively defined as follows:

(1) Every direct derivation $L \Rightarrow_{r,g} R$ with surjective g is a forward closure.

(2) A derivation $G \Rightarrow^+ H \Rightarrow M$ is a forward closure if $G \Rightarrow^+ H$ is an instance of a forward closure, and $H \Rightarrow M$ is a direct derivation such that

- $G \Rightarrow^+ H$ and $H \Rightarrow M$ are dependent,
- $New(H) \subseteq Redex(H \Rightarrow M)$.

Note that the prefix $G \Rightarrow^+ H$ of $G \Rightarrow^+ H \Rightarrow M$ is not a forward closure in general. This happens only in the special case where $G \Rightarrow^+ H$ has been taken as an instance of itself.

Example 3.8 Consider the following two rules:

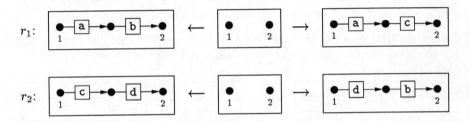

This system has (up to isomorphism) five forward closures:

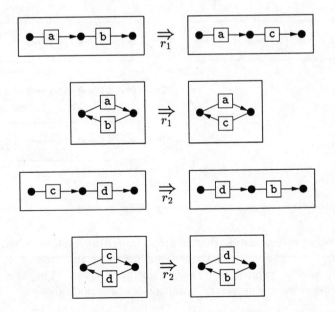

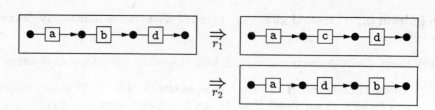

The last forward closure is constructed by applying r_2 to an instance of the first forward closure.

Definition 3.9 (infinite forward closure) An *infinite forward closure* is an infinite derivation $G_0 \Rightarrow G_1 \Rightarrow G_2 \Rightarrow \ldots$ that contains a forward closure as a prefix. That is, there is some $n \geq 1$ such that $G_0 \Rightarrow^+ G_n$ is a forward closure.

If each rule removes at least one item (and only such systems will matter in the following, see below), then every extension $G_0 \Rightarrow^+ G_{n+i}$ of the prefix $G_0 \Rightarrow^+ G_n$ is also a forward closure. This is because for every $p \geq n$, the step $G_p \Rightarrow G_{p+1}$ removes an item in the image of the right-hand side of some rule in $G_0 \Rightarrow^+ G_p$.

Example 3.10 Suppose that the rule r_2 of Example 3.8 is modified by exchanging the labels d and b in the right-hand side, yielding the following rule:

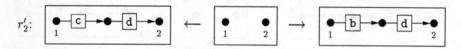

Now the modified system admits an infinite forward closure:

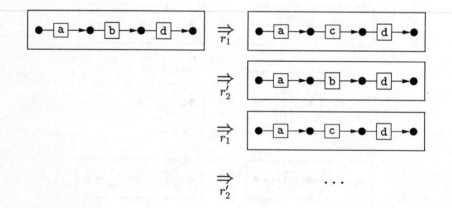

It is important to notice that a graph rewriting system is trivially non-terminating if it contains a rule that does not remove anything. This is because such a rule can be applied to its own right-hand side. Thus, the absence of non-removing rules is a necessary condition for termination.

Definition 3.11 A graph rewriting system is *consumptive* if each rule $(L \leftarrow K \rightarrow R)$ has a non-surjective morphism $K \rightarrow L$.

Theorem 3.12 (main theorem) *A graph rewriting system is terminating if, and only if, it is consumptive and does not admit an infinite forward closure.*

Example 3.13 The system in Example 3.8 is terminating as it is consumptive and possesses only finite forward closures. Note that for this system, termination cannot be proved by comparing the sizes of hypergraphs or the numbers of labels before and after rule applications. Both rules are size-preserving, and after two consecutive steps with r_1 and r_2, the numbers of a's and b's in a hypergraph is the same as before.

3.3 Proof of the main theorem

This subsection contains the lemmas and additional definitions needed for the proof of Theorem 3.12. The essence of the proof is provided by Lemmas 3.17 and 3.19. The former shows that every infinite derivation in which each step depends on previous steps can be transformed into an infinite forward closure. Lemma 3.19 shows that every infinite derivation of a consumptive system can be reordered such that each step depends on previous steps.

Definition 3.14 (dependence chain) A *dependence chain* is a derivation $G_0 \Rightarrow \ldots \Rightarrow G_n$ such that $G_0 \Rightarrow^+ G_i$ and $G_i \Rightarrow G_{i+1}$ are dependent for $i = 1, \ldots, n-1$. An *infinite dependence chain* is an infinite derivation $G_0 \Rightarrow G_1 \Rightarrow \ldots$ such that $G_0 \Rightarrow^+ G_n$ is a dependence chain for each $n \geq 1$.

The idea is to transform an infinite dependence chain into an infinite forward closure by cutting off all items that neither occur in any redex nor are generated by the rules. To formalize this, a partial hypergraph morphism *track* is introduced which maps the initial hypergraph of a derivation into the final one.

Definition 3.15 (track) Given a direct derivation $G \Rightarrow H$, the partial hypergraph morphism $track_{G \Rightarrow H} \colon G \rightarrow H$ is defined by

$$track_{G \Rightarrow H}(x) = \begin{cases} c'(c^{-1}(x)) & \text{if } x \in c(D), \\ \text{undefined} & \text{otherwise,} \end{cases}$$

where $c \colon D \rightarrow G$ and $c' \colon D \rightarrow H$ are the morphisms in the lower row of Figure 2. (The morphism c is injective by pushout properties and hence there is an inverse morphism $c^{-1} \colon c(D) \rightarrow D$.) For a derivation $G = G_0 \Rightarrow G_1 \Rightarrow \ldots \Rightarrow G_n = H$, $track_{G \Rightarrow^+ H}$ is defined by

$$track_{G \Rightarrow^+ H} = track_{G_{n-1} \Rightarrow G_n} \circ \ldots \circ track_{G_0 \Rightarrow G_1}.$$

The above described "clipping" of a derivation amounts to remove from the initial hypergraph all items that will never occur in a redex, and to remove from the subsequent hypergraphs all descendants of these items.

Lemma 3.16 (clipping lemma) *Let $\Delta: G_0 \Rightarrow G_1 \Rightarrow \ldots$ be a finite or infinite derivation and C_0 be the subhypergraph of G_0 consisting of $Redex(G_0 \Rightarrow G_1)$ and all items x such that $track_{G_0 \Rightarrow +G_i}(x) \in Redex(G_i \Rightarrow G_{i+1})$ for some $i \geq 1$. Then there is a derivation $Clip(\Delta): C_0 \Rightarrow C_1 \Rightarrow \ldots$ such that Δ is an instance of $Clip(\Delta)$.*

Proof Straightforward generalization of the corresponding proof for finite derivations in [Kre77]. □

Lemma 3.17 *If a graph rewriting system admits an infinite dependence chain, then it admits an infinite forward closure.*

Proof Let $\Delta: G_0 \Rightarrow G_1 \Rightarrow \ldots$ be an infinite dependence chain. Then $Clip(\Delta)$: $C_0 \Rightarrow C_1 \Rightarrow \ldots$ is also a dependence chain since clipping preserves dependence and independence of each two subsequent steps in Δ.

Claim. For each $i \geq 1$, $Clip(C_0 \Rightarrow^+ C_i)$ is a forward closure.

Proof. By definition of clipping, $Clip(C_0 \Rightarrow C_1)$ is a step $L \Rightarrow R$ with $L = Redex(C_0 \Rightarrow C_1)$. Clearly, this is a forward closure. Now suppose, as induction hypothesis, that $C_0 \Rightarrow^+ C_n$ satisfies the claim for some $n \geq 1$. Let $Clip(C_0 \Rightarrow^+ C_n \Rightarrow C_{n+1}) = (B_0 \Rightarrow^+ B_n \Rightarrow B_{n+1})$. Then $B_0 \Rightarrow^+ B_n$ is an instance of $Clip(C_0 \Rightarrow^+ C_n)$, and $New(B_n) = track_{C_0 \Rightarrow +C_n}(X)$ where X consists of all x whose descendants occur in no redex of $C_0 \Rightarrow^+ C_n$ but in $Redex(C_n \Rightarrow C_{n+1})$. Since $Redex(C_n \Rightarrow C_{n+1}) = Redex(B_n \Rightarrow B_{n+1})$, it follows by induction hypothesis that $B_0 \Rightarrow^+ B_n \Rightarrow B_{n+1}$ is a forward closure ($B_0 \Rightarrow^+ B_n$ and $B_n \Rightarrow B_{n+1}$ are dependent because $C_0 \Rightarrow^+ C_n$ and $C_n \Rightarrow C_{n+1}$ are so). □

By the claim, it suffices to show that there is some k with $Clip(C_0 \Rightarrow^+ C_k) = (C_0 \Rightarrow^+ C_k)$. Since $C_0 \Rightarrow C_1 \Rightarrow \ldots$ results from clipping, each item in C_0 occurs in some redex. Hence there is some $k \geq 1$ such that each item occurs in a redex of $C_0 \Rightarrow^+ C_k$. But then, obviously, $Clip(C_0 \Rightarrow^+ C_k)$ is $C_0 \Rightarrow^+ C_k$. □

By the virtue of Lemma 3.17, proving that a non-terminating consumptive graph rewriting system has an infinite forward closure reduces to proving that there is an infinite dependence chain. The latter is achieved by using a transformation process on infinite derivations which involves the shifting of rule applications. Given a rule application that is independent from all previous steps in a derivation, the *Shift* operation defined next moves this application to the beginning of the derivation.

Definition 3.18 (shift)

(1) $Shift(G \Rightarrow^* M \Rightarrow_r N) = (G \Rightarrow_r N)$ if $G \Rightarrow^* M$ has length 0.

(2) $Shift(G \Rightarrow^* H \Rightarrow_q M \Rightarrow_r N) = (G \Rightarrow^* M' \Rightarrow_q N)$ if $G \Rightarrow^* H \Rightarrow_q M$ and $M \Rightarrow_r N$ are independent and $(G \Rightarrow^* M') = Shift(G \Rightarrow^* H \Rightarrow_r M')$, where $H \Rightarrow_r M' \Rightarrow_q N$ result from interchanging $H \Rightarrow_q M \Rightarrow_r N$ such that $G \Rightarrow^* H \Rightarrow_r M'$ are independent.

Lemma 3.19 *Every non-terminating consumptive graph rewriting system admits an infinite dependence chain.*

Proof Let $G_0 \Rightarrow G_1 \Rightarrow \ldots$ be an infinite derivation. There is some $n \geq 1$ such that $track_{G_0 \Rightarrow^+ G_j}(G_0)$ has the same size for each $j \geq n$ (since the size of $track_{G_0 \Rightarrow^+ G_i}(G_0)$ does not increase for growing i). As the given system is consumptive, this implies the following:

For each $j \geq n$, $G_j \Rightarrow G_{j+1}$ removes an item generated by $G_0 \Rightarrow^+ G_j$. $\qquad (*)$

(An item in G_j is *generated* by $G_0 \Rightarrow^+ G_j$ if it is not in $track_{G_0 \Rightarrow^+ G_j}(G_0)$.) Now consider the n steps $G_0 \Rightarrow \ldots \Rightarrow G_n$. Each of these steps is (trivially) a dependence chain. An infinite transformation process is used to show that at least one of these chains can be extended to an infinite dependence chain. Each step of the process transforms a derivation $G_0 \Rightarrow^+ G_j$ composed from n dependence chains into a derivation $G_0 \Rightarrow^+ G_{j+1}$ that again consists of n dependence chains. Thus, at least one of the chains must grow ad infinitum. Each step of the process is realized by applying two transformation rules. In the following description of these rules, derivations on top of the bar are transformed into those below the bar, and framed derivations represent dependence chains.

$$
\textbf{extend:} \quad \frac{\boxed{X \Rightarrow^+ Y} \Rightarrow_r Z}{\boxed{X \Rightarrow^+ Z}} \quad \text{where } X \Rightarrow^+ Y \text{ and } Y \Rightarrow_r Z \text{ are dependent}
$$

$$
\textbf{shift:} \quad \frac{\boxed{X \Rightarrow^+ Y} \Rightarrow_r Z}{X \Rightarrow_r \boxed{X' \Rightarrow^+ Z}} \quad \begin{array}{l} \text{where } X \Rightarrow^+ Y \text{ and } Y \Rightarrow_r Z \text{ are independent} \\ \text{and } Shift(X \Rightarrow^+ Y \Rightarrow_r Z) = (X \Rightarrow_r X' \Rightarrow^+ Z) \end{array}
$$

It remains to show that each derivation $G_0 \Rightarrow^+ G_j \Rightarrow_r G_{j+1}$, where $G_0 \Rightarrow^+ G_j$ consists of n dependence chains, can be transformed into a derivation $G_0 \Rightarrow^+ G_{j+1}$ consisting of n dependence chains. Starting with the nth dependence chain, repeated application of **shift** leads to a segment $X \Rightarrow^+ Y \Rightarrow_r Z$ with $X \Rightarrow^+ Y$ being the mth dependence chain, $1 \leq m \leq n$, and such that **extend** can be applied. For, by the property $(*)$, the step $G_j \Rightarrow_r G_{j+1}$ removes an item generated by one of the dependence chains and hence cannot be shifted beyond this chain. $\qquad \square$

Proof of Theorem 3.12 A graph rewriting system that is not consumptive or that admits an infinite forward closure is clearly non-terminating. Conversely, Lemmas 3.17 and 3.19 show that a non-terminating consumptive system admits an infinite forward closure. $\qquad \square$

4 Application to term graph rewriting

In this section the application of Theorem 3.12 to term graph rewriting is sketched. Details of the approch used here can be found in [HP91, Plu93]. (For related approaches, see [BvEG+87, CR93] and the collection [SPvE93].)

Let Σ be a signature such that $type(f) \geq 1$ for each $f \in \Sigma$. Such an f is considered as function symbol of arity $type(f) - 1$ (0-ary symbols are constants). Given a hypergraph G over Σ and a hyperedge e with $type(lab_G(e)) = n$, the nodes at positions 1 to $n-1$ in $att_G(e)$ are the *argument nodes* of e while the node at position n is the *result node*. G is a *jungle* if there are no cycles and if each node is the result node of at most one hyperedge. Jungles represent terms similar to trees but allow shared subterms. For example, the hypergraph in Figure 1 is a jungle representing the term f(a,b,b) with a shared constant b.

A term rewrite rule $l \rightarrow r$ consists of two terms l and r that may contain variables. The variables in r must also occur in l, and l must not be a single variable. Each such rule is translated into a graph rule $(L \leftarrow K \rightarrow R)$, where L and R are jungles representing l and r. K is obtained from L by removing the hyperedge representing the leftmost function symbol in l. Variables are represented as non-result nodes, where multiple occurrences of a variable in l or r are always shared in L and R.

Example 4.1 The term rewrite rules

$$\begin{aligned}
\text{f(a,b,x)} &\rightarrow \text{f(x,x,x)} \\
\text{b} &\rightarrow \text{a}
\end{aligned}$$

(where x is a variable) are translated into the following graph rewrite rules:

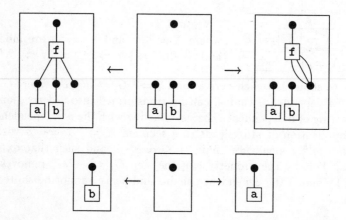

The graph rules resulting from the translation of term rewrite rules are called *evaluation rules*. The set of all evaluation rules for a given term rewriting system is denoted by $\mathcal{E}$. Evaluation rules preserve the jungle structure; the rewrite relation on jungles induced by $\mathcal{E}$ is denoted by $\Rightarrow_{\mathcal{E}}$.

To test the relation $\Rightarrow_{\mathcal{E}}$ for termination, only forward closures starting from jungles need to be considered since $\Rightarrow_{\mathcal{E}}$ is a relation on jungles. Moreover, evaluation rules are consumptive as each application removes a hyperedge. Hence Theorem 3.12 can be refined as follows, where an *infinite jungle forward closure* is an infinite forward closure starting from a jungle.

Theorem 4.2 *The relation $\Rightarrow_{\mathcal{E}}$ is terminating if, and only if, $\mathcal{E}$ does not admit an infinite jungle forward closure.*

Example 4.3 It is easy to check that the two evaluation rules in Example 4.1 induce only jungle forward closures of at most two steps. There are three one-step forward closures resulting from the first rule (the variable node can be identified with each of the two result nodes of the hyperedges for **a** and **b**) and one one-step forward closure resulting from the second rule. The three former closures can be extended by applying the second rule, yielding the only possible two-step closures. Obviously, these cannot be extended. Hence, by Theorem 4.2, the relation $\Rightarrow_{\mathcal{E}}$ is terminating. In contrast, the given term rewriting system is non-terminating due to the following infinite rewrite sequence:

$$\texttt{f(a,b,b)} \rightarrow \texttt{f(b,b,b)} \rightarrow \texttt{f(a,b,b)} \rightarrow \ldots$$

Thus Theorem 4.2 can be used to prove the termination of evaluation rules even if term rewriting is non-terminating (vice versa it holds that $\Rightarrow_{\mathcal{E}}$ is terminating whenever term rewriting is [HP91]).

It is worth noting that the above rules constitute a non-terminating system if they are used on arbitrary hypergraphs. The reason is that the first rule can be applied to the hypergraph that is obtained from the left-hand side by fusing the three argument nodes of the **f**-hyperedge. This application yields the same hypergraph again (which is not a jungle since the two constant hyperedges have the same result node), hence there is a cyclic derivation.

The jungle evaluation relation $\Rightarrow_{\mathcal{E}}$ is not adequate for non-left-linear term rewriting systems, i.e. systems where some rules have repeated variables in their left-hand sides. This is because for such a system, an evaluation rule need not be applicable to a jungle although the underlying term rewrite rule is applicable to the represented term. This problem is overcome in [HP91, Plu93] by adding "folding rules" which fuse hyperedges with identical labels and argument nodes. The termination of systems of evaluation and folding rules can still be characterized as in Theorem 4.2. However, the folding rules cause infinitely many (finite) jungle forward closures. Future research has to show whether this problem can be solved by further restricting the class of jungle forward closures.

References

[BvEG$^+$87] Hendrik Barendregt, Marko van Eekelen, John Glauert, Richard Kennaway, Rinus Plasmeijer, and Ronan Sleep. Term graph rewriting. In *Proc. Parallel Architectures and Languages Europe*, pages 141–158. Springer Lecture Notes in Computer Science 259, 1987.

[CR93] Andrea Corradini and Francesca Rossi. Hyperedge replacement jungle rewriting for term rewriting systems and logic programming. *Theoretical Computer Science*, 109:7–48, 1993.

[Der81]   Nachum Dershowitz. Termination of linear rewriting systems (pre-liminary version). In *Proc. Automata, Languages, and Programming*, pages 448–458. Springer Lecture Notes in Computer Science 115, 1981.

[Der87]   Nachum Dershowitz. Termination of rewriting. *Journal of Symbolic Computation*, 3:69–116, 1987.

[Ehr79]   Hartmut Ehrig. Introduction to the algebraic theory of graph grammars. In *Proc. Graph-Grammars and Their Application to Computer Science and Biology*, pages 1–69. Springer Lecture Notes in Computer Science 73, 1979.

[EK76]   Hartmut Ehrig and Hans-Jörg Kreowski. Parallelism of manipulations in multidimensional information structures. In *Proc. Mathematical Foundations of Computer Science*, pages 284–293. Springer Lecture Notes in Computer Science 45, 1976.

[Geu89]   Oliver Geupel. Overlap closures and termination of term rewriting systems. Report MIP-8922, Fakultät für Mathematik und Informatik, Universität Passau, Passau, Germany, 1989.

[Hab92]   Annegret Habel. Hypergraph grammars: Transformational and algorithmic aspects. *Journal of Information Processing and Cybernetics*, 5:241–277, 1992.

[HP91]   Berthold Hoffmann and Detlef Plump. Implementing term rewriting by jungle evaluation. *RAIRO Theoretical Informatics and Applications*, 25(5):445–472, 1991.

[Kre77]   Hans-Jörg Kreowski. Manipulationen von Graphmanipulationen. Dissertation, Technische Universität Berlin, 1977.

[Plu93]   Detlef Plump. Evaluation of functional expressions by hypergraph rewriting. Dissertation, Universität Bremen, Fachbereich Mathematik und Informatik, 1993.

[RS85]   Neil Robertson and P.D. Seymour. Graph minors — a survey. In Ian Anderson, editor, *Surveys in Combinatorics 1985*, pages 153–171. London Mathematical Society, 1985.

[SPvE93]   Ronan Sleep, Rinus Plasmeijer, and Marko van Eekelen, editors. *Term Graph Rewriting: Theory and Practice*. John Wiley, 1993.

A uniform approach to graph rewriting : the pullback approach [*]

Michel Bauderon

Laboratoire Bordelais de Recherche en Informatique
Université Bordeaux I
33405 Talence Cedex, France
bauderon@labri.u-bordeaux.fr

Abstract. Most of the works in the theory of graph rewriting can be put into two main categories : edge (or hyperedge) rewriting and node rewriting. Each has been described by a specific formalism, both have given rise to many significant developments and many works have been devoted to the comparison of both approaches. In this paper, we describe a new categorical formalism, which provides a common framework to both approaches and makes their comparison much clearer.

1 Introduction

Graphs can roughly be considered from two points of view, either as sets of vertices linked by edges or as sets of edges glued by vertices, each point of view leadimg to a different kind of graph rewriting systems, basically *edge rewriting* (or hyperedge replacement, shortly HR, see [5, 13] and *node rewriting* (or vertex replacement, shortly VR, see [9, 15]).

In both cases the basic ingredients are given by specifying what is to be replaced, how it is linked to the rest of the graph, by what it will be replaced and how the replacing part will be connected to the remaining part of the original graph, the main difference between both types of rewriting being probably that node rewriting may create new edges in an unpredictable way (cf [15]), while edge rewriting does not create anything, but simply unites into a single object already existing items.

This has made a big difference when trying to develop a more abstract setting for graph rewriting. Indeed, considering the graph to be rewritten as "embedded" in the big graph (in a sense which we shall not make more precise), it quickly appeared that the categorical generalisations of union and equivalence relation, namely coproduct and coequalizer where enough to give a good description edge-oriented rewriting. This gave rise to the well known double-pushout approach extensively developed by the Berlin school under the name of algebraic theory of graph grammars [11], which we would rather call categorical. Unfortunately, this approach was absolutely unable to describe the creation of edges and thus was not applicable to node rewriting.

---

[*] This work has been supported by the Esprit BRA "Computing with graph transformations"

Other approaches to graph grammars have been developed since, providing a genuine algebraic description of graph rewriting in the usual sense [5, 6] and an extensive logical theory of graphs. The table in figure 1 summarises the mains characteristics of both types of graph rewriting together with the main developments.

	Hyperedge Replacement	Vertex replacement
Nature of the graphs	set of edges glued by nodes	set of nodes linked by edges
Substitution	replaces an edge	replaces a node
Interface	a family of nodes	a set of edges
Connection	glues interface nodes	creates interface edges
Algebraic framework	Available	Available
Categorical approach	Available	None
Logical Theory	Available	Available

Fig. 1. Hyperedge vs. Vertex replacement

Trying to extend earlier works on infinite hypergraphs generated by systems of recursive equations (c.f. [1, 2]) from the edge rewriting to the node rewriting context, we had to develop a new theory which would provide us with a notion of approximation for these infinite graphs. We knew from [1] that a categorical framework was necessary and it turned out that it was a complete category instead of a cocomplete one that provided the good one. Our first results in that direction have been presented in [3].

Further investigation of this new framework then showed that it was much richer than expected, and that provided one accepted to change one's point of view on graphs, it was not only encompassing both the vertex (NLC) and edge (HR) replacement systems, but also richer systems as described by the double pushout approach or the NCE grammars [16].

As a concluding remark for this introduction, let us simply suggest that the surprising fact is not that a mechanism based on product proves to be so effective, but rather that union and pushout have taken such a proeminent part in the theory of graph rewriting : we have the feeling that throughout computer science, product is usually more popular than union (product types are more widely used than union types, and SQL natural join is clearly a pullback).

This short paper tries to describes how this new framework unifies the description of those rewriting mechanisms and can provide a clear setting to compare them. As far as was possible, we have tried to give enough examples to support the intuition of the reader and to convince him/her of the interest of this new approach. Unfortunately, this did not leave much room for a thorough discussion of some of its properties. More details will be found in forthcoming works.

Section 2 describes the general framework that we shall use, section 3 and 4 the single and double pullback approaches and show how they encompass both edge and node rewriting.

2   Basic Properties of Graphs

Let N be the set of integers. In this paper we shall consider simple, undirected graphs, possibly with loops and shall use indistinctly the words node and vertex. For more details on category theory we refer the reader to standard textbooks such as [17].

Definition 1. A *graph* is a pair $G = \langle V, E \rangle$ where V is the set of vertices and $E \subseteq V \times V$ is the set of edges. An edge between vertices u and v will be denoted by $[u, v]$. A node v in V is reflexive if $[v, v] \in E$. If S is any set, the complete graph K_S over S is the graph $< S, S \times S >$. A *hypergraph* is simply a bipartite graph.

As usual, considering G as a hypergraph means that its set of nodes is split into the set of nodes and the set of hyperedges of the corresponding hypergraph. We shall call their elements respectively nodes and hyperedges. The arity of a hyperedge will be the number of its neighbours.

Definition 2. A *graph morphism* $h : G \longrightarrow G'$ is a pair $h =< h_V, h_E >$ of mappimgs with $h_V : V \longrightarrow V$ and $h_E : E \longrightarrow E'$ such that $h_E([u, v]) = [h_V(u), h_V(v)]$. If G and G' are hypergraphs, h will be a hypergraph morphism if it respects the bipartitions of G and G' (we do not ask that h preserves arities of the hyperedges).

It is well known that the good properties of graph morphisms turn the set of graphs into a category that we shall denote by $\mathcal{G}$. Clearly, hypergraphs and hypergraph morphisms form a subcategory of $\mathcal{G}$ whose properties are not very interesting and we shall most often have to consider hypergraphs as graphs. We shall always distinguish between the words graph and hypergraph.

Proposition 3. *The category $\mathcal{G}$ has arbitrary products and equalizers. The graph with one vertex and one edge is a terminal object simply denoted by $\odot$. It is a neutral element for the product.*

Proof. We shall not give the proof of this result which is classic, but simply enumerate the description of the objects that such a proof would build. If G_1 and G_2 are two graphs, their product $G_1 \times G_2$ is classically defined by its sets of vertices V and edges E, in the following way :

- $V = V_1 \times V_2$
- $E = \{[u_1 u_2, v_1 v_2] / [u_1, v_1] \in E_1 \wedge [u_2, v_2] \in E_2\}$.

The definition of the corresponding projections $\pi_i : G \longrightarrow G'$ is quite obvious.

Also known as the *Kronecker product* [12], the product we have just defined is different from the cartesian product of graph used in some other areas of

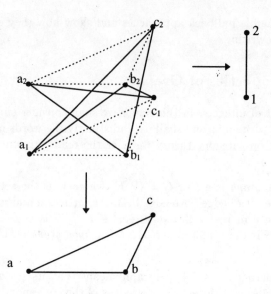

Fig. 2. The product of $K_3 \times K_2$

computer science (we refer the reader to some contributions in this volume) and which does not have the appropriate universal property to be a categorical product. A simple example, that of the categorical product of K_3 by K_2, is shown on figure 2, where the dotted triangles do not belong to the product but are drawn to help the reader understand the construction. We think the notations are self explaining.

The equalizer of two morphisms f and g is simply given by the inclusion of the subgraph where $f = g$. The only arrow from any graph G to $\odot$ sends all nodes and all edges of G respectively into the unique node and edge of $\odot$. It is now easily checked from the definition of the product that $\odot$ is a unit. Note that the unique vertex of $\odot$ creates all the vertices of the product, while the unique edge creates all the edges. $\qquad \square$

Corollary 4. *The category $\mathcal{G}$ has arbitrary limits (is complete). In particular, $\mathcal{G}$ has pullbacks.*

Proof. It is a standard result that a category has limits if and only if it has both products and equalizers. Let us simply indicate here that the pullback of two graph morphisms $f_i : G_i \longrightarrow F, i = 1, 2$ is a pair of arrows $h_i : H \longrightarrow G_i, i = 1, 2$ where H is the subgraph of the product consisting of exactly those items (nodes

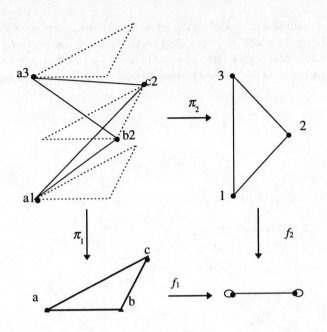

Fig. 3. A simple example of pullback

and vertices) on which $f_i \circ \pi_i$ coincide. Note also that if F is $\odot$, the pullback is equal to the product.   □

As an example, figure 3 describes the pullback of a pair of morphisms (f_1, f_2), from K_3 to K_2 (the complete reflexive graph on two vertices) defined :

  – on the vertices by $f_1(a) = f_2(1) = f_2(3)$ and $f_1(b) = f_1(c) = f_2(2)$,
  – on the edges by coherence with the definition of a morphism.

3   Single Pullback Rewriting

3.1   General Rewriting

A basic framework for graph rewriting could be the following :

Definition 5. A production rule is a graph morphism $p : R \longrightarrow L$, where L is the left-hand side of the rule, while R is the right-hand side. An occurrence of the left-hand side of p in the graph G is a morphism $x : G \longrightarrow L$. The rewriting of G by p at occurrence x is the pullback of x and p.

As a simple example and justification of this statement, let us consider the pullback in figure 4, which is but a slight variation on the previous one : we have merely added two loop edges, one on the node a, an other one on the node 2. Then, computation of the pullback creates two more edges, $[a_1, a_3]$ and $[b_2, c_2]$.

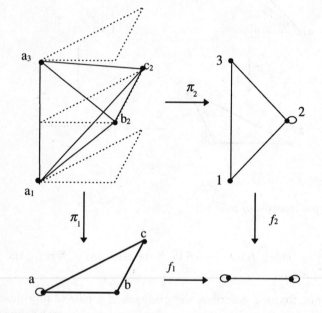

Fig. 4. A slight variation

Although our interpretation might look like being imposed upon the reader, it is clear that this new pullback can be interpreted (for instance, since it is symmetric) as the rewriting of node 2 by the graph K_2 (based on the vertices b and c) with a connection relation stating that all vertices of the right hand side will be linked to all nodes which where linked to the vertex 2 in the original graph : this is an example of what litterature calls an NLC rewriting.

Of course, we do not pretend that this example is a good model of NLC rewriting since for instance, it is not able to rewrite a node in a larger context, but we thing that it is sufficient to justify the idea of using pullback as a generic rewriting mechanism. In the next sections, we will show how it can actually specialise to describe the usual vertex and edge rewritings.

3.2 Node Rewriting

A vertex replacement rewriting rule is usually defined by giving *separately* the graph to be substituted to a node with a certain label and a connection relation which specifies the way its nodes will be linked to the neighbours of the rewritten node. Trying to integrate both part of the rule in a unique setting has only given rise so far to quite awkward formalism (cf. [8]). We now translate this rewriting mechanism into our new setting where all the items of the traditional NLC (*Node labelled Controlled*) mechanism will be integrated within the rewriting rule itself, in what we consider to be a quite elegant way. In the following definition, totally reflexive means that all nodes are reflexive (cf definition 1).

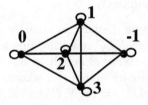

Fig. 5. The alphabet graph A_3

Definition 6. Let A be the infinite totally reflexive graph with vertices $\{-1\} \cup N$ and edges $\{[-1, n]/n \in N\} \cup \{[0, n]/n \in N\} \cup \{[m, n]/n, m \in \{-1\} \cup N\}$. A will be called the *alphabet graph*.

In other words, A is obtained by taking the countable complete graph K_ω and two extra nodes called 0 and -1 and linking both of them to all the nodes of K_ω. For convenience, we shall call 0 the context, -1 the unknown and all other nodes the letters. Considering the graph A will allow us to take into account an arbitrary number of distinct letters, but if we only need a finite number m of such letters, we can restrict to its subgraph A_m, where ω is replaced by $[1, m]$ (see figure 5 for the graph A_3). As a matter of fact, we shall not really need letters or labels, since the labelling of nodes will be provided by morphisms into A, but we shall sometimes use colors or various kinds of ellipsis to make the drawings more intuitive.

Definition 7. Let G be a graph and u be a vertex of G. A *label* a on u is a morphism $a : G \longrightarrow A$ such that $a^{-1}(-1) = \{u\}$ and for each $i \in N^*$, either $a^{-1}(i)$ is empty or it consists of immediate neighbours of u. An *unknown* is a label on a reflexive node.

Intuitively, a label on u distinguishes between u, its immediate neighbours which are mapped to the letters in A and the rest of the graph, mapped onto the context in A.

Definition 8. A VR-rule is a morphism $r : R \longrightarrow A$ where $\#r^{-1}(0) = 1$ and all the $r^{-1}(i), i \in N$ have at most a single element. A *production* is a pair (a, r) where a is an unknown and r is a VR- rule.

Intuitively, $r^{-1}(-1)$ is the graph to be substituted to the rewritten node, and the edges describe the connection relation of the node rewriting rule, $a^{-1}(-1)$ is the node x to be rewritten, $a^{-1}(n)$ for $n \in A$ are those neighbours which will be connected to the rewritten graph according to the connection described by $r^{-1}(n)$, and $a^{-1}(0)$ are the nodes of G which will not be affected by the rewriting (this is why we talk about *the context* of x).

Definition 9. The application of r to G at a is the pullback of r and a. Let $\underline{G}$ denote the graph built as a pullback.

We now show that our rewriting mechanism actually encompasses NLC rewriting. We first describe how any NLC rewriting rule in the sense of [15] can be described by a VR-rule in the sense of definition 8.

We first let $\underline{A}$ be an "ordinary alphabet", whoses letters we consider to be enumerated : $a_1, \ldots, a_n, \ldots$. Let $\rho = (a, X, C)$ be an NLC rule, where $a \in \underline{A}$ is the label of the nodes where ρ can be applied, X is the right hand side of the rule and C the connection relation over the alphabet A. We let R be the following graph :

- vertices : those of X, one i for each letter $a_i \in A$, an extra one u,
- edges : those of X, one from each i to a vertex v labelled by b in X iff $(a, b) \in C$, one from each i to u.

The morphism is defined :

- on the nodes by : $r(u) = 0, r(i) = i$ and $r(v) = -1$ for each node v coming from X,
- on the edges simply by being a morphism.

Definition 10. The rule $r : R \longrightarrow A$ is the VR-rule associated with the NLC rule ρ.

A very simple example is provided by the NLC rule whose rewriting part and connection rule are represented on the left side of figure 6. The corresponding VR-rule is represented on the right side of the same figure 6. Once again, let us insist on the fact that we do not need any label for our rewriting mechanism and that the integers used in the coding are simply the names of the vertices which are explicitly indicated merely to make the drawing clearer (at least do we hope). A "pullback encoded" NLC rewriting step is described on figure 7 where we have removed all the integers and used ellipsis to indicate the way edges are mapped onto each other.

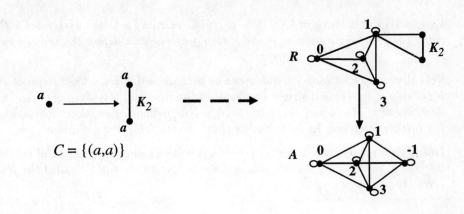

$$C = \{(a,a)\}$$

Fig. 6. Coding a simple *NLC* rule

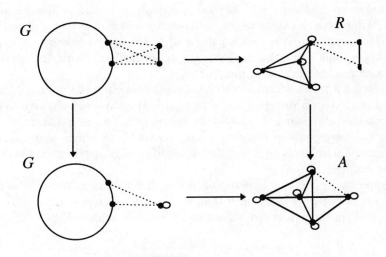

Fig. 7. An *NLC* rewriting step

Proposition 11. *Let $\rho = (X, C)$ be an NLC rule and r be the associated VR-rule. Then the application of ρ and r to a graph G both define the same graph $\underline{G}$.* □

Note that conversely, any VR-rule r can be decomposed into the basic items of an NLC rule :$r^{-1}(-1)$ is the right hand side of the rule, and the edges between the elements of $r^{-1}(-1)$ and those of:$r^{-1}(i), i \in N$ define the connection relation. For further reference, let us formalise that into the following definition.

Definition 12. For any VR-rule r, $r^{-1}(-1)$ will be called the *right hand side* of the rule r, $r^{-1}(0)$ its *left hand side* and the $r^{-1}(i), i \in N$ will be called the *link vertices*.

3.3 Edge Rewriting

Let us quickly indicate how edge - or rather hyperedge - rewriting can be simulated by pullback rewriting.

The simplest way to describe hyperedge rewriting in a categorical framework is probably to consider hypergraphs with n sources (or n-hypergraphs [5]), i.e, arrows $\underline{n} \xrightarrow{d} D$, where $\underline{n}$ is the discrete hypergraph with n vertices.

An unknown x of type p in the n-hypergraph $\underline{n} \xrightarrow{g} G$ is also an arrow $\underline{p} \xrightarrow{x} G$. Alternately, x can be seen as a hyperedge and $x(p)$ as its set of vertices.

An HR-rewriting rule is then a pair (x, d) where x is an unknown of type p in an n-hypergraph g and d is a p-hypergraph. In this framework, the rewriting of g by d at x is the pushout of the pair (x, d).

We now need to transform all the basic ingredients into the pullback framework and convert embeddings x and d into projections. We simply have to consider both the unknown and sources as hyperedges, i.e., as extra nodes and edges in the corresponding bipartite graph, replace both source and unknown morphisms by morphisms on A and define the hyperedge rewriting as the corresponding pullback.

More precisely, if G is a hypergraph, we shall define an unknown x of type p as a morphism $x : G \longrightarrow A$, such that $x^{-1}(0)$ is a reflexive hyperedge (i.e., a reflexive node in the hyperedges subset of V) and for each $n in N$,

$$\#x^{-1}(n) = \begin{cases} 1 \text{ if } 1 \le n \le p \\ 0 \ \text{ otherwise.} \end{cases} \tag{1}$$

A *hypergraph with p sources* H will simply be a distinguished unknown $d : H \longrightarrow A$. A *production rule* will again be a pair (x, d) and the rewriting of G by d at x the pullback of the pair (x, d).

Proposition 13. *Let $H \xleftarrow{h} \underline{G} \xrightarrow{g} G$ be the pullback of the pair (x, d). Then $\underline{G} \xrightarrow{doh} A$ is a p-hypergraph.* □

With these definitions, it is now clear how with any HR-rule (x, d) with pushout hypergraph G' we can associate a production rule (x, d) with pullback hypergraph $\underline{G}$.

Proposition 14. *Let (x, d) be an HR-rule and (x, d) be the corresponding production rule. The hypergraphs G' and $\underline{G}$ are isomorphic.* □

3.4 Comparison

Many works have been devoted to the study of the relative powers of vertex and hyperedge rewriting yielding very deep comparison results (see for instance [6, 8, 9, 10, 18]). Although we can not go here into great details, we believe that the previous results clearly show why VR-rewriting is strictly more powerful than HR rewriting. This could be summarised by the 'equation' :

$$VR = HR + \text{multiple source}$$

Indeed a rewriting rule as defined in paragraph 3.3 is clearly a rewriting rule as defined in paragraph 3.2, and conversely. But the main difference lies in the rewriting site : in the HR case, all neighbours of the rewritten node (the hyperedge) are sent to different vertices in the alphabet graph A, while in the VR case, several of them can be sent to the same node ('have the same label'). Then, depending on the number of such nodes, the pullback mechanism will create the necessary edges to link the new nodes to the old ones. This flexibility of pullback rewriting will appear even more strikingly when considering parallel rewriting [4]. Note that unlike [5], this formalism does not allow for multiple edges hence for multiple sources on the same node.

4 Double Pullback Rewriting

4.1 NCE Rewriting

Double pullback rewriting can be defined exactly like double pushout rewriting as a means to substitue a graph to a full subgraph of a given graph. Simply, as shown in 8 all the arrows will be the other way round and the occurrence of the left-hand side of the rule within the graph to rewrite will not be an embedding but a projection, a different which shall be significant.

Definition 15. Let $L \xrightarrow{r} A$ be a VR-rule and G be a graph. An occurrence of r in G is a morphism $G \xrightarrow{p} L$ such that the restriction of p to $(p \circ r)^{-1}(-1)$ is one to one.

In other words, using the terminology of definition 12, the right hand side of r occurs exactly in G, p projects its occurrence bijectively into L, its neighbours somehow on the link vertices in L, and the rest of G on the left hand side of L.

Proposition 16. *There exists a graph D and two morphisms $D \xrightarrow{s} A$ and $G \xrightarrow{h} D$ such that (p, h) is the pullback of (s, r).*

Proof. Immediate from the definition of an occurrence. It is easily seen that D is obtained from L by simply reducing the occurrence of the right hand side of r to a graph $\odot$, linked uniformly to its former neighbour vertices. □

Double Pushout Rewriting

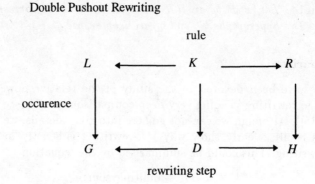

rule

occurence

rewriting step

Double Pullback Rewriting

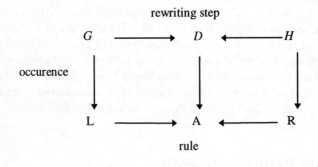

rewriting step

occurence

rule

Fig. 8. Double Pushout vs. Double Pullback

Remark. Unlike the case of the double pushout approach, there are no application conditions. This is due to the deep difference between embedding and projection : projection does not allow edges crossing the boundary between the context and the graph to be rewritten, since it would imply the existence of an edge between the vertices 0 and -1 in the alphabet graph.

Definition 17. A production rule is a pair $L \xleftarrow{l} A \xrightarrow{r} R$ of morphisms with target the alphabet graph.

Proposition 18. *A production rule in the sense of definition 17 is exactly what is called a NCE (Neighboorhood Controlled Embedding) production in the litterature. It can be applied to any of its occurrences.* □

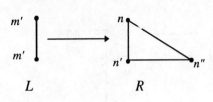

$$(m,n, f(v)) = 1 \qquad (m,n'', f(v)) = 0$$
$$(m,n, f(v'')) = 1 \qquad (m,n'', f(v'')) = 1$$
$$(m',n, f(v)) = 0 \qquad (m',n'', f(v)) = 0$$
$$(m',n, f(v'')) = 1 \qquad (m',n'', f(v'')) = 1$$
$$(m,n, f(v)) = 1$$
$$(m,n, f(v'')) = 1$$
$$(m',n, f(v)) = 0$$
$$(m',n, f(v'')) = 1$$

(a) The rule   *(b) The connection relation*

Fig. 9. An NCE rewriting rule

Giving a proof of this result would imply restating several definitions and would not fit within the space requirements for this paper. Let us simply give the example of the rule described in figure 9, which we hope will fully convince the reader.

It describes a standard NCE rewriting rule as defined in [16], where the left part shows the graph L to be rewritten by R, and where the right part describe the connection relation : a node n in the copy of R which is added to $G - L$ will be linked to a node v in $G - L$ if and only if there exists a node m in L linked to v and if the label $f(v)$ is such that $(m, n, f(v)) = 1$ (we refer the reader to [16] for more extensive details on NCE rewriting).

Now, let us simply claim without further justification but the evidence of the picture that the right column of figure 10 actually encodes this rule and that the double pullback described in the right column does actually describe the effect of NCE rewriting.

4.2 Subgraph Rewriting : The Double Pushout Approach

The previous results indicate clearly that the double pushout approach to graph rewriting can be emulated by a double-pullback rewriting and therefore that it is a a special case of NCE rewriting.

4.3 Conclusion

This paper introduces a new approach to vertex replacement which has been developed while trying to give a rigorous definition of the generation of infinite

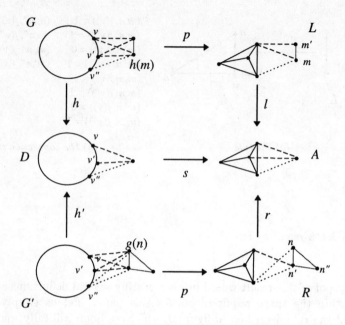

Fig. 10. An NCE rewriting step

graphs generated by vertex replacement [3]. This new framework is quite interesting *per se* since it eventually closes a gap between the formalisms describing HR and VR system : both could be described either by set theoretical definitions describing the substitution mechanism or by an algebraic formalism giving a clean description of the generated languages, but only HR systems could be described within a categorical framework. Pullback rewriting has the double merit of filling the hole and of providing a uniform description of graph rewriting.

Further work will show how this formalism can be used to provide :

- a detailed description of infinite graphs generated by systems of recursive equations (see [3] for first results),
- a description of the simultaneous rewriting of a set of nodes or hyperedges by a single rule [4] by simply collecting several rewriting sites into a single one (provided of course that some compatibility conditions are satisfied),
- description of other rewriting systems such as label rewriting, node rewriting in hypergraphs or single pullback rewriting of partial morphisms (rewritings a la single push-out).

References

1. M. Bauderon, Infinite Hypergraphs I. Basic properties, *Theor. Comput. Sci* 82 (1991) 177-214
2. M. Bauderon, Infinite Hypergraphs II. Systems of recursive equations, *Theor. Comput. Sci* 103 (1993) 165-190
3. M. Bauderon, A categorical approach to vertex replacement : the generation of infinite graphs, *5th International Conference on Graph Grammars and their applications to Computer Science, Williamsburg, November 1994, To appear, Lect. Notes in Comp. Sci.*
4. M. Bauderon, Parallel rewriting through the pullback approach, in SEGRA-GRA'95, to appear, *Elect. Notes in Theor. Comp. Sci.*
5. M. Bauderon, B. Courcelle, Graph expressions and graph rewriting, *Math. Systems Theory* 20 (1987), 83-127
6. B. Courcelle, An axiomatic definition of context-free rewriting and its applications to NLC graph grammars, *Theor. Comput. Sci* 55 (1987), 141-181.
7. B. Courcelle, Graph rewriting : an algebraic and logic approach, in *J. Van Leeuwen, ed, Handbook of Theoretical Computer Science*, Vol. B (Elsevier, Amsterdam, 1990) 193-242
8. J. Engelfriet, G. Rozenberg, A comparison of boundary graph grammars and context-free hypergraph grammars, *Inf. Comp.* 84, 1990, 163-206
9. J. Engelfriet, G. Rozenberg, Graph grammars based on node rewriting : an introduction to NLC grammars, *Lect. Notes in Comp Sci* N 532, 1991, 12-23.
10. J. Engelfriet, L. Heyker, G. Leih, Context free graph languages of bounded degree are generated by apex graph grammars, *Acta Informatica* 31, 341-378 (1994)
11. H. Ehrig, Introduction of the algebraic theory of graph grammars, in Graph Grammars and their applications to Computer Science, *Lect. Notes in Comp Sci* N 73, 1979, 1-69
12. M. Farzan, D.A. Waller, Kronecker products and local joins of graphs, *Can. J. Math.* Vol. XXIX, No 2 1977, 255-269
13. A. Habel, Hyperedge Replacement : Grammars and Languages, *Lect. Notes in Comp Sci* N 643, 1992.
14. D. Janssens, G. Rozenberg, On the structure of node-label-controlled graph languages, *Inform. Sci.* 20 (1980), 191-216.
15. D. Janssens, G. Rozenberg, Graph grammars with node label controlled rewriting and embedding, *Lect. Notes in Comp. Sci.* 153 (1982), 186-205.
16. D. Janssens, G. Rozenberg, Graph grammars with neighboorhood-controlled embedding, *Theor. Comp. Sci.* 21 (1982), 55-74.
17. S. McLane, *Categories for the working mathematician*, Springer, Berlin, 1971.
18. W. Vogler, On hyperedge replacement and BNLC graph grammars, *Discrete Applied Mathematics* 46 (1993) 253-273.

Visualizing Two- and Three-Dimensional Models of Meristematic Growth

F. David Fracchia

Graphics and Multimedia Lab, School of Computing Science
Simon Fraser University, Burnaby, B.C. V5A 1S6 CANADA

Abstract. The role that cell division patterns play on the shape of an organism is quite significant. The relationship between such division patterns and meristem development has been extensively studied in the literature. This paper presents several two- and three-dimensional map and cellwork L-system models that simulate patterns described by Lück and Lück [11]. The advantage of these models over their predecessors is that, through the use of a geometric model to determine cell shape, they can be translated into a visual form quite easily in order to animate (and validate) the growth of the meristem.

1 Introduction

The study of cell division patterns plays a significant role in the area of developmental biology. The spatial and temporal organization of cell divisions are believed to be the main factors determining the form of a tissue [1]. This paper presents two-dimensional map L-system and three-dimensional cellwork L-system models for the visualization of meristematic patterns, such as those exhibited by the division of apical cells in ferns and mosses.

Meristematic and phyllotactic patterns have been studied extensively by Lück et al [8, 9, 10, 11]. Specifically, they have classified such patterns based on the number of distinct division planes exhibited by an apical cell, which in turn determines the number of walls comprising apical and merophyte cells. Furthermore, they have presented double-wall map and cellwork L-systems which describe these patterns. However, as detailed as their models are, they lack the functionality necessary to automatically generate realistic cell shapes.

Marker based map and cellwork L-systems with precise and realistic geometric interpretation have been developed by Fracchia et al [4, 5] and used to model a variety of plants at the early developmental stage. In this paper, the specific models proposed by Lück and Lück [11] are reformulated in terms of marker based systems, and visualized using computer graphics techniques.

The motivation is two-fold. First, as a research tool, the graphical simulations allow one to directly study the effect of meristematic division patterns on organism shape, and verify the models proposed by Lück and Lück. Second, as a method for visualization, the simulations provide a tool for highlighting and presenting significant features. For example, pseudocolouring may be used to distinguish between successive merophytes generated by the apical cell. Furthermore, the models serve as a testbed for further experiments.

The paper is organized as follows. Section 2 briefly describes map L-systems and cellwork L-systems. Section 3 presents four models of meristematic growth derived from the double-wall cellworks and illustrations of Lück and Lück [11], each in their respective two- and three-dimensional representations, and the resulting visualizations. Problems open for future research are outlined in Section 4.

2 Map and Cellwork L-systems

This section briefly describes marker based map and cellwork L-systems. The reader is referred to [4, 5] for more details. The underlying mathematical framework consists of two components. First, the meristematic cell division patterns are expressed, at a topological level, using the formalism of two-dimensional map L-systems and three-dimensional cellwork L-systems. Next, cell geometry is modelled using a physically-based approach that takes into account the osmotic pressure inside the cells and the tension of the walls.

2.1 Two-Dimensional Map L-systems

Map L-systems, first proposed by Lindenmayer and Rozenberg [7], operate on a map which corresponds to a microscopic view of a cellular layer where regions represent cells, and edges represent cell walls perpendicular to the plane of view.

A marker based map L-system is composed of a starting map, or *axiom*, and production rules of the form: $A \to \alpha$, where the directed (with arrow) or neutral (without) edge $A \in \Sigma$ is called the *predecessor*, and the string α, composed of symbols from Σ and special symbols $[,], +, -$, is called the *successor*. Arrows placed above edge symbols indicate whether the successor edges have directions consistent with, or opposite to, the predecessor edge.

Pairs of matching brackets [and] delimit *markers*, which specify possible attachment sites for region-dividing walls. The symbols $+$ or $-$ within the brackets indicate whether the marker is placed to the left or to the right of the predecessor edge. The next symbol in the brackets is the marker label. There can be only one marker within a set of brackets. The left arrow over a marker indicates that the marker is directed towards the predecessor edge, and the right arrow indicates that it is oriented away from that edge. If no arrow is present, the marker is neutral. Figure 1 gives two examples of production rules.

A derivation step consists of two phases (Figure 2): First, each edge in the map is replaced by successor edges and markers using the corresponding edge production. If more than one production has the same predecessor, the first one encountered is chosen (the user specifies order in the input). Second, each region is scanned for matching markers; those that have the same label and proper orientation. If more than two matching markers exists in a region, then a pair is chosen arbitrarily; these will create a new edge which will split the region. However, the pair must not belong to the same production in order to avoid degenerate walls. The remaining markers are discarded.

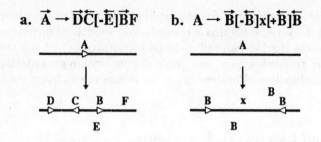

a. $\vec{A} \to \vec{D}\overleftarrow{C}[\text{-}\overleftarrow{E}]\vec{B}F$ **b.** $A \to \vec{B}[\text{-}\overleftarrow{B}]x[\text{+}\overleftarrow{B}]\overleftarrow{B}$

Fig. 1. Examples of map L-system productions: (a) directed and (b) neutral.

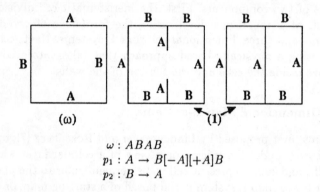

$$\omega : ABAB$$
$$p_1 : A \to B[-A][+A]B$$
$$p_2 : B \to A$$

Fig. 2. The phases of a map L-system derivation step.

2.2 Three-Dimensional Cellwork L-systems

Cyclic cellwork L-systems were first introduced by Lindenmayer [6] and then modified to become marker based by Fracchia and Prusinkiewicz [4]. Cellwork L-systems operate on a three-dimensional extension of a map. Cells are polyhedral volumes bounded by polygon regions (walls) composed of edges.

A marker based cellwork L-system is defined by a finite alphabet of edge labels Σ, a finite alphabet of wall labels Γ, a starting cellwork ω, and a finite set of edge productions P. The initial cellwork is specified as a list of walls and their bounding edges. Edges may be directed or neutral. Each production is of the form $A : \beta \to \alpha$, where the edge $A \in \Sigma$ is the *predecessor*; the string $\beta \in \Gamma^+ \cup \{",", "*"\}$ is a list of *applicable walls* (* denotes all walls and a list of several walls is separated by commas); and the string α, composed of edge labels from Σ, wall labels from Γ, and symbols [and], is the *successor*. Figure 3 gives an example of a production.

Pairs of matching brackets [and] delimit *markers* which specify possible attachment sites for new edges and walls. Arrows indicate the directions of the successor edges and markers with respect to the predecessor edge. The list β

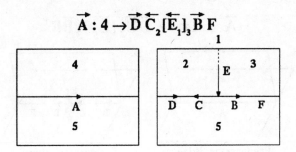

Fig. 3. Example of cellwork L-system production (directed).

contains all walls into which a marker should be inserted. In addition to the labels for edges and markers, a successor specifies the labels of walls which may be created as a result of production application.

A derivation step in a cellwork L-system consists of three phases, illustrated in Figure 4 for the rule: $\overrightarrow{A}{:}1,4 \rightarrow \overrightarrow{D}\ \overleftarrow{C}_2[\overrightarrow{E}_5]_3\overrightarrow{B}\ F$. First, each edge in the cellwork is replaced by successor edges and markers using one or more productions. The rule above applies to the edge A if it belongs to one or more walls labelled 1 or 4 (Figure 4a), and subdivides the edge into four edges D, C, B and F as well as introduces marker E into all walls of type 1 or 4 which share edge A (Figure 4b). If no production exists for an edge, the edge remains unchanged. Second, each wall is scanned for matching markers. If a match inducing a consistent labelling of daughter walls is found, the wall is subdivided. In Figure 4c the E markers have been matched with similar markers inserted in walls 1 or 4 by other assumed productions to create new edges which split the walls. The daughter wall created before the matched marker in the direction of the predecessor edge A will be labelled 2, and the wall formed after the marker will be labelled 3. Third, each cell is scanned for a circular sequence of new division edges having the same wall label. If such a sequence is found, it is used to bound the new wall which will divide the cell into two daughter cells. In order for this to occur, all markers involved must specify the same label for the new wall; 5 in this example (Figure 4d).

The same rules outlined previously for map L-systems as to the choice of matching pairs of markers apply with the following exception. A wall may be subdivided more than once as long as new division edges do not intersect and a consistent labelling of daughter walls is possible. In contrast, a cell may be divided only once in any derivation step.

The limitation of the scope of a production to specific walls may create a consistency problem while rewriting edges. That is, an edge incident to two differently labelled walls may be the predecessor of two separate productions, each with its own distinct successor. If the two successors do not match a conflict arises as to which to choose. We assume the cellwork L-systems under consideration are free of such inconsistencies. However, this does not preclude the possibility

$$\vec{A} : 1{,}4 \rightarrow \overset{\leftrightarrow}{D}\overset{\leftrightarrow}{C}_2\,[\vec{E}_5\,]_3\,\vec{B}F$$

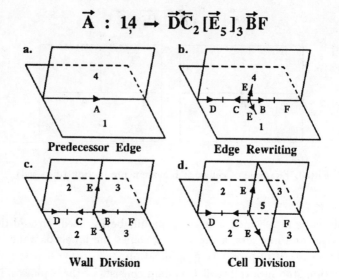

Fig. 4. The phases of a cellwork L-system derivation step.

of applying several productions simultaneously to the same edge. Such is the case when several productions apply to one successor edge but only differ in the markers specified within the successor (the successor edges are consistent).

2.3 Geometric Model

Maps and cellworks are topological objects without inherent geometric properties. In order to visualize them, some method for assigning geometric interpretation must be applied. We take a physically-based approach and assume that the shape of cells and thus the shape of the entire organism result from the action of forces.

The cellular structure is represented as a two-(three-)dimensional network of masses corresponding to cell vertices, connected by springs which correspond to cell walls (edges). These springs are assumed to be always straight and obey Hooke's law. The motivation for their use is that they model surface tension. Furthermore, they have a rest length of zero which would cause the entire structure to collapse inwards.

To counteract the compression of springs the cells exert pressure on their bounding walls; motivated by osmotic pressure. The pressure on a wall is directly proportional to the wall length (area) and inversely proportional to the cell area (volume), and is divided evenly between the wall vertices,

The position of each vertex, and thus the shape of the structure, is computed by first summing the forces at each vertex due to incident springs and the pressure exerted on cell walls (edges). Then, using Newton's second law of motion, we obtain a system of $2N$ differential equations, where N is the total number of vertices, which can be solved numerically using the forward (explicit) Euler method. This procedure is repeated until the change in all vertex positions from the previous to current iteration falls below a threshold value. This indicates that the equilibrium state has been approximated to the desired accuracy. The next derivation step is then performed and the method is applied once again. In such a way, the developmental process is simulated as periods of continuous cell expansion, delimited by instantaneous cell divisions. A more detailed mathematical description appears in Fracchia et al [4, 5].

3   Meristem Models

The apices of most fern and moss meristems result from repeated divisions of the apical cell into merophytes. These divisions usually proceed in a helicoidal manner. For the purposes of this paper and to remain consistent throughout all the models, the apical cell is assumed to be triangular (tetrahedral in three dimensions) in shape and the divisions are assumed to proceed in a counter-clockwise fashion when viewed from the top of the meristem (for map and cellwork L-systems 3 and 4). In reality, such divisions may proceed in a clockwise direction in some organisms, and may even alternate (these would be simple changes to the models).

The four models below represent the "theoretical elementary shoots" which lie on the diagonal of the meristimatic group table in Lück and Lück [11]. Each model will be accompanied by illustrations showing the topological changes associated with the derivation steps and rendered sequences of the cells in which the geometric model is applied. The polyhedral cells are rendered as Gouraud shaded spheres centred at the midpoint of the minimum and maximum bounding values of the polyhedron, with a radius of the average distance between vertices of the polyhedron and the computed centre. The choice of spheres was for aesthetic reasons (they simply looked better than shaded polyhedra).

The map and cellwork L-systems were derived by observing the repetitive nature of the cell divisions patterns. Dividing apical cells were distinguished from nondividing merophytes. Apical cells were then further classified based on the orientation of their division wall. Finally, the somewhat tedious task of mapping these divisions to wall (edge) labels and production rules was performed. Each model was akin to solving a small geometric puzzle.

3.1   Two-Dimensional Models

In map L-system 1 (Figure 5), productions p_1 and p_2 subdivide the apical cell into a merophyte and new apical cell. The division is such that only one merophyte is a neighbour of the apical cell (a filament). In this case the apical cell repeats the same division pattern (same division wall orientation) every derivation step. Therefore the productions must, assuming a compact set of productions, maintain the same sequence of wall labels for each new apical cell. Specifically the labels A and B are assigned to the two outer walls of the apical cell and productions are designed to divide the cell and regenerate these same labelled walls. Only one wall label could have been used but the production rule would have generated markers outside of the cell; a scenario we have attempted to avoid in order to produce "cleaner" models.

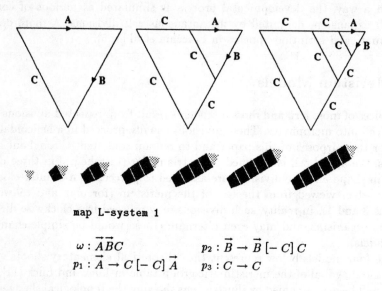

map L-system 1

$$\omega : \vec{A}BC \qquad\qquad p_2 : \vec{B} \rightarrow \vec{B}\,[-\,C]\,C$$
$$p_1 : \vec{A} \rightarrow C\,[-\,C]\,\vec{A} \qquad p_3 : C \rightarrow C$$

Fig. 5. Meristimatic pattern defined by map L-system 1: topological view of the first two cell divisions and a rendered sequence with the geometric model applied for 6 divisions.

In map L-system 2 (Figure 6), productions p_1-p_4 subdivide the apical cell into a merophyte and new apical. The division is such that it alternates between one merophyte generated to the left (p_1,p_2) and then one to the right (p_3,p_4). In this case, two merophytes are neighbours of the apical cell. Since apical cell divisions alternate another set of two labels and productions were used to distinguish the division wall orientations. The first set of productions generated walls with labels of the second set and vice-versa in order to repeat the dividion pattern. Different colours are used to indicate the apical cell and the alternating merophytes.

In map L-system 3 (Figure 7), productions p_1-p_6 subdivide the apical cell into a merophyte and new apical. The division is such that it alternates between three merophytes generated in a counterclockwise fashion. In order to account for the third division in the pattern (viewing this as an extension of map L-system 2) a set of two more labels and productions were introduced. Each set transformed the cell walls to the proper orientation and labels for the next set. It is important to note that the cells begin to self-intersect as the number of steps exceed fifteen, since the outer cells cannot move very much relative to the inner cells (they are constrained by the number of walls).

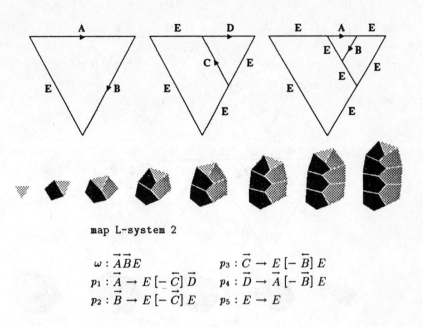

map L-system 2

$$\omega : \vec{A}\vec{B}E$$
$$p_1 : \vec{A} \rightarrow E\,[-\,\vec{C}]\,\vec{D} \qquad p_3 : \vec{C} \rightarrow E\,[-\,\vec{B}]\,E$$
$$p_2 : \vec{B} \rightarrow E\,[-\,\vec{C}]\,E \qquad p_4 : \vec{D} \rightarrow \vec{A}\,[-\,\vec{B}]\,E$$
$$\qquad\qquad\qquad\qquad\quad p_5 : E \rightarrow E$$

Fig. 6. Meristimatic pattern defined by map L-system 2: 7 divisions.

In map L-system 4 (Figure 8), the previous map L-system has been modified and rules added to divide the merophyte walls (p_{11}) as they age (controlled by a delay from rules p_9-p_{10}). Each cell is modified such that it has one more wall than the next youngest cell. This increases the flexibility of cells (with respect to how much bending can occur) so that the geometric model produces a more pleasing shape. Furthermore, it remains robust at further divisions.

3.2   Three-Dimensional Models

The derivation of cellwork L-systems 1-4 parallel those of map L-systems 1-4 with the exception that the divisions must be handled within three-dimensional cells. The puzzle remained the same: observe the repeated patterns and determine the number of distinct changes in the apical cell and the division wall orientation. The task of then determining the appropriate labels and productions was more difficult because of the added spatial dimension. For example, the top face of the three-dimensional models described by cellwork L-systems 1-3 behave in the same manner as map L-systems 1-3 respectively. The difference lies in determining the productions to handle the remaining faces.

Cellwork L-system 1 (Figure 9), like map L-system 1, generates only one merophyte which neighbours on the apical cell. The initial walls and associated edges are shown but the specific three-dimensional vertex positions are omitted for the sake of brevity. In the first derivation step, production p_1 divides walls labelled 1. The inserted edges form a cycle that divides the cell with a new wall labelled 2.

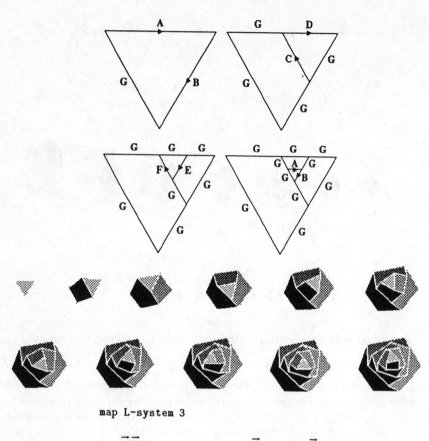

map L-system 3

$$\omega : \overrightarrow{A} B G \qquad p_4 : \overrightarrow{D} \to G \, [- \overleftarrow{E}] \, G$$

$$p_1 : \overrightarrow{A} \to G \, [- \overline{C}] \, \overrightarrow{D} \qquad p_5 : \overrightarrow{E} \to G \, [- \overleftarrow{A}] \, \overrightarrow{B}$$

$$p_2 : \overrightarrow{B} \to G \, [- \overrightarrow{C}] \, G \qquad p_6 : \overrightarrow{F} \to G \, [- \overrightarrow{A}] \, G$$

$$p_3 : \overrightarrow{C} \to G \, [- \overleftarrow{E}] \, \overrightarrow{F} \qquad p_7 : G \to G$$

Fig. 7. Meristimatic pattern defined by map L-system 3: 10 divisions.

Cellwork L-system 2 (Figure 10), like map L-system 2, generates two alternating merophytes in a cell layer. Both merophytes remain neighbours of the apical cell until the next derivation generates a new neighbour.

Cellwork L-system 3 (Figure 11), like map L-system 3, generates three alternating merophytes in a cell layer. The productions result in the apical cell dividing such that alternating merophytes are generated counterclockwise around the apical cell.

In cellwork L-system 4 (Figure 12), as in cellwork L-system 3, the apical cell generates three alternating merophytes, however, the merophytes increase

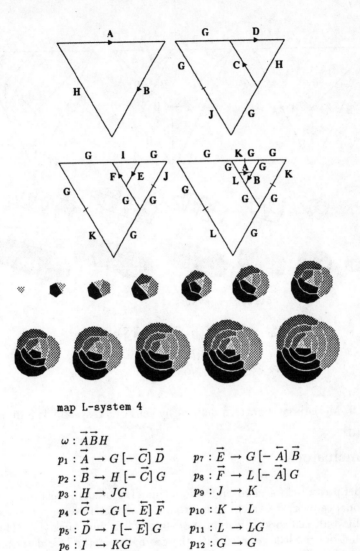

map L-system 4

$$\omega : \vec{A}\vec{B}H$$
$$p_1 : \vec{A} \rightarrow G\,[-\,\bar{C}]\,\vec{D}$$
$$p_2 : \vec{B} \rightarrow H\,[-\,\vec{C}]\,G$$
$$p_3 : H \rightarrow JG$$
$$p_4 : \vec{C} \rightarrow G\,[-\,\bar{E}]\,\vec{F}$$
$$p_5 : \vec{D} \rightarrow I\,[-\,\vec{E}]\,G$$
$$p_6 : I \rightarrow KG$$

$$p_7 : \vec{E} \rightarrow G\,[-\,\bar{A}]\,\vec{B}$$
$$p_8 : \vec{F} \rightarrow L\,[-\,\bar{A}]\,G$$
$$p_9 : J \rightarrow K$$
$$p_{10} : K \rightarrow L$$
$$p_{11} : L \rightarrow LG$$
$$p_{12} : G \rightarrow G$$

Fig. 8. Meristimatic pattern defined by map L-system 4: 11 divisions.

in the number of walls with age (in this case, in conjunction with divisions in the apical cell) like map L-system 4. Notice that the structure rotates itself after each division.

In all examples, it is quite apparent that the shape of the meristem is directly associated with the division patterns. Furthermore, the shapes resulting from the geometric model coincide with the hand-drawn illustrations in [11].

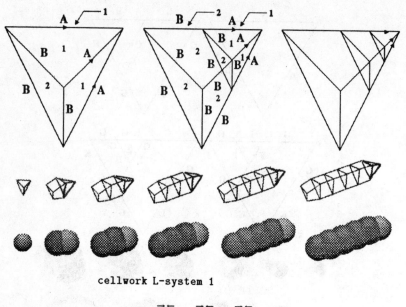

cellwork L-system 1

$$\omega : 1112 \quad \overrightarrow{A}\overleftarrow{A}B \quad \overrightarrow{A}\overleftarrow{A}B \quad \overrightarrow{A}\overleftarrow{A}B \quad BBB$$
$$p_1 : \overrightarrow{A}: * \to B_2[B_2]_1 \overrightarrow{A}$$
$$p_2 : B: * \to B$$

Fig. 9. Meristimatic pattern defined by cellwork L-system 1: 5 divisions.

4 Conclusions

This paper presented four two-dimensional map L-system models and their three-dimensional cellwork L-system counterparts of meristem development based on the double-wall cellwork L-system models of Lück and Lück [11]. The advantage is that our models, coupled with the geometric method for determining cell shape, provide a direct translation into a visual sequence of images animating the growth of the meristem. The position of the division walls and the resulting cell expansion are clearly visible, and significantly aid in understanding the relationship between apical cell topology and meristematic patterns.

However, there is room for further improvements. For example, the geometric model used to determine cell shape involves many arbitrary assumptions, such as the equal distribution of pressure between the wall vertices, and reduction of surface tension to forces acting along walls. However, biological observations which would provide a solid basis for such refinements do not appear in the literature.

Also, it would have been more convenient and compact to specify the division patterns using cell systems [2, 3], however cell systems only allow for walls to divide if intersected by a new division wall. This would have excluded the map

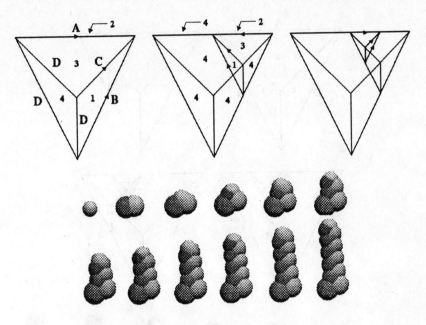

cellwork L-system 2

$$\omega : 1234 \quad \overrightarrow{B}\overleftarrow{C}D \quad \overrightarrow{A}\overleftarrow{B}D \quad \overrightarrow{C}\overleftarrow{A}D \quad DDD$$

$p_1 : \overrightarrow{A}: 2 \rightarrow D_4[\overrightarrow{B}_1]_2 \overrightarrow{A} \qquad p_6 : \overrightarrow{B}: * \rightarrow DD$

$p_2 : \overrightarrow{A}: 3 \rightarrow D_4[\overrightarrow{C}_1]_3 \overrightarrow{A} \qquad p_7 : \overrightarrow{C}: 1 \rightarrow D_4[D_1]_4 \, D$

$p_3 : \overrightarrow{A}: * \rightarrow D\overrightarrow{A} \qquad\qquad p_8 : \overrightarrow{C}: 3 \rightarrow D_4[\overrightarrow{C}_1]_3 \, D$

$p_4 : \overrightarrow{B}: 1 \rightarrow D_4[D_1]_4 \, D \qquad p_9 : \overrightarrow{C}: * \rightarrow DD$

$p_5 : \overrightarrow{B}: 2 \rightarrow D_4[\overrightarrow{B}_1]_2 \, D \qquad p_{10} : \overrightarrow{D}: * \rightarrow D$

Fig. 10. Meristimatic pattern defined by cellwork L-system 2: 11 divisions.

and cellwork L-system 4 models. Three-dimensional cell systems are currently under development.

The fact that the meristem data was available in a form readily translatable into map and cellwork L-systems made this specific research possible. Currently, more complex models of three-dimensional meristem development are being developed, such as branching patterns (multiple apical cells) and merophyte divisions. However, the lack of data required to derive the cellwork L-systems is proving to be a challenging problem.

Acknowledgments

Many thanks to Drs. Jacqueline and Hermann Lück for inspiring discussions and sharing their insights into meristematic growth. Thanks to Sheelagh Carpendale

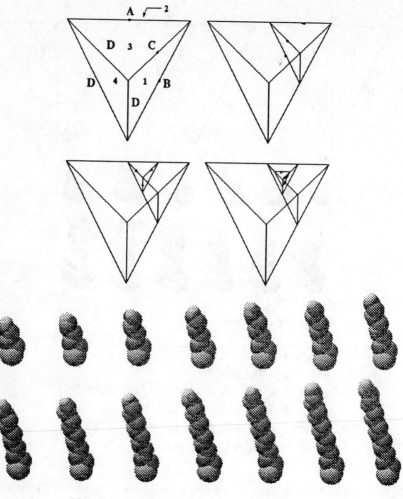

cellwork L-system 3

$$\omega : 1234 \quad \overrightarrow{B}\overleftarrow{C}D \quad \overrightarrow{A}\overleftarrow{B}D \quad \overrightarrow{C}\overleftarrow{A}D \quad DDD$$

$p_1 : \overrightarrow{A}: 2 \rightarrow D_4[\overleftarrow{B_2}]_1 \ \overleftarrow{C}$ $\qquad p_6 : \overrightarrow{B}: * \rightarrow DD$

$p_2 : \overrightarrow{A}: 3 \rightarrow D_4[\overrightarrow{A_2}]_3 \ \overleftarrow{C}$ $\qquad p_7 : \overrightarrow{C}: 1 \rightarrow D_4[D_2]_4 \ D$

$p_3 : \overrightarrow{A}: * \rightarrow D\overleftarrow{C}$ $\qquad p_8 : \overrightarrow{C}: 3 \rightarrow D_4[\overrightarrow{A_2}]_3 \ D$

$p_4 : \overrightarrow{B}: 1 \rightarrow D_4[D_2]_4 \ D$ $\qquad p_9 : \overrightarrow{C}: * \rightarrow DD$

$p_5 : \overrightarrow{B}: 2 \rightarrow D_4[\overrightarrow{B_2}]_1 \ D$ $\qquad p_{10} : D: * \rightarrow D$

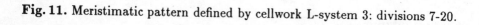

Fig. 11. Meristimatic pattern defined by cellwork L-system 3: divisions 7-20.

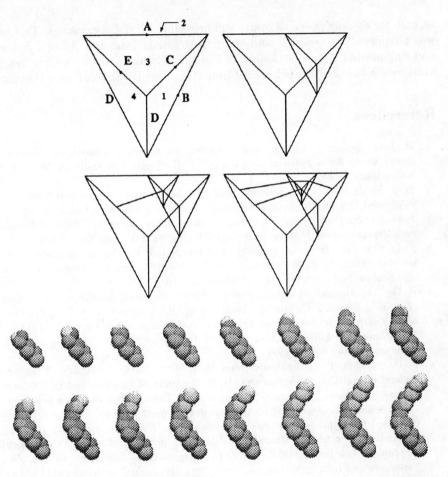

cellwork L-system 4

ω : 1234 $\quad \vec{B}\overleftarrow{C}D \quad \vec{A}\overleftarrow{B}D \quad \vec{C}\overleftarrow{A}E \quad \vec{E}DD$

$p_1 : \vec{A}: 2 \to D_5[\vec{B}_2]_1 \, \overleftarrow{C}$

$p_2 : \vec{A}: 3 \to D_5[\vec{A}_2]_3 \, \overleftarrow{C}$

$p_3 : \vec{A}: 5 \to D_5[D_5]_5 \, \overleftarrow{C}$

$p_4 : \vec{A}: * \to D\overleftarrow{C}$

$p_5 : \vec{B}: 1 \to D_5[D_2]_4 \, D$

$p_6 : \vec{B}: 2 \to D_5[\vec{B}_2]_1 \, D$

$p_7 : \vec{B}: 5 \to D_5[D_5]_5 \, D$

$p_8 : \vec{B}: * \to DD$

$p_9 : \vec{C}: 1 \to D_5[D_2]_4 \, \overleftarrow{E}$

$p_{10} : \vec{C}: 3 \to D_5[\vec{A}_2]_3 \, \overleftarrow{E}$

$p_{11} : \vec{C}: 5 \to D_5[D_5]_5 \, \overleftarrow{E}$

$p_{12} : \vec{C}: * \to D\overleftarrow{E}$

$p_{13} : D: * \to D$

$p_{14} : \vec{E}: * \to \vec{F}$

$p_{15} : \vec{F}: 5 \to D_5[D_5]_5 \, G$

$p_{16} : \vec{F}: * \to DG$

$p_{17} : G: 5 \to D_5[D_5]_5 \, D$

$p_{18} : G: * \to DD$

Fig. 12. Meristimatic pattern defined by cellwork L-system 4: divisions 8-23.

as well, for reading previous drafts and providing helpful suggestions. This work was supported by research and equipment grants from the Natural Sciences and Engineering Research Council of Canada. Thanks also to the Graphics and Multimedia Lab and the School of Computing Science, Simon Fraser University.

References

1. M. J. M. de Boer. *Analysis and computer generation of division patterns in cell layers using developmental algorithms.* PhD thesis, University of Utrecht, the Netherlands, 1989.

2. M. J. M. de Boer, F. D. Fracchia, and P. Prusinkiewicz. A model for cellular development in morphogenetic fields. In G. Rozenberg and A. Salomaa, editors, *Lindenmayer Systems. Impacts on Theoretical Computer Science, Computer Graphics, and Developmental Biology*, pages 351–370. Springer-Verlag, Berlin, 1992.

3. F. D. Fracchia. Integrating lineage and interaction for the visualization of cellular structures. In *Fifth International Workshop on Graph Grammars and Their Application to Computer Science*, pages 241–246, 1994.

4. F. D. Fracchia and P. Prusinkiewicz. Physically-based graphical interpretation of marker cellwork L-systems. In H. Ehrig, H. J. Kreowski, and G. Rozenberg, editors, *Graph grammars and their application to computer science; Fourth International Workshop*, Lecture Notes in Computer Science 532, pages 363–377. Springer-Verlag, Berlin, 1991.

5. F. D. Fracchia, P. Prusinkiewicz, and M. J. M. de Boer. Animation of the development of multicellular structures. In N. Magnenat-Thalmann and D. Thalmann, editors, *Computer Animation '90*, pages 3–18, Tokyo, 1990. Springer-Verlag.

6. A. Lindenmayer. Models for plant tissue development with cell division orientation regulated by preprophase bands of microtubules. *Differentiation*, 26:1–10, 1984.

7. A. Lindenmayer and G. Rozenberg. Parallel generation of maps: Developmental systems for cell layers. In V. Claus, H. Ehrig, and G. Rozenberg, editors, *Graph grammars and their application to computer science; First International Workshop*, Lecture Notes in Computer Science 73, pages 301–316. Springer-Verlag, Berlin, 1979.

8. H. B. Lück and J. Lück. De la theorie des automates vers la phyllotaxie. In B. Millet, editor, *Mouvements, rythmes et irritabilité chez les végétaux, Hommage à Lucien Baillaud*, pages 85–94. Université de Franche-Comté, Besançon, 1993.

9. J. Lück, A. Lindenmayer, and H. B. Lück. Models for cell tetrads and clones in meristematic cell layers. *Botanical Gazette*, 149:1127–141, 1988.

10. J. Lück and H. B. Lück. Double-wall cellwork systems for plant meristems. In H. Ehrig, H. J. Kreowski, and G. Rozenberg, editors, *Graph grammars and their application to computer science; Fourth International Workshop*, Lecture Notes in Computer Science 532, pages 564–581. Springer-Verlag, Berlin, 1991.

11. J. Lück and H. B. Lück. Parallel rewriting dw-cellwork systems for plant development. In J. Demongeot and V. Capasso, editors, *Mathematics applied to biology and medicine*, pages 461–466. Wuerz Publishing Ltd., Winnipeg, 1992.

Graph-theoretical Methods to Construct Entity-Relationship Databases

Sven Hartmann

Fachbereich Mathematik
Universität Rostock
18051 Rostock, Germany

Abstract. Within the recent years, the entity-relationship approach has become one of the most popular methods in high-level database design. In this approach data are modelled as entity and relationship types. Usually relationship types come along with certain restrictions that influence the structure of databases. Cardinality constraints are the most commonly used class of constraints used to model such restrictions. Database instances satisfying given cardinality constraints are said to be valid. The aim of this paper is to show how to use methods from graph theory to determine the class cardinalities of entity and relationship types in valid databases. We develop algorithms for this purpose and a number of variations of the problem, namely to construct databases of minimum size and linear ternary databases.

1 Introduction

Database design is the process of determining the organization of a database. The entity-relationship approach introduced by Chen [4] is one of the most popular database design models. This approach has given rise to various algebraic investigations, for a survey see Thalheim [11] or Batini et.al. [2]. In recent research combinatorial aspects have been considered, too (cf. Thalheim [10], Lenzerini and Nobili [8], and Engel and Hartmann [5]).

This paper shall contribute to the analysis of the mathematical and, in particular, combinatorial foundations of the entity-relationship approach. After summarizing the basic concepts, where we have followed [11] (although other terminologies exist, too), the subsequent sections contain a number of results achieved by methods from combinatorial optimization and graph theory adopted for this purpose.

Let $\mathcal{E}$ be a non-empty, finite set. In the context of the entity-relationship approach the elements are called *entity types*. With each entity type E we associate a set $\mathcal{C}(E)$ called the *domain* of E. Members of the sets $\mathcal{C}(E)$ are *entity instances* or *entities*, for short.

Intuitively, entities can be seen as real-world objects, which are of interest for any application or purpose. By classifying them and specifying their significant properties (called attributes), we obtain entity types which are frequently used to model the objects in their domains. Note, that entity types are considered to be static, while their domains may be dynamic.

A *relationship type* R is associated with a non-empty, finite subset $\mathcal{N}(R)$ of $\mathcal{E}$. For a given relationship type R, a *relationship instance* or *relationship r* is defined as the image of R under a function assigning with each E in $\mathcal{N}(R)$ an element from $\mathcal{C}(E)$. The cardinality of $\mathcal{N}(R)$ is the order of R. Relationship types of order 2 are said to be *binary*.

Relationship types are used to model associations between real world objects, i.e. entities. Again, relationship types are seen to be static, while the set of relationships modeled by a relationship type R (called *population* or *relationship set* $\mathcal{C}(R)$) may be dynamic.

Usually relationship types come along with certain restrictions that limit the possible combinations among the entities participating in their instances. In the context of the entity-relationship approach these restrictions are said to be *integrity constraints*. The most commonly used kind of these restrictions are *cardinality constraints*. They specify the number of relationship instances that an entity instance of a given entity type in a relationship set participates in.

For a relationship type R and an entity type $E \in \mathcal{N}(R)$ the cardinality constraint

$$\text{comp}(R, E) = (m, n)$$

specifies that an instance e from $\mathcal{C}(E)$ appears in $\mathcal{C}(R)$ at least m and at most n times.

Example. Let us consider a relationship type for a simple workshop model covering a catalogue of sessions. Each session can be characterized by a triple of entities: a subject, a time and a location. Hence we have three entity types which shall be denoted by B (subject type), T (time type) and L (location type). The relationship type S to model the sessions can be associated now with the set $\{B, T, L\}$. (For a graphical representation see Figure 1.)

The sessions in the workshop may come along with certain restrictions. Each subject shall be discussed at least two and at most three times, there shall always be five sessions in parallel, and each location shall be used three or four times. Obviously these restrictions are cardinality constraints, since they limit the number of sessions a subject, a time or a location shall be involved in:

$$\text{comp}(S, B) = (2, 3), \quad \text{comp}(S, T) = (5, 5), \quad \text{comp}(S, L) = (3, 4).$$

Cardinality constraints are already discussed over a longer period. While, for binary relationship types, their properties are rather well understood, there is still a need for further investigations for relationship types of larger order.

Given a set of cardinality constraints for a relationship type R, a relationship set $\mathcal{C}(R)$ is said to be *valid*, if the cardinality constraints hold in this relationship set.

The considerations presented so far can be extended to collections $R_1, \ldots, R_s$ of different relationship types, which are called *database schemes*. Of course, the sets $\mathcal{N}(R_1), \ldots, \mathcal{N}(R_s)$ in such a database scheme must not be disjoint. But however, entity types participating in several relationship types will always have the same domains.

Let a *database instance* or *database*, for short, be a collection of relationship sets $\mathcal{C}(R_1), \ldots, \mathcal{C}(R_s)$. We call it *valid* if each $\mathcal{C}(R_i)$, $i = 1, \ldots, s$, is valid for the relationship type R_i. We call a database scheme *binary* or *ternary* iff all the relationship types R_i are of order at most 2 or 3, respectively.

When dealing with cardinality constraints in database schemes there are at least three main problems that have to be considered:

- Which database schemes have valid instances?
- How to use cardinality constraints to construct valid instances?
- How to verify that a given instance is valid?

The first problem has been considered for example by Lenzerini and Nobili [8] and Thalheim [10], while the third problem is known to be NP-complete [7]. The major part of this paper is devoted to the second question.

Example (continued). Consider a database scheme containing the session type S introduced above to model the sessions in a workshop. When organizing such a workshop we have to arrange the subjects to be discussed and the available times and locations in such a way, that all restrictions are satisfied. Thus, we are looking for valid instances of our database scheme.

This paper is organized as follows. In Section 2 we discuss a graph-theoretical representation of database schemes. In subsequent sections we will show how to construct valid database instances (called realizers). Here we consider variations of the original problem, namely realizers of minimum size and linear realizers, too.

2 The graph-theoretical model

From the graph-theoretical point of view, a database scheme $\mathfrak{D}$ is a bipartite digraph $\mathfrak{G}$ with vertex set $\mathcal{V}$ and edge set $\mathcal{A}$, together with two functions $\alpha_D : \mathcal{A} \to \mathbb{N} \cup \{0\}$ and $\beta_D : \mathcal{A} \to \mathbb{N} \cup \{\infty\}$ defined on the edge set of $\mathfrak{G}$.

The two partition classes of the vertex set are just the sets $\mathcal{E}$ of entity types and $\mathcal{R}$ of relationship types. A relationship type R is connected by an edge (R, E) to an entity type E iff E is participating in R. The set of all entity types connected to an $R \in \mathcal{R}$ shall be denoted by $\mathcal{N}(R)$, the set of all relationship types connected to an $E \in \mathcal{E}$ by $\mathcal{N}(E)$.

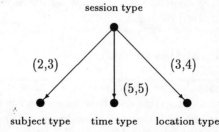

Fig. 1 The database scheme for the workshop model from Section 1

The functions α_D and β_D represent the cardinality constraints in the database scheme. For any relationship type R and a participating entity type E we have

$$comp(R, E) = (\alpha_D((R, E)), \beta_D((R, E))).$$

Now a *database instance* can be defined as a bipartite digraph $\mathfrak{R}$ with vertex set $\mathcal{V}_R$ and edge set $\mathcal{A}_R$ such that there exist two mappings $\phi : \mathcal{V}_R \to \mathcal{V}$ and $\psi : \mathcal{A}_R \to \mathcal{A}$ satisfying the following conditions:

(R1) $\psi(a) = (\phi(r), \phi(e))$ for all edges $a = (r, e)$ in $\mathfrak{R}$,

(R2) $\phi^{-1}(T) \neq \emptyset$ for all types T in $\mathfrak{G}$,

(R3) $|\phi^{-1}(E) \cap \mathcal{N}(r)| = 1$ for all edges $A = (R, E)$ in $\mathfrak{G}$ and all pre-images r in $\mathfrak{R}$ of the relationship type R.

According to condition (R1) the database scheme is the homomorphic image of its instances under the mapping ϕ. The pre-images of any entity type E or relationship type R are just the instances of this type. Hence, $\phi^{-1}(E)$ is the domain of E and $\phi^{-1}(R)$ the relationship set of R in the database instance.

Let T be an arbitrary type in the database scheme. In the sequel, we shall use the term *class* to refer to the set $\mathcal{C}(T) = \phi^{-1}(T)$ of instances of T. According to condition (R2), there are no empty classes in a database instance. The number of instances of a type T shall be called the *class cardinality* of the type T and denoted by $x_R(T)$. Additionally, the function $x_R : \mathcal{V} \to I\!N$ is called the *class cardinality function* of the database instance $\mathfrak{R}$.

In the database scheme, each relationship type R is connected to the set $\mathcal{N}(R)$ of participating entity types. Similary, each instance r of R in the database instance is connected to a set $\mathcal{N}(r)$ of entities, where by condition (R3) $\mathcal{N}(r)$ contains exactly one instance of each entity type in $\mathcal{N}(R)$.

We call an instance $\mathfrak{R}$ of a database scheme $\mathfrak{D}$ *valid* or a *realizer* of $\mathfrak{D}$ iff the following condition holds, too:

(R4) $\alpha_D(A) \leq |\phi^{-1}(R) \cap \mathcal{N}(e)| \leq \beta_D(A)$ for all edges $A = (R, E)$ in $\mathfrak{G}$ and all instances e in $\mathfrak{R}$ of the entity type E.

A database scheme is said to be *satisfiable* iff it has at least one realizer. The inequalities in condition (R4) are specified by the cardinality constraints of the database scheme. They give us a lower and an upper bound on the number of instances of the relationship type R that an instance e of the participating entity type E is involved in.

3 Class cardinalities in realizers

In this section we consider the problem of constructing realizers of given database schemes. Let $\mathfrak{D}$ be a satisfiable database scheme with realizer $\mathfrak{R}$. Further let $A = (R, E)$ be an arbitrary edge in $\mathfrak{G}$. Conditions (R3) and (R4) imply

$$\alpha_D(A)x(E) \leq x(R) \leq \beta_D(A)x(E) \tag{1}$$

by summing up over all pre-images of E. Hence, to obtain a realizer of an database scheme $\mathfrak{D}$, we are required to find class cardinalities that are *feasible* in that inequality (1) is satisfied.

In the sequel, we use the term *class cardinality function for* $\mathfrak{D}$ to refer to a positive integral function defined on the vertex set of $\mathfrak{G}$. Additionally, we call it feasible iff the system of inequalities given by (1) holds.

This leads to a two-phase approach for constructing realizers: In a first phase, one looks for feasible class cardinalities for the entity and relationship types. This provides the vertex set of a realizer, i.e. the instances of the types. The edges are constructed afterwards in a second phase.

For the second phase, Lenzerini and Nobili [8] proved, that for every feasible class cardinality function x there is a realizer $\mathfrak{R}$ such that the classes $\mathcal{C}(T)$ in $\mathfrak{R}$ have the cardinalities $x(T)$. They proposed an efficient algorithm for the second phase using congruences of numbers.

For the first phase, K. Engel and the author [5] described a polynomial-time algorithm using shortest-path methods.

For the database scheme $\mathfrak{D}$ we consider a new digraph $\mathfrak{G}'$ which can be obtained from $\mathfrak{G}$ by adding to each edge $A = (R, E)$ its reverse edge $A^{-1} = (E, R)$.

On the edge set of $\mathfrak{G}'$ we define a length function $l_D : \mathcal{A} \cup \mathcal{A}^{-1} \to \mathbb{R} \cup \{\infty\}$ by

$$l_D(A) := \begin{cases} \infty & \text{if } \alpha_D(A) = 0, \\ -\log \alpha_D(A) & \text{otherwise} \end{cases}$$

and

$$l_D(A^{-1}) := \begin{cases} \infty & \text{if } \beta_D(A) = \infty, \\ \log \beta_D(A) & \text{otherwise} \end{cases}$$

where $A \in \mathcal{A}$.

A subset $\mathcal{P} = \{A_1, A_2, \ldots, A_s\}$ of edges in $\mathfrak{G}'$ with $A_i = (T_{i-1}, T_i)$, $i = 1, \ldots, s-1$, for some types T_i in the digraph $\mathfrak{G}'$, $i = 1, \ldots, s$, is called a *directed path* and, if additionally $T_0 = T_t$, a *directed cycle* in $\mathfrak{G}'$. We define the length of a directed path $\mathcal{P}$ by

$$l_D(\mathcal{P}) := \sum_{A \in \mathcal{P}} l_D(A).$$

A directed cycle $\mathcal{C}$ is said to be *critical* for the database scheme $\mathfrak{D}$ if it has negative length with respect to l_D, i.e. iff $l_D(\mathcal{C}) < 0$.

Let x be an arbitrary class cardinality function and put $p(T) = \log x(T)$ for all types T. Then (1) is equivalent to

$$-l_D(A^{-1}) \le p(E) - p(R) \le l_D(A) \text{ for all edges } A = (R, E) \in \mathcal{A},$$

i.e. to

$$p(T) - p(U) \le l_D(A) \text{ for all edges } A = (U, T) \in \mathcal{A} \cup \mathcal{A}^{-1}. \tag{2}$$

So we are lead to the problem of finding a function $p : \mathcal{V} \to \mathbb{N} \cup \{0\}$ satisfying (2). Such a function is called a *potential function* in $\mathfrak{G}'$ with respect to l_D.

The existence of a potential function is known to be ensured for a digraph with strongly connected components and finite length function iff it contains no directed cycles of negative length. (The necessity is clear and the sufficiency follows from the correctness of the Ford-Bellman algorithm described below.)

4 Normalization of database schemes

Unfortunately, the length function l_D defined above may not even be finite. Therefore, a certain normalization of database schemes is proposed in [5]. Let us consider the scheme $\mathfrak{N}$ consisting of the digraph $\mathfrak{G}$ and the functions $\alpha_N : \mathcal{A} \to I\!N \cup \{1/\mu\}$ and $\beta_N : \mathcal{A} \to I\!N$ defined by

$$\alpha_N(A) = \begin{cases} 1/\mu & \text{if } \alpha_D(A) = 0, \\ \alpha_D(A) & \text{otherwise} \end{cases}$$

and

$$\beta_N(A) = \begin{cases} \mu & \text{if } \beta_D(A) = \infty, \\ \beta_D(A) & \text{otherwise} \end{cases}$$

where $A \in \mathcal{A}$ and

$$\mu = \prod_{\substack{A \in \mathcal{A}, \\ \alpha_D(A) \neq 0}} \alpha_D(A) \tag{3}$$

or $\mu = 1$ if $\alpha_D(A) = 0$ for all $A \in \mathcal{A}$. (In this case the problem becomes trivial: the digraph $\mathfrak{G}$ itself is a realizer of $\mathfrak{D}$.) We call $\mathfrak{N}$ the *normalized scheme* of $\mathfrak{D}$.

As above, we define a length function l_N on the edge set of $\mathfrak{G}'$ by $l_N(A) := -\log \alpha_N(A)$ and $l_N(A^{-1}) := \log \beta_N(A)$ for all $A \in \mathcal{A}$. This time, the length function l_N is always finite, and the critical directed cycles in $\mathfrak{G}'$ with respect to l_N are the same as with respect to l_D.

Lemma 1 *Let $\mathfrak{D}$ be a database scheme and $\mathfrak{N}$ its normalized scheme. Then a class cardinality function x is feasible for $\mathfrak{D}$ if it is feasible for $\mathfrak{N}$.*

Proof. This is an easy consequence of the inequalities $\alpha_D(A) \leq \alpha_N(A)$ and $\beta_N(A) \leq \beta_D(A)$ for all edges $A = (R, E)$ in $\mathfrak{G}$. ∎

5 Computing feasible class cardinalities

The usual way for computing a potential function in a digraph with strongly connected components and finite length function would be the following: For every component choose an arbitrary vertex U and put $p(U) := 0$. The potential of a vertex T would be the length of a shortest directed path from the fixed vertex U in the same component to T. It follows from properties of shortest directed paths that the resulting potential function p satisfies (2).

The following algorithm uses this idea in order to construct a feasible class cardinality function of a given database scheme $\mathfrak{D}$. We assume that the normalized scheme $\mathfrak{N}$ is given as input to the algorithm. In addition, let $\mathfrak{G}'_1, \ldots, \mathfrak{G}'_s$ be the components of $\mathfrak{G}'$ and $\mathcal{V}_1, \ldots, \mathcal{V}_s$ their vertex sets which can be determined with the help of breath first search or the union-find algorithm (cf. Jungnickel [6] or Sedgewick [9]). Further let μ be the number introduced above.

Algorithm 1

```
 1. for T ∈ V do x(T) ← ∞ od;
 2. for i = 1 to s do
 3.     choose U ∈ Vᵢ;
 4.     x(U) ← μ;
 5.     repeat
 6.         for T ∈ Vᵢ do x^old(T) ← x(T) od;
 7.         for E ∈ Vᵢ ∩ E do
 8.             x(E) ← min{x(E), x(R)/αN(A) : A = (R, E) for all A ∈ A};
 9.         od;
10.         for R ∈ Vᵢ ∩ R do
11.             x(R) ← min{x(R), βN(A)x(E) : A = (R, E) for all A ∈ A};
12.         od;
13.     until x(T) = x^old(T) for all T ∈ Vᵢ;
14. od.
```

Theorem 2 *Let $\mathfrak{D}$ be an database scheme without critical directed cycles. Then Algorithm 1 terminates and finds a feasible class cardinality function x.*

Sketch of the proof. (For a detailed proof see [5].) The length function l_N defined on the edge set of $\mathfrak{G}'$ is finite. Hence, we can use the Ford-Bellman algorithm [3] for computing a potential function in $\mathfrak{G}'$. In steps 3 to 13 we formulate this algorithm directly for the original problem, i.e. for determining a class cardinality function satisfying (1). ∎

Example (continued). In our example μ would be 30. If we choose the session type S as the starting vertex U in the algorithm we obtain the feasible class cardinalities $x(B) = 15, x(T) = 6, x(L) = 10$ and $x(S) = 30$. Hence a realizer of $\mathfrak{D}$ may contain 15 subjects $b^0, \ldots, b^{14}$, 6 times $t^0, \ldots, t^6$, 10 locations $l^0, \ldots, l^9$ which are arranged in 30 sessions $s^0, \ldots, s^{29}$. An example of such a realizer is shown in the following table, where each column corresponds to a session and contains the subject, time and location connected to this session in the realizer.

s^0	s^1	s^2	s^3	s^4	s^5	s^6	s^7	s^8	s^9	s^{10}	s^{11}	s^{12}	s^{13}	s^{14}	s^{15}	s^{16}	s^{17}	s^{18}	s^{19}	s^{20}	s^{22}	s^{22}	s^{23}	s^{24}	s^{25}	s^{26}	s^{27}	s^{28}	s^{29}
b^0	b^1	b^2	b^3	b^4	b^5	b^6	b^7	b^8	b^9	b^{10}	b^{11}	b^{12}	b^{13}	b^{14}	b^0	b^1	b^2	b^3	b^4	b^5	b^6	b^7	b^8	b^9	b^{10}	b^{11}	b^{12}	b^{13}	b^{14}
t^0	t^0	t^0	t^0	t^0	t^1	t^1	t^1	t^1	t^1	t^2	t^2	t^2	t^2	t^2	t^3	t^3	t^3	t^3	t^3	t^4	t^4	t^4	t^4	t^4	t^5	t^5	t^5	t^5	t^5
l^0	l^0	l^0	l^1	l^1	l^1	l^2	l^2	l^2	l^3	l^3	l^3	l^4	l^4	l^4	l^5	l^5	l^5	l^6	l^6	l^6	l^7	l^7	l^7	l^8	l^8	l^8	l^9	l^9	l^9

Fig. 2 A realizer of the database scheme

Example. Consider again the workshop model from Section 1. Assume a participant intends to discuss each subject at least once. Hence, we introduce a new relationship type P associated to the set $\{B, T\}$ containing the subject type B and the time type T. Of course, at a time he can only attend one session. Therefore cardinality constraints are given by

$$comp(P, B) = (1, \infty) \quad \text{and} \quad comp(P, T) = (0, 1).$$

By adding this new relationship type to $\mathfrak{D}$ we obtain a new database scheme $\mathfrak{D}'$.

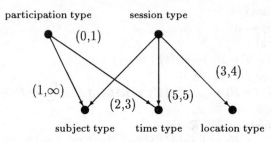

Fig. 3 The database scheme with the new participation type

However, the new database scheme $\mathfrak{D}'$ is not satisfiable. It contains a critical directed cycle containing the edges $(P, B), (B, S,), (S, T)$ and (T, P) where again S denotes the session type, B the subject type and T the time type. And of course, it is easy to check that the database scheme $\mathfrak{D}'$ has no realizers since

$$\alpha_{D'}((S, T))x(T) = 5x(T) \leq x(S) \leq \alpha_{D'}((S, B))x(B) = 3x(B) < 5x(B)$$

implies $x(T) < x(B)$, while the new cardinality constraints

$$\alpha_{D'}((P, B))x(B) = x(B) \leq x(P) \leq \beta_{D'}((P, T))x(T) = x(T)$$

give us $x(B) \leq X(T)$. Hence, we obtain a contradiction, i.e there exists no feasible class cardinality function for the new database scheme $\mathfrak{D}'$.

6 Realizers of minimum size

For practical purposes we are looking for realizers of small size. This gives us a lower bound for the space needed to store the information collected in an instance of a given database. In addition, small realizers are of special interest since it is always possible to obtain large realizers from small ones as the following statement shows.

Proposition 3 *Let $\mathfrak{D}$ be a satisfiable database scheme, and x and y feasible class cardinality functions for $\mathfrak{D}$. Then the class cardinality functions $x + y$, λx (for any $\lambda \in \mathbb{N}$) and $\frac{1}{\xi}x$ (where ξ denotes the greatest common divisor of the class cardinalities $x(T)$ of the types T in the database scheme) are feasible, too.*

When asking for realizers of small size we may want to minimize the following parameters:

- the number of instances in a fixed class,
- the number of instances in the largest class or
- the total number of instances in the realizer.

To understand the relations between these problems, we investigate the structure of the set of feasible class cardinality functions for a given database scheme. We shall show that this set forms a distributive lattice. The necessary definitions can be found e.g. in [1].

Given two class cardinality functions x and y, we say that y *dominates* x and write $x \leq y$ iff, for all types T, $x(T) \leq y(T)$. Henceforth, we denote by $\mathcal{X}$ the set of all feasible class cardinality functions. It is easy to see, that $\mathcal{X}$ forms a poset together with the dominance relation defined above. We shall denote this poset by $\mathcal{X}$, too.

Theorem 4 *The poset $\mathcal{X}$ is a distributive lattice.*

Sketch of the proof. (For a detailed proof see [5].) Let $\mathcal{C}$ be the chain $(1 < 2 < 3 < \ldots)$. Obviously, X is an induced subposet of the direct product $\mathcal{P} = \mathcal{C}^n$, where n is the number of types in the database scheme $\mathcal{D}$.

It remains to show that $\mathcal{X}$ is a sublattice of $\mathcal{P}$ (cf. [1, Theorem 2.8]). This can be done by verifying that, for any two elements of $\mathcal{X}$, also their meet and join (in $\mathcal{P}$) belong to $\mathcal{X}$. ∎

Example (continued). Consider again the database scheme containing the entity types B, T, L and the relationship type S. The following figure shows the sublattice of class cardinality functions x of all realizers with at most 20 sessions. Each node is labeled by the tuple $(x(B), x(T), x(L), x(S))$ giving the numbers of sessions, subjects, times and locations of the corresponding feasible class cardinality function.

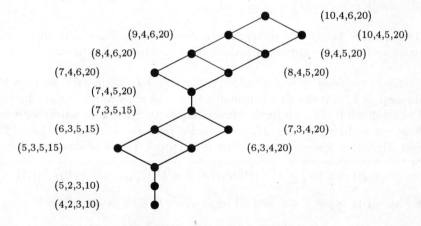

Fig. 4 A sublattice of feasible class cardinality functions

In addition to Proposition 3, Theorem 5 provides a second way of constructing new feasible class cardinality functions from given ones simply by determining the meet and the join of any two of them.

7 Computing minimum class cardinalities

According to Proposition 3 we find arbitrary large feasible class cardinality functions. On the other hand, each class cardinality function dominates the function $y : \mathcal{V} \to I\!N$ given by $y(T) = 1$ for every type. Hence, there is always a minimum feasible class cardinality function (with respect to the dominance relation defined in Section 6). Algorithm 1 usually does not find this one.

In order to determine the minimum feasible class cardinality function, we present the following algorithm. We assume, that the database scheme $\mathfrak{D}$ is given as input to the algorithm. As usually, $\lceil x \rceil$ denotes the smallest integer not smaller than x.

Algorithm 2

1. $k \leftarrow 0$;
2. for $T \in \mathcal{V}$ do $x^0(T) \leftarrow 1$ od;
3. repeat
4. $k \leftarrow k + 1$;
5.   for $E \in \mathcal{E}$ do
6. $x^k(E) \leftarrow \max\{x^{k-1}(E), \lceil x^{k-1}(R)/\beta_D(A) \rceil : A = (R, E) \in \mathcal{A}\}$
7.   od;
8.   for $R \in \mathcal{R}$ do
9. $x^k(R) \leftarrow \max\{x^{k-1}(R), \alpha_D(A)x^{k-1}(E) : A = (R, E) \in \mathcal{A}\}$
10.   od;
11. until $x^k(T) = x^{k-1}(T)$ for all $T \in \mathcal{V}$;
12. $s \leftarrow k$
13. for $T \in \mathcal{V}$ do $x(T) \leftarrow x^s(T)$ od.

Theorem 5 *Let $\mathfrak{D}$ be a satisfiable database scheme. Then Algorithm 2 terminates and finds the minimum feasible class cardinality function x.*

Sketch of the proof. (For a detailed proof see [5].) To begin with, we show by induction on k, that the class cardinality functions x^k, $k = 1, \ldots, s$ are dominated by each feasible class cardinality function y of $\mathfrak{D}$. Since $\mathfrak{D}$ is satisfiable it has at least one realizer $\mathfrak{R}$. Thus, $x(T)$ is bounded from above by $x_R(T)$ for all $T \in \mathcal{V}$ and Algorithm 2 terminates. It remains to prove, that x is feasible. We have

$$x(E) = x^s(E) \geq x^{s-1}(R)/\beta_D(A) = x^t(R)/\beta_D(A) = x(R)/\beta_D(A)$$

for all entity types $E \in \mathcal{E}$ and all edges $A = (R, E) \in \mathcal{A}$ as well as

$$x(R) = x^t(R) \geq \alpha_D(A)x^{t-1}(E) = \alpha_D(A)x^t(E) = \alpha_D(A)x(E)$$

for all relationship types $R \in \mathcal{R}$ and all edges $A = (R, E) \in \mathcal{A}$. This implies (1), i.e. x is a feasible class cardinality function. ∎

A realizer of a given database scheme whose class cardinality function is just the minimum one contains the minimum number of instances of each entity type or relationship type in the database scheme. Hence it has the minimum total number of instances, too.

8 Linear realizers

We call an instance $\mathfrak{R}$ of a database scheme $\mathfrak{D}$ *linear* iff for two instances r and r' of a given relationship type R there is at most one entity participating in both, r and r', i.e.

$$|\mathcal{N}(r) \cap \mathcal{N}(r')| \le 1.$$

In this section we are interested in realizers having this property.

Example (continued). In the realizer represented by Figure 2 there exist sessions s^0 (associated to the entities b^0, t^0 and l^0) and s^1 (associated to the entities b^1, t^0 and l^0). Hence there are two sessions to be hold at the same time and same location. Of course, this is impossible in the sense of the model. To avoid this situation we ask for linear realizers.

Lemma 6 *Let $\mathfrak{D}$ be a satisfiable database scheme. If $\mathfrak{D}$ has a linear realizer with the class cardinality function x then*

$$\sum_{A=(R,E)} \alpha_D(A) - |\mathcal{N}(R)| + 1 \le x(R) \le \min_{E, E' \in \mathcal{N}(R)} \{x(E) x(E')\} \qquad (4)$$

holds for each relationship type R in $\mathfrak{D}$.

Proof. The second part of the inequality is trivial. To show the first part, let $\mathfrak{R}$ be a linear realizer of the database scheme and x its class cardinality function. Further, let R be an arbitrary relationship type of $\mathfrak{D}$. Denote by σ the number of triples (r, r', e) consisting of two instances r and r' of R and an entity e participating in both, r and r'. Obviously,

$$\sigma = \sum_{E \in \mathcal{N}(R)} \sum_{e \in \mathcal{C}(E)} \binom{|\mathcal{N}(e)|}{2}.$$

By the concavity of the function $f(z) = \binom{z}{2}$ and Jensen's inequality we obtain

$$\sigma \ge \sum_{E \in \mathcal{N}(R)} x(E) \binom{x(R)/x(E)}{2}.$$

Since $\mathfrak{R}$ is linear, we have

$$\sigma \le \binom{x(R)}{2}$$

and thus

$$\sum_{E \in \mathcal{N}(R)} x(E) \binom{x(R)/x(E)}{2} \leq \binom{x(R)}{2},$$

$$x(R) \sum_{E \in \mathcal{N}(R)} \frac{1}{x(E)} - |\mathcal{N}(R)| \leq x(R) - 1. \tag{5}$$

By condition (R4) we know $\alpha_D(A) \leq x(R)/x(E)$ for all edges $A = (R, E)$, which implies the first part of the stated inequality. ∎

Example (continued). Obviously, inequality (4) does not hold for the minimum feasible class cardinality function of the workshop scheme, since $x(S) = 10$ while $x(T)x(L) = 6$. Hence, there is no linear realizer with these class cardinalities.

Let x be a feasible class cardinality function for the database scheme $\mathfrak{D}$ and R a relationship type in $\mathfrak{D}$. Denote the entity types participating in R by $E_1, \ldots, E_n$ such that $x(E_1) \leq x(E_2) \leq \ldots \leq x(E_n)$. If $x(E_1) = 1$ or $x(E_2) \geq n - 1$ then the right part of inequality (4) immediately implies the left part. In the first case, we have

$$x(R)(\frac{1}{x(E_1)} + \ldots + \frac{1}{x(E_n)} - 1) \leq x(E_2)(1 + \frac{1}{x(E_2)} + \ldots + \frac{1}{x(E_n)} - 1)$$
$$\leq n - 1$$

and, in the second case,

$$x(R)(\frac{1}{x(E_1)} + \ldots + \frac{1}{x(E_n)} - 1) \leq x(E_1)x(E_2)(\frac{1}{x(E_1)} + \ldots + \frac{1}{x(E_n)} - 1)$$
$$\leq x(E_1)x(E_2)(\frac{1}{x(E_1)} + \frac{n-1}{x(E_2)} - 1)$$
$$= x(E_1)(n - 1) - (x(E_1) - 1)x(E_2)$$
$$\leq x(E_1)(n - 1) - (x(E_1) - 1)(n - 1) = n - 1.$$

Since $n = |\mathcal{N}(R)|$ this implies (5), from which we derived the left part of (4). Note, that for relationship types of order $n = 3$ as in our example, this condition always holds, since otherwise $1 < x(E_1) \leq x(E_2) < 2 = n - 1$ causes a contradiction.

Additionally, for database schemes whose relationship types all have order at most 3, we shall show that the conditions given in Lemma 6 are sufficient, too.

Theorem 7 *Let $\mathfrak{D}$ be satisfiable ternary database scheme with a feasible class cardinality function x. $\mathfrak{D}$ has a linear realizer iff*

$$x(R) \leq \min_{E, E' \in \mathcal{N}(R)} \{x(E)x(E')\} \tag{6}$$

holds for every relationship type R in $\mathfrak{D}$.

Proof. The necessity is given by Lemma 6. To verify the sufficiency, let R be an arbitrary relationship type of $\mathfrak{D}$. Denote the entity types connected to R by E_1, E_2 (if R is binary or ternary) and E_3 (if R is ternary).

For each typ T in the database scheme, let $\mathcal{C}(T) = \{t^0, \ldots, t^{x(T)-1}\}$ be the class of T and $\mathcal{V}$ the union of the classes of all types in $\mathfrak{D}$. Further, put $\delta = \lfloor x(R)/x(E_1) \rfloor$ and $\zeta = x(R) \bmod x(E_1)$. Thus, $\alpha_D((R, E_1)) \leq \delta \leq \beta_D((R, E_1))$ holds by (1) and ζ is a number between 0 and $x(E_1) - 1$.

For $k = 0, \ldots, x(R) - 1$ we connect the relationship instance r^k to the instance $e_1^{\lfloor k/(\delta+1) \rfloor}$ if $k < (\delta+1)\zeta$ and otherwise to the instance $e_1^{\lfloor (k-\zeta)/\delta \rfloor}$ of the entity type E_1. Then, the entities $e_1^0, \ldots, e_1^{\zeta-1}$ participate in exactly $\delta + 1$ instances of R, while the other ones participate in exactly δ instances of R. Note, that for $\zeta > 0$ we have $\delta + 1 = \lceil x(R)/x(E_1) \rceil$.

If R is binary or ternary, we connect each relationship r^k, $k = 0, \ldots, x(R) - 1$, also to the instance $e_2^{k \bmod x(E_2)}$ of the entity type E_2. Hence, each of the entities participates in at least $\lfloor x(R)/x(E_2) \rfloor$ and at most $\lceil x(R)/x(E_2) \rceil$ instances of R.

In addition, if R is ternary, we connect each r^k, $k = 0, \ldots, x(R) - 1$ to the instance $e_3^{(k + \lfloor k/\gamma \rfloor) \bmod x(E_3)}$, where γ is the lowest common multiple of $x(E_2)$ and $x(E_3)$. Again, each of the entities participates in at least $\lfloor x(R)/x(E_3) \rfloor$ and at most $\lceil x(R)/x(E_3) \rceil$ instances of R.

The digraph $\mathfrak{R}$ with vertex set $\mathcal{V}$ and edge set $\mathcal{A}$, which we obtain as described above, is obviously a realizer of $\mathfrak{D}$. It remains to prove, that $\mathfrak{R}$ is linear. Let R be a relationship type of the database scheme and r^p and r^q ($p < q$) two different instances of R.

To begin with, we are looking for entities participating in both, r^p and r^q. Obviously, r^p and r^q are connected to the same instance e_1^z of E_1 iff

$$\sum_{s=0}^{z-1} |\mathcal{N}(e_1^s) \cap \mathcal{C}(R)| \leq p < q < \sum_{s=0}^{z} |\mathcal{N}(e_1^s) \cap \mathcal{C}(R)| \tag{7}$$

(Note, that $|\mathcal{N}(e_1^s) \cap \mathcal{C}(R)|$ is the number of instances of R connected to the entity e_1^s.) In particular, this implies

$$q - p \leq |\mathcal{N}(e_1^z) \cap \mathcal{C}(R)| - 1 < x(R)/x(E_1). \tag{8}$$

Further, r^p and r^q are connected to the same instance of E_2 iff there is an integer $\lambda_2 \in \mathbb{N}$ such that

$$q - p = \lambda_2 x(E_2), \tag{9}$$

and r^p and r^q are connected to the same instance of E_3 iff there is an integer $\lambda_3 \in \mathbb{N}$ such that

$$q - p = \lambda_3 x(E_3) - \lfloor q/\gamma \rfloor + \lfloor p/\gamma \rfloor. \tag{10}$$

Now, $\mathfrak{R}$ is linear iff r^p and r^q are not connected to the same instances of two different entity types. Assume, r^p and r^q are connected to the same instances of both, E_1 and E_2. By (8) and (9), we obtain $x(E_2) \leq \lambda_2 x(E_2) = q - p < x(R)/x(E_1)$, i.e. $x(E_1)x(E_2) < x(R)$ contradicting (6).

Assume, r^p and r^q are connected to the same instance e_1^z of E_1 and the same instance of E_3. By (8) and (6) we have $q - p < x(R)/x(E_1) \leq x(E_3) \leq \gamma$. Hence it holds $q/\gamma < p/\gamma + 1$, i.e.

$$\lfloor q/\gamma \rfloor \leq \lfloor p/\gamma \rfloor + 1. \tag{11}$$

This implies by (10), (8) and (6)

$$
\begin{aligned}
x(E_3) - 1 &\leq \lambda_3 x(E_3) - 1 \leq \lambda_3 x(E_3) - \lfloor q/\gamma \rfloor + \lfloor p/\gamma \rfloor = q - p \\
&\leq |\mathcal{N}(e_1^z) \cap \mathcal{C}(R)| - 1 < x(R)/x(E_1) \leq x(E_3).
\end{aligned}
$$

Therefore, we obtain $q - p = x(E_3) - 1$, $x(E_3) = |\mathcal{N}(e_1^z) \cap \mathcal{C}(R)|$ and

$$\lfloor q/\gamma \rfloor - \lfloor p/\gamma \rfloor = 1. \tag{12}$$

If $x(E_3) = |\mathcal{N}(e_1^z) \cap \mathcal{C}(R)| = \delta$, we have $\delta = x(R)/x(E_1)$ and $\zeta = 0$. Hence, all entities e_1^s, $s = 0, \ldots, z$, are connected to the same number (either $\delta + 1$ or δ) of instances of R. Thus, (7) is equivalent to

$$
\begin{aligned}
z x(E_3) &\leq p & < \quad q < (z+1) x(E_3) \\
z &\leq p/x(E_3) & < \quad q/x(E_3) < z + 1 \\
\lfloor p/\gamma \rfloor &= \lfloor q/\gamma \rfloor
\end{aligned}
$$

contradicting (12).

Finally, assume that r^p and r^q are connected to the same instances of both types, E_2 and E_3. Then (9) and (10) imply

$$q - p = \lambda_2 x(E_2) = \lambda_3 x(E_3) - \lfloor q/\gamma \rfloor + \lfloor p/\gamma \rfloor.$$

Let ξ denote the greatest common divisor of $x(E_2)$ and $x(E_3)$. Thus ξ divides $\lfloor q/\gamma \rfloor - \lfloor p/\gamma \rfloor$. By $0 \leq \lfloor q/\gamma \rfloor - \lfloor p/\gamma \rfloor < x(R)/\gamma \leq x(E_2) x(E_3)/\gamma = \xi$, this implies

$$\lfloor p/\gamma \rfloor = \lfloor q/\gamma \rfloor. \tag{13}$$

Therefore, we obtain $q - p = \lambda_2 x(E_2) = \lambda_3 x(E_3)$, and γ divides $q - p$. Now, $p < q$ implies $q - p \geq \gamma$ while (13) implies $q - p < \gamma$. This gives a contradiction and concludes the proof. ∎

Example (continued). For the feasible class cardinalities $x(B) = 7, x(T) = 4, x(L) = 6$ and $x(S) = 20$, we would obtain a realizer whose sessions correspond to the columns in the following table. Note that this time session s^0 is the only one which takes place at time t^0 and location l^0. Obviously, the new realizer is linear.

s^0	s^1	s^2	s^3	s^4	s^5	s^6	s^7	s^8	s^9	s^{10}	s^{11}	s^{12}	s^{13}	s^{14}	s^{15}	s^{16}	s^{17}	s^{18}	s^{19}
b^0	b^0	b^0	b^1	b^1	b^1	b^2	b^2	b^2	b^3	b^3	b^3	b^4	b^4	b^4	b^5	b^5	b^5	b^6	b^6
t^0	t^1	t^2	t^3	t^0	t^1	t^2	t^3	t^0	t^1	t^2	t^3	t^0	t^1	t^2	t^3	t^0	t^1	t^2	t^3
l^0	l^1	l^2	l^3	l^4	l^5	l^0	l^1	l^2	l^3	l^4	l^5	l^1	l^2	l^3	l^4	l^5	l^0	l^1	l^2

Fig. 5 A linear realizer of the database scheme

9 Conclusion

We have developed algorithms for determining the number of instances of entity and relationship types in valid databases. They show, how to use cardinality constraints in database design for the solution of several problems. In particular, we considered the construction of databases of minimum size and linear databases. Databases of minimum size minimize both, the number of instances of a fixed entity or relationship type as well as the total number of instances. Linear databases do not contain entities which commonly participate in more than one instance of a fixed relationship type. This property is desired in several applications as our example shows. Our concepts can be encorporated into design systems and then be used in the design process.

Acknowledgement. I am thanking Konrad Engel for helpful discussions and useful comments while preparing this paper.

References

[1] M. Aigner, Combinatorial Theory (Springer, Berlin, 1979).

[2] C. Batini, S. Ceri and S. Navathe, Conceptual database design, An entity-relationship approach (Benjamin Cummings, Redwood, 1992).

[3] R.E.Bellman. On a routing problem. Quart. Appl. Math. 16 (1958), 87-90.

[4] P. Chen, The Entity-Relationship Model: Towards a unified view of data. ACM TODS 1,1 (1984), 9-36.

[5] K. Engel and S. Hartmann, Constructing realizers of semantic entity-relationship schemes (1994), submitted to Discr.Appl.Math.

[6] D. Jungnickel, Graphen, Netzwerke und Algorithmen (BI-Wissenschaftsverlag, Mannheim, 1990).

[7] S. Hartmann, Schemes and Satisfiablity, Preprint (1995).

[8] M. Lenzerini and P. Nobili, On the satisfiability of dependency constraints in Entity-Relationship schemata, Information Systems Vol.15, 4 (1990) 453-461.

[9] R. Sedgewick, Algorithms (Addison-Wesley, Reading, Mass., 1988).

[10] B. Thalheim, Foundations of Entity-Relationship Modeling, Annals of Mathematics and Artificial Intelligence 6 (1992).

[11] B. Thalheim, Fundamentals of Entity-Relationship Models (Springer, Berlin, 1993).

[12] B. Thalheim, A survey on Database Constraints, Reihe Informatik I-8, Universität Cottbus (1994).

An Approximation Algorithm for 3-Colourability

Ingo Schiermeyer

Lehrstuhl für Diskrete Mathematik und Grundlagen
der Informatik, Technische Universität Cottbus
D-03013 Cottbus, Germany
e-mail:schierme@math.tu-cottbus.de

Abstract. We present a polynomial time approximation algorithm to colour a 3-colourable graph G with $3f(n)$ colours, if G has minimum degree $\delta(G) \geq \alpha n/f(n)$, where $\Omega(1) \leq f(n) \leq O(n)$ and α is a positive constant. We also discuss $NP - completeness$ and $\#P - completeness$ of restricted k-Colourability problems.

Key words: Graph, k-colouring, exact and approximation algorithm, complexity

1 Introduction

We use Bondy & Murty [2] for terminology and notation not defined here and consider simple graphs only.

An assignment of k colours to the vertices of a graph G is called a *k-colouring* of G if adjacent vertices are coloured by different colours. A graph G is *k-colourable* if there exists an *l-colouring* of G, where $l \leq k$. The minimum k for which a graph G is *k-colourable* is called the *chromatic number* of G and is denoted by $\chi(G)$. A graph G is *k-chromatic* if $\chi(G) = k$.

Let *k-Colourability* denote the decision problem whether a given graph G is *k-colourable*. It is well-known that *k-Colourability* is *NP-complete* for $k \geq 3$ and solvable in polynomial time for $k = 1$ and $k = 2$ (cf. [7]). Even the problem of determining the chromatic number of an arbitrary graph to within a given factor $M < 2$ has also been shown to be NP-complete [6]. Moreover, Lund and Yannakakis (cf. [11]) recently proved the following remarkable result for k-Colourability.

Theorem 1.1 *There is an $\epsilon > 0$ such that no polynomial-time approximation algorithm can have worst-case ratio growing as $O(|V|^\epsilon)$ unless $P = NP$.*

Therefore, analyzing the worst-case running time of exponential-time algorithms, which have already been studied intensively for the *Maximum Independent Set Problem* (cf. [14], [17]) and for the *k-Satisfiablity Problem* (cf. [12], [15]), is of increasing interest.

2 Exact Colouring Algorithms

Lawler [9] showed that determining the chromatic number of an arbitrary graph can be solved by the algorithm of Christofides [3] with worst-case running time of $O(mn \cdot (1 + 3^{1/3})^n)$, where m is the number of edges in the graph and n is the number of vertices, respectively (note: $1 + 3^{1/3} \approx 2,4422$).

The ideas of Christofides algorithm can be described as follows: An independent subset I of vertices is *maximal* if it is not a proper subset of any other independent subset I'. It is well-known that if a graph is k-colourable then there is a partition of its vertex set into k independent subsets where at least one of these independent subsets is maximal. Now computing all maximal independent sets of a given graph G and repeating this computation for the remaining graphs the chromatic number of G can be determined.

As suggested by Lawler [9], to test a graph for 3-Colourability, one can generate all maximal independent sets in time of $O(mn \cdot 3^{n/3})$ and then check the induced subgraph on each complementary set of vertices for bipartiteness. It follows that such a test can be made in $O(mn \cdot 3^{n/3})$ time, where $3^{1/3} \approx 1,4422$. This bound is sharp since graphs on n vertices may have up to $3^{n/3}$ maximal independent sets as has been shown by Moon & Moser [13].

Recently, we established an algorithm which decides 3-Colourability of an arbitrary graph in less than $O(1,398^n)$ steps [16]. Choosing a vertex u of maximum degree $\Delta(G)$ there are at most $3^{(n-1-\Delta(G))/3}$ maximal independent sets containing u. Exploiting the structure of the graph and a very sophisticated case analysis in the generation process of all maximal independent sets containing u leads to the improvement of the constant to $1,398$. Moreover, in several stages the algorithm can verify that a given graph G is not 3-colourable without generating all maximal independent sets containing u.

In [16] we extended the concept of maximal independent sets to the concept of maximal bipartite graphs. We proved that a graph on n vertices has no more than $10^{n/5}$ maximal bipartite graphs. Based on this concept we obtained three improved algorithms for deciding the 4-, 5- and 6-Colourability Problem in less than $O(1,585^n), O(1,938^n)$ and $O(2,155^n)$ steps, respectively.

3 Decision and Enumeration Problems

In [4] Edwards considers k-Colourability with respect to the minimum degree $\delta(G)$ of a given graph G.

Definition 3.1 *For each fixed integer k and rational number α with $0 \leq \alpha < 1$, we define the problem $\Pi(k, \delta, \alpha)$ as follows:*
    Input: *Graph $G = (V, E)$ with $|V| = n$ and $\delta(G) \geq \alpha n$.*
    Question: *Is G k-colourable?*

For example, the problem $\Pi(3, \delta, 0)$ is the problem 3-Colourability.

Theorem 3.2 *[4] Let k be an integer ≥ 3. Then*
(1) if $0 \leq \alpha \leq (k-3)/(k-2)$, $\Pi(k, \delta, \alpha)$ is NP-complete;
(2) if $\alpha > (k-3)/(k-2)$, $\Pi(k, \delta, \alpha) \in P$.

For $k = 3$ part (1) is just the statement that 3-Colourability is NP-complete. However, he also proved the following lemma.

Lemma 3.3 *[4] Let $C, \eta > 0$ be fixed. Then the following problem is NP-complete.*
Input: Graph $G = (V, E)$ with $|V| = n$ and $\delta(G) \geq Cn^{1-\eta}$.
Question: Is G 3-colourable?

We now consider k-Colourability with respect to the number of edges $m = |E(G)|$ of a given graph G. First observe that lemma 3.3 can be extended to the larger class of all graphs with $m \geq \frac{C}{2} \cdot n^{2-\eta}$ edges. Next we relax the hypothesis '$\delta(G) \geq \alpha n$' to '$m(G) \geq \alpha n^2$'.

Definition 3.4 *For each fixed integer k and rational number α with $0 \leq \alpha < 1$, we define the problem $\Pi(k, m, \alpha)$ as follows:*
Input: Graph $G = (V, E)$ with $|V| = n$ and $m \geq \alpha n^2$.
Question: Is G k-colourable?

For example, the problem $\Pi(3, m, 0)$ is the problem 3-Colourability.

Theorem 3.5 *Let k be an integer ≥ 3. Then*
(1) if $\alpha \geq 1/k$, $\Pi(k, m, \alpha) \in P$;
(2) if $0 \leq \alpha < 1/k$, $\Pi(k, m, \alpha)$ is NP-complete.

Proof:
(1) We may assume that G is a complete k-partite graph with vertex sets $V_1 \cup V_2 \cup \ldots \cup V_k = V$ of cardinalities $n_1 \geq n_2 \geq \ldots \geq n_k$. Then m is maximal if $n_k \geq n_1 - 1$ (arithmetic-geometric mean). Hence, if $m > n^2/k$, then G is not k-colourable. If $m = n^2/k$, then $G \cong K_{n/k, \ldots, n/k}$, which is (uniquely) k-colourable.
(2) We reduce 3-Colourability to this problem.
Let $\beta := \sqrt{k\alpha}$, $M := \frac{\beta n}{1-\beta}$. We now construct a graph $G' = (V', E')$ on $n' := n + kM$ vertices as the union of G and a complete balanced k-partite graph $K_{M, \ldots, M}$ on kM vertices. Then

$$m'(G') = \frac{k(k-1)}{2} \left(\frac{\beta}{1-\beta}\right)^2 \left(\frac{n}{k}\right)^2 + m \geq \frac{1}{k} \cdot \left(\frac{\beta n}{1-\beta}\right)^2 = \frac{\beta^2}{k} \cdot \left(\frac{n}{1-\beta}\right)^2 = \alpha \cdot n'^2.$$

Finally, G' is k-colourable if and only if G is k-colourable and it is clear that G' can be constructed in polynomial time.

$\square$

We now turn to the corresponding enumeration problems.

Definition 3.6 *For each fixed integer k and rational number α with $0 \leq \alpha < 1$, we define the problem $\#\Pi(k, \delta, \alpha)$ as follows:*
   Input: *Graph $G = (V, E)$ with $|V| = n$ and $\delta(G) \geq \alpha n$.*
   Question: *How many k-colourings of G are there?*

Theorem 3.7 *[4] Let k be an integer ≥ 3. Then*
   (1) if $\alpha > (k-2)/(k-1)$, $\#\Pi(k, \delta, \alpha) \in P$;
   (2) if $0 \leq \alpha < (k-2)/(k-1)$, $\#\Pi(k, \delta, \alpha)$ is $\#P$-complete.

Remark: Thus, for α satisfying

$$\frac{k-3}{k-2} < \alpha < \frac{k-2}{k-1},$$

the decision problem is solvable in polynomial time, whereas the corresponding enumeration problem is $\#P - complete$ and so probably intractable.

Definition 3.8 *For each fixed integer k and rational number α with $0 \leq \alpha < 1$, we define the problem $\#\Pi(k, m, \alpha)$ as follows:*
   Input: *Graph $G = (V, E)$ with $|V| = n$ and $m \geq \alpha n^2$.*
   Question: *How many k-colourings of G are there?*

With lemma 3.5 we obtain the following corollary.

Corollary 3.9 *Let k be an integer ≥ 3. Then*
   (1) if $\alpha \geq 1/k$, $\#\Pi(k, m, \alpha) \in P$;
   (2) if $0 \leq \alpha < 1/k$, $\#\Pi(k, m, \alpha)$ is $\#P$-complete.

4   An Approximation Algorithm for 3-Colourability

Recently Khanna, Linial and Safra [8] showed that 3-Colourability is not 4/3-approximable, i.e., there is no polynomial time algorithm to colour a 3-colourable graph with at most 4 colours unless $P = NP$.

Welsh in the 80's asked whether 3-Colourability is M-approximable for a (large) constant M, say $M = 100$, which is still open.

In [1] Blum presents an algorithm to colour any 3-colourable graph with $O(n^{3/8}polylog(n))$ colours in polynomial time, which improves several previous approximation algorithms for 3-colourable graphs (cf. [1] for a short survey).

Our approximation algorithm is based on the *greedy cover algorithm* (cf. [10]), which may be described as follows:

A subset U of V is called a *dominating set* of G if every vertex is either in U or is adjacent to some vertex in U. The greedy cover algorithm then constructs a dominating set of G as follows: Let v_1 be a vertex of maximum degree $\Delta = d(v_1)$. Suppose that $v_1, \ldots, v_k$ have been selected but they do not form a dominating set for G. Then let v_{k+1} be a vertex which covers a maximum number of (so far uncovered) vertices. This way we obtain a dominating set $U = \{v_1, v_2, \ldots\}$ of G.

150

Theorem 4.1 *Let G be a 3-colourable graph satisfying $\delta(G) \geq \alpha n / f(n)$, where $\Omega(1) \leq f(n) \leq O(n)$ and α is a positive constant. Then G can be coloured with $3f(n)$ colours in polynomial time.*

Remark: For $f(n) = 1$ this is part (2) of theorem 3.2.

Proof:

We follow the approach of Edwards [4] to show (2) of 3.2. From [10] it follows that

$$|U| \leq \frac{(1 + log(\delta + 1))n}{\delta + 1}$$

which leads to

$$|U| \leq C \cdot log\, n \cdot f(n),$$

where C is a constant only depending on α. We now partition U into $f(n)$ subsets U_i. Then $|U_i| \leq C\, log\, n$. For each U_i let $H_i := G[U_i \cup N(U_i)]$. Then each H_i is 3-colourable since G is 3-colourable. As described in [4] to determine whether or not H_i is 3-colourable, we can try all possible assignments of the colours $3i - 2, 3i - 1, 3i$ to the elements of U_i, and whenever a (proper) colouring l of these vertices is obtained we can test in time bounded by a fixed polynomial $p(n)$ whether or not the colouring l can be extended to a 3-colouring of the whole H_i. This extension given in [4] can be summarized as follows: For a proper colouring l of U_i there are at most two possible colours for each vertex $v \in V(H_i) - U_i$. Now an instance of 2-Sat can be constructed such that the colouring l can be extended to a 3-colouring of H_i if and only if this instance of 2-Sat has a satisfying truth assignment. Since 2-Satisfiability can be solved in polynomial time (cf. [5]), say $p(n)$, and since there are at most $3^{C\, logn}$ possible assignments of 3 colours to U_i, we can find a 3-colouring of H_i in polynomial time.

Now $V(G)$ is the (not necessarily disjoint) union of $V(H_1), \ldots, V(H_{f(n)})$. Hence, repeating the above described algorithm for each H_i, we can find in polynomial time a colouring of G with $3 \cdot f(n)$ colours $1, 2, \ldots, 3f(n)$.

$\square$

References

[1] A. Blum, *New Approximation Algorithms for Graph Colouring*, J. ACM 41 (1994) 470 - 516.

[2] J. A. Bondy and U. S. R. Murty, *Graph Theory with Applications* (Macmillan, London and Elsevier, New York, 1976).

[3] N. Christofides, *An Algorithm for the Chromatic Number of a Graph*, Computer J. 14 (1971) 38 - 39.

[4] K. Edwards, The Complexity of Colouring Problems on Dense Graphs, Theoretical Computer Sciene 43 (1986) 337 - 343.

[5] S. Even, A. Itai and A. Shamir, *On the complexity of timetable and multicommodity flow problems*, SIAM J. Comput. 5 (1976) 691 - 703.

[6] M. R. Garey and D. S. Johnson, *The complexity of Near-Optimal Graph Coloring*, J. ACM 23 (1976) 43 - 49.

[7] M. R. Garey and D. S. Johnson, *Computers and Intractability, A Guide to the Theory of NP-Completeness*, W. H. Freeman and Company, San Francisco, 1979.

[8] S. Khanna, N. Linial and S. Safra, *On the hardness of approximating the chromatic number*, Proc. of the 2nd Israel Symp. on the Theory of Computing and Systems, IEEE Computer Society, 250 - 260.

[9] E. L. Lawler, *A Note on the Complexity of the Chromatic Number Problem*, Inform. Process. Lett. 5 (1976) 66 - 67.

[10] L.Lovász, *On the Ratio of Optimal and Fractional Covers*, Discrete Math. 13 (1975) 383 - 390.

[11] C. Lund and M. Yannakakis, *On the hardness of approximating minimization problems*, Proc. of the Annual ACM Symposium on the Theory of Computing, Vol. 25 (1993) 286 - 293.

[12] B. Monien and E. Speckenmeyer, *Solving Satisfiability in less than 2^n Steps*, Discrete Appl. Math. 10 (1985) 287 - 295.

[13] J. W. Moon and L. Moser, *On Cliques in Graphs*, Israel J. of Math. 3 (1965) 23 - 28.

[14] J. M. Robson, *Algorithms for Maximum Independent Sets*, J. of Alg. 7 (1986) 425 -440.

[15] I. Schiermeyer, *Solving 3-Satisfiability in less than $1,579^n$ Steps*, Lecture Notes in Computer Science 702 (1993) 379 - 394.

[16] I. Schiermeyer, *Fast exact Colouring Algorithms*, RWTH Aachen, preprint 1993, Journal Tatra Mountains Mathematical Publications, in print.

[17] R. E. Tarjan and A. E. Trojanowski, *Finding a Maximum Independent Set*, SIAM J. Comput., Vol. 6, No. 3, September 1977, 537 - 546.

The Malleability of TSP_{2Opt}

Sophie Fischer and Leen Torenvliet

University of Amsterdam, Department of Computer Science, Plantage Muidergracht 24, 1018 TV Amsterdam. Supported (in part) by grants NWO/SION 612316801 and HC&M grant ERB4050PL93-0516

Abstract. We prove that the local search optimization problem TSP_{2Opt}, though not known to be PLS-complete, shares an important infeasibility property with other PLS-complete sets.

1   Introduction

NP-Optimization problems form a central object of study in operations research. Since at present no technique is known to find optimal solutions for these problems in reasonable time, nor is such a technique likely to be uncovered, practical algorithms tend to give near optimal solutions, where "near optimal" means different things in different contexts.

One of the techniques, recently introduced by Johnson, Papadimitriou and Yannakakis[JPY88], is called "Local Search." Instead of striving for a local optimum for the problem, one is content with a solution that is the best solution in a region. The technique starts out with an initial solution as its current solution and then keeps repeating the following steps. It considers some neighbors of the current solution to see if one of these is a better solution than the current one. If it can find such a solution it makes this solution the current solution. Otherwise it decides that the current solution is the best solution among the neighbors, stops and outputs the current solution as its current optimum. Problems with the property that an initial solution can be construed in polynomial time, the neighborhood can be inspected for a better solution in polynomial time, and the conclusion that a given solution is locally optimal can be made in polynomial time, are called "polynomial local search problems." The class that captures these problems is called PLS.

Among the problems in PLS, some have the property that finding a local optimum is "as hard as" finding a local optimum for any problem in PLS. That is, for these problems a translation mechanism, the PLS reduction exists, that translates local optima back to local optima in the problem reduced from. These problems are called PLS-complete. In the paper [PSY90] also a stronger form of the PLS reduction is defined—the tight PLS reduction. This reduction again requires that a local optimum can be translated back. Furthermore, the longest augmenting path of the problem reduced from is related to an augmenting path in the problem reduced to. The related path in the second problem must be at least as long as the path in the first problem.

Hence if one can prove that for a certain problem the path leading to a local optimum is long or otherwise computationally hard to construct then so will the path be in a search problem to which this problem can be reduced using a tight PLS reduction. Papadimitriou, Shaeffer and Yannakakis demonstrated a local search problem for which the question whether a specified local optimum that is reachable from a given initial solution is complete for $PSPACE$. Since this was the only local optimum reachable from the initial solution, it follows that computing a local optimum from a given solution is $PSPACE$-hard for this problem. They finally showed that for the problem CircuitFlip this question is $PSPACEF$-complete. for various other local search problems this question is also $PSPACEF$-complete, see for instance [K89], [K90], [SY] and [P90].

The traveling salesperson problem with the 2Opt neighborhood is defined as an optimization problem on a complete weighted graph. An initial Hamiltonian circuit in the graph is given. In each step the local search algorithm may select two edges (x_1, x_2), (y_1, y_2), remove these from the tour, and add the edges (x_1, y_2) and (y_1, x_2). It is presently unknown whether this problem is complete for PLS under PLS reductions.

The main result of this paper is to show that TSP_{2Opt} shares the computational improbability of determining a reachable local optimum with the PLS-complete problems, and we will do so in Section 4. In Section 3 we construct the reduction on which our main result is based. In the next section we will give definitions and notations and in particular discuss the nature of TSP_{2Opt}.

2 Preliminaries

In this section we give the definitions used throughout this paper. We also give some earlier results.

2.1 Local search

Definition 2.1 ([JPY88]) *A* Polynomial Local Search *(PLS) problem A is a five tuple $\langle I_A, \mathcal{FS}_A, f_A, opt_A, N_A \rangle$, where I_A is the set of instances of A, for each $I \in I_A$ the set $\mathcal{FS}_A(I)$ is the set of feasible solutions of I, f_A assigns to every feasible solution an integer value, opt_A is either min or max depending on the optimization nature of the problem, and for each $I \in I_A, s \in \mathcal{FS}_A(I)$, $N_A(I, s)$ is the set of neighbors of s, i.e., a set of feasible solutions associated with solution s.*
It holds that $I_A \in P$. Also, it is required that for each $I \in I_A$ an initial solution in $\mathcal{FS}_A(I)$ can be computed in polynomial time. Finally, for all $s \in \mathcal{FS}_A(I)$ it must be possible to decide in polynomial time whether s is locally optimal, i.e., whether s has a better cost than every $s' \in N_A(I, s)$, and if s is not locally optimal it must also be possible to compute in polynomial time a neighbor of s with a better cost.

The set of feasible solutions of a local search problem, together with a neighborhood structure, can be interpreted as a directed graph. By (a, b) we denote the edge directed from a to b.

Definition 2.2 *[PSY90] Let $A = \langle I_A, \mathcal{FS}_A, f_A, opt_A, N_A \rangle \in PLS$. For $I \in I_A$, define the local search graph $G_A(I) = (V_A(I), E_A(I))$ as follows:*
$$V_A(I) = \{s \mid s \in \mathcal{FS}_A(I)\}.$$
$$E_A(I) = \{(s, s') \mid s' \in N_A(I, s) \text{ and } f_A(I, s') > f_A(I, s)\}.$$

Let $s, s' \in \mathcal{FS}_A(I)$. We say that feasible solution s' is reachable from feasible solution s, if there is a directed path from s to s' in $G_A(I)$. Such a directed path is called an augmenting *path.*

Let A be a problem in PLS. The *standard local search algorithm*, given $I \in I_A$, will first compute an initial solution $s \in \mathcal{FS}_A(I)$ in polynomial time. Then the following step is repeated, until a locally optimal solution is found.

The standard local search algorithm decides in polynomial time whether s is locally optimal. If s is not locally optimal, it computes a solution $s' \in N_A(I, s)$ with a better cost than s and sets s to s'.

The standard local search algorithm will only go from one feasible solution to another with a strictly better cost. Consider the local search graph G_A. The standard local search algorithm walks along the paths of this graph. One arc in G_A denotes one step of the standard local search algorithm.

2.2 Space Machines

Definition 2.3 *[BDG88] A Turing machine with one tape is a five-tuple $M = \langle Q, \Sigma, \delta, q_0, F \rangle$, where*

1. *Q is a finite set of internal states*
2. *Σ is the tape alphabet*
3. *$q_0 \in Q$ is the initial state*
4. *F is a set of accepting final states*
5. *$\delta : Q \times \Sigma \to Q \times \Sigma \times \{R, L, N\}$ is a partial function called the transition function of M.*

Definition 2.4 *Let M be a Turing machine, a configuration of M is a description of the contents of the tape of M, the state of M and the tapecel that is scanned by the head of M.*

A computation of M is described by a sequence $c_0, c_1, \ldots, c_f$ of configurations, where c_0 is the initial configuration of M on x, c_f is the final configuration of M on x and c_i can be obtained from c_{i-1} by applying the transition function of M once.

Without loss of generality, in this paper we assume that $F = \{q_f\}$ and $\Sigma = \{0, 1\}$. If a Turing machine M runs in polynomial space, there exists a polynomial p, such that each computation of M on a string x uses at most $p(|x|)$ tapecels of the tape. A finite computation of M consists of at most $2^{p'(|x|)}$ computation steps, p' a polynomial. Furthermore, at the end of a computation, the first $p(|x|)$ tapecels on the tape of M contain the symbol 0 and the head of M scans the $(p(|x|) + 1)$'st tapecel.

2.3 Problems

We consider the local search problem TSP with the $2Opt$-neighborhood. Instances for TSP are complete, weighted graphs. The feasible solution set of an instance G contains all tours in G. The cost of a tour in G is the sum of all the weights on its edges. We try to find a tour in G with a minimal cost. Let T be a tour in G. The set of neighbors of T contains all tours T', that can be obtained by deleting two edges from T and inserting two new edges in T.

The decision problem $RDV(TSP_{2Opt})$ is based on the local search problem TSP_{2Opt}.

Definition 2.5 *Let G be a complete, weighted graph and T, T_f two tours in G. Assume that T_f is locally optimal. The decision problem $RDV(TSP_{2Opt})$ is defined as follows.*
Given G, T and T_f
Question *Is T_f reachable from T by a sequence of improving $2Opt$-changes.*

We also consider the following problem for Turing machines.

Definition 2.6 Polynomial Space Acceptance
Given *A polynomial space bounded Turing machine M and string x.*
Question *Does M accept x?*

It is well-known that *Polynomial Space Acceptance* is $PSPACE$-complete. For instance, see [GJ79] for a proof that *Linear Space Acceptance* is $PSPACE$-complete.

3 Constructing a Complete, Weighted Graph

In this section we construct a complete, weighted graph G based on a deterministic Turing machine M. Then we define special tours in G and augmenting paths in $G_{TSP_{2Opt}}(G)$. A computation step of M can be simulated by a unique sequence of improving $2Opt$-changes. In the last subsection we give some intuition for this simulation.

3.1 Constructing G

The vertices in G Graph G has three types of vertices, the *normal*-vertices, the *guide*-vertices and the *dummy*-vertices. The normal-vertices are used to represent configurations together with a time indication by tours in G. The guide-vertices are used to simulate computations of M by improving $2Opt$-changes. The dummy-vertices are used to guarantee that there do not exist different short augmenting paths.

For each tapecel i of M, G contains a subset V_i of normal-vertices. This subset V_i again can be divided into subsets, each having its own function.

1. Subset $H^i \subset V_i$ contains the vertices $h_1^i, h_2^i, h_3^i, h_4^i$. These vertices are used to indicate whether the head of M is on tapecel i or not.

2. Subset $R^i \subset V_i$ contains the vertices $r_1^i, r_2^i, r_3^i, r_4^i$. These vertices are used to indicate whether the symbol in tapecel i is 0.

3. Subset $S^i \subset V_i$ contains the vertices $s_1^i, s_2^i, s_3^i, s_4^i$. These vertices are used to indicate whether the symbol in tapecel i is 1.

For each state $q_j \in Q$, subset Q^j is a subset of normal-vertices in G and contains the vertices $q_1^j, q_2^j, q_3^j, q_4^j$. These vertices indicate whether M is in state q_j or not. The bitstring σ consists of $p'(|x|)$ bits and is called the *time bitstring*. It will keep track of at which point in the computation we are. For each bit i' in σ, G contains a subset $T^{i'} = \{t_1^{i'}, t_2^{i'}, t_3^{i'}, t_4^{i'}\}$ and a subset $U^{i'} = \{u_1^{i'}, u_2^{i'}, u_3^{i'}, u_4^{i'}\}$ of normal-vertices. These vertices are used to represent that the value of the i''th bit in the time bitstring is 0 or 1, respectively. Finally, G contains a subset D of normal vertices, containing d_1, d_2, d_3, d_4.

We now turn to the guide-vertices. The guide-vertices are also grouped in subsets. Graph G contains subset H of guide-vertices. Subset H contains the guide-vertices η_0, η_1, η_2 and $\delta\eta_1, \delta\eta_2$. For each tapecel i, a subset A^i of guide-vertices is in G. Subset A^i contains the guide vertices $\alpha_0^i, \alpha_1^i, \alpha_2^i$ and $\delta\alpha_1^i, \delta\alpha_2^i$. For each symbol $b \in \Sigma$ and tapecel i, a subset $A^{i,b}$ of guide-vertices is in G. Subset $A^{i,b}$ contains the guide-vertices $\alpha_0^{i,b}, \alpha_1^{i,b}, \alpha_2^{i,b}$ and $\delta\alpha_1^{i,b}, \delta\alpha_2^{i,b}$. For each bit i' in the time bitstring, subsets $\Theta^{i'}, \Lambda^{i',j}$ and $M^{i',j}$, $j \in \{1,2\}$ of guide-vertices are in G. Subset $\Theta^{i'}$ contains the vertices $\vartheta_0^{i'}, \vartheta_1^{i'}, \vartheta_2^{i'}$ and $\delta\vartheta_1^{i'}, \delta\vartheta_2^{i'}$. Subset $\Lambda^{i',j}$ contains the vertices $\lambda_0^{i',j}, \lambda_1^{i',j}, \lambda_2^{i',j}$ and $\delta\lambda_1^{i',j}, \delta\lambda_2^{i',j}$. Subset $M^{i',j}$ contains the vertices $\mu_0^{i',j}, \mu_1^{i',j}, \mu_2^{i',j}$ and $\delta\mu_1^{i',j}, \delta\mu_2^{i',j}$, $j \in \{1,2\}$.

The last sets of guide-vertices are used to simulate transitions of M. For each tapecel i, each symbol b, each state q_j and for each $k \in \{1,2,3,4,5,6\}$, we introduce the subsets $A^{i,b,q_j,k}$ of guide-vertices. A subset $A^{i,b,q_j,k}$ contains the vertices $\alpha_0^{i,b,q_j,k}, \alpha_1^{i,b,q_j,k}, \alpha_2^{i,b,q_j,k}$ and $\delta\alpha_1^{i,b,q_j,k}, \delta\alpha_2^{i,b,q_j,k}$.

For each subset S of normal-vertices or guide-vertices we define a subset $\Delta(S)$ of dummy-vertices in G containing the vertices $\Delta_1(S)$ and $\Delta_2(S)$.

Weights in G We will define the weights on the edges of G. The subsets of guide vertices can be divided into the layers L_v, $1 \leq v \leq 5p'(|x|) + 9$. Let $1 \leq i \leq p(|x|)$, $1 \leq i' \leq p'(|x|)$, $b \in \{0,1\}$, $q_j \in Q$ and $k \in \{1,2,3,4,5,6\}$. Layer L_1 contains subset H. Layer L_2 contains the subsets A^i. Layer L_3 contains the subsets $A^{i,b}$. Layer L_{3+k} contains the subsets $A^{i,b,q_j,k}$. Layer $L_{9+i'}$ contains the subset $\Theta^{i'}$. Layer $L_{9+p'(|x|)+2i'-1}$ contains the subset $\Lambda^{i',1}$. Layer $L_{9+p'(|x|)+2i'}$ contains the subset $\Lambda^{i',2}$. Layer $L_{9+5p'(|x|)-2i'+1}$ contains the subset $M^{i',1}$. Layer $L_{9+5p'(|x|)-2i'+2}$ contains the subset $M^{i',2}$.

Now we construct *triples* consisting of two subsets with guide-vertices and one subset with normal-vertices. Figure 1 gives all triples that exist among the subsets of vertices in G. In the first part of the figure, the triples that always exist among the subset are denoted. In the second part, the triples that depend on the transition function of M are denoted.

By carefully checking the triples we can prove that each triple is determined by exactly two of its members.

Let $1 \leq i \leq p(|x|)$, $1 \leq i' \leq p'(|x|)$, $b \in \{0,1\}$ and $q_j \in Q$.

$[\![H, A^i, H^i]\!]$	H in L_1	A^i in L_2
$[\![A^i, A^{i,0}, R^i]\!]$	A^i in L_2	$A^{i,0}$ in L_3
$[\![A^i, A^{i,1}, S^i]\!]$	A^i in L_2	$A^{i,1}$ in L_3
$[\![A^{i,b}, A^{i,b,q_j,1}, Q^j]\!]$	$A^{i,b}$ in L_3	$A^{i,b,q_j,1}$ in L_4
$[\![\Theta^{i'}, \Lambda^{i',1}, T^{i'}]\!]$	$\Theta^{i'}$ in $L_{9+i'}$	$\Lambda^{i',1}$ in $L_{9+p'(|x|)+2i'-1}$
$[\![\Theta^{i'}, \Theta^{i'+1}, U^{i'}]\!]$	$\Theta^{i'}$ in $L_{9+i'}$	$\Theta^{i'+1}$ in $L_{9+i'+1}$
$[\![\Lambda^{i',1}, \Lambda^{i',2}, U^{i'}]\!]$	$\Lambda^{i',1}$ in $L_{9+p'(|x|)+2i'-1}$	$\Lambda^{i',2}$ in $L_{9+p'(|x|)+2i'}$
$[\![\Lambda^{i',2}, M^{i'-1,1}, T^{i'}]\!]$	$\Lambda^{i',2}$ in $L_{9+p'(|x|)+2i'}$	$M^{i'-1,1}$ in $L_{9+5p'(|x|)-2i'-1}$
$[\![M^{i',1}, M^{i',2}, T^{i'}]\!]$	$M^{i',1}$ in $L_{9+5p'(|x|)-2(i'+1)-1}$	$M^{i',2}$ in $L_{9+5p'(|x|)-2(i'+1)}$
$[\![M^{i',2}, M^{i'-1,1}, U^{i'}]\!]$	$M^{i',2}$ in $L_{9+5p'(|x|)-2(i'+1)}$	$M^{i'-1,1}$ in $L_{9+5p'(|x|)-2i'-1}$
$[\![M^{1,2}, H, U^1]\!]$	$M^{1,2}$ in $L_{9+5p'(|x|)}$	H in L_1

$[\![A^{i,b,q_j,1}, A^{i,b,q_j,2}, H^{i+1}]\!]$	if M moves on symbol b and state q_j its head one place to the right $A^{i,b,q_j,1}$ in L_4, $A^{i,b,q_j,2}$ in L_5
$[\![A^{i,b,q_j,1}, A^{i,b,q_j,2}, H^{i-1}]\!]$	if M moves on symbol b and state q_j its head one place to the left $A^{i,b,q_j,1}$ in L_4, $A^{i,b,q_j,2}$ in L_5
$[\![A^{i,b,q_j,1}, A^{i,b,q_j,3}, D]\!]$	if M does not move on symbol b and state q_j its head $A^{i,b,q_j,1}$ in L_4, $A^{i,b,q_j,3}$ in L_6
$[\![A^{i,b,q_j,2}, A^{i,b,q_j,3}, H^i]\!]$	if M moves on symbol b and state q_j its head $A^{i,b,q_j,2}$ in L_5, $A^{i,b,q_j,3}$ in L_6
$[\![A^{i,0,q_j,3}, A^{i,0,q_j,4}, R^i]\!]$	if M on symbol 0 and in state q_j changes the symbol $A^{i,b,q_j,3}$ in L_6, $A^{i,b,q_j,4}$ in L_7
$[\![A^{i,1,q_j,3}, A^{i,1,q_j,4}, R^i]\!]$	if M on symbol 1 and in state q_j changes the symbol $A^{i,b,q_j,3}$ in L_6, $A^{i,b,q_j,4}$ in L_7
$[\![A^{i,b,q_j,3}, A^{i,b,q_j,5}, D]\!]$	if M does not change on symbol b and state q_j the value of the symbol $A^{i,b,q_j,3}$ in L_6, $A^{i,b,q_j,5}$ in L_8
$[\![A^{i,0,q_j,4}, A^{i,0,q_j,5}, S^i]\!]$	if M on symbol 0 and in state q_j changes the symbol $A^{i,b,q_j,4}$ in L_7, $A^{i,b,q_j,5}$ in L_8
$[\![A^{i,1,q_j,4}, A^{i,1,q_j,5}, S^i]\!]$	if M on symbol 1 and in state q_j changes the symbol $A^{i,b,q_j,4}$ in L_7, $A^{i,b,q_j,5}$ in L_8
$[\![A^{i,b,q_j,5}, A^{i,b,q_j,6}, Q^j]\!]$	if M changes on symbol b and in state q_j its state to $q_{j'}$ $A^{i,b,q_j,5}$ in L_8, $A^{i,b,q_j,6}$ in L_9
$[\![A^{i,b,q_j,5}, \Theta^1, D]\!]$	if M does not change on symbol b and state q_j its state $A^{i,b,q_j,5}$ in L_8, Θ^1 in L_{9+1}
$[\![A^{i,b,q_j,6}, \Theta^1, Q^j]\!]$	if M changes its state on symbol b and in state q_j $A^{i,b,q_j,6}$ in L_9, Θ^1 in L_{9+1}

Fig. 1. The triples among the subsets of vertices in G

Property 1 *Let Ξ, Ψ, Ξ' and Ψ' be subsets of guide vertices and let X and X' be subsets of normal vertices. Suppose that $[\![\Xi, \Psi, X]\!]$ and $[\![\Xi', \Psi', X']\!]$ are two triples in Figure 1. Then*

$$|\{\Xi, \Psi, X\} \cap \{\Xi', \Psi', X'\}| \neq 2$$

Using the triples of Figure 1 and the layer division on the subsets of guide vertices we define the weights on the edges of G. Weights of the edges in G are denoted by 2-tuples. The tuple-number $\langle x, y \rangle$ indicates the value $x\mathcal{R} + y$, where $\mathcal{R} = 2^{6p'(|x|)}$. The value of the first component of a weight is sometimes denoted by a quinary string of $p'(|x|)$ symbols of the set $\{0, 1, 2, 3, 4\}$. The value $val(\omega)$

of quinary string $\omega = \omega_{p'(|x|)}\omega_{p'(|x|)-1}\omega_{p'(|x|)-2}\ldots\omega_1$ is defined as

$$val(\omega) = \sum_{i'=0}^{p'(|x|)} \omega_{i'} 5^{i'-1}$$

To describe which weights are assigned to which edges, we use the subsets Ξ and Ψ with guide vertices, the subset X with normal vertices and the subset $\Delta(S)$ with dummy vertices. Let $\Xi = \{\delta\xi_1, \delta\xi_2, \xi_0, \xi_1, \xi_2\}$, $\Psi = \{\delta\psi_1, \delta\psi_2, \psi_0, \psi_1, \psi_2\}$, $X = \{x_1, x_2, x_3, x_4\}$ and $\Delta(S) = \{\Delta_1(S), \Delta_2(S)\}$. The weights to the edges of G are defined as follows.

1. The edges (x_1, x_2), (x_2, x_3), (x_3, x_4), (x_1, x_3) and (x_2, x_4) have weight $\langle 0, 1\rangle$, unless $X = U^{i'}$. Then edge $(u_1^{i'}, u_3^{i'})$ has weight $\langle 0^{p(|x|)-i'} 20^{i'-1}, 0\rangle$.

2. The edges $(\delta\xi_1, \delta\xi_2)$, $(\delta\xi_2, \xi_0)$, $(\delta\xi_1, \xi_1)$, $(\xi_1, \delta\xi_2)$, (ξ_0, ξ_1) and (ξ_1, ξ_2) have weight $\langle 0, 1\rangle$.

3. Edge $(\Delta_1(S), \Delta_2)$ has weight $\langle 0, 1\rangle$.

4. Edge $(x_4, \Delta_1(X))$ has weight $\langle 0, 1\rangle$. Edge $(\xi_2, \Delta_1(\Xi))$ has weight $\langle 0, 1\rangle$. Suppose that after subset $\Delta(S)$, tour T_0 traverses subset $\tilde{S}$. If $\tilde{S}$ is a normal subset, edge $(\Delta_2(S), \tilde{s}_1)$ has weight $\langle 0, 1\rangle$. If $\tilde{S}$ is a guide subset, edge $(\Delta_2(S), \delta\tilde{s}_1)$ has weight $\langle 0, 1\rangle$.

5. Suppose that Ξ belongs to layer L_v, $1 \leq v \leq 9 + 5p'(|x|)$. Then the second component of the weight of (ξ_0, ξ_2) is equal to $(11 + 5p'(|x|) - i')9$. If $1 \leq i' \leq 9 + p'(|x|)$ or if $\Xi = \Lambda^{i',1}$, the first component of the weight of (ξ_0, ξ_2) is equal to 0. If $\Xi = \Lambda^{i',2}$, then the first component of the weight of (ξ_0, ξ_2) is equal to $0^{p'(|x|)-i'} 20^{i'-1}$. If $\Xi = M^{i',1}$ or $\Xi = M^{i',2}$, then the first component of the weight of (ξ_0, ξ_2) is equal to $0^{p'(|x|)-(i'+1)} 20^{i'}$.

6. To enforce sequences of improving 2Opt-changes, assign to edge (ξ_0, x_2) weight $w(\xi_0, \xi_2) - \langle 0, 1\rangle$, unless $\Xi = \Lambda^{i',1}$. Then assign to edge (ξ_0, x_2) weight $w(\xi_0, \xi_2) - \langle 0, 1\rangle + \langle 0^{p'(|x|)-i'} 20^{i'-1}, 0\rangle$.
   If Ξ is in layer L_v, $4 \leq v \leq 9$, layer $L_{9+p'(|x|)+2i'-1}$ or layer $L_{9+3p'(|x|)+2i'}$, then assign to

edge	weight
(ξ_0, x_3)	$w(\xi_0, x_2) - \langle 0, 1\rangle$
(ξ_0, ψ_1)	$w(\xi_0, x_2) - \langle 0, 2\rangle$
$(\psi_1, \delta\xi_2)$	$w(\xi_0, x_2) - \langle 0, 3\rangle$
$(\delta\xi_2, \delta\psi_2)$	$w(\xi_0, x_2) - \langle 0, 4\rangle$
$(\delta\psi_2, \xi_1)$	$w(\xi_0, x_2) - \langle 0, 5\rangle$
$(\delta\psi_2, x_1)$	$w(\xi_0, x_2) - \langle 0, 6\rangle$
(ψ_2, x_1)	$w(\xi_0, x_2) - \langle 0, 7\rangle$
(ξ_2, x_1)	$\langle 0, 1\rangle$
(ψ_0, x_3)	$\langle 0, 1\rangle$

Otherwise, assign to

edge	weight
(ξ_0, ψ_1)	$w(\xi_0, x_2) - \langle 0, 2 \rangle$
$(\psi_1, \delta\xi_2)$	$w(\xi_0, x_2) - \langle 0, 3 \rangle$
$(\delta\xi_2, \delta\psi_2)$	$w(\xi_0, x_2) - \langle 0, 4 \rangle$
$(\delta\psi_2, \xi_1)$	$w(\xi_0, x_2) - \langle 0, 5 \rangle$
$(\delta\psi_2, x_1)$	$w(\xi_0, x_2) - \langle 0, 6 \rangle$
(ψ_2, x_1)	$w(\xi_0, x_2) - \langle 0, 7 \rangle$
(ξ_2, x_1)	$\langle 0, 1 \rangle$
(ψ_0, x_2)	$\langle 0, 1 \rangle$

7. All edges not mentioned so far have a huge weight, which we will denote with ∞.

Consider set Ξ. If the edges (ξ_0, ξ_1), (ξ_1, ξ_2), $(\delta\xi_2, \xi_0)$ and $(\delta\xi_1, \delta\xi_2)$ are on a tour, we say that the vertices in subset Ξ are traversed in their *restorder*. If the edges (ξ_0, ξ_2), $(\delta\xi_1, \xi_1)$, $(\delta\xi_2, \xi_0)$ and $(\xi_1, \delta\xi_2)$ are on a tour, we say that the restorder of the vertices in subset Ξ is disturbed.

We have one important property on the weights of the edges in G.

Property 2 *Let Ξ, Ψ be two different sets of guide-vertices. Let X be a set of normal vertices. All edges between X and Ψ or all edges between Ξ and Ψ have weight ∞ if and only if $[\![\Xi, \Psi, X]\!]$ is not a triple in Figure 1.*

3.2 Tours in G

In this section we associate with a configuration c of M and a timestamp t, $1 \le t \le 2^{p'(|x|)}$, a tour $T_{\langle c,t \rangle}$ in G. We will describe in which order $T_{\langle c,t \rangle}$ traverses the subsets of G by describing in which order the pairs S, $\Delta(S)$ are traversed by $T_{\langle c,t \rangle}$, see Figure 2. Here S is a subset of normal or guide-vertices and $\Delta(S)$ is its associated dummy subset. Tour $T_{\langle c,t \rangle}$ always first traverses the vertices in S and then in $\Delta(S)$. Suppose that $\Delta(S)$ contains the vertices $\Delta_1(S)$ and $\Delta_2(S)$. Then $T_{\langle c,t \rangle}$ first visits $\Delta_1(S)$ and then $\Delta_2(S)$.

We first show how tour $T_{\langle c,t \rangle}$ traverses the subsets of normal vertices. Tour $T_{\langle c,t \rangle}$ traverses all subsets V_i of normal vertices in order. So first the vertices in the subset belonging to tapecel 1 are traversed, then all vertices in the subset belonging to tapecel 2 and so on, until all vertices in the subset belonging to tapecel $p(|x|)$ have been traversed. We assume that the subsets in V_i are traversed in the following order. First the pair of subsets H^i is traversed, then the pair of subsets R^i and then the pair of subsets S^i. After traversing all subsets V_i, tour $T_{\langle c,t \rangle}$ traverses the pair of subsets Q^j, $q_j \in Q$, in order of index j. To traverse the subsets of vertices associated with the bits of the time bitstring, we start traversing the subsets associated with the least significant bit and we end by traversing the subsets associated with the most significant bit. For each bit i', we first traverse the pair of subsets $T^{i'}$ and then the pair of subsets $U^{i'}$.

The order in which the vertices in a subset V_i are traversed indicates information about which tapecel the head of M scans and the contents of each

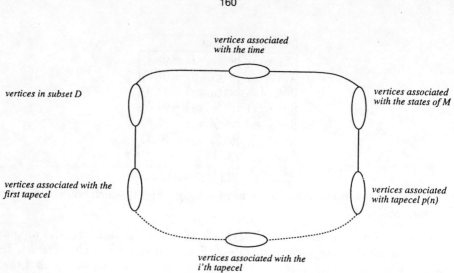

Fig. 2. The order of traversing the subsets of normal vertices in G

tapecel of the worktape of M. The fact that the head of M is on tapecel i is encoded in a tour by traversing the vertices in H^i in order $h_1^i, h_3^i, h_2^i, h_4^i$. Otherwise, the vertices are traversed in order $h_1^i, h_2^i, h_3^i, h_4^i$. If tapecel i contains symbol 0, this is encoded by traversing the vertices in R^i and S^i in order $r_1^i, r_3^i, r_2^i, r_4^i$ and $s_1^i, s_2^i, s_3^i, s_4^i$. If tapecel i contains symbol 1, this is encoded by traversing the vertices in R^i and S^i in order $r_1^i, r_2^i, r_3^i, r_4^i$ and $s_1^i, s_3^i, s_2^i, s_4^i$. The state of M is represented by the order in which the vertices of the subsets Q^j are traversed. If M is in state q_j, the vertices in subset Q^j are traversed in order $q_1^j, q_3^j, q_2^j, q_4^j$. Otherwise the vertices are traversed in order $q_1^j, q_2^j, q_3^j, q_4^j$.

We also encode a value of the time bitstring in tours of G. If bit i' has value 0 the vertices in $T^{i'}$ and $U^{i'}$ are traversed in order $t_1^{i'}, t_3^{i'}, t_2^{i'}, t_4^{i'}$ and $u_1^{i'}, u_2^{i'}, u_3^{i'}, u_4^{i'}$. If bit i' has value 1 the vertices in $T^{i'}$ and $U^{i'}$ are traversed in order $t_1^{i'}, t_2^{i'}, t_3^{i'}, t_4^{i'}$ and $u_1^{i'}, u_3^{i'}, u_2^{i'}, u_4^{i'}$.

The pairs of subsets of guide-vertices and their associated dummy subset are traversed in the following order. Let Ξ be a subset of guide-vertices. All subsets Ξ, where Ξ is in layer L_v, are traversed before the subsets Ψ, where Ψ belongs to layer L_{v+1}, $1 \le v \le 9 + 5p'(|x|) - 1$. In L_2, the subsets are traversed in the order of their index. So subset $A^{i'}$ is traversed before subset $A^{i'+1}$, $1 \le i' \le p'(|x|) - 1$. In L_3, the subset $A^{i',b}$ is traversed in lexicographical order on $\langle i', b \rangle$. In L_v, $4 \le v \le 9$, the subset $A^{i',b,q_j,v}$ in traversed in lexicographical order on $\langle i', b, j, v \rangle$. For each subset $\Xi \ne H$ of guide-vertices, the vertices are traversed in order $\delta\xi_1, \delta\xi_2, \xi_0, \xi_1, \xi_2$ The vertices of subset H are traversed in order $\delta\eta_1, \eta_1, \delta\eta_2, \eta_0, \eta_2$.

3.3 Augmenting paths in $G_{TSP_{2Opt}}(G)$

In this subsection we state the key-lemma in the proof of our main result. This lemma deals with the existence of unique short augmenting paths from a specified tour to another specified tour.

Because of space limitations we do not prove the lemma here. A full proof can be found in [Fis]. First two definitions.

Definition 3.1 *Let G be the complete, weighted graph as defined in the previous section. Let T be a tour in G. Let $\Xi = \{\delta\xi_1, \delta\xi_2, \xi_0, \xi_1, \xi_2\}$ be a subset of guide-vertices in G. Assume that*

1. *T traverses all vertices in one subset of normal, guide or dummy-vertices consecutively, i.e. one subset is traversed after the other.*
2. *T traverses all vertices in the guide-subsets in their restorder, except for guide-subset Ξ, which is traversed in disturbed order.*
3. *all edges on T, except (ξ_0, ξ_2) have weight $\langle 0, 1 \rangle$.*

Then T is called a useful *tour.*

Definition 3.2 *Let Ξ, Ψ be subsets of guide vertices and let X be a set of normal vertices, such that $[\![\Xi, \Psi, X]\!]$ is a triple in Figure 1. Suppose that tour $T \in \mathcal{T}$ contains edge (x_1, x_2). Subset X is active wrt Ξ, if (ξ_0, x_2) has a weight smaller than ∞. Otherwise, X is passive wrt Ξ.*

The key-lemma is as follows.

Lemma 3.1 *Let Ξ, Ψ be subsets of guide vertices and let X be a subset of normal vertices, such that $[\![\Xi, \Psi, X]\!]$ is a triple in Figure 1. Let T be a useful tour in G. Let $1 \le i \le p(|x|)$, $1 \le i' \le p'(|x|)$, $b \in \{0, 1\}$, $l \in \{1, 2\}$, $q_j \in Q$ and $k \in \{1, 2, 3, 4, 5, 6\}$. Suppose that T traverses all guide-subsets in their restorder, except for Ξ. Suppose that X is active wrt Ξ in T. Then there is a useful tour T' reachable from T with*

1. *The restorder of Ξ is restored and the restorder of Ψ is disturbed on T'.*
2. *If $\Xi \in \{H, A^i, A^{i,b}, \Theta^{i'}\}$ the length of the augmenting path from T to T' is 8 and the traversal-order of X is not changed.*
3. *If $\Xi \in \{A^{i,b,q_j,k}, \Lambda^{i,j'}, M^{i,j'}\}$ the length of the augmenting path from T to T' is 9 and the traversal-order of X is changed.*
4. *There is at most one augmenting path from T to T' of length 8 or 9.*

In Figure 3 we depict the $2Opt$-changes in a unique sequence of 8 improving local search steps.

3.4 Intuition

In this subsection we give the ideas used in the many-one reduction from *PSA* to $RDV(Min TSP_{2Opt})$. Let M be a deterministic polynomial-space bounded Turing machine, let x be a string and let σ be the associated time bitstring. Let G be

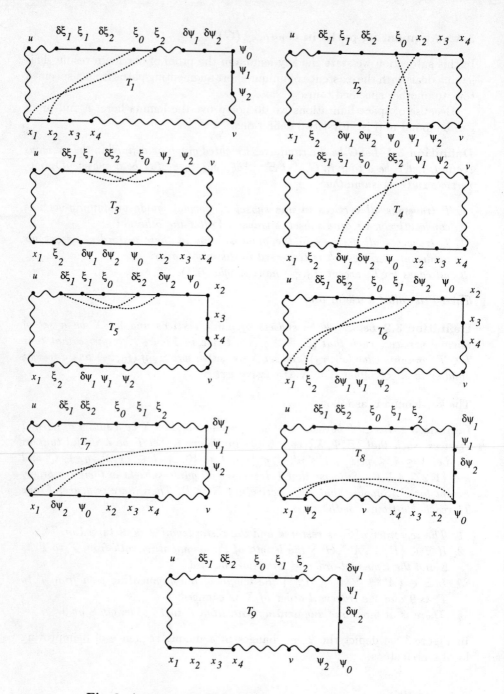

Fig. 3. A sequence of 8 improving $2Opt$-changes

based on M, x and σ as before. Each of the three kinds of subsets in G play their own role. The dummy-subsets are used for overhead purposes. We will not consider them here. The normal subsets are used to encode configurations of M and values to σ. Consider a directed supergraph SG. A vertex in SG is called a supervertex. Each supervertex contains the normal subsets associated with the configurations of M and the bits of σ. With each normal subset in a supervertex a traversal-order is associated. (Note here that the SG has a number of nodes that is exponential in the number of nodes of G.) In Subsection 3.2 two traversal-orders are indicated for normal subsets. One of these traversal-orders is assigned to a normal subset in a supervertex.

Let $\mathcal{V}_1$, $\mathcal{V}_2$ be two supervertices in SG. Suppose that configuration c_α and time bitstring value σ_α are encoded in $\mathcal{V}_\alpha$, $\alpha \in \{1, 2\}$. There is a superarc from $\mathcal{V}_1$ to $\mathcal{V}_2$ if and only if M can go in 1 step from c_1 to c_2 and $\sigma_1 = \sigma_2 + 1$. With each supervertex in SG a tour in G is associated. This tour traverses the normal subsets as indicated by the supervertex. Let T_1 and T_2 be the tours associated with respectively $\mathcal{V}_1$ and $\mathcal{V}_2$. The graph G and the tours T_1 and T_2 are constructed in such a way that there is a unique path from T_1 to T_2. This is the only augmenting path leaving T_1.

The augmenting path leaving T_1 is induced by the guide subsets. The augmenting path consists of several augmenting subpaths of 8 or 9 improving $2Opt$-changes. Each of these augmenting subpaths accomplishes something. In order

1. It is decided which tape cel is scanned by the head of M. Assume that tape cel i is scanned by the head.
2. The symbol in de cel scanned by the head of M is determined. Assume that tape cel i contains symbol 0.
3. The state of M is determined. Assume that M is in state q_j.
4. Suppose that based on 0 and q_j, M scans in the next configuration tape cel h_k, writes symbol b in tape cel i and changes the state of M to q_l.
   (a) The traversal-order of the normal subset H^k is changed. If $k = i$ this step is not performed.
   (b) The traversal-order of the normal subset H^i is changed. If $k = i$ this step is not performed.
   (c) The traversal-order of the normal subset R^i is changed. If $b = 0$ this step is not performed.
   (d) The traversal-order of the normal subset S^i is changed. If $b = 0$ this step is not performed.
   (e) The traversal-order of the normal subset Q^l is changed. If $j = l$ this step is not performed.
   (f) The traversal-order of the normal subset Q^j is changed. If $j = l$ this step is not performed.
5. The least significant 1-bit in the time bitstring is determined. First the traversal-order of $T^{p'(|x|)}$ is considered. Then the traversal-order of $T^{p'(|x|)-1}$, and so on. Assume that bit i' is the least significant 1-bit in the time bitstring.
6. The traversal-order of the normal subset $U^{i'}$ is changed.

7. The traversal-order of the normal subset $T^{i'}$ is changed.
8. The traversal-orders of the normal subsets associated with the lesser significant bits are changed. First the traversal-order of the T-subset associated with the bit is changed. Then the traversal-order of the U-subset.

On each subpath two guide subsets are used. One guide subset is traversed in rest-order at the beginning of the subpath. This guide subset is also used on the next augmenting subpath. Then its rest-order is disturbed. The restorder of the other guide subset is disturbed. At the end of the subpath this guide subset is traversed in rest-order.

The costs of tours in G consist of 2 components. During a computation step the second component is decreased. After a computation step the first component is decreased.

4 Main Result

With all the precomputation we did in the previoius section we come to our main theorem. We use the following two lemma's in the mainresult. Proofs of these lemma's can be found in [Fis].

Lemma 4.1 *Let c, c' be two configurations of M, and let $1 \le t \le 2^{p'(|x|)}$. There is an augmenting path from $T_{\langle c,t \rangle}$ to $T_{\langle c',t-1 \rangle}$ if and only if M can go in one step from c to c'. Furthermore, if there is an augmenting path from $T_{\langle c,t \rangle}$ to $T_{\langle c',t-1 \rangle}$, then this augmenting path is unique.*

Lemma 4.2 *There is at most one augmenting path leaving $T_{\langle c,t \rangle}$.*

Theorem 4.1 *Polynomial Space Acceptance $\le_m^p RDV(TSP_{2Opt})$.*

Proof The many one reduction computes from M and x, the graph G, and the tours $T_{\langle c_0, 2^{p'(|x|)} \rangle}$ and $T_{\langle c_f, 0 \rangle}$, where G, $T_{\langle c_0, 2^{p'(|x|)} \rangle}$ and $T_{\langle c_f, 0 \rangle}$ are as in the previous section. Suppose that the computation of M on x consists of the configurations $c_0, c_1, \ldots, c_{2^{p'(|x|)}}$, where M can go in one step from $c_{i'-1}$ to $c_{i'}$, $1 \le i' \le 2^{p'(|x|)}$. By Theorem 4.1, there exists a unique path from $T_{\langle c_{i'-1}, 2^{p'(|x|)} - i' + 1 \rangle}$ to $T_{\langle c_{i'}, 2^{p'(|x|)} - i' \rangle}$ By Corrolary 4.2, this is the only augmenting path leaving $T_{\langle c_{i'-1}, 2^{p'(|x|)} - i' + 1 \rangle}$. So there is a unique augmenting path from $T_{\langle c_0, 2^{p'(|x|)} \rangle}$ to $T_{\langle c_{2^{p'(|x|)}}, 0 \rangle}$, and this is the only augmenting path leaving $T_{\langle c_0, 2^{p'(|x|)} \rangle}$. It then follows that $c_{2^{p'(|x|)}} = c_f$ if and only if $\langle G, T_{\langle c_0, 2^{p'(|x|)} \rangle}, T_{\langle c_f, 0 \rangle} \rangle \in RDV(TSP_{2Opt})$.
$\square$

Corollary 1 *The standard local optimum version of TSP_{2Opt} is PSPACEF-complete.*

5 Conclusions

A very interesting question in the theory of local search problems is whether TSP with the $2Opt$-neighborhood is PLS-complete or not. To show this one must PLS-reduce a PLS-complete problem to TSP_{2Opt}. Krentel, see [K89], proved that $MaxSat$ with the Flip-neighborhood is PLS-complete. In the $MaxSat$ problem Boolean formulas in CNF are given, with on each clause a weight. Goal is to find an assignment that maximizes the sum of the weights of the satisfied clauses. The set of neigbors of an assignment τ contains all assignments that assign to all variables but one, the same value as τ.

To PLS-reduce $MaxSat$ to TSP_{2Opt}, one must map an instance ϕ of $MaxSat$ to a complete graph G. Furthermore, each locally optimal tour of G must be mapped in polynomial time to a locally optimal assignment of ϕ.

With the techniques developed in this paper, we can only partially achieve this goal. We can map ϕ to a complete graph G, such that each locally optimal solution not containing an edge of weight ∞ can be mapped to a locally optimal assignment to ϕ. Just use a local search variant of a polynomial space bounded Turing machine problem, such that for each assignment to ϕ, there is exactly one locally optimal solution reachable. Then simulate a computation of the Turing machine by an augmenting path in $G_{TSP_{2Opt}}(G)$. Unfortunately, we are neither able to prove that tours containing an edge of weight ∞ can not be locally optimal nor are we able to map a locally optimal tour containing an edge of weight ∞ to a locally optimal assignment.

With the presently available tools, the best result we can get is the $PSPACE$-hardness proof given in this paper. With the techniques presented, we hope to further interrelate TSP_{2Opt} with other PLS-complete problems.

References

[BDG88] J.L. Balcázar, J. Diáz and J. Gabarró. Structural Complexity 1. Springer-Verlag, 1988.

[Fis] S.T. Fischer. *The Solution Sets of Local Search Problems*. PhD thesis, University of Amsterdam, Department of Computer Science, Amsterdam, The Netherlands, 1995.

[GJ79] M. Garey and D. Johnson. Computers and Intractability: A Guide to the Theory of NP-Completeness. W. H. Freeman, San Francisco, 1979.

[JPY88] D. Johnson, C. Papadimitriou, and M. Yannakakis. How easy is local search? *J. Comput. System Sci.*, 37:79–100, 1988.

[K89] M. Krentel. Structure in Locally Optimal Solutions. Proc. 30th Annual IEEE Symposium on Foundations of Computer Science, Research Triangle Park, NC, 1989, pp 216-221.

[K90] M. Krentel. On Finding and Verifying Locally Optimal Solutions. *Siam J. Comput.*, 19:742-749, August 1990.

[P90] C.H. Papadimitriou. The complexity of the Lin-Kernighan heuristic for the traveling salesman problem. Proc. 22nd Annual ACM Symposium on Theory of Coomputing, Baltimore, MD, 1990, pp84-94.

[PSY90] C.H. Papadimitriou, A.A. Shäffer and M. Yannakakis. On the complexity of local optimality. Proc. 22nd Annual ACM Symposium on Theory of Coomputing, Baltimore, MD, 1990, pp84-94.

[SY]   A.A. Shäffer and M. Yannakakis. Simple local search problems that are hard to solve. *Siam J. Comput.*, 20:56-87, February 1991.

Non-Oblivious Local Search for Graph and Hypergraph Coloring Problems *

Paola Alimonti

Dipartimento di Informatica e Sistemistica, Università di Roma "La Sapienza",
via Salaria 113, 00198 Roma, Italia
E-mail: alimon@dis.uniroma1.it

1 Introduction

Local search is a general technique that is extensively used to solve difficult combinatorial optimization problems [12]. The general local search algorithm starts from an initial feasible solution and repeatedly tries to improve the value of the current solution searching for a better solution in its neighborhood.

Recently, the complexity, the computational power and the limits of local search has been studied from more theoretical points of view. Namely, in [9] a class of NP-hard problems for which local optimality can be verified in polynomial time has been introduced, and in [5] the quality of the local optima has been considered to provide a characterization of the problems.

Successively, in [2], and independently in [11], a new local search technique, that generalizes the standard technique and enlarge the power of this general paradigm, has been presented. This new approach and the standard technique have been called, respectively, *non-oblivious* and *oblivious* local search [11]. Informally speaking, the non-oblivious technique differs from the oblivious one since the search for optimality is based on an auxiliary objective function, instead of being based on the objective function of the problem.

In [2, 11] it has been shown that, by means of the non-oblivious technique, it is possible both to achieve better performance ratio for some problems approximable by means of the oblivious techniques, and to approximate problems not approximable by means of the oblivious technique.

In this paper we propose approximation algorithms based on the non-oblivious local search technique, in order to solve NP-hard hypergraph colorability problems [7].

Directed and undirected hypergraphs are generalizations of directed and undirected graphs, and has been widely used in computer science to represent concepts and structures from different areas. Classical examples of combinatorial structures that have been represented by hypergraphs are: functional and join dependencies in relational databases, Horn formulas in propositional calculus, implication in problem solving, Petri nets [1, 3, 4, 14].

* Work partially supported by: the ESPRIT Basic Research Action No.7141 (ALCOM II); the Italian Project "Algoritmi, Modelli di Calcolo e Strutture Informative", Ministero dell'Università e della Ricerca Scientifica e Tecnologica.

Directed and undirected hypergraphs consist of a finite set of nodes and a set of hyperarcs defined over the nodes. In undirected case, they are usually simply called hypergraphs [6], and hyperarcs are defined as arbitrary non-empty subsets of the set of nodes; in the directed case [3], a hyperarc is defined as a pair $<X, y>$, where X is a nonempty subset of nodes and y is a node. Clearly directed and undirected graphs are special cases of directed and undirected hypergraphs.

In this work we consider several hypergraph colorability problems for undirected and directed hypergraphs that are generalization of many well known NP-hard optimization problems, such as Max Cut, Max Directed Cut, h-Colorability, and Hypergraph 2-Colorability (also known as Set Splitting) [7].

In particular, by means of this approach we will obtain the first approximation algorithms for several colorability problems on directed and undirected hypergraphs. In one case (that is the 2-colorability for undirected hypergraphs or Set Splitting) we will be able to achieve an approximation ratio better than the previously known one (although the same ratio as been independently obtained by Kann et al. [10] by means of different techniques.) Finally, in another case (that is the 2-colorability for directed graphs or Max Directed Cut) our approximation ratio is worse than the one achieved in [8]. Nevertheless, the last result remains interesting due to the simplicity of the technique and on the fact that it holds also for hypergraphs. We also observe that our approximation ratios are better than that obtained by another simple technique, that is the greedy technique [13].

Finally, we believe that, apart from the specific results achieved for colorability problems, this work give a contribution in the investigation of the computational power of the local search paradigm in the approximation of NP-hard optimization problems.

The paper is organized as follows. In the next section, we describe the non-oblivious local search technique and we report the main results from [2]. In section 3 we deal with colorability problems for undirected hypergraphs, and then in section 4 we consider the case of directed hypergraphs. Finally, in the appendix we prove a tight bound on the performance ratio of non-oblivious local search.

2 Non-Oblivious Local Search

In this section we briefly describe the non-oblivious local search technique and we quote the main results from [2].

In order to apply local search to obtain approximation algorithms, the non-oblivious technique is based on the same assumptions on the problems and on the neighborhood structures of the oblivious technique, that guarantee to develop polynomial time algorithms [5, 9]. In particular, with regard to the neighborhood structure, it make use of a definition based on the fact that, for NP-hard optimization problems, a solution can be represented by a set of items; then the neighborhood mapping is defined as the set of solutions obtained by changing a bounded number of items of the current solution.

The difference between oblivious and non-oblivious local search consists on the fact that, while in the oblivious technique the search for optimality is based on the objective function of the problem, in the non-oblivious technique this search is led by an auxiliary objective function. In [2, 11], it has been shown that, by means of the non-oblivious technique it is possible: 1) to achieve better performance ratio for some problems approximable by means of the oblivious technique; 2) to approximate problems not approximable by means of the oblivious technique. The reason why non-oblivious local search is more powerful than the oblivious one is due to the fact that the introduced objective function better represents the features of the problems, and in the search for optimality is allowed to move to solutions of decreasing real value.

In [2] the technique has been introduced with respect to the approximation of logical disjunctive formulas, all of whose disjuncts are conjunctions containing up to k literals, for some constant $k \geq 1$.

Given a logical formula $F = \{c_1, ..., c_m\}$ of m conjunctive form clauses, formed by n Boolean variables $Y = \{y_1, ..., y_n\}$, all of whose conjunct contains up to k literals, find the truth assignment for Y that satisfies the maximum number m^* of clauses.

Given a truth assignment T for Y, a bounded neighborhood mapping for T is defined as the set of solutions obtained by changing the truth value of a bounded number of variables.

In the oblivious technique the search for optimality is based on the objective function of the problem, and thus it is expressed in terms of the number of satisfied clauses. In [2] it has been shown that by means of this technique the problem cannot be approximated; that is, for any approximation ratio, whatever neighborhood size we allow, there are local optimal solutions, with respect to such neighborhood that do not guarantee the ratio.

In the non-oblivious technique: 1) the search for optimality is led by a weighted sum of the clauses in the formula, where the weight of each clause depends on the number of true literals in the clause; 2) the neighborhood of a solution is the set of solutions obtained by changing the value of at most one variable (1-bounded neighborhood).

Let $T = \{p_1, ..., p_n\}$ be a truth assignment for Y, where, either $p_i = y_i$, or $p_i = \overline{y}_i$. Let S_j be the set of clauses where j literals are false in T and let $|p_i|_{S_j}$ ($|\overline{p_i}|_{S_j}$) be the cardinality of the literal p_i ($\overline{p_i}$) in S_j ($1 \leq i \leq n, 0 \leq j \leq k$).

$$|S_0| = \sum_{i=1}^{n} \frac{|p_i|_{S_0}}{k} \qquad |S_j| = \sum_{i=1}^{n} \frac{|p_i|_{S_j}}{k-j} = \sum_{i=1}^{n} \frac{|\overline{p_i}|_{S_j}}{j} \qquad |S_k| = \sum_{i=1}^{n} \frac{|\overline{p_i}|_{S_k}}{k}$$

Let $W = \sum_{j=0}^{k} L_j \cdot |S_j|$ be the weighted sum of the cardinalities of the S_j, where

$$L_j = \frac{1 + k \cdot L_{j+1} - (k - j - 1) \cdot L_{j+2}}{j+1} \quad (0 \leq j \leq k - 1) \qquad L_j = 0 \quad (j \geq k)$$

Then the following theorem holds:

Theorem 2.1 *[2] Let F be a logical formula of m conjunctive clauses of up to k literals, over a set Y of n boolean variables, and T ⊆ Y be an independently obtained truth assignment for Y. If T is a local optimum for a 1-bounded neighborhood and for the objective function W, then T satisfies at least $m \cdot 2^{-k}$ clauses.*

3 Undirected Hypergraphs

Undirected hypergraphs, usually simply called hypergraphs [6], are defined as follows:

Definition 3.1 *A hypergraph $\mathcal{H}$ is a pair $< V, H >$ where V is the set of nodes, H is the set of hyperarcs, and for each $X \in H$, $X \subseteq V$.*

Although there is a unique definition of graph h-colorability, if we generalize and we deal with hypergraphs, we have several possible different problems, depending on the considered definition of coloring.

In fact, while in the case of a graphs an arc is in the coloring if and only if its extremes have different colors, there are many possible definition of being in the coloring for hyperarcs. If we consider, for instance, the hypergraph 2-colorability problem, we can say that a hyperarc is "well colored": either if there are at least two node with different colors in the hyperarc (as in [7]); or, in a more general way, if there is at least a given fraction of the nodes in the hypergraph for each one of the two colors.

In this paper we consider two different versions of hypergraph 2-colorability. In the former one, hyperarcs are in the coloring, if they contains at least one node with each one of the two colors, that is we refer to the definition of Hypergraph 2-Colorability problem given in [7]. In the latter one, that we call Hypergraph 2-Perfect-Colorability, hyperarcs are in the coloring if they have, either the same number of nodes, or the same number of nodes but one for each one of the two colors.

Definition 3.2 *Let $\mathcal{H} = < V, H >$ be a hypergraph. A subset of hyperarcs $H' \subseteq H$ is* 2-Colorable *if there exists a coloring such that each hyperarc $X \in H'$ contains at least one node for each one of the two colors. The* Hypergraph 2-Colorability *problem is the problem of finding a 2-Colorable subset $H' \subseteq H$, that maximizes the cardinality $|H'|$.*

Definition 3.3 *Let $\mathcal{H} = < V, H >$ be a hypergraph. A subset of hyperarcs $H' \subseteq H$ is* 2-Perfect-Colorable *if there exists a coloring such that each hyperarc $X \in H'$ contains, either the same number of nodes, or the same number of nodes but one, for each one of the two colors. The* Hypergraph 2-Perfect-Colorability *problem is the problem of finding a 2-Perfect-Colorable subset $H' \subseteq H$, that maximizes the cardinality $|H'|$.*

Finally we consider the problem of the Hypergraph h-Colorability, according to the definition for which hyperarcs are in the coloring, if they contain at least one node for each one of the h colors.

Definition 3.4 *Let* $\mathcal{H}=< V,H >$ *be a hypergraph. A subset of hyperarcs* $H' \subseteq H$ *is* h-*Colorable if there exists a coloring such that each hyperarc* $X \in H'$ *contains at least one node for each one of the* h *colors. The* Hypergraph h-Colorability *problem is the problem of finding a* h-*Colorable subset* $H' \subseteq H$, *that maximizes the cardinality* $|H'|$.

The main idea underlying the proposed algorithms is based on representing instances of hypergraph colorability problems as instances of the satisfiability problem. Then a solution is determined by means of non-oblivious local search within some approximation ratio that depends on combinatorial arguments on the structure of the problems.

Given a Hypergraph h-Colorability problem over a hypergraph $\mathcal{H}=< V,H >$, with $|V| = n$ and $|H| = m$, consider a representation of the h colors by means of the truth value of $\lceil \log h \rceil$ boolean variables. Then we construct a logical formula $F_{\mathcal{H},h}$ in the following way:

a) we associate to each node $v_i \in V$ a number equal to $\lceil \log h \rceil$ of boolean variables $y_{i,j}$ $(1 \le i \le n, 1 \le j \le \lceil \log h \rceil)$;

b) we associate to each hyperarc $X = \{v_1, \ldots, v_s\} \in H$, with size $k = |X|$ a number equal to $2^{\lceil \log h \rceil \cdot k}$ of conjunctive clauses of $k \cdot \lceil \log h \rceil$ literals ($p_{1,1} \wedge \ldots \wedge p_{1,\lceil \log h \rceil} \wedge \ldots \wedge p_{k,1} \wedge \ldots \wedge p_{k,\lceil \log h \rceil}$), where, either $p_{i,j} = y_{i,j}$, or $p_{i,j} = \overline{y}_{i,j}$ $(1 \le i \le k, 1 \le j \le \lceil \log h \rceil)$.

A truth assignment for the variables $x_{i,j}$ $(1 \le i \le n, 1 \le j \le \lceil \log h \rceil)$ corresponds to a coloring of the nodes in V.

To illustrate this consider a hypergraph $\mathcal{H} =< V, H >$, where $V = \{v_i : i = 1, \ldots, 7\}$ and $H = \{\{v_1, v_2, v_3\}, \{v_1, v_5, v_7\}, \{v_3, v_4, v_6, v_7\}\}$. We construct the formula $F_{\mathcal{H},2}$ in the following way:

a) we associate to V a set of boolean variables $Y = \{y_i : 1 \le i \le 7\}$, and

b) we associate to H a set of conjunctive clauses $C = \{\{\overline{y}_1 \wedge \overline{y}_2 \wedge \overline{y}_3\}, \{\overline{y}_1 \wedge \overline{y}_2 \wedge y_3\}, \{\overline{y}_1 \wedge y_2 \wedge \overline{y}_3\}, \{\overline{y}_1 \wedge y_2 \wedge y_3\}, \{y_1 \wedge \overline{y}_2 \wedge \overline{y}_3\}, \{y_1 \wedge \overline{y}_2 \wedge y_3\}, \{y_1 \wedge y_2 \wedge \overline{y}_3\}, \{y_1 \wedge y_2 \wedge y_3\}, \{\overline{y}_1 \wedge \overline{y}_5 \wedge \overline{y}_7\}, \ldots, \{y_1 \wedge y_5 \wedge y_7\}, \{\overline{y}_3 \wedge \overline{y}_4 \wedge \overline{y}_6 \wedge \overline{y}_7\}, \ldots, \{y_3 \wedge y_4 \wedge y_6 \wedge y_7\}\}$.

In the following we will discuss the case of hypergraphs for which all hyperarcs have size k. However similar results can be proved for the general case by introducing dummy variables in the formula associated to the problems, that allows to achieve approximation ratio that depends on the minimal size of the hyperarcs in the hypergraph.

3.1 2-Colorability

The first problem we consider is the Hypergraph 2-Colorability problem, based on the classical definition of 2-coloring of hyperarcs (i.e. a hyperarc is in the coloring if it contains at least one node for each color). This problem is also called Set Splitting.

The next theorem shows that, given a hypergraph with m hyperarcs, in which the size of the hyperarcs is k, our algorithm determines a solution such that at least $m \cdot \frac{2^{k-1}-1}{2^{k}-1}$ hyperarcs are in the coloring, and then it approximates the problem within $\frac{2^{k-1}-1}{2^{k}-1}$.

Theorem 3.1 *Let $\mathcal{H}=< V, H >$ be a hypergraph, with $|H| = m$ hyperarcs of size k, then the Hypergraph 2-Colorability problem of $\mathcal{H}$ can be solved within approximation ratio $\frac{2^{k-1}-1}{2^{k}-1}$.*

Proof. Let $\mathcal{H}=< V, H >$ be a hypergraph, with $|H| = m$, for which the size of each hyperarc $X \in H$ is k. Consider now the logical formula F obtained from $F_{\mathcal{H},2}$ by removing the clauses in which all literals are positive and the clauses in which all literals are negative. This corresponds to remove $2 \cdot m$ clauses, that is, for each hyperarc $X \in H$ the 2 clauses satisfied by truth assignments representing coloring in which all nodes in X have the same color. Then we have a logical formula of $M = (2^k - 2) \cdot m$ clauses, in which each clause is a conjunction of k literals. Therefore, by means of non-oblivious local search, we can determine a truth assignment that satisfies at least $2^{-k} \cdot M$ clauses in F. Furthermore, no more than one clause corresponding to the same hyperarc can be simultaneously satisfied. Hence, the number of hyperarcs in the coloring is at least $2^{-k} \cdot M = 2^{-k} \cdot (2^k - 2) \cdot m = \frac{2^{k-1}-1}{2^{k-1}} \cdot m \geq \frac{2^{k-1}-1}{2^{k-1}} \cdot m^*$.

Now we consider the 2-Perfect-Colorability problem. We will show that given a hypergraph $\mathcal{H}=< V, H >$ with $|H| = m$ hyperarcs of size k, the problem can be solved within approximation ratio $\sqrt{\frac{2}{\pi \cdot k}}$ if k is even, and $2 \cdot \sqrt{\frac{2}{\pi \cdot (k+1)}}$ if k is odd.

Theorem 3.2 *Let $\mathcal{H}=< V, H >$ be a hypergraph, with $|H| = m$ hyperarcs of size k, then the Hypergraph 2-Perfect-Colorability problem of $\mathcal{H}$ can be solved within approximation ratio $\sqrt{\frac{2}{\pi \cdot k}}$ if k is even and $2 \cdot \sqrt{\frac{2}{\pi \cdot (k+1)}}$ if k is odd.*

Proof. Let $\mathcal{H}=< V, H >$ be a hypergraph, with $|H| = m$, for which the size of each hyperarc $X \in H$ is k. We consider two cases: k is even and k is odd.

a) If k is even a hyperarc is in the coloring if it contains the same number of nodes for each one of the two colors. Consider now the logical formula F obtained by considering the clauses of $F_{\mathcal{H},2}$ in which there are exactly $\frac{k}{2}$ positive literals and $\frac{k}{2}$ negative literals. It contains $\frac{k!}{\frac{k}{2}! \cdot \frac{k}{2}!}$ clauses for each hyperarcs in $\mathcal{H}$. Then we have a logical formula of $M = \frac{k!}{\frac{k}{2}! \cdot \frac{k}{2}!} \cdot m$ clauses, in which each clause is a conjunction of k literals. Therefore, by means of non-oblivious local search, we can determine a truth assignment that satisfies at least $2^{-k} \cdot M$ clauses in F. Furthermore, no more than one clause corresponding to the same hyperarc can be simultaneously satisfied. Therefore, the number of hyperarcs in the coloring is at least $2^{-k} \cdot M = 2^{-k} \cdot \frac{k!}{\frac{k}{2}! \cdot \frac{k}{2}!} \cdot m \geq 2^{-k} \cdot \frac{k!}{\frac{k}{2}! \cdot \frac{k}{2}!} \cdot m^*$.

Hence, from Stirling approximation the approximation ratio is $\sqrt{\frac{2}{\pi \cdot k}}$.

b) If k is odd a hyperarc is in the coloring if it contains the same but one number of nodes for each one of the two colors. Consider now the logical formula F obtained by considering the clauses of $F_{\mathcal{H},2}$ in which there are $\frac{k-1}{2}$ positive literals and $\frac{k+1}{2}$ negative literals, and the clauses of $F_{\mathcal{H},2}$ in which there are $\frac{k+1}{2}$ positive literals and $\frac{k-1}{2}$ negative literals. It contains exactly $\frac{(k+1)!}{\frac{(k+1)!}{2}!\cdot\frac{(k+1)!}{2}!}$ clauses for each hyperarcs in $\mathcal{H}$. Therefore, by similar arguments than in the even case, we achieve approximation ratio $2\cdot\sqrt{\frac{2}{\pi\cdot(k+1)}}$.

3.2 h-Colorability

Finally, we consider the problem of the Hypergraph h-Colorability in which the size of the hyperarcs is $k = h$, that is the problem of determining a coloring for $\mathcal{H}$ in which no nodes in a hyperarc have the same color.

Theorem 3.3 *Let $\mathcal{H}=< V, H >$ be a hypergraph, with $|H| = m$ hyperarcs of size $k = h$, then the Hypergraph h-Colorability problem of $\mathcal{H}$ can be solved within approximation ratio $\frac{\sqrt{2\cdot\pi\cdot k}}{e^k} \cdot (\frac{k}{k+1})^k$.*

Proof. Let $\mathcal{H}=< V, H >$ be an hypergraph, with $|H| = m$, for which the size k of each hyperarc $X \in H$ is h. The formula F associated to this problem can be obtained by considering all the clauses in $F_{\mathcal{H},k}$ corresponding to h-coloring of the hyperarcs, and then by considering for each hyperarc in $\mathcal{H}$ a number of clauses equal to $h!$, that is the number of permutation of h elements. Hence the total number of clauses in F is $M = h! \cdot m$, and each clause is the conjunction of $h \cdot \lceil \log h \rceil$ literals. Therefore, by means of non-oblivious local search, we can determine a truth assignment that satisfies at least $2^{-h\cdot\lceil\log h\rceil} \cdot M \geq 2^{-h\cdot\log(h+1)} \cdot M = (h+1)^{-h} \cdot M$ clauses in F. Furthermore, no more than one clause corresponding to the same hyperarc can be simultaneously satisfied. Therefore, from Stirling approximation, the number of hyperarcs in the coloring is at least $(h + 1)^{-h} \cdot M = \frac{h!}{(h+1)^{-h}} \cdot m \geq \frac{\sqrt{2\cdot\pi\cdot h}}{e^h} \cdot (\frac{h}{h+1})^h \cdot m = \frac{\sqrt{2\cdot\pi\cdot k}}{e^k} \cdot (\frac{k}{k+1})^k \cdot m \geq \frac{\sqrt{2\cdot\pi\cdot k}}{e^k} \cdot (\frac{k}{k+1})^k \cdot m^*$.

4 Directed Hypergraphs

In this section we deal with the 2-Colorability problem for directed hypergraphs.

Definition 4.1 *A directed hypergraph $\mathcal{H}$ is a pair $< V, H >$ where V is the set of nodes, H is the set of directed hyperarcs, and for each $< X, v >\in H$, $X \subseteq V$ and $v \in V$.*

Definition 4.2 *Given a directed hypergraph $\mathcal{H}=< V, H >$, the Directed Hypergraph 2-Colorability problem is the problem of finding a partition in two sets S and $\bar{S}$ of V, that maximize the cardinality $|H'|$, where $H' \subseteq H$ and for each hyperarc $< X, v >\in H'$, $X \subseteq S$ and $v \in \bar{S}$.*

Note that if $\mathcal{H}$ is a graph this problem is Max Directed Cut.

4.1 Limits to the Oblivious Local Search

In this section we show that the Directed Hypergraph 2-Colorability problem cannot be approximated by means of oblivious local search, even if we restrict to the case of directed graphs (and then if we consider Max Directed Cut). We will show that, for any approximation ratio r, whatever neighborhood size b we allow (i.e. whatever number b of nodes we allow to change color), there are local optima with respect to such neighborhood that do not guarantee the ratio.

To illustrate this consider the following examples:

a) Consider a directed graph $G = (V, E)$, $V = \{v_i : i = 1, \ldots, 20\}$ $E = \{(v_i, v_j) : (i = 1, 2) \wedge (j = 3, 4)\} \cup \{(v_i, v_j) : (i = 3, 4) \wedge (j = 8 \cdot i - 19, \ldots, 8 \cdot i - 12)\}$ $(n = |V| = 20, m = |E| = 20)$ and a solution $(S, \bar{S})$, $S = \{v_i : i = 1, 2, 5, \ldots, 20\}$, $\bar{S} = \{v_i : i = 3, 4\}$.
The number of edges in the cut is equal to 4, the optimal value m^* is equal to 16, and the number of nodes to move to improve the solution is equal to 4. Then $(S, \bar{S})$ is a local optimum for $r = \frac{1}{4}$, and $h = 3$.

b) Consider a directed graph $G = (V, E)$, $V = \{v_i : i = 1, \ldots, 30\}$ $E = \{(v_i, v_j) : (i = 1, 2, 3) \wedge (j = 4, 5, 6)\} \cup \{(v_i, v_j) : (i = 4, 5, 6) \wedge (j = 8 \cdot i - 25, \ldots, 8 \cdot i - 18)\}$ $(n = |V| = 30, m = |E| = 30)$ and a solution $(S, \bar{S})$, $S = \{v_i : i = 1, 2, 3, 7, \ldots, 30\}$, $\bar{S} = \{v_i : i = 4, 5, 6\}$.
The number of edges in the cut is equal to 6, the optimal value m^* is equal to 24, and the number of nodes to move to improve the solution is equal to 5. Then $(S, \bar{S})$ is a local optimum for $r = \frac{1}{4}$, and $h = 4$.

c) Consider a directed graph $G = (V, E)$, $V = \{v_i : i = 1, \ldots, 24\}$ $E = \{(v_i, v_j) : (i = 1, 2) \wedge (j = 3, 4)\} \cup \{(v_i, v_j) : (i = 3, 4) \wedge (j = 10 \cdot i - 25, \ldots, 10 \cdot i - 16)\}$ $(n = |V| = 24, m = |E| = 24)$ and a solution $(S, \bar{S})$, $S = \{v_i : i = 1, 2, 5, \ldots, 24\}$, $\bar{S} = \{v_i : i = 3, 4\}$.
The number of edges in the cut is equal to 4, the optimal value m^* is equal to 20, and the number of nodes to move to improve the solution is equal to 4. Then $(S, \bar{S})$ is a local optimum for $r = \frac{1}{5}$, and $h = 3$.

The previous examples show that the number of nodes it needs to move to improve a solution and the fraction of arcs not in the cut can be arbitrarily increased with the size of the problem. The next theorem generalize the result shown in the previous examples. The proof is omitted.

Theorem 4.1 *For any constant $\delta > 0$, and for any constant r, $0 \leq r \leq 1$, there exists a directed graph $G = (V, E)$, a cut $(S, \bar{S})$, and a solution $(S, \bar{S})$, such that the number of edges in the cut is less than $m^* \cdot r$, and $(S, \bar{S})$ is a local optimum with respect to a δ-bounded neighborhood.*

4.2 Non-Oblivious Local Search

In the remaining of this section we prove that the Directed Hypergraphs 2-Colorability problem can be solved within approximation ratio $\frac{1}{2^k - 1}$ by means of non oblivious local search.

This result is based on the following theorem that shows that non-oblivious local search achieves better performance ratio, in the approximation of the maximum satisfiability problem of formula of conjunctive clauses, with respect to the results of [2, 11]; we obtain a tight performance ratio by considering the approximation with respect to the value m^* of the optimal solutions, instead of considering the approximation with respect to the total number m of the clauses in the formula. A detailed proof of the theorem is given in Appendix A.

Theorem 4.2 *Let F be a logical formula of m conjunctive clauses of up to k literals, over the set Y of n Boolean variables, and $T \subseteq Y$ be an independently obtained truth assignment for Y. If T is a local optimum for a 1-bounded neighborhood and for the objective function W, then T satisfies at least $\frac{m^*}{2^k-1}$ clauses.*

With this result we are able to prove the following theorem.

Theorem 4.3 *Let $\mathcal{H} = < V, H >$ be a directed hypergraph, with $|H| = m$ hyperarcs of size k, then the Directed Hypergraphs 2-Colorability problem of $\mathcal{H}$ can be solved within approximation ratio $\frac{1}{2^k-1}$.*

Proof. Given a directed hypergraph $\mathcal{H} = < V, H >$, with $|V| = n$ and $|H| = m$, we construct a logical formula $F_{\mathcal{H}}$ of m clauses in the following way:

a) we associate to each node $v_i \in V$ a boolean variables y_i;
b) we associate to each hyperarc $< X, v_k > \in H$, $X = \{v_1, \ldots, v_{k-1}\}$ a conjunctive clauses of k literals $(y_1 \wedge \ldots \wedge y_{k-1} \wedge \overline{y}_k)$.

A truth assignment $T = \{p_1, \ldots, p_n\}$ for the variables y_i ($1 \le i \le n$) corresponds to a coloring of the nodes in V, such that, if $p_i = y_i$ then the node $v_i \in S$, and if $p_i = \overline{y}_i$ then $v_i \in \bar{S}$. Furthermore, given a solution T, a hyperarc is in the corresponding coloring $(S, \bar{S})$ if and only if the associated clause is satisfied by T. From theorem 4.2 we can determine a truth assignment that satisfies at least $\frac{m^*}{2^k-1}$ clauses, and then a 2-coloring for $\mathcal{H}$ of at least $\frac{m^*}{2^k-1}$ hyperarcs.

Acknowledgments: I would like to thank Giorgio Ausiello and Pierluigi Crescenzi for many helpful comments and suggestions.

References

1. P. Alimonti, and E. Feuerstein, Petri Nets, Hypergraphs and Conflicts, *Proc. of the 18th International Workshop on Graph-Theoretic Concept in Computer Science*, LNCS 657, 293-309, 1992.
2. P. Alimonti, New Local Search Approximation Techniques for Maximum Generalized Satisfiability Problems, *Proc. 2nd Italian Conference on Algorithms and Complexity*, LNCS 788, 40-53, 1994.
3. G. Ausiello, A. D'Atri, D. Saccà, Graph Algorithms for Functional Dependency Manipulation, *Journal of ACM*, 30, 752–766, 1983.

4. G. Ausiello, and G.F. Italiano, On-Line Algorithms for Polynomially Solvable Satisfiability Problems, *Journal of Logic Programming*, 10, 69-90, 1991.

5. G. Ausiello, and M. Protasi, Local Search, Reducibility and Approximability of NP Optimization Problems, *Information Processing Letters*, 1995 (to appear).

6. C. Berge, *Graphs and Hypergraphs*, North Holland, Amsterdam, 1973.

7. M. Garey, and D. Johnson, *Computers and Intractability: a Guide to the Theory of NP-Completeness*, Freeman, San Francisco, 1979.

8. M. Goemans, and D.P. Williamson, .878-Approximation Algorithms for MAX CUT and MAX 2SAT, *Proc. of the 35th Annual IEEE Conference on Foundations of Computer Science*, 1994.

9. D.S. Johnson, C.H. Papadimitriou, and M. Yannakakis, How Easy Is Local Search?, *Journal of Computer and System Sciences*, 37, 79-100, 1988.

10. V. Kann, J. Lagergren, A. Panconesi, Approximability of Maximum Splitting of k-Sets, *Unpublished Manuscript*, 1995.

11. S. Khanna, R. Motwani, M. Sudan, and U. Vazirani, On Syntactic versus Computational Views of Approximability, *Proc. of the 35th Annual IEEE Conference on Foundations of Computer Science*, 1994.

12. C. Papadimitriou, and K. Steiglitz, *Combinatorial Optimization Algorithms and Optimization*, Prentice-Hall, Englewood Cliffs, New Jersey, 1982.

13. C. Papadimitriou, and M. Yannakakis, Optimization, Approximation, and Complexity Classes, *Journal of Computer and System Sciences*, 43, 425-440, 1991.

14. D. Saccà, Closure of Database Hypergraphs, *Journal of ACM*, 32, 774-803, 1985.

A   Performance Guarantee by Non-Oblivious Local Search

Now we prove that non-oblivious local search achieves better approximations ratio with respect to the results given in [2, 11]. More precisely, we consider the maximum satisfiability problem of a logical formula of m conjunctive clauses of k literals and we obtain a tight a tight performance ratio of $\frac{1}{2^k-1}$ by considering the number of clauses satisfied by means of non-oblivious local search, with respect to the value m^* of the global optimal solutions, while the previous result was given with respect to the total number m of the clauses in the formula.

Let F be a logical formula of m conjunctive clauses of up to k literals, over a set $Y = \{y_1, \ldots, y_n\}$ of Boolean variables, and let $T = \{p_1, \ldots, p_n\}$ be a truth assignment for Y, where $p_i = y_i$ or $p_i = \bar{y}_i$. Let S_j be the set of clauses where j literals are false in T and let $|p_i|_{S_j}$ ($|\bar{p}_i|_{S_j}$) be the cardinality of the literal p_i ($\bar{p}_i$) is in S_j ($1 \le i \le n, 0 \le j \le k$).

$$|S_0| = \sum_{i=1}^{n} \frac{|p_i|_{S_0}}{k} \qquad |S_j| = \sum_{i=1}^{n} \frac{|p_i|_{S_j}}{k-j} = \sum_{i=1}^{n} \frac{|\bar{p}_i|_{S_j}}{j} \qquad |S_k| = \sum_{i=1}^{n} \frac{|\bar{p}_i|_{S_k}}{k}$$

Let $W = \sum_{j=0}^{k} L_j \cdot |S_j|$ be the weighted sum of the cardinalities of the S_j, where

$$L_j = \frac{1 + k \cdot L_{j+1} - (k - j - 1) \cdot L_{j+2}}{j+1} \quad (0 \le j \le k-1) \qquad L_j = 0 \quad (j \ge k)$$

$$(1)$$

The following lemmas show some properties of the coefficients L_j of the objective function W.

Lemma 1. $L_0 - L_1 = \frac{2^k - 1}{k}$.

Proof. Since from 1, for $(0 \leq j \leq k - 1)$

$$L_j - L_{j+1} = \frac{1 + (k - j - 1) \cdot (L_{j+1} - L_{j+2})}{j + 1}$$

we have

$$L_0 - L_1 = \sum_{i=1}^{k} \frac{(k-1)!}{i! \cdot (k - 1 - i)!} = \frac{2^k - 1}{k}.$$

Lemma 2. $L_j - L_{j+1} > L_{j+1} - L_{j+2}$, for $0 \leq j \leq k - 1$.

Proof. The lemma is proved by induction. Since from 1 $L_j = 0$ for $j \geq k$, and $L_{k-1} = \frac{1}{k}$, we have $L_{k-1} - L_k > L_k - L_{k+1}$.

Suppose now that $L_{j+1} - L_{j+2} > L_{j+2} - L_{j+3}$ for some $0 \leq j \leq k - 2$; we will show that $L_j - L_{j+1} > L_{j+1} - L_{j+2}$. Therefore, from 1, we have

$$L_j - L_{j+1} = \frac{1 + k \cdot L_{j+1} - (k - j - 1) \cdot L_{j+2}}{j + 1} - \frac{1 + k \cdot L_{j+2} - (k - j) \cdot L_{j+3}}{j + 2} >$$

$$\frac{1 + k \cdot L_{j+1} - (k - j - 1) \cdot L_{j+2} - 1 - k \cdot L_{j+2} + (k - j) \cdot L_{j+3}}{j + 1} =$$

$$\frac{k \cdot (L_{j+1} - L_{j+2}) - (k - j - 1) \cdot L_{j+2} + (k - j) \cdot L_{j+3}}{j + 1} >$$

$$\frac{k \cdot (L_{j+1} - L_{j+2}) - (k - j - 1) \cdot (L_{j+2} - L_{j+3})}{j + 1} >$$

$$\frac{k \cdot (L_{j+1} - L_{j+2}) - (k - j - 1) \cdot (L_{j+1} - L_{j+2})}{j + 1} = L_{j+1} - L_{j+2}$$

Lemma 3. $L_j - L_{j+1} \geq \frac{1}{j+1}$, for $0 \leq j \leq k - 1$.

Proof. It follows immediately, from 1, that, for $0 \leq j \leq k - 1$

$$L_j - L_{j+1} = \frac{1 + (k - j - 1) \cdot (L_{j+1} - L_{j+2})}{j + 1} \geq \frac{1}{j + 1}.$$

Now we prove the thesis of theorem 4.2. More precisely in the next theorem the case in which clauses contain exactly k literals is considered; but, since this does not affect the study of the worst case, the stated result also holds for clauses containing up to k literals.

Theorem A.1 *Let F be a logical formula of m conjunctive clauses of k literals, over the set Y of $n = r + s$ Boolean variables, and $T \subseteq Y$ be an independently obtained truth assignment for Y. If T is a local optimum for a 1-bounded neighborhood and for the objective function W, then T satisfies at least $\frac{m}{2^k - 1}$ clauses.*

Proof. Let $T = \{p_1, \ldots, p_r, q_1, \ldots, q_s\}$ be the truth assignment determined by the algorithm, and S (U) be the set of clauses satisfied (unsatisfied) by T. Let m^* be the cardinality of the global optima, and consider a truth assignment $T^* = \{p_1, \ldots, p_r, \overline{q}_1, \ldots, \overline{q}_s\}$ corresponding to a global optimal solution. Let S^* (U^*) be the set of clauses in S (U) satisfied by T^*, and S' (U') be the set of clauses in S (U) not satisfied by T^*. Then the cardinality of the global optima is $m^* = m_S^* + m_U^*$, where $m_S^* = |S^*|$ $(m_U^* = |U^*|)$.

Since T is a local optimum, we have $|S| \geq \frac{|S|+|U|}{2^k}$, and therefore

$$(2^k - 1) \cdot (m_S^* + |S'|) \geq m_U^* + |U'| \qquad (2)$$

Consider now the sets S^*, S', U^*, and U'.

a) S^* is formed by clauses $(p_{i_1} \wedge \ldots \wedge p_{i_k})$;
b) S' is formed by clauses $(p_{i_1} \wedge \ldots \wedge p_{i_{k-1}} \wedge q_{i_k}), \ldots, (p_{i_1} \wedge q_{i_2} \wedge \ldots \wedge q_{i_k}), (q_{i_1} \wedge \ldots \wedge q_{i_k})$ in S_0.

$$|S'| \geq \sum_q \frac{1}{k} \cdot |q|_{S_0} \qquad (3)$$

c) U^* is formed by clauses $(p_{i_1} \wedge \ldots \wedge p_{i_{k-1}} \wedge \overline{q}_{i_k})$ in S_1, $(p_{i_1} \wedge \ldots \wedge p_{i_{k-2}} \wedge \overline{q}_{i_{k-1}} \wedge \overline{q}_{i_k})$ in S_2, $\ldots$, $(p_{i_1} \wedge \overline{q}_{i_2} \wedge \ldots \wedge \overline{q}_{i_k})$ in S_{k-1}, $(\overline{q}_{i_1} \wedge \ldots \wedge \overline{q}_{i_k})$ in S_k;

$$|U^*| = m_U^* \leq \sum_{j=1}^{k} \left(\frac{1}{j} \cdot \sum_q |\overline{q}|_{S_j} \right) \qquad (4)$$

d) U' is formed at least by all the clauses containing literals q in the sets S_i for $i \geq 1$.

$$|U'| \geq \sum_{j=1}^{k-1} \left(\frac{1}{k-j} \cdot \sum_q |q|_{S_i} \right) \qquad (5)$$

Since T is a local optimal solution no literal q exists such that $\Delta W(q) > 0$.

$$\Delta W(q) = \sum_{j=0}^{k} L_j \cdot \Delta |S_j|(q) =$$

$$= L_0 \cdot (|\overline{q}|_{S_1} - |q|_{S_0}) + \sum_{j=1}^{k-1} L_j \cdot (|q|_{S_{j-1}} - |\overline{q}|_{S_j} + |\overline{q}|_{S_{j+1}} - |q|_{S_j}) =$$

$$= -(L_0 - L_1) \cdot |q|_{S_0} - (L_1 - L_2) \cdot |q|_{S_1} - \ldots - L_{k-1} \cdot |q|_{S_{k-1}} +$$
$$+ (L_0 - L_1) \cdot |\overline{q}|_{S_1} + (L_1 - L_2) \cdot |\overline{q}|_{S_2} + \ldots + L_{k-1} \cdot |\overline{q}|_{S_k} =$$

$$- \sum_{j=0}^{k-1} [(L_j - L_{j+1}) \cdot |q|_{S_j}] + \sum_{j=1}^{k} [(L_{j+1} - L_j) \cdot |\overline{q}|_{S_j}] \leq 0$$

Then

$$\sum_{j=0}^{k-1}[(L_j - L_{j+1}) \cdot |q|s_j] \geq \sum_{j=1}^{k}[(L_{j+1} - L_j) \cdot |\overline{q}|s_j]$$

hence, doing the sum over all literals q

$$\sum_{j=0}^{k-1}[(L_j - L_{j+1}) \cdot \sum_q |q|s_j] \geq \sum_{j=1}^{k}[(L_{j+1} - L_j) \cdot \sum_q |\overline{q}|s_j]$$

From 3, and lemma 1

$$(L_0 - L_1) \cdot \sum_q |q|s_0 = (2^k - 1) \cdot \frac{1}{k} \cdot \sum_q |q|s_0 = (2^k - 1) \cdot |S'| \qquad (6)$$

Furthermore, from lemmas 1 and 2

$$(L_i - L_{i+1}) \leq (L_1 - L_2) = \frac{(L_0 - L_1) - 1}{k - 1} = \frac{2^k - 1 - k}{k \cdot (k-1)} \quad (j \geq 1)$$

then, for $i \geq 1$

$$(L_i - L_{i+1}) \cdot \sum_q |q|s_i \leq \frac{2^k - 1 - k}{k \cdot (k-1)} \cdot \sum_q |q|s_i = \frac{2^k - 1 - k}{k} \cdot \frac{1}{k-1} \cdot \sum_q |q|s_i$$

hence, from 5

$$\sum_{j=1}^{k-1}[(L_j - L_{j+1}) \cdot \sum_q |q|s_j] \leq \frac{2^k - 1 - k}{k} \cdot |U'| \qquad (7)$$

Moreover, from Lemma 3, for $0 \leq j \leq k - 1$

$$(L_j - L_{j+1}) \cdot \sum_q |\overline{q}|s_{j+1} \geq \frac{1}{j+1} \cdot \sum_q |\overline{q}|s_{j+1}$$

then, from 4

$$\sum_{j=0}^{k-1}[(L_j - L_{j+1}) \cdot \sum_q |\overline{q}|s_j \geq m_U^* \qquad (8)$$

Therefore, from 6, 7, and 8

$$(2^k - 1) \cdot |S'| + \frac{2^k - 1 - k}{k} \cdot |U'| \geq m_U^* \qquad (9)$$

Suppose now $(2^k - 1) \cdot |S| = (2^k - 1) \cdot (|S'| + m_S^*) < m^*$. Since from 2 $|U'| + m_U^* < m^*$, then $|U'| < m_S^*$. Hence, from 9, $(2^k - 1) \cdot (|S'| + m_S^*) > (2^k - 1) \cdot |S'| + \frac{2^k - 1}{k} \cdot |U'| + m_S^* \geq m_S^* + m_U^* = m^*$, a contradiction.

Finally, we will show that the result of the previous theorem is tight, that is there exists an instance of the problem for which the algorithm calculates a solution with size exactly $\frac{m^*}{2^k-1}$.

Let be $F = \{(y_1 \wedge y_2), (y_3 \wedge y_4), (\overline{y}_1 \wedge \overline{y}_2), (\overline{y}_1 \wedge \overline{y}_3), (\overline{y}_1 \wedge \overline{y}_4), (\overline{y}_2 \wedge \overline{y}_3), (\overline{y}_2 \wedge \overline{y}_4)(\overline{y}_3 \wedge \overline{y}_4), \}$. The solution $T = \{y_1, y_2, y_3, y_4\}$ is a local optimum for a 1-bounded neighborhood and $W = 2 \cdot |S_0| + \frac{1}{2} \cdot |S_1|$, the number of clauses satisfied by T is equal to 2, and m^* is equal to 6.

On Interval Routing Schemes and Treewidth*

Hans L. Bodlaender[1], Richard B. Tan[1,2], Dimitris M. Thilikos[3] and Jan van Leeuwen[1],

[1] Department of Computer Science, Utrecht University,
P.O. Box 80.089, 3508 TB Utrecht, the Netherlands
{hansb,jan,rbtan}@cs.ruu.nl

[2] Department of Computer Science, University of Sciences and Arts of Oklahoma
Chickasha, Oklahoma 73018, USA

[3] Computer Technology Institute, P.O. Box 1122, 26110 Patras, Greece, and
Department of Computer Engineering and Informatics, University of Patras
26500 Rio, Patras, Greece
sedthilk@cti.gr

Abstract. In this paper, we investigate which processor networks allow k-label Interval Routing Schemes, under the assumption that costs of edges may vary. We show that for each fixed $k \geq 1$, the class of graphs allowing such routing schemes is closed under minor-taking in the domain of connected graphs, and hence has a linear time recognition algorithm. This result connects the theory of compact routing with the theory of graph minors and treewidth.

We also show that every graph that does not contain $K_{2,r}$ as a minor has treewidth at most $2r - 2$. In case the graph is planar, this bound can be lowered to $r + 2$. As a consequence, graphs that allow k-label Interval Routing Schemes under dynamic cost edges have treewidth at most $4k$, and treewidth at most $2k + 3$ if they are planar.

Similar results are shown for other types of Interval Routing Schemes.

1 Introduction

A common problem in processor networks is that messages that are sent from one processor to another processor must be routed through the network. The classical solution is to give each processor a routing table, with an entry for each destination specifying over which link the message must be forwarded. A disadvantage of this method is that these tables grow with network size, and may become too large for larger processor networks.

* This research was partially supported by the ESPRIT Basic Actions Program of the EC under contract No. 8141 (project ALCOM II). The research of the second author was also partially supported by the Netherlands Organization for Scientific Research (NWO) under contract NF 62-376 (NFI project ALADDIN: *Algorithmic Aspects of Parallel and Distributed Systems*). The research of the third author was also partially supported by the European Union ESPRIT Basic Research Project GEPPCOM (contract No. 9072). Correspondence on this paper to the first author.

Several different routing methods have been proposed that do not have this disadvantage. One such method is the interval routing method, together with its generalization k-label interval routing and variants of these. An overview of these and other compact routing methods can be found in [22].

Interval routing was introduced by Santoro and Khatib [21] and van Leeuwen and Tan [15]. Several well-known classes of networks, such as meshes, rings and hypercubes, allow interval routing schemes that are optimal, in the sense that messages always follow the shortest path to their destination. The method was applied in the C104 Router Chip, used in the INMOS T9000 Transputer design [13].

Frederickson and Janardan [12] considered interval routing in the setting of dynamic cost links (i.e., in the case that the cost of edges is variable). Actually, they considered a variant of interval routing, called *strict* interval routing. For these, they gave a precise characterization of the graphs with dynamic cost links which allow optimum strict interval routing schemes: these are exactly the *outerplanar* graphs. Bakker, Tan and van Leeuwen [3] obtained a similar result for general interval routing: a graph with dynamic cost links has an optimum interval routing scheme if and only if it is outerplanar or K_4. Another restriction of interval routing was introduced by Bakker, Tan and van Leeuwen in [2]: linear interval routing. It has also been applied in concrete networks. Here, also a precise characterization exists of the graphs which allow optimum linear interval routing schemes with dynamic cost links.

All of the interval routing schemes assumes that each link has one unique label, which is a (possibly cyclic) interval of processor names. All can be generalized to multi-label schemes, where each link has a number of labels. We consider the k-label schemes: each link has at most k labels. The issue we study in this paper is: *which graphs allow k-label interval routing schemes in the setting of dynamic cost links*.

Surprisingly, new and deep graph theoretical results on graph minors of Robertson and Seymour (see Section 2.1) can be used for the analysis of this problem. With the help of these results, we show *non-constructively* the existence of finite characterizations of which graphs allow certain routing schemes. Also, we give a non-constructive proof of the *existence* of linear time algorithms that check whether a desired routing scheme exists for a given graph. These algorithms heavily depend on the use of tree-decompositions. We also show that graphs, allowing a k-label interval routing scheme (in the setting of dynamic cost links) have treewidth at most $4k$. This not only gives a partial characterization of the graphs which have such routing schemes, but also, as the hidden constant factor of these algorithms is exponential in the treewidth of the tree-decomposition, it helps to decrease the running time of algorithms that would test the property.

As a main lemma, we show that every graph either contains $K_{2,r}$ as a minor, or has treewidth at most $2r - 2$. This can be seen as a special case of a result of Robertson and Seymour [18]: every planar graph H has an associated constant c_H, such that any graph G either contains H as a minor or has treewidth at most c_H. The best general bound for c_H known is $20^{2(2|V_H|+4|E_H|^5)}$ [20]. Our result gives a much better bound in the case of graphs of the form $K_{2,r}$. Similar results for other specific graphs can be found in [4] (trees), [11] (cycles and subgraphs of cycles), [7] (disjoint copies of K_3), and [6] (graphs that are minor of a circus graph and $(2 \times k)$-grid). The result of this main lemma can be seen as an additional result, fitting into this framework. Applied to the routing problem, it gives the first graph-theoretic complexity bound on the graphs that admit optimal k-label interval routing schemes. Another consequence we discuss is that 'most' random graphs (even 'sparse random graphs') do not allow k-label interval routing schemes under the dynamic cost edges assumption, for small values of k. Additionally, we give variants of the results when the graphs are restricted to be planar. [4]

The paper is organized as follows. The next section introduces some definitions and relevant background results. In section 3, we show the result that graphs with k-interval routing schemes is closed under minor-taking. Section 4 contains the main result on treewidth and a variant of the result for the planar graphs. Finally we close with some open problems.

2 Definitions and preliminary results

In this section, we introduce the most important definitions and mention some known results. In Section 2.1, we introduce graph-theoretic notions and results, and in Section 2.2, concepts and results from interval routing and its variants.

2.1 Graph theoretic definitions and preliminary results

All graphs in this paper will be assumed to be undirected, simple and finite. The number of vertices of a graph $G = (V, E)$ will be denoted by $n = |V|$. The notion of treewidth was introduced by Robertson and Seymour [18].

Definition 1. A *tree-decomposition* of a graph $G = (V, E)$ is a pair (X, T) with $T = (I, F)$ a tree and $X = \{X_i \mid i \in I\}$ a family of subsets of V, one for each node of T, such that

- $\bigcup_{i \in I} X_i = V$.
- for all edges $(v, w) \in E$, there exists an $i \in I$ with $v \in X_i$ and $w \in X_i$.
- for all $i, j, k \in I$: if j is on the path from i to k in T, then $X_i \cap X_k \subseteq X_j$.

The *treewidth* of a tree-decomposition $(\{X_i \mid i \in I\}, T = (I, F))$ is $\max_{i \in I} |X_i| - 1$. The treewidth of a graph G is the minimum treewidth over all possible tree-decompositions of G.

---

[4] Recently, Andreas Parra [16] obtained an improvement on our result: every graph either contains $K_{2,r}$ as a minor, or has treewidth at most $2r - 4$, $r \geq 4$.

There are several well known equivalent characterizations of the notion of treewidth; for instance, a graph has treewidth at most k, if and only if it is a partial k-tree, or a subgraph of a chordal graph with maximum clique size at most $k + 1$ (see [14]).

A graph $G = (V, E)$ is said to be a *minor* of a graph $H = (W, F)$, if G can be obtained from H by a series of vertex deletions, edge deletions, and edge contractions; where an edge contraction is the operation that takes two adjacent vertices v and w, and replaces it by a new vertex, adjacent to all vertices that were adjacent to v or w. A class of graphs $\mathcal{G}$ is said to be *closed under taking of minors*, if for every $G \in \mathcal{G}$, every minor H of G belongs to $\mathcal{G}$. For classes of graphs $\mathcal{G}, \mathcal{H}$, we say that $\mathcal{G}$ is *closed under taking of minors in the domain* $\mathcal{H}$, if for every graph $G \in \mathcal{G} \cap \mathcal{H}$, every minor H of G with $H \in \mathcal{H}$ belongs to $\mathcal{G}$.

In a long series of papers, Robertson and Seymour proved their famous graph minor theorem (formerly 'Wagner's conjecture'):

Theorem 2 See [17]. *For every class of graphs $\mathcal{G}$, that is closed under taking of minors, there exists a finite set of graphs, called the obstruction set of $\mathcal{G}$, $ob(\mathcal{G})$, such that for all graphs H, $H \in \mathcal{G}$, if and only if there is no graph G in the obstruction set of $\mathcal{G}$ that is contained in H as a minor.*

Fellows and Langston [10] derived the following consequence and variant of this result.

Theorem 3. *Let $\mathcal{G}$ be a class of graphs, closed under taking of minors in the domain $\mathcal{H}$, with $\mathcal{G} \subseteq \mathcal{H}$. There exists a finite set of graphs, the obstruction set of $\mathcal{G}$ in $\mathcal{H}$, $ob_{\mathcal{H}}(\mathcal{G})$, such that for all graphs $H \in \mathcal{H}$, $H \in \mathcal{G}$, if and only if there is no graph $G \in ob_{\mathcal{H}}(\mathcal{G})$ that is contained in H as a minor.*

It should be noted that the proofs of these results are (inherently) non-constructive. As for every fixed graph H, there exists an $O(n^3)$ time algorithm that tests whether H is a minor of a given graph G with n vertices [19], it follows that every minor-closed class of graphs has a cubic recognition algorithm, and every minor-closed class of graphs in a domain $\mathcal{H}$ has a cubic algorithm that tests whether graphs from $\mathcal{H}$ belong to $\mathcal{G}$. However, as the proof of Theorem 2 is non-constructive, we only know the algorithm exists, but we do not have the algorithm itself.

In several cases, faster algorithms exist.

Theorem 4 [18]. *For every planar graph H, there exists a constant c_H, such that for every graph G, either H is a minor of G, or the treewidth of G is at most c_H.*

Moreover, for every fixed integer k and graph H, there exists a linear time algorithm, such that when given a graph $G = (V, E)$ with a tree-decomposition of treewidth at most k, decides whether H is a minor of G, using standard methods for graphs with bounded treewidth (see e.g. [1].) As such tree-decompositions can be found in linear time [5], when existing, the following result holds:

Theorem 5. *Let $\mathcal{G}$ be a class of graphs that is closed under taking of minors, and that does not contain all planar graphs. Then there exists a linear time algorithm that tests whether a given graph G belongs to $\mathcal{G}$.*

Proof. (This proof is basically taken from [10], but we now use the algorithm of [5] for finding tree-decomposition of small treewidth.) Suppose H is a planar graph that does not belong to $\mathcal{G}$. First test whether the treewidth of input graph G is at most c_H. If not, we can safely conclude that $H \notin \mathcal{G}$. Otherwise, find a tree-decomposition of G of treewidth at most c_H with the algorithm of [5], and use this tree-decomposition to test whether a graph in $ob(\mathcal{G})$ is a minor of G. □

Theorem 6. *Let $\mathcal{G}$ be a class of graphs that is closed under taking of minors in the domain $\mathcal{H}$, $\mathcal{G} \subseteq \mathcal{H}$. Suppose there is at least one planar graph that belongs to $\mathcal{H}$ but not to $\mathcal{G}$. Then there exists a linear time algorithm that tests whether a given graph $G \in \mathcal{H}$ belongs to $\mathcal{G}$.*

Proof. (We again use an only slightly modified variant of a proof from [10].) Suppose H is a planar graph with $H \in \mathcal{H}$, $H \notin \mathcal{G}$. If $G \in \mathcal{G}$, then G does not contain H as a minor, hence has treewidth at most c_H. So, again we can first test whether the treewidth of G is at most k. If not, we are done. Otherwise, we compute a tree-decomposition of G with treewidth at most c_H, and then use this tree-decomposition to test in linear time whether G contains a graph in $ob_{\mathcal{H}}(\mathcal{G})$ as a minor. □

The constant factor of the linear time algorithms mentioned above is exponential in the treewidth of the tree-decomposition used, i.e., in c_H, H a planar graph not in $\mathcal{G}$ (but in $\mathcal{H}$). The constant factor in the original result of Robertson and Seymour was 'astronomically large'. In a later paper, Robertson et al. [20] improved this result, and obtained a constant factor of $20^{2(2|V_H|+4|E_H|^5)}$.

Still in most, if not all, practical cases, this constant factor is much too large, and makes the algorithm practically infeasible. This is the motivation, why we looked for much smaller values of c_H for graphs of the form $K_{2,r}$, as these graphs are planar, connected and can be shown to be 'outside' the considered classes of graphs.

2.2 Definitions and preliminary results on interval routing

Unless stated otherwise, intervals will be assumed to be 'cyclic' in the set $\{0, 1, \ldots, n-1\}$, ($n = |V|$); thus if $a > b$ then the interval $[a, b]$ denotes the set $\{a, a+1, \ldots, n-1, 0, \ldots, b-1\}$.

The shortest distance from vertex $u \in V$ to a vertex $v \in V$ in a graph $G = (V, E)$ when edges have costs given by edge cost function $c : E \to \mathbf{R}$, is denoted by $d_{G,c}(u, v)$. When G and/or c are clear from the context, we drop them from the subscript. The cost of a path p under edge cost function c is denoted by $c(p)$.

A node labelling of a graph $G = (V, E)$ is a bijective mapping $nb : V \to \{0, 1, \ldots, n-1\}$. An interval labelling scheme (ILS) of a graph $G = (V, E)$ is a node labelling nb of G, together with a labelling l, mapping each link (outgoing edge) to an interval $[a, b)$, $a, b \in \{0, 1, \ldots, n-1\}$, such that for every vertex v, the set of all labels of links outgoing from v partitions the set $\{0, 1, \ldots, n-1\}$.

Given an ILS, routing is done as follows. Each message contains, amongst others, the node label $nb(w)$ of its destination node w. When a node x receives a message with destination-label $dest$, it first looks whether $nb(x) = dest$. If so, the message has reached its destination, and is not routed any further. Otherwise, the message is transferred over the link with label $[a, b)$ such that $dest \in [a, b)$. An ILS is *valid*, if for all nodes v, w, messages sent from v to w eventually reach w by this procedure. An *interval routing scheme* (in short: IRS) is a valid ILS.

The notion of strict interval labelling schemes is obtained in a similar way: modify the definition of ILS in the sense that all labels of links associated with nodes v must partition the set $\{0, 1, \ldots, n-1\} - \{nb(v)\}$, i.e., the label of v may not appear in the labels of any of its outgoing links. A *linear interval labelling scheme* is an ILS where no interval label 'wraps' around, i.e., for all interval labels $[a, b)$ $a < b$. Strict linear interval labelling schemes, strict interval routing schemes (SIRS), linear interval routing schemes (LIRS), and strict linear interval routing schemes (SLIRS) are defined in the obvious way.

For each of these notions, we also define k-label variants. Here, each link is labelled with at most k (cyclic) intervals. All (cyclic) intervals associated with links of a node v must together partition $\{0, 1, \ldots, n-1\}$ (or $\{0, 1, \ldots, n-1\} - \{nb(v)\}$, in the case of strict labellings.) Again, a message is transferred over the link e for which one of its labels is an interval that contains the destination-number. k-label interval routing schemes, k-label linear interval routing schemes, etc., are defined as can be expected, and abbreviated as k-IRS, k-LIRS, etc. Note that an IRS is an 1-IRS, etc.

A routing scheme is *optimal* for a graph $G = (V, E)$, together with an assignment of non-negative costs to each edge $e \in E$, if whenever a message is sent from node v to node w, the path taken by this message is a minimum cost path from v to w.

We say that graph $G = (V, E)$ with dynamic cost links has an optimum k-IRS, if there exists a node labelling nb of G, such that for all assignments of non-negative costs to edges of E, there exists an IRS (nb, l) that is optimal for this cost assignment.

The class of graphs $k - IRS$ is defined as the set of all graphs G that have an optimum k-IRS with dynamic cost links. In the same way, we define classes k-LIRS, k-SIRS, k-SLIRS. See [22] for an overview of several results on these classes. We have the following relationships.

Theorem 7. *(i) (Frederickson, Janardan [12]) k-IRS $\subset (k+1)$-IRS.*
(ii) (Bakker, Tan, van Leeuwen [2]) k-LIRS $\subset (k+1)$-LIRS.
(iii) (Bakker, Tan, van Leeuwen [2]) k-IRS $\subset (k+1)$-LIRS.
(iv) (Bakker, Tan, van Leeuwen [3]) k-SIRS $\subset k$-IRS $\subset (k+1)$-SIRS.
(v) (Bakker, Tan, van Leeuwen [3]) k-SLIRS $\subset k$-LIRS $\subset k$-IRS.

Interestingly, one can prove that if we do not require optimal schemes, but allow paths to be at most some fixed constant $\alpha > 1$ larger than optimal, then we still cannot handle larger classes of graphs: the worst case number of intervals per link needed over all cost assignments equals that of the usual case $\alpha = 1$.

3 Closedness under minor taking

In this section we prove that for each fixed integer $k \geq 1$, each of the classes k-IRS, k-LIRS, k-SIRS and k-SLIRS is closed under taking of minors in the domain of connected graphs. The reason that this result is interesting is that it enables us to apply results from the theory of graph minors and of graphs of bounded treewidth to the theory of interval routing. We first prove a lemma which will be used later.

Lemma 8. *Let $G = (V, E)$ be a graph with edge costs $c : E \to \mathbf{R}^+ \cup \{0\}$. There exists an edge cost function $c' : E \to \mathbf{Z}^+$, such that for all $u, v \in V$: each shortest path p from u to v in G under edge costs c' is also a shortest path from u to v in G under edge costs c.*

Proof. Let $\mathcal{P}$ be the set of all simple paths in G. Define

$$\epsilon = \min\{|c(p) - c(p')| \mid p, p' \in \mathcal{P}, \ c(p) \neq c(p')\}$$

Note that $\epsilon > 0$. Define $c' : E \to \mathbf{Z}^+$ by taking for all $e \in E$:

$$c'(e) = \left\lfloor \frac{2|V|c(e)}{\epsilon} \right\rfloor$$

Suppose p is a shortest path from u to v under edge costs c', but not under edge costs c'. Let p' be another path from u to v with $c(p') < c(p)$. By definition of ϵ, we have $c(p') < c(p) - \epsilon$. Let $n = |V|$. Now

$$c'(p') = \sum_{e \in p}\left(\left\lfloor \frac{2n \cdot c(e)}{\epsilon} \right\rfloor + 1\right) \leq n + \frac{2n}{\epsilon}c(p') \leq n + \frac{2n}{\epsilon}(c(p) - \epsilon)$$

$$< \frac{2n}{\epsilon}c(p) = \sum_{e \in p}\frac{2n \cdot c(e)}{\epsilon} \leq \sum_{e \in p}\left(\left\lfloor \frac{2n \cdot c(e)}{\epsilon} \right\rfloor + 1\right) = c'(p)$$

So, $c'(p') < c'(p)$, hence p was not a shortest path from u to v under edge cost c', contradiction. □

Theorem 9. *Let $k \in \mathbf{N}$ be a fixed constant. k-IRS, k-SIRS, k-LIRS, k-SLIRS are closed under minor taking in the domain of connected graphs.*

Proof. We only prove the theorem for k-IRS, the other cases are similar. It is sufficient to prove, that if a connected graph $G = (V', E')$ is obtained from a graph $H = (V, E) \in k$-IRS by one of the following operations: removal of a vertex, removal of an edge, contraction of an edge, then $G \in k$-IRS. Suppose $H \in k$-IRS; let nb be a vertex labelling, such that for any cost assignment, there exists a k-label interval routing scheme (nb, l) for H.

First, suppose that G is obtained from H by removing an edge e_0. Use the same numbering nb for G. For any cost assignment $c : E_G \to \mathbf{R}^+ \cup \{0\}$, consider the cost assignment $c' : E_H \to \mathbf{R}^+ \cup \{0\}$, with for all $e \in E_G : c'(e) = c(e)$, and take $c(e_0) = 1 + \sum_{e \in E_G} c(e)$, i.e., the cost of e_0 is chosen so large that no minimum cost path will ever use the edge e_0. Hence, any k-label interval routing scheme (nb, l) for H with costs c' will also be a k-label interval routing scheme for G with costs c.

Next, suppose that G is obtained from H by removing a vertex $v \in V$ and all of its adjacent edges. By first removing all edges adjacent to v but one, as in the previous case, it follows that we may assume v has degree 1. Now, no shortest path between two vertices w and x, $x \neq v$, $x \neq w$ uses v. Label the vertices in V' as follows: if $nb(w) < nb(v)$, then take $nb'(w) = nb(w)$, and if $nb(w) > nb(v)$, then $nb'(w) = nb(w) - 1$. For any edge cost function c on G, we can make an IRS as follows: consider the same edge cost function c on H, giving the unique edge from v some arbitrary cost, and find an IRS (nb, l) for this function on H. Applying the same relabelling (decrease all labels larger than $nb(v)$ by one) on labelling l, we obtain a labelling l' such that (nb', l') is an IRS for G with edge costs c.

Finally, suppose G is obtained from H by contracting the edge $(v, w) = e_0 \in E_H$ to a vertex v'. Let $nb' : V \to \{0, 1, \ldots, |V_H| - 1\}$ be the function, obtained by taking for all $x \in V_G - \{v\}$, $nb'(x) = nb(x)$. Actually, there is a 'gap' in nb': there is no vertex x with number $nb'(x) = nb(v)$. This is resolved by decreasing all labels larger than $nb(v)$ by one, as in the case of removing a vertex.

Let $c : E_G \to \mathbf{R}^+ \cup \{0\}$ be a cost assignment for H. By Lemma 8, there exists a cost assignment $c' : E_G \to \mathbf{N}^+$, such that all shortest paths under cost assignment c' are shortest paths under cost assignment c. Let $\alpha = 1 + \sum_{e \in E_G} c'(e)$, a forbidding weight. Now let $c'' : E_H \to \mathbf{R}^+$ be defined as follows: for all edges $(x, y) \in E_H$ with $x, y \notin \{v, w\}$, let $c''(x, y) = c'(x, y)$. For $y \neq w$, if $(v, y) \in E_H$, take $c''(v, y) = c'(v', y)$. For $y \neq v$, if $(w, y) \in E_H$, then if $(v, y) \in E_H$, then let $c''(w, y) = \alpha$, otherwise let $c''(w, y) = c'(v', y) + 1/4$. Finally, we let $c''(v, w) = 1/8$.

Let (nb, l) be a k-IRS for H with cost c''. We can use l to build a k-IRS (nb, l) for G with cost c'. First note that H without the edges of cost α is still connected. So, no shortest path takes an edge of cost α, and all links corresponding to these edges have an empty label. For every link $(x, (x, y))$ with $x \notin \{v, w\}$, take in l' the same labels as in l. For a link $(v', (v', y))$, take in l' the union of the labels of links $(v, (v, y))$ and $(w, (w, y))$. Note that one of these links is either non-existing or empty, so this label will not consist of more than k intervals. Also, note that for every node x, the shortest path from v to x does not use w, if and only if the shortest path from w to x uses v. The same holds with roles of v and w reversed. It follows that no vertex label will appear in more than one label of a link outgoing from v'. We now have shown that l' is a k-ILS.

It remains to be shown that l' gives shortest paths in G. Consider nodes x and y in V_G. Let p be a shortest path in H between nodes x and y following links as directed by l'. (If $x = v'$, then take $x = v$ in H. Similar, if $y = v'$.) Note that if both v and w appear in p, then they must occur as consecutive nodes on this path, as all edges except (v, w) have cost at least 1. Let p' be the path in G, obtained from p by replacing a possible occurrence of v, w or both by one occurrence of v'. Observe that l' will direct a message from x to y via path p'. Finally, observe that p' is a shortest path from x to y in G with costs c', hence also with costs c. $\square$

Theorem 10. *(i) (Frederickson, Janardan [12]) $K_{2,2k+1} \notin k\text{-}SIRS$.*
(ii) $K_{2,2k+1} \notin k\text{-}IRS$.
(iii) (Bakker, Tan, van Leeuwen [3]) $K_{2,2k} \notin k\text{-}LIRS$ or $k\text{-}SLIRS$.

Proof. (i) is shown in [12]. The proof of (ii) is very similar. $\square$

It follows now from Theorem 3 that for each fixed $k \geq 1$, the classes k-IRS, k-LIRS, k-SIRS, and k-SLIRS have a finite characterization in terms of obstruction sets. Combining Theorem 10, Theorem 9 and Theorem 6 gives the following result.

Corollary 11. *For each fixed $k \in \mathbf{N}$, there exists a linear time algorithm that decide whether given a graph $G = (V, E)$ belongs to the class $k\text{-}IRS$ (or: $k\text{-}SIRS$, $k\text{-}LIRS$, $k\text{-}SLIRS$).*

It should be noted that this result is *non-constructive*: we know the algorithm exists, but to write down the algorithm, we must know the corresponding finite obstruction set, which we do not know. Unfortunately, we only know of much slower versions of these results.

Theorem 12. *For any fixed $k \geq 1$, one can construct algorithms that test, for a given graph $G = (V, E)$ with a given node labelling nb, whether for all costs assignments $c : E \to \mathbf{R}^+ \cup \{0\}$, there exists an $k\text{-}IRS$ (or: $k\text{-}SIRS$, $k\text{-}LIRS$, $k\text{-}SLIRS$) (nb, l) for G with costs c, in $O(n^{2k+3})$ time.*

The algorithm is based on the fact that yes-instances have bounded treewidth, a tree-decomposition is used to verify for each link whether there exists a cost assignment that needs too many labels for that link.

Corollary 13. *One can construct an algorithm that tests whether for a given integer $k \in \mathbf{N}$, and graph $G = (V, E)$, $G \in k\text{-}IRS$ (or: $G \in k\text{-}SIRS$, $G \in k\text{-}LIRS$, $G \in k\text{-}SLIRS$).*

4 The treewidth of graphs with k-label interval routing schemes

The main object of this section is to prove the following result.

Theorem 14. *Every graph $G = (V, E)$ contains $K_{2,r}$ as a minor or has treewidth at most $2r - 2$, $r \geq 1$.*

We also have a variant of this result with a slightly sharper bound for the case that G is planar. [5]

As the result trivially holds for $r = 1$, we suppose $r \geq 2$ in the remainder of this section.

Given a graph $G = (V, E)$ and a set $S \subseteq V$, let $\partial S = \{v \in V - S \mid \exists u \in S, \{u, v\} \in E\}$ (i.e., the neighbors of vertices in S that do not belong to S.)

Definition 15. A set $S \subseteq V$ is an *s-t-separator* in $G = (V, E)$ ($s, t \in V$), if s and t belong to different connected components of $G[V - S]$. S is a *minimal s-t-separator*, if it does not contain another s-t-separator as a proper subgraph. S is a *minimal separator*, if there exist vertices $s, t \in V$ for which S is a minimal s-t-separator.

Note that minimal separators can contain other minimal separators as proper subgraph. We will use in fact a different property of minimal separators, as given in the following lemma, which is easy to proof.

Lemma 16. *A non-empty set S is a minimal separator in G if and only if there are at least two components, G_1, G_2 of $G[V - S]$ such that $S \subseteq \partial V(C_i)$, $i = 1, 2$ (i.e. each vertex in S has a neighbor in both G_1 and G_2). We call two such components* separated components.

Lemma 17. *If G contains a minimal separator S, with $|S| \geq r$, then $K_{2,r}$ is a minor of G.*

Proof. Let S be a minimal separator and consider two separated components G_A and G_B of $G[V - S]$. Remove any vertex from any other component and $|S| - r$ vertices from S. If we now contract all edges in G_A and G_B that are not incident with a vertex in S, we obtain $K_{2,r}$. $\square$

Definition 18. For given $r \geq 2$, let $\mathcal{D}_r$ be the class of all graphs $G = (V_0 \cup V_1 \cup V_2 \cup V_3, E)$, such that

- V_0, V_1, V_2, V_3 are disjoint sets.
- $V_0 = \{v_0\}$. v_0 is adjacent to all vertices in V_1 and no vertices in $V_2 \cup V_3$.
- $|V_1| < r$. Every vertex in V_1 is adjacent to at least one vertex in V_2 and to no vertex in V_3.
- Every vertex in V_2 is adjacent to at least one vertex in V_1.
- Every vertex in V_3 is adjacent to less than r vertices in V_2, and is not adjacent to vertices in $V_0 \cup V_1 \cup V_3$.

Write $V_3 = \{v_3^1, \ldots, v_3^m\}$.

Definition 19. Let $G = (V, E)$ be a graph and $\mathcal{S}$ a collection of subsets of $V(G)$. Denote by $\mathrm{CL}(G, \mathcal{S})$ the graph obtained from G by making every set $S_i \in \mathcal{S}$ into a clique, i.e., $\mathrm{CL}(G, \mathcal{S}) = (V, E \cup \{(v, w) \mid v \neq w, \exists S_i \in \mathcal{S} : v, w \in S_i\})$.

---

[5] Recently, Andreas Parra improved Theorem 14, and obtained a bound of $2r - 4$, for $r \geq 4$.

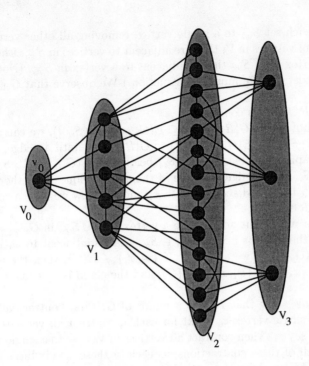

Fig. 1. Example of a graph in D_6.

Definition 20. Let G be a graph and $\mathcal{S}$ a collection of subsets of $V(G)$. Denote by $\mathrm{EX}(G, \mathcal{S})$ the graph obtained from G by adding to every set $S_i \in \mathcal{S}$ a new vertex $v_{\mathrm{new},i}$ which is adjacent to all vertices in S_i. (In case $|\mathcal{S}| = 1$, we denote the "new" vertex as v_{new}).

Lemma 21. (See [6, 4].) *For any graph $G = (V, E)$, either $K_{1,r}$ is a minor of G or treewidth$(G) \leq r - 1$.*

Lemma 22 See e.g., [8]. *Let $(\{X_i, i \in I\}, T)$ be a tree-decomposition of graph $G = (V, E)$. For any clique K of G, there exists an $i \in I$ with $V(K) \subseteq X_i$.*

Lemma 23. *For any graph $G \in \mathcal{D}_r$, either $K_{2,r}$ is a minor of G or G has a tree-decomposition of treewidth $\leq 2r - 2$ which is also a tree-decomposition of $\mathrm{CL}(G, \{\partial\{v_0\}, \partial\{v_3^1\}, \ldots, \partial\{v_3^m\}\})$.*

Proof. Let $G_{\mathrm{clique}} = \mathrm{CL}(G, \{\partial\{v_0\}, \partial\{v_3^1\}, \ldots, \partial\{v_3^m\}\})$, $V(K_{r,1}) = \{w_0, w_1, \ldots, w_r\}$ and $E(K_{r,1}) = \{\{w_0, w_1\}, \ldots, \{w_0, w_r\}\}$. By Lemma 21, either $K_{1,r}$ is a minor of $G_{\mathrm{clique}}[V_2]$ or treewidth$(G_{clique}[V_2]) \leq r - 1$. We consider these cases separately.

Case 1. $K_{1,r}$ is a minor of $G_{\mathrm{clique}}[V_2]$: We can identify disjoint sets $S_{w_i} \subseteq V_2$, $0 \leq i \leq k$, such that each set induces a connected subgraph, and we get $K_{1,r}$ by contracting each set S_{w_i} to a single vertex, removing all other vertices. Denote by R the set of vertices in V_3 that are adjacent to vertices in S_{w_0}. Choose for each i, $1 \leq i$, a vertex $w_i \in S_{w_i}$ that is adjacent to a vertex in S_{w_0}. (Note that these vertices w_i exist, by the minor construction.) We observe that $G_{\text{clique}}[R \cup S_{w_0}]$ is connected.

Claim I: $G[R \cup S_{w_0}]$ is connected.

Suppose not. As $E(G[R \cup S_{w_0}]) \subseteq E(G_{clique}[R \cup S_{w_0}])$, we can add edges in $E(G_{clique}[R \cup S_{w_0}]) - E(G[R \cup S_{w_0}])$ to $G[R \cup S_{w_0}]$ until an edge, say $\{x_1, x_2\}$ makes the graph connected. As $\{x_1, x_2\}$ belongs to $E(G_{clique}[R \cup S_{w_0}])$, but not to $E(G[R \cup S_{w_0}])$, the edge is in one of the added cliques, i.e., there must be a vertex $x_3 \in V_3$ that is adjacent to both x_1 and x_2. Now we have a contradiction, as $x_3 \in R \subseteq R \cup S_{w_0}$.

Claim II: For all i, w_i is adjacent to a vertex in $R \cup S_{w_0}$ in G.

For all i, there exists a vertex $x_i \in S_{w_0}$ that is adjacent to w_i in G_{clique}. If $\{w_i, x_i\} \in E(G)$, then we are done. If $\{u_i, x_i\} \notin E(G)$, then there is a vertex $x_i^3 \in V_3$ with $\{x_i, x_i^3\}$, $\{w_i, x_i^3\} \in E(G)$, and the claim is true, as w_i is adjacent to $x_i^3 \in R$.

We can now show that $K_{2,r}$ is a minor of G. First contract all vertices in $R \cup S_{w_0}$ to a single vertex z_0. Next for each i, contract all vertices in S_{w_i} to a single vertex, say z_i. Then contract all vertices in $V_0 \cup V_1$ to a single vertex z_{r+1}. (We can do all of these contractions, as each of these sets induces a connected subgraph of G.) By claim II, z_0 is adjacent to each vertex in $\{z_1, \ldots, z_r\}$. Also for each i, as $S_{w_i} \subseteq V_2$ and each vertices in V_2 is adjacent to at least one vertex in V_1, z_{r+1} is adjacent to each z_i. This yields the graph $K_{2,r}$.

Case 2. Treewidth$(G_{clique}[V_2]) \leq r - 1$: We show that treewidth$(G_{\text{clique}}) \leq 2r - 2$. Take a tree-decomposition $(\{X_i \mid i \in I\}, T = (I, F))$ of $G_{\text{clique}}[V_2]$ with treewidth $\leq r - 1$. Observe that $(\{X_i \cup V_1 \mid i \in I\}, T)$ is a tree-decomposition of $G_{\text{clique}}[V_1 \cup V_2]$ with treewidth at most $r - 1 + |V_1| \leq 2r - 2$. Using this tree-decomposition, we can build a tree-decomposition of G_{clique} of treewidth $\leq 2r - 2$, as follows. Add nodes $j_0, j_3^1, \ldots, j_3^m$ to I, with $X_{j_0} = \{v_0\} \cup \partial\{v_0\}$ and $X_{j_3^i} = \{v_3^i\} \cup \partial\{v_3^i\}$. From Lemma 22 there exists for each $i \in I$ a node $j_i' \in I$, with $\partial\{v_3^i\} \subseteq X_{j_i'}$. We can make j_i adjacent to this node j_i'. Finally, make j_0 adjacent to an arbitrary node $j_0' \in I$. We now have a tree-decomposition of G of treewidth at most $2r - 2$. $\square$

Definition 24. A *terminal graph* is a triple $G = (V, E, S)$ where (V, E) is a graph and $S \subseteq V$ is an ordered subset of its vertices. We call S the *terminal set* of G.

Definition 25. Consider two terminal graphs $G_i = (V_i, E_i, S_i)$, $i = 1, 2$ such that $|S_1| = |S_2|$. Define $G_1 \oplus G_2$ as the graph obtained by taking the disjoint union of G_1 and G_2 and then identifying the corresponding terminal vertices in S_1 and S_2.

Lemma 26. *Consider two terminal graphs $G_i = (V_i, E_i, S_i)$, $i = 1, 2$ such that $|S_1| = |S_2|$. Suppose that for $i = 1, 2$, $G_i[S_i]$ is a clique. If treewidth$(G_i) \leq$*

k_i, $i = 1, 2$, *then there is a tree-decomposition of $G_1 \oplus G_2$ with treewidth at most* $\max\{k_1, k_2\}$.

Proof. Take tree-decompositions $\{X_j^i \mid j \in I^i\}, T^i = (I^i, F^i)$ of G^i of treewidth at most k_i, $i = 1, 2$. By Lemma 22, there are $j_0^i \in I^i$ with $S_i \subseteq X_{j_0^i}^i$, $i = 1, 2$. Taking the disjoint union of the two tree-decompositions and connecting nodes j_0^1 and j_0^2 yields the desired tree-decomposition: one easily verifies that $(\{X_j^1 \mid j \in I^1\} \cup \{X_j^2 \mid j \in I_2\} \mid T = (I^1 \cup I^2, F^1 \cup F^2 \cup \{(j_0^1, j_0^2)\}))$ is a tree-decomposition of $G_1 \oplus G_2$ with treewidth at most $\max\{k_1, k_2\}$. □

We are now ready to prove Theorem 14. In fact, we prove the following, slightly stronger result.

Theorem 27. *Let $G = (V, E)$ be a graph. Then, for any $r \geq 1$, either $K_{2,r}$ is a minor of G or for any minimal separator S where $|S| < r$, G has a tree-decomposition with treewidth $\leq 2r - 2$ that is also a tree-decomposition of* $CL(G, \{S\})$.

Proof. The result trivially holds when G is complete. Suppose G is not complete. We use induction on $|V|$. The theorem clearly holds for $|V| = 3$. Assume that the theorem holds for any graph with less than n vertices. Let $G = (V, E)$ be a graph with n vertices and let S be a minimal separator with $|S| < r$ (in the case where $|S| \geq r$, we have from Lemma 17 that $K_{2,r}$ is a minor of G). Let $G_i = (V_i, E_i)$, $i = 1, \ldots, m$, be the connected components of $G[V - S]$ and $\bar{G}_i = EX(G[V_i \cup \partial V_i], \{\partial V_i\})$. We denote the corresponding "new" nodes as v_{new}^i. Notice that each graph $\bar{G}_i$ has $\partial\{v_{new}^i\}$ as a minimal separator and $|\partial\{v_{new}^i\}| < r$. We consider two cases.

Case 1. $|V(\bar{G}_i)| < n$ for all i, $1 \leq i \leq m$: From the induction hypothesis, either $K_{2,r}$ is a minor of $\bar{G}_i$ for some i, or for all i $\bar{G}_i$ has a tree-decomposition of treewidth $\leq 2r - 2$ that is also a tree-decomposition of $CL(\bar{G}_i, \{\partial\{v_{new}^i\}\})$. In the first instance, since $\bar{G}_i$ is a minor of G, $K_{2,r}$ is also a minor of G.

So suppose that for all i, $\bar{G}_i$ has a tree-decomposition of treewidth $\leq 2r - 2$ which is also a tree-decomposition of $CL(\bar{G}_i, \{\partial\{v_{new}^i\}\})$. We now construct a tree-decomposition of $CL(G, \{S\})$ with treewidth $\leq 2r - 2$. Let H_i be the graph obtained from $CL(\bar{G}_i, \{\partial\{v_{new}^i\}\})$ by removing the "new" vertex v_{new}^i. Clearly any graph H_i, $i = 1, \ldots, m$ has a tree-decomposition of treewidth $\leq 2r - 2$ and is a subgraph of $CL(G, \{S\})$. Consider now the graphs H_1 and H_2. We have that $S_{1,2} = V(H_1) \cap V(H_2) \subseteq S$ and thus $S_{1,2}$ induces a clique in H_1 and H_2. Make H_1 and H_2 into terminal graphs with terminal set $S_{1,2}$. From Lemma 26 the graph $H_{1,2} = H_1 \oplus H_2$ has also a tree-decomposition of treewidth $\leq 2r - 2$. Notice that $H_{1,2}$ is a subgraph of $CL(G, \{S\})$ and $S_{1,2,3} = V(H_{1,2}) \cap V(H_3) \subseteq S$, thus $S_{1,2,3}$ induces a clique in $H_{1,2}$ and H_3. Now make $H_{1,2}$ and H_3 terminal graphs with terminal set $S_{1,2,3}$ and apply again Lemma 26 to obtain a tree-decomposition of $H_{1,2,3} = H_{1,2} \oplus H_3$ with treewidth $\leq 2r - 2$. In this manner, by repeatedly applying Lemma 26, we can merge all the tree-decompositions of the graphs H_i, $i = 1, \ldots, m$ and thus construct a tree-decomposition of $CL(G, \{S\})$ that has treewidth $\leq 2r - 2$.

Case 2. If case 1 does not hold, then there are only two components in $G[V - S]$ and at least one of them contains only one vertex v_0. Consider the set $D = \partial S - \{v_0\}$. and assume that it does not contain a minimal separator of cardinality $\geq r$ (if it does, then by Lemma 17, $K_{2,r}$ is a minor of G).

Let G_i, $1 \leq i \leq m$ be the components of $G[V - D - S - \{v_0\}]$ and let $N_i = \partial V(G_j) \subseteq D$, $1 \leq i \leq m$. Notice that, as D does not contain minimal separators of cardinality $\geq r$, $|N_i| < r$ for each i, $1 \leq i \leq m$. Let $\bar{G}_i = \text{EX}(C_i, \{N_i\})$, $1 \leq i \leq m$, and denote the corresponding "new" vertices as v_{new}^i. As each $\bar{G}_i$ has less than n vertices, by induction hypothesis, either $K_{2,r}$ is a minor of $\bar{G}_i$ for some i or $\bar{G}_i$ has a tree-decomposition of treewidth $\leq 2r - 2$ which is also tree-decomposition of $\text{CL}(\bar{G}_i, \{N_i\})$ for each i.

In the first case, $\bar{G}_i$ is a minor of G and thus $K_{2,r}$ is a minor of G. In the second case, observe that $F = \text{EX}(G[D \cup S], \{S, N_1, \ldots, N_m\})$ is a member of D_r. From Lemma 23, either $K_{2,r}$ is a minor of F which implies that $K_{2,r}$ is a minor of G, as F is a minor of G, or F has a tree-decomposition of treewidth $\leq 2r - 2$ which is also a tree-decomposition of $\text{CL}(F, \{S, N_1, \ldots, N_m\})$. Suppose the latter. We now construct a tree-decomposition of $\text{CL}(G, \{S, N_1, \ldots, N_m\})$ with treewidth $\leq 2r - 2$. For each i, $1 \leq i \leq m$, let H_i be the graph obtained from $\text{CL}(\bar{G}_i, \{N_i\})$ if we remove the "new" vertex v_{new}^i. Also let F_0 be the graph obtained from F by removing the "new" vertices $v_{\text{new},i}$ corresponding to the sets $N_1, \ldots, N_m$. Clearly each H_i and F_0 have tree-decompositions of treewidth $\leq 2r - 2$. We observe that for F_0 and H_1, $V(F_0) \cap V(H_i) = N_1$ induces a clique in F_0 and H_1. Make F_0 and H_1 into terminal graphs with terminal set N_1. By Lemma 26, the graph $F_1 = F_0 \oplus H_1$ has also a tree-decomposition of treewidth $\leq 2r - 2$. Now N_2 induces a clique on F_1 and H_2, so we can make these graphs terminal graphs with terminal set N_2, and apply Lemma 26 again to obtain a tree-decomposition of $F_2 = F_1 \oplus N_2$ of width $\leq 2r - 2$. Continuing in this manner we can merge all the tree-decompositions of the graphs H_i, $i = 1, \ldots, r$ with the tree-decomposition of F_0, and thus construct a tree-decomposition of $\text{CL}(G, \{S, N_1, \ldots, N_m\})$ with treewidth $\leq 2r - 2$. As $\text{CL}(G, \{S\})$ is a subgraph of $\text{CL}(G, \{S, N_1, \ldots, N_m\})$, this completes the proof of the theorem. $\square$

Corollary 28. *(i) Every graph in k-IRS and k-SIRS has treewidth at most $4k$. (ii) Every graph in k-LIRS and k-SLIRS has treewidth at most $4k - 2$.*

Proof. If $G \in k$-SIRS or $G \in k$-IRS, then $K_{2,2k+1}$ is not a minor of G hence G has treewidth at most $2(2k+1) - 2 = 4k$. Similarly for the second assertion use $K_{2,2k}$ to obtain a bound of $4k - 2$. $\square$

These results can be seen as partial characterizations of graphs which allow k-label interval routing schemes (with dynamic edge costs). The result also indicates a limitation of the interval routing method: as 'most graphs have large treewidth' (see e.g. [14], chapter 5), the set of graphs in k-IRS only covers a small part of all graphs (or even of all sparse graphs, see [14]).

In the case when $k = 1$, more precise bounds are known. As 1-SIRS equals the class of connected outerplanar graphs [12], and outerplanar graphs have treewidth at most 2, every graph in 1-SIRS has treewidth at most 2. Similarly, the characterizations of 1-LIRS in [2] and of 1-SIRS in [3] show that every graph in 1-LIRS has treewidth (and even pathwidth) at most 2, and every graph in 1-IRS has treewidth at most 3, and has treewidth at most 2 if not equal to K_4.

The results also have consequences for *random graphs*. We mention some results, obtained by Kloks [14]. Let $G_{n,m}$ denote a random graph with n vertices and m edges. For a precise meaning of the term 'almost every' we refer to [9].

Theorem 29 Kloks[14]. *(i) Let $\delta > 1.18$. Then almost every graph $G_{n,m}$ with $m \geq \delta n$ has treewidth $\Theta(n)$.*
(ii) For all $\delta > 1$ and $0 < \epsilon < (\delta - 1)/(\delta + 1)$, almost every graph $G_{n,m}$ with $m \geq \delta n$ has treewidth at least n^ϵ.

Corollary 30. *(i) Let $\delta > 1.18$. Then almost every graph $G_{n,m}$ with $m \geq \delta n$, the smallest k for which $G_{n,m} \in k-IRS$ is of size $\Theta(n)$.*
(ii) Let $\delta > 1$ and $0 < \epsilon < (\delta - 1)/(\delta + 1)$. Then almost every graph $G_{n,m}$ with $m \geq \delta n$, the smallest k for which $G_{n,m} \in k-IRS$ fulfills $k \geq n^\epsilon$.

Finally, we mention that better bounds can be derived if G is required to be planar. Details can be found in the full journal version.

Theorem 31. *If G is planar, then either G contains $K_{2,r}$ as a minor or the treewidth of G is at most $r + 2$.*

Corollary 32. *(i) Every planar graph in k-IRS and k-SIRS has treewidth at most $2k + 3$.*
(ii) Every planar graph in k-LIRS and k-SLIRS has treewidth at most $2k + 2$.

5 Conclusions

In this paper, we made a perhaps somewhat surprising and interesting connection between the theory of compact routing schemes, and the theory of graph minors and treewidth of graphs. Several angles of this connection are still left unexplored.

As main open problems, we like to mention several issues that deal with constructivity. Is it possible to *construct* linear time algorithms that test whether a given graph belongs to k-IRS or one of its variants, for a fixed k . In several other cases, a non-constructive proof of a linear or small degree polynomial time bound was only the first step towards a fully constructive solution (e.g., [5]). Will our Corollary 11 also be such a first step? But even if we know that a graph belongs to k-IRS (or a related class), we do not have a corresponding node labelling. How much time does it cost to construct such a node labellings? And, given a node labelling, how much time does it cost to verify that it has a k-label IRS (or variant) for every edge cost assignment? We also suspect that our bounds of the treewidth of graphs that do not contain $K_{2,r}$ as a minor, or that belong to k-(S)(L)IRS are not sharp. More related open problems are mentioned e.g. in [22].

References

1. S. Arnborg, J. Lagergren, and D. Seese. Easy problems for tree-decomposable graphs. *J. Algorithms*, 12:308–340, 1991.

2. E. M. Bakker, R. B. Tan and J. van Leeuwen. Linear interval routing schemes. *Algorithms Review*, 2:45–61, 1991.

3. E. M. Bakker, R. B. Tan and J. van Leeuwen. Manuscript, 1994.

4. D. Bienstock, N. Robertson, P. D. Seymour, and R. Thomas. Quickly excluding a forest. *J. Comb. Theory Series B*, 52:274–283, 1991.

5. H. L. Bodlaender. A linear time algorithm for finding tree-decompositions of small treewidth. In *Proceedings of the 25th Annual Symposium on Theory of Computing*, pages 226–234, New York, 1993. ACM Press. To appear in SIAM J. Comput.

6. H. L. Bodlaender. On linear time minor tests with depth first search. *J. Algorithms*, 14:1–23, 1993.

7. H. L. Bodlaender. On disjoint cycles. *Int. J. Found. Computer Science*, 5(1):59–68, 1994.

8. H. L. Bodlaender and R. H. Möhring. The pathwidth and treewidth of cographs. *SIAM J. Disc. Math.*, 6:181–188, 1993.

9. B. Bollobas. *Random Graphs*. Academic Press, London, 1985.

10. M. R. Fellows and M. A. Langston. Nonconstructive tools for proving polynomial-time decidability. *J. ACM*, 35:727–739, 1988.

11. M. R. Fellows and M. A. Langston. On search, decision and the efficiency of polynomial-time algorithms. *J. Comp. Syst. Sc.*, 49:769–779, 1994.

12. G. N. Frederickson and R. Janardan. Designing networks with compact routing tables. *Algorithmica*, 3:171–190, 1988.

13. Inmos. *The T9000 Transputer Products Overview Manual*, 1991.

14. T. Kloks. *Treewidth. Computations and Approximations*. Lecture Notes in Computer Science, Vol. 842. Springer Verlag, Berlin, 1994.

15. J. van Leeuwen and R. B. Tan. Computer networks with compact routing tables. In G. Rozenberg and A. Salomaa, editors, *The Book of L*, pages 298–307. Springer-Verlag, Berlin, 1986.

16. A. Parra. Personal communication, 1995.

17. N. Robertson and P. D. Seymour. Graph minors — a survey. In I. Anderson, editor, *Surveys in Combinatorics*, pages 153–171. Cambridge Univ. Press, 1985.

18. N. Robertson and P. D. Seymour. Graph minors. II. Algorithmic aspects of tree-width. *J. Algorithms*, 7:309–322, 1986.

19. N. Robertson and P. D. Seymour. Graph minors. XIII. The disjoint paths problem. *J. Comb. Theory Series B*, 63:65–110, 1995.

20. N. Robertson, P. D. Seymour, and R. Thomas. Quickly excluding a planar graph. Technical Report TR89-16, DIMACS, 1989.

21. N. Santoro and R. Khatib. Labelling and implicit routing in networks. *Computer Journal*, 28:5–8, 1985.

22. R. B. Tan and J. van Leeuwen. Compact routing methods: A survey. Technical Report UU-CS-1995-05, Department of Computer Science, Utrecht University, Utrecht, 1995.

Highly Fault-Tolerant Routings and Diameter Vulnerability for Generalized Hypercube Graphs

Koichi WADA*, Takaharu IKEO, Kimio KAWAGUCHI and Wei CHEN

Nagoya Institute of Technology,
Gokiso-cho, Syowa-ku, Nagoya 466, JAPAN
email:(wada,tikeo,kawaguchi,chen)@elcom.nitech.ac.jp

Abstract. Consider a communication network G in which a limited number of link and/or node faults F might occur. A routing ρ for the network(a fixed path between each pair of nodes) must be chosen without knowing which components might become faulty. The diameter of the surviving route graph $R(G,\rho)/F$, where the surviving route graph $R(G,\rho)/F$ is a directed graph consisting of all nonfaulty nodes in G with a directed edge from x to y iff there are no faults on the route from x to y, could be one of the fault-tolerant measures for the routing ρ. In this paper, we show that we can construct efficient and highly fault-tolerant routings on a k-dimensional generalized d-hypercube $C(d,k)$ such that the diameter of the surviving route graph is bounded by constant for the case that the number of faults exceeds the connectivity of $C(d,k)$.

1 Introduction

We consider the problem of constructing efficient and fault-tolerant routings in a computer (or communication) network. As usual, a network is modeled as a graph, with nodes corresponding processors (or switching offices) and edges corresponding communication links. A *routing* assigns to any pair of distinct nodes in the graph a fixed path between them. This specified path is called *route* between them. We assume that the network communication protocol has no information about the topology of the network. Thus, all communication must be done by routes in a fixed routing table for the given network.

When nodes and/or links failures occur, routes that utilize the failed elements are disconnected. However, as long as the network is connected, communication between the endpoints of a disconnected route can be still possible through a sequence of alternate unfailed routes. Under the assumption of the fixed routing,

---

* This author was partially supported by the Okawa Institute of Information and Telecommunication(94-11) and the Telecommunications Advancement Foundation.

it has been proposed that we should estimate the diameter of a directed graph called *surviving route graph* as a measure of the network reliability [1, 2].

Given a graph G, routing ρ and a set of faults F, the surviving route graph is defined to be a directed graph consisting of all nonfaulty nodes in G, with a directed edge from x to y iff the route from x to y is survival. In a network with a fixed routing, the time required to send a message along a route is often dominated by the message processing time at the endpoints of the route [2]. Under the assumption, the total message transmission time is proportional to the diameter of the surviving route graph. Therefore, we need routings that minimizes this diameter. Many results have been obtained for the diameter of the surviving route graph [5, 8, 10].

The diameter of the surviving route graph can be considered as a generalization of the diameter vulnerability. The diameter vulnerability of G is how much the diameter of G increases in case of failures in G. So it is considered as the diameter of the surviving route graph for the routing on which only edges are defined as routes. Thus, the diameter vulnerability can be also discussed in the surviving route graph model.

When we examine the diameter of the surviving route graph for networks, we usually assume that the number of faults is less than the (node and/or edge) connectivity of their corresponding graphs, which ensures that the graphs are connected even if their faults occur. However, there are a lot of situations that graphs are still connected when the number of faults is greater than their connectivity. Several work has been done about fault-tolerance of networks and their measures when the number of faults is greater than their connectivity [3, 4, 7, 11, 9].

In this paper, we consider the diameter of the surviving route graph for a k-dimensional generalized d-hypercube $((d, k)$-cube for short) $C(d, k)$ when the number of faults is greater than the connectivity. The (d, k)-cube has d^k nodes, which are represented by d-ary numbers of length k and two nodes are adjacent if their corresponding numbers differ in exactly one position. Thus, the $(2, k)$-cube is the well known Boolean cube. Like Boolean cubes, (d, k)-cubes possess rich recursive structures and symmetry properties and so they are versatile networks.

Dolev, Halpern, Simons and Strong have shown in [2] that for the $(2, k)$-cube $(k \geq 1)$, if the number of faults is less than k, the diameter of the surviving route graph for any minimal routing(i.e. a routing in which the route between any two nodes is assigned to one of the shortest paths between them) on $C(2, k)$ is bounded by three, and there exists a minimal routing on $C(2, k)$ such that the diameter of the surviving route graph is bounded by two.

In this paper, we first show that we can extend the results to (d, k)-cube $C(d, k)(d \geq 3)$. That is, if the number of faults is less than $(d-1)k$, the diameter of the surviving route graph for any minimal routing on $C(d, k)$ is at most three and there exists a minimal routing on $C(d, k)$ such that the diameter of the surviving route graph is bounded by two. Then using the results we evaluate the diameter of the surviving route graph for $C(d, k)$ when the number of faults is less than $2(d-1)k - d$ and the remaining graph is still connected. As far as we know, this is the first result for the diameter of the surviving route graph in the case that the number of faults exceeds the connectivity of the graph. We obtain the following results.

When the number of faults F is less than $2(d-1)k - d$ in $C(d, k)$ and $C(d, k) - F$ is connected,

(1) the diameter of $C(d, k) - F$ increases at most by 2,
(2) the diameter of the surviving route graph for any minimal routing on $C(d, k)$ is at most 6 and
(3) there exists a minimal routing on $C(d, k)$ such that the diameter of the surviving route graph is bounded by 4.

The rest of the paper is organized as follows. In Section 2 the necessary definitions are given. Section 3 contains the results on $C(d, k)$ when the number of faults is less than the connectivity. In Section 4 the diameter of the surviving route graph for $C(d, k)$ is discussed when the number of faults exceeds the connectivity.

2 Preliminaries

In this section, we give definitions and terminology. We refer the reader to [6] for basic graph terminology.

2.1 Surviving route graphs

Unless otherwise stated, we deal with an undirected graph $G = (V, E)$ that corresponds to a network. For a node $v \in V$ and a subgraph $G' = (V', E')$ of G $N_{G'}(v) = \{u | (v, u) \in E'\}$. If $G' = G$, $N_{G'}(v)$ is simply denoted by $N(v)$. The *distance* between nodes x and y in G is the length of the shortest path between x and y and is denoted by $dis_G(x, y)$. The *diameter* of G is the maximum of $dis_G(x, y)$ over all pairs of nodes in G and is denoted by $D(G)$.

Let $G = (V, E)$ be a graph. Define $P_G(x, y)$ to be the set of all simple paths between the nodes x and y in G, and $P(G)$ to be the set of all simple paths in G. A *routing* is a partial function $\rho : V \times V \to P(G)$ such that $\rho(x, y) \in P_G(x, y)$. The path specified to be $\rho(x, y)$ is called the *route from x to y*. The length of the route $\rho(x, y)$ is denoted by $|\rho(x, y)|$. Let ρ be a routing on G, and let $G' = (V', E')$ be a subgraph of G. The *routing ρ on the subgraph G'* is a subfunction of ρ such that the domain of ρ is limited to $V' \times V'$.

For a graph $G = (V, E)$, let $F \subseteq V \cup E$ be a set of nodes and edges called the set of *faults*. $F \cap V(= F_V)$ and $F \cap E(= F_E)$ are called the set of *node faults* and the set of *edge faults* respectively. In what follows, we consider the node faults only. If an object such as route or path does not contain any element of F, the object is called *fault free*.

For a graph $G = (V, E)$, a routing ρ on G and a set of faults $F(= F_V \cup F_E)$, the *surviving route graph*, $R(G, \rho)/F$, is a directed graph with node set $V - F_V$ and edge set $E(G, \rho, F) = \{<x, y> | \rho(x, y) is\ defined\ and\ fault\ free\}$.

A routing ρ is a *partial* routing if $\rho(x, y)$ is undefined for some nodes $x \neq y$, otherwise ρ ia a *total* routing.

A routing ρ is a *minimal* routing if it holds that $|\rho(x, y)| = dis_G(x, y)$ for any node pair (x, y) defined by ρ. A routing ρ is an *edge-routing* if $\rho(x, y)$ is defined for every edge $(x, y) \in E$.

A routing ρ is a *bidirectional* routing if it holds that $\rho(x, y) = \rho(y, x)$ for any pair of distinct nodes x and y defined by ρ. Note that if the routing ρ is bidirectional, the surviving route graph $R(G, \rho)/F$ can be represented as an undirected graph and $dis_{R(G, \rho)/F}(x, y) = dis_{R(G, \rho)/F}(y, x)$ for any pair of distinct nodes x and y.

Let G be a graph and ρ be a routing on G. For a pair of distinct nodes x and y in G, x and y is (d, f)−*tolerant with respect to* ρ if for every set of F of f faults in G such that $G - F$ is connected, $dis_{R(G, \rho)/F}(x, y) \leq d$. If the routing ρ is apparent in the context, we simply say x and y is (d, f)−tolerant. A *routing ρ on a graph G is (d,f)-tolerant* if x and y is (d, f)−tolerant for any pair of distinct nodes x and y. A *graph G is called (d, f)−tolerant* if there exists a (d, f)−tolerant routing ρ on G.

2.2 Generalized (d, k)-cube

Given two graphs $G = (V_G, E_G)$ and $H = (V_H, E_H)$, *their Cartesian product* $G \times H$ is a graph (V, E), where $V = V_G \times V_H$ and $((i, j), (k, l)) \in E$ if and only if both (i, j) and (k, l) are nodes in V and either $i = k$ and $(j, l) \in E_H$ or $j = l$ and $(i, k) \in E_G$. The *H plane defined by i* (*G plane defined by j*) in $G \times H$ is the subgraph of $G \times H$ determined by all nodes having the first (second) coordinate equal to i (j).

The *k-dimensional generalized d-hypercube graph* ((d, k)-*cube* for short) $C(d, k) = (V(d, k), E(d, k))$ is defined recursively as follows:

1. $C(d, 1)$ is the complete graph of d nodes, denoted by K_d,
2. $C(d, k) = C(d, k - 1) \times K_d$.

From the definition of $C(d, k)$, $|V(d, k)| = d^k$, a node of $C(d, k)$ can be represented by d−tuples $a_1, a_2, ..., a_k$ where $a_j \in \{0, 1, ..., d - 1\}$ and two nodes are connected by an edge in $E(d,k)$ if there corresponding d-tuples differ in exactly one coordinate.

Figure 1 shows an example of generalized (d,k)-cubes.

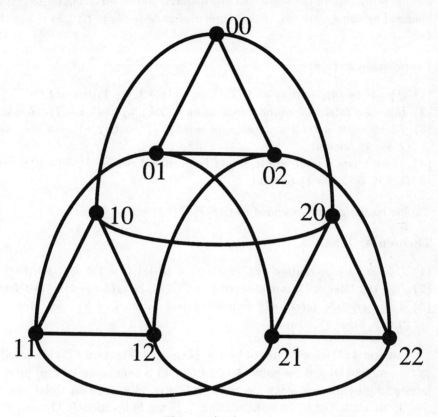

Fig. 1. $C(3,2)$.

For $C(d,k)$ and two nodes x and y, let $Sc(x,y)$ be the subgraph induced by all nodes in minimal length paths between x and y. The (d,k)-cube $C(d,k)$ has the following properties, which are easily verified.

Lemma 1. *Let $d \geq 2$. For any pair of nodes x and y in $C(d,k)$, $Sc(x,y)$ is isomorphic to $C(2, dis_{C(d,k)}(x,y))$.*

Lemma 2. *Between any two nodes x and y in $C(d,k)$, there exist exactly $(d-1)k$ node-disjoint paths, in which $dis_{C(d,k)}(x,y)$ of these paths are of length $dis_{C(d,k)}(x,y)$, $dis_{C(d,k)}(x,y)(d-2)$ of these paths are of length $dis_{C(d,k)}(x,y)+1$ and the remaining $(d-1)(k-dis_{C(d,k)}(x,y))$ paths are of length $dis_{C(d,k)}(x,y)+2$.*

3 Fault-tolerant routings for $C(d, k)$

It is shown in [2] that every minimal routing on the Boolean cube $C(2, k)$ has good performance in the sense that the diameter of the surviving route graph is bounded by some constant, if the number of node faults in $C(2, k)$ is less than k.

Proposition 3. [2, 8]

(1) If ρ_k is an edge-routing on $C(2, k)$, ρ_k is $(k + 1, k - 1)$-tolerant.
(2) If ρ_k is a total and minimal routing on $C(2, k)$, ρ_k is $(3, k - 1)$-tolerant.
(3) There exists a total and minimal routing λ_k on $C(2, k)$ such that λ_k is $(2, k - 1)$-tolerant.
(4) There exists a total, minimal and bidirectional routing λ'_k on $C(2, k)$ such that λ'_k is $(2, k - 1)$-tolerant.

The results can be extended to $C(d, k)(d \geq 3)$ except Theorem 3 (4).

Theorem 4. Let $d \geq 3$.

(1) If ρ_k is an edge-routing on $C(d, k)$, ρ_k is $(k + 1, (d - 1)k - 1)$-tolerant.
(2) If ρ_k is a total and minimal routing on $C(d, k)$, ρ_k is $(3, (d-1)k-1)$-tolerant.
(3) There exists a total and minimal routing λ_k on $C(d, k)$ such that λ_k is $(2, (d - 1)k - 1)$-tolerant.

Theorem 4 (1) is easily proved by using Lemma 2. Theorem 4 (3) is a corollary to the result of [1] and the routing λ_k on $C(d, k)$ is defined as $\lambda_k(x, y)$ proceeds from x to y by moving along the the coordinates on which they differ one at a time from left to right. Thus the routing λ_k is not bidirectional. Theorem 4 (2) follows from the next lemma which has been proved for $d = 2$ [2].

We define a pair of nodes x and y to be *safe with respect to a set of faults* F iff every minimal length path from x to y is fault-free. A sequence of nodes $x_1, \ldots, x_m$ is safe with respect to F if each consecutive pair of nodes in the sequence is safe with respect to F [2].

Lemma 5. Let $d \geq 3$. If $|F| < (d - 1)k$, then for any pair of nonfaulty nodes x and y in $C(d, k)$ there are nodes u and v $(u \neq v)$ such that the sequence x, u, v, y is safe with respect to F

Proof. We prove the lemma by induction on k. The argument for $k = 1$ is straightforward. We assume the induction hypothesis for $k - 1(k \geq 2)$. Let $x = x_1 x_2 \ldots x_k$ and $y = y_1 y_2 \ldots y_k$ be non-faulty nodes in $C(d, k)$. There are two cases.

(Case 1: $dis_{C(d,k)}(x,y) < k$**)**

There is some coordinate on which the nodes x and y have the same value. Without loss of generality, $x_1 = y_1 = 0$. Let $x = 0x'$ and $y = 0y'$. Define sets of node faults F_0 and F' as

$F_0 = \{f | f = f_1 f_2 \ldots f_k \in F$ and $f_1 = 0\}$ and $F' = F - F_0$.

If $|F_0| < (d-1)(k-1)$, we can apply the induction hypothesis to the induced subgraph by nodes of which first coordinates are 0 and there are nodes $u = 0u', v = 0v'$ such that x', u', v', y' is safe with respect to F_0. Then the sequence x, u, v, y is safe with respect to F.

Otherwise($|F_0| \geq (d-1)(k-1)$), since $|F'| \leq d-2$. there is a fault-free subcube $C^i(d, k-1)(1 \leq i \leq d-1)$. Thus the sequence x, ix', iy', y is safe with respect to F.

(Case 2: $dis_{C(d,k)}(x,y) = k$**)**

Without loss of generality, we assume that $x = 0^k$ and $(d-1)^k$. Define sets of node faults F_s and F' as

$F_s = \{f | f \in Sc(0^k, (d-1)^k)$ and $f \in F\}$ and
$F' = F - F_s$.

If $|F_s| \geq k$, then $|F'| < (d-2)k$. There are $(d-2)k$ sequences
$x = 0^k, u(a,i) = 0^i a 0^{k-1-i}, v(a,i) = (d-1)^i a(d-1)^{k-1-i}, y = (d-1)^k$
$(1 \leq a \leq d-1, 0 \leq i \leq k-1)$ such that if $a \neq b$ or $i \neq j$ then
  $Sc(x, u(a,i)) \cup Sc(u(a,i), v(a,i)) \cup Sc(v(a,i), y)$ and
  $Sc(x, u(b,j)) \cup Sc(u(b,j), v(b,j)) \cup Sc(v(b,j), y)$
are node-disjoint except x and y. Thus, there is a pair (a,i) such that $x = 0^k, u(a,i) = 0^i a 0^{k-1-i}, v(a,i) = (d-1)^i a(d-1)^{k-1-i}, y = (d-1)^k$ is safe with respect to F.

If $|F_s| < k$, since $Sc(0^k, (d-1)^k)$ is isomorphic to $C(2,k)$ from Lemma 1, the result for $d = 2$ can be used and this case is proved in [2].

4 Highly fault-tolerant routings for $C(d, k)$

Since the node-connectivity of $C(d,k)$ is $(d-1)k$ from Lemma 2, it may be disconnected if the number of node faults exceeds $(d-1)k - 1$. In this section, we consider a set F of node faults whose cardinality is about twice the node-connectivity of $C(d,k)$ under the assumption that $C(d,k) - F$ is connected. It is known that if a node fault F^+ in $C(d,k)$ satisfies $|F^+| \leq 2(d-1)k - d - 1$ and $N(v) \not\subseteq F^+$ for any node of $C(d,k)$ (that is, all neighbours of v are not faulty), $C(d,k) - F^+$ is connected [11]. This result can be obtained from the fact that there are $2(d-1)k - d$ node-disjoint paths between any two non-adjacent edges in $C(d,k)$ [11].

Theorem 6. *Let $d \geq 3$ Let ρ_k be a minimal routing on $C(d,k)$ and be $(\delta(k), (d-1)k-1)$-tolerant, where $\delta(k)$ is a monotone non-decreasing function whose range is non-negative integers, $\delta(0) = 0$ and $\delta(k) \geq 2(k \geq 2)$. If F^+ is any set of node faults in $C(d,k)$ such that $|F^+| \leq 2(d-1)k-d-1$ and $C(d,k)-F^+$ is connected, then $D(R(C(d,k), \rho_k)/F^+) \leq \delta(k-1)+3$.*

Proof. We show the theorem by induction on k. It can be verified for $k = 1$ and $k = 2$. We assume that the theorem holds for any integer less than k. Let $C^i(0 \leq i \leq d-1)$ denote $d-1$ $C(d,k-1)$ in $C(d,k)$. Let $H = R(C(d,k), \rho_k)/F^+$ and let (x,y) be any ordered pair of distinct nodes in H. Let $F_i(0 \leq i \leq d-1)$ denote the set of node faults in C^i. The cases are divided according as $dis_{C(d,k)}(x,y) < k$ or not. If $dis_{C(d,k)}(x,y) < k$, there exists i such that x and y are in C^i.

(**Case 1:** x and y in C^i) Let $x = ix'$ and $y = iy'$.

(**Subcase 1-1:**$|F_i| \leq 2(k-1)(d-1)-d-1$:)

(**Subcase 1-1-1**) If $|F_i| \leq 2(k-1)(d-1)-d-1$ and there are $u \in N_{C^i}(x)$ and $v \in N_{C^i}(y)$ such that both u and v are non-faulty, x and y are connected in C^i [11] we have

$$dis_H(x,y) \leq \delta(k-2)+3 \leq \delta(k-1)+3$$

using the induction.

(**Subcase 1-1-2**) If $|F_i| \leq 2(k-1)(d-1)-d-1$ and ($N_{C^i}(x) \subseteq F_i$ or $N_{C^i}(y) \subseteq F_i$), x and y are disconnected in C^i. We assume that $N_{C^i}(x) \subseteq F_i$. There are two cases that $dis_{C^i}(x,y) \geq 3$ and $dis_{C^i}(x,y) = 2$.

Note that if $dis_{C^i}(x,y) \geq 3$, $N_{C^i}(x) \cap N_{C^i}(y) = \phi$. Let f_r be the number of faults in the $d-1$ subcubes $C^j(j \neq i) - \{jx'|(j \neq i)\}$ and let ℓ be the number of non-faulty nodes $jx'(j \neq i)$ in the $d-1$ subcubes C^j. Since x is not isolated in $C(d,k)$, it holds that $1 \leq \ell \leq d-1$. Let f_y be the number of faulty nodes in $N_{C^i}(y)$, where $0 \leq f_y \leq (k-1)(d-1)-d-1$. Since

$$f_r \leq (2k(d-1)-d-1) - (k-1)(d-1) - (d-1-\ell) - f_y$$
$$\leq (k-1)(d-1) + (\ell-2) - f_y,$$

there is a subcube C^a such that ax' is non-faulty and $|F_a| \leq \frac{1}{\ell} f_r$. Then it holds that $|F_a| \leq (k-1)(d-1)-1$. Considering the set of $(k-1)(d-1)-f_y$ paths $\{(y = iy', iu', au')|iu' \in N_{C^i}(y) - F_i\}$, there exists a fault-free path $(y = iy', iu', au')$ because it holds that $((k-1)(d-1)-f_y) - |F_a| \geq 1$. We have

$$dis_H(x,y) \leq \delta(k-1)+3.$$

If $dis_{C^i}(x,y) = 2$, then $N_{C^i}(x) \cap N_{C^i}(y) \neq \phi$ and $|N_{C^i}(x) \cap N_{C^i}(y)| = 2$. By using the same argument for the former case, for a subcube C^a with $ax' \notin F_a$ if $ay' \notin F_a$ or there is a fault-free path (iy', iu', au'), it holds that $dis_H(x,y) \leq \delta(k-1)+2$ or $dis_H(x,y) \leq \delta(k-1)+3$, respectively. Otherwise, there is a non-faulty node $iv' \in N_{C^i}(y)$ because $|F_i| \leq 2(k-1)(d-1)-d-1$ and $N_{C^i}(x) \subseteq F_i$. Furthermore, there are $2-p$ non-faulty nodes $aw' \in N_{C^a}(ax') \cap N_{C^a}(ay')$. There are at most $d-2-p$ faults in $d-2$ subcubes $C^j(j \neq i,a)$. Since if $p \geq 1$ or there is a subcube with at least 2 faults, we can find a subcube with no faults and we have $dis_H(x, y) \leq \delta(k-1) + 2$, we may consider the case that there is a non-faulty nodes $aw' \in N_{C^a}(ax') \cap N_{C^a}(ay')$ and each subcube C^b has one fault. If bx' and by' are non-faulty, $dis_H(x, y) \leq \delta(k-1) + 2$. If bx' is non-faulty but by' is faulty, $dis_H(x, y) \leq \delta(k-1) + 3$, because the path (bv', iv', y) is fault-free. If bx' is faulty, there is a fault-free path $(x, ax', aw', bw', by', y)$ with length 5. Since $\delta(k-1) \geq 2$ we have

$dis_H(x, y) \leq \delta(k-1) + 3.$

(Subcase 1-2) If $|F_i| \geq 2(k-1)(d-1) - d$, since $|F^+| \leq 2k(d-1) - d - 1$, remaining faults not in C^i is at most $2d - 3$.

(Subcase 1-2-1: There exists a s.t. $ax', ay' \notin F_a$)

Noting that $(k-1)(d-1) - 1 \geq 2d - 3$, since $dis_H(ax', ay') \leq \delta(k-1)$, we have

$dis_H(x, y) \leq \delta(k-1) + 2.$

(Subcase 1-2-2: There does not exist a s.t. $ax', ay' \notin F_a$)

Since $|F^+| - |F_i| \leq 2d - 3$ and there are $d - 1$ subcubes except C^i, there is a subcube C^a such that $|F_a| \leq 1$. Since either ax' or ay' is faulty, we can assume that $ax' \in F_a$ and $ay' \notin F_a$. Since x can not be isolated in $C(d, k)$, there is a node $z \in N_{C(d,k)}(x)$. If $z \in C^i$, $az' \notin F_a$ because $|F_a| \leq 1$ and $ax' \in F_a$. Thus,

$dis_H(x, y) \leq \delta(k-1) + 3.$

Otherwise(there does not exist $z \in C^i$) $|N_{C^i}(x) \cap F_i| = (k-1)(d-1)$ and there exists $b' (\neq a)$ such that $z = bx' \in C^b$.

If $dis_{C^i}(x, y) \geq 3$, $N_{C^i}(x) \cap N_{C^i}(y) = \phi$ and $|N_{C^i}(y)| = (k-1)(d-1)$. Since there are $(k-1)(d-1)$ node-disjoint paths $\{(bw', iw', iy' = y)|iw' \in N_{C^i}(y)\}$ and there are at most $(k-1)(d-1) - 2$ faults on them, there exists at least one fault-free path $(bw', iw', iy' = y)$. Furthermore, $|F_b| \leq (k-1)(d-1) - 1$. Thus,

$dis_H(x, y) \leq \delta(k-1) + 3.$

If $dis_{C^i}(x, y) = 2$, $|N_{C^i}(x) \cap N_{C^i}(y)| = 2$. For this case considering $(k-1)(d-1) - 2$ paths

$\{(bw', iw', iy' = y)|iw' \in N_{C^i}(y), iw' \notin F_i\}$

and two paths

$\{(bw', aw', ay', iy' = y)|iw' \in N_{C^i}(y), iw' \in F_i\},$

which are node-disjoint, we can obtain a fault-free path with length $\delta(k-1) + 3$ or 5. Since $\delta(k-1) \geq 2(k \geq 3)$, it holds that $dis_H(x, y) \leq 5 \leq \delta(k-1) + 3$.

(Case 2: $dis_{C(d,k)} = k$) Let $x = ix'$ and $y = jy'(i \neq j)$.

(Subcase 2-1: $|F_i| \geq (k-1)(d-1)$ and $|F_j| \geq (k-1)(d-1))$

Since $|F^+| - (|F_i| + |F_j|) \leq d - 3$, there is a subcube C^a with no faults. In this case we can get $dis_H(x, y) \leq \delta(k-1) + 2 \leq \delta(k-1) + 3$.

(Subcase 2-2: $|F_i| \leq (k-1)(d-1) - 1$ or $|F_j| \leq (k-1)(d-1) - 1)$

(Subcase 2-2-1: There is a subcube $C^a (a \neq i, j)$ such that $|F_a| \leq 1)$

We can assume that $|F_i| \leq (k-1)(d-1) - 1$.

For the cases that
$$iy' \notin F_i,$$
$$ay' \notin F_a \text{ or}$$
$$N_{C^j}(jy') \nsubseteq F_j,$$
it is easily shown that
$$dis_H(x,y) \leq \delta(k-1) + 1,$$
$$dis_H(x,y) \leq \delta(k-1) + 3 \text{ or}$$
$$dis_H(x,y) \leq \delta(k-1) + 3,$$
respectively. We consider the case that $iy' \in F_i$, $ay' \in F_a$ and $N_{C^j}(jy') \subseteq F_j$. Let f_r be the number of faults in the $d-3$ subcubes $C^p(p \neq i, j, a) - \{py'|(p \neq 0, 1, a)$ and let ℓ be the number of non-faulty nodes $py'(p \neq i, j, a)$ in the $d-3$ subcubes C^p. Since y is not isolated it holds that $1 \leq \ell \leq d-3$. Let f_x be the number of faulty nodes in $N_{C^i}(x)$, where $0 \leq f_x \leq (k-1)(d-1) - 2$ because iy' is faulty. Since
$$f_r \leq (2k(d-1) - d - 1) - (k-1)(d-1) - 2 - (d-3-\ell) - f_x$$
$$\leq (k-1)(d-1) + (\ell-2) - f_x,$$
there is a subcube C^b such that by' is non-faulty and $|F_b| \leq \frac{1}{\ell} f_r$. Then it holds that $|F_b| \leq (k-1)(d-1) - 1$. By considering the set of $(k-1)(d-1) - f_x$ paths $\{(x = ix', iu', bu')|iu' \in N_{C^i}(x) - F_i\}$, there exists a fault-free path $(x = ix', iu', bu')$ because it holds that $((k-1)(d-1) - f_x) - |F_b| \geq 1$. Since $|F_b| \leq (k-1)(d-1) - 1$, we have $dis_H(x,y) \leq \delta(k-1) + 3$.

(**Subcase 2-2-2**: For any subcube $C^a(a \neq i, j)$ $|F_a| \geq 2$)

We can assume that $|F_j| \leq (k-1)(d-1) - 1$. The number of faults in $C^i \cup C^j$ is at most
$$(2k(d-1) - d - 1) - \Sigma_{a \neq i,j}|F_a| \leq 2k(d-1) - 3d + 3.$$
For the case that there is a non-faulty node $u \in N_{C^i}(x)$, since
$$|N_{C^i}(x) \cup N_{C^i}(u)| = 2(k-1)(d-1) - (d-2) = 2k(d-1) - 3d + 4,$$
there exists a path (x, jx'), (x, iv', jv') or (x, u, iw', jw') that does not include any fault in F^+, where $iv' \in N_{C^i}(x)$ and $iw' \in N_{C^i}(u) - \{x\}$. Thus, we have
$$dis_H(x,y) \leq \delta(k-1) + 3.$$
For the case that there are no non-faulty nodes $u \in N_{C^i}(x)$, that is, $N_{C^i}(x) \subseteq F_i$, it can be shown similar to the proof of the latter case of Subcase 2-2-1.

Theorem 6 does not hold for $C(2,k)$. However all cases except the latter case of Subcase 1-1-2 can be proved under the conditions of Theorem 6 and if some conditions are added stated below the remaining case can be proved.

Theorem 7. *Let ρ_k be a minimal routing on $C(2,k) = (V_k, E_k)$ and be $(\delta(k), k-1)$-tolerant, where $\delta(k)$ is a monotone non-decreasing function whose range is non-negative integers, $\delta(0) = 0$ and $\delta(k) \geq 2(k \geq 2)$. Moreover, we assume that ρ_k is total or $\delta(k) \geq 3(k \geq 2)$. If F^+ is any set of node faults in $C(2,k)$ such that $|F^+| \leq 2k - 3$ and $C(2,k) - F^+$ is connected, then $diam(R(C(2,k), \rho_k)/F^+) \leq \delta(k-1) + 3$.*

The following corollary is directly obtained from Theorem 6 and 7.

Corollary 8. *Let $d \geq 2$.*

(1) *If ρ_k is an edge-routing on $C(d, k)$, ρ_k is $(k + 3, 2k(d - 1) - d - 1)$-tolerant.*

(2) *If ρ_k is a total and minimal routing on $C(d, k)$, ρ_k is $(6, 2k(d - 1) - d - 1)$-tolerant.*

(3) *There exists a total and minimal routing λ_k on $C(d, k)$ such that λ_k is $(5, 2k(d - 1) - d - 1)$-tolerant.*

In [9], it is shown that if ρ_k is an edge-routing on $C(2, k)$ then $(k + 2, 2k - 3)$-tolerant. We can extend the result to $C(d, k)(d \geq 3)$ using the proof of Theorem 6 and Lemma 2.

Theorem 9. *Let $d \geq 2$. If ρ_k is an edge-routing on $C(d, k)$, then $(k + 2, 2k(d - 1) - d - 1)$-tolerant.*

We can also improve the result of Corollary 8 (3).

Let λ_k be any $(2, (d - 1)k - 1)$-tolerant routing on $C(d, k)$. Define a routing π_k on $C(d, k)$ by using λ_{k-1} as follows:

$$
\pi_k(x, y) = \begin{cases} \lambda_{k-1}(x', y') & if \ (x = ax' \ and \ y = ay')(0 \leq a \leq d - 1) \\ \lambda_{k-1}(x', y')(ay', by') & if \ x = ax' \ and \ y = by'(a < b) \\ (bx', ax')\lambda_{k-1}(x', y') & if \ x = bx' \ and \ y = ay'(a < b) \end{cases}
$$

We can show that the routing π_k is $(4, 2k(d - 1) - d - 1)$-tolerant. The proof can be done by using the proof of Theorem 6.

Theorem 10. *Let $d \geq 2$. There exists a total and minimal routing π_k on $C(d, k)$ such that π_k is $(4, 2k(d - 1) - d - 1)$-tolerant.*

5 Concluding remarks

We have shown highly fault-tolerant routings on generalized (d, k)-cubes. Although faults considered in this paper are restricted to nodes, the results shown in this paper still hold for faults of nodes and edges. Interesting open problems are to evaluate the diameter of the surviving route graph of $C(d, k)$ for the case that the number of faults exceeds $2(d - 1)k - d - 1$, or to construct a $(3, 2(d - 1)k - d - 1)$-tolerant routing on $C(d, k)$.

References

1. A.Broder, D.Dolev, M.Fischer and B.Simons, "Efficient fault tolerant routing in network," *Information and Computation*75,1,pp.52–64(Oct. 1987).
2. D.Dolev, J.Halpern , B.Simons and H.Strong, "A new look at fault tolerant routing, " *Information and Computation*72,3,pp.180–196(Oct. 1987).
3. A.-H.Esfahanian, "Generalized measures of fault tolerance with application to n-cube networks," *IEEE Trans. on Computers*38, 11 ,pp.1585–1591(1989).
4. A.-H.Esfahanian and S.L.Hakimi, "On computing a conditional edge-connectivity of a graph," *Information Processing Letters*, 27, pp.195–199(1988).
5. P.Feldman, "Fault tolerance of minimal path routing in a network," *Proc. of ACM 17th STOC*,pp.327–334(May 1985).
6. F.Harary, Graph theory, *Addison-Wesley*, Reading, MA (1969).
7. F.Harary, "Conditional connectivity," *Networks*, 13, pp.346–357(1983).
8. K.Kawaguchi and K.Wada, "New Results in Graph Routing," *Information and Computation*106, 2,pp.203–233(1993).
9. S.Latifi, "Combinatorial analysis of the fault-diameter of the n-cube," *IEEE Trans. on Computers*42, 1 ,pp.27–33(1993).
10. D.Peleg and B.Simons, "On fault tolerant routing in general graph," *Information and Computation*74,1,pp.33–49(Jul. 1987).
11. K.Wada, K.Kawaguchi and H.Fujishima, "A New Measure of Fault-Tolerance for Interconnection Networks," *Proceeding of 1990 BILKENT International Conference on New Trends in Communication, Control and Signal Processing*, pp.111–118(1990).

Hot-Potato Routing on Multi-Dimensional Tori*

Friedhelm Meyer auf der Heide and Matthias Westermann

Department for Mathematics and Computer Science
and Heinz Nixdorf Institute
University of Paderborn
33098 Paderborn, Germany

Abstract. We consider the hot-potato routing problem. The striking feature of this form of packet routing is that there are no buffers at the nodes. Thus packets are always moving.

A probabilistic hot-potato routing protocol is presented that routes random functions on the (n, d)-torus. If at most $\frac{d}{88} n^d$ packets, evenly distributed among the processors, have to be routed, they all have reached their destinations in $dn + O\left(d^3 \log n\right)$ steps, with high probability, if $3 \le d = O\left(n^\epsilon\right)$ with $\epsilon \in \left(0, \frac{1}{2}\right)$. This improves upon previous results where similar time bounds are only obtained for constant d and n^d packets.

1 Introduction

Hot-potato routing or deflection routing is a variant of packet routing where the packets are always moving, i.e. they are treated as hot potatoes. In each step a processor must send out all packets it received at the beginning of the step. If two or more packets want to use the same link at the same time, only one packet is allowed to do so. The other packets are sent out over other free links, i.e. they may be forced to move farther away from their destinations.

The advantage of hot-potato routing is that packets are not stored between time steps, as in the store-and-forward routing model. Thus buffers are not required except for input and output buffers in each processor. Input buffers contain the packets before they start to move, output buffers contain them after they have reached their destinations. Packets in transit are never stored in buffers. This keeps the hardware cheap and routing cycles very fast. Because of these reasons hot-potato routing is especially useful in optical networks [AS92, HK90, S90], where buffering involves the packets to be stored in electronic media.

1.1 The Routing Model

A *processor network* is represented by a directed graph whose nodes are processors and whose edges are the unidirectional communication links.

---

* This work is partially supported by the DFG-Forschergruppe "Effiziente Nutzung massiv paralleler Systeme" and the DFG Leibnitz Grant

Definition 1 (n, d)**-torus.** Let $n \geq 1$ and $d \geq 1$. The d-dimensional torus with egde length n, the (n, d)-torus, is a directed graph with node set $[n]^d$ ($[n] := \{0, \ldots, n-1\}$). For each dimension $i \in [d]$, each node $(v_0, \ldots, v_{d-1})$ has one edge to node $(v_0, \ldots, (v_i - 1) \bmod n, \ldots, v_{d-1})$ and one edge to node $(v_0, \ldots, (v_i + 1) \bmod n, \ldots, v_{d-1})$. Thus the (n, d)-torus has n^d nodes and $2dn^d$ edges.

Each node has one *input* and one *output buffer*. Initially the packets are stored in the input buffers. If a packet reaches its destination, it is stored in the output buffer. Packets in transit are never stored in buffers.

Routing is performed in discrete, synchronous time steps. At the beginning of each time step, a node may receive at most one packet along each incoming edge. At the end of the time step, the node sends out each incoming packet along one of the outgoing edges, at most one packet per edge. This is always possible, since in the (n, d)-torus the in-degree of each node is equal to its out-degree, so that always each incoming packet can find a free outgoing edge.

We route functions from $F(n^d, p)$.

- If $p < 1$, routing $f \in F(n^d, p) := \{f : [\lfloor pn^d \rfloor] \longrightarrow [n^d]\}$ means that initially $\lfloor pn^d \rfloor$ packets are randomly distributed in the network, at most one per node, the i-th of which has to be sent to node $f(i)$.
- If $p \geq 1$, routing $f \in F(n^d, p) := \{f : [n^d] \times [\lfloor p \rfloor] \longrightarrow [n^d]\}$ means that initially each node $\bar{v}$ has $\lfloor p \rfloor$ packets, the i-th of which has to be sent to node $f(\bar{v}, i)$.

As soon as a packet has left its starting node, it moves in each step until it reaches its destination.

For a hot-potato routing protocol it might occur that a set of packets is routed in such a way that they mutually deflect each other in an infinite loop from which they never recover. This situation is called *livelock*. Our routing protocol is constructed in a way that a livelock cannot appear.

1.2 Known Results about Hot-Potato Routing

In the following the term "*with high probability*" means "with probability of at least $1 - N^{-\alpha}$", where N is the number of the nodes and $\alpha \geq 1$ is a constant.

Experimental results on simulations of hot-potato routing on various networks are documented in [AS92, GG93, GH92, HK90, M89, S90]. In several of these papers a probabilistic analysis of simple protocols is presented, but various independence assumptions are made to make the analysis tractable.

Feige and Raghavan [FR92] present a simple deterministic hot-potato routing protocol that routes a random $f \in F(n^2, 1)$ on the $(n, 2)$-torus in $2n + O(\log n)$ steps, with probability at least $1 - o\left(\frac{1}{n}\right)$. Additionally they present a simple deterministic routing protocol that routes a random $f \in F(2^d, 1)$ on the $(2, d)$-torus (d-dimensional hypercube) in $O(d)$ steps, with high probability. They also show that these protocols are in the worst case asymptotically optimal.

Kaklamanis, Krizanc and Rao [KKR93] present a deterministic hot-potato routing protocol that routes a random $f \in F\left(n^d, 1\right)$ on the (n, d)-torus, d constant, in $\frac{dn}{2} + O(\log^2 n)$ steps, with high probability. The restrictions to constant d and n^d packets are inherent in their analysis. They also present a routing protocol that routes a random $f \in F\left(n^2, 1\right)$ on the $(n, 2)$-mesh in $2n + O(\log^2 n)$ steps, with high probability.

Hajek [H91] considers hot-potato routing protocols in which priority is given to packets closer to their destinations. He shows that routing k packets to arbitrary destinations can be done in $d + 2k$ steps on the $(2, d)$-torus (d-dimensional hypercube).

The routing protocols in [FR92, H91, KKR93] are free of livelocks.

1.3 New Result

In this paper a probabilistic hot-potato routing protocol is presented that routes a random $f \in F\left(n^d, \frac{\lfloor d/2 \rfloor}{44}\right)$ on the (n, d)-torus in $dn + O\left(d^3 \log n\right)$ steps, with high probability, if $3 \le d = O\left(n^\epsilon\right)$ with $\epsilon \in \left(0, \frac{1}{2}\right)$. The protocol is free of livelocks. It is the first asymptotically optimal hot-potato routing protocol on tori with non-constant dimensions.

The paper is organised as follows: Section 2 presents the new routing protocol. Section 3 contains its analysis.

2 The Routing Protocol

Let $n \ge 2$ and $d \ge 3$. Our protocol is based on the following partition of the (n, d)-torus. In the following is $(a \tilde{+} b) := ((a + b) \bmod d)$ and $(a \tilde{-} b) := ((a - b) \bmod d)$. For each $i \in [d]$ is defined:

- *i-plane*: If $i \ne d - 1$, for each $\bar{a} \in [n]^i$ and $\bar{b} \in [n]^{d-i-2}$ exists an i-plane with node set $\{\bar{a}\} \times [n]^2 \times \{\bar{b}\}$. Otherwise for each $\bar{a} \in [n]^{d-2}$ exists an i-plane with node set $[n] \times \{\bar{a}\} \times [n]$. An i-plane has all edges that connect its nodes in the (n, d)-torus.

  An i-plane is isomorphic to a $(n, 2)$-torus. It uses only edges of dimension i and $i \tilde{+} 1$ of the (n, d)-torus. For fixed i, the (n, d)-torus can be partitioned into n^{d-2} different i-planes, which are all node and edge disjoint. An i-plane and an i'-plane are edge disjoint, if $i' \notin \{i \tilde{-} 1, i, i \tilde{+} 1\}$.

- *i-routing cycle*: For each $\bar{a} \in [n]^i$ and $\bar{b} \in [n]^{d-i-1}$ exists an i-routing cycle with node set $\{\bar{a}\} \times [n] \times \{\bar{b}\}$. For each $x \in [n]$ it has one edge from node $(a_0, \ldots, a_{i-1}, x, b_0, \ldots, b_{d-i-2})$ to $(a_0, \ldots, a_{i-1}, (x+1) \bmod n, b_0, \ldots, b_{d-i-2})$, if i is even and $\sum_{j=0}^{i-1} a_j + \sum_{j=0}^{d-i-2} b_j$ is odd, or if i is odd and $\sum_{j=0}^{i-1} a_j + \sum_{j=0}^{d-i-2} b_j$ is even. Otherwise it has one edge from node $(a_0, \ldots, a_{i-1}, x, b_0, \ldots, b_{d-i-2})$ to $(a_0, \ldots, a_{i-1}, (x - 1) \bmod n, b_0, \ldots, b_{d-i-2})$.

  An i-routing cycle is isomorphic to a $(n, 1)$-torus, in which the cycle of edges is unidirectional. It uses only edges of dimension i of the (n, d)-torus. For fixed i, the (n, d)-torus can be partitioned into n^{d-1} different i-routing cycles, which are all node and edge disjoint. All routing cycles are edge disjoint.

— *i-block* and *i-parking orbit*: Let $c > 0$ be an integer that only depends on d. In each i-plane the nodes are divided along the dimension i into strips of length $4c$ and along the dimension $i\tilde{+}1$ into strips of length 6. For that purpose $4c$ and 6 must be divisors of n. Each i-plane is now divided into blocks of size $4c \times 6$. Such a block is called *i-block*.

The routing cycles occupy only half of the outgoing edges of each node of the (n, d)-torus. With the other half of the edges a directed cycle of edges is formed in each i-block, as in Fig. 1. Such a directed cycle of edges is called *i-parking orbit*.

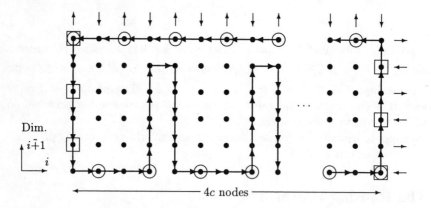

Fig. 1. *i*-parking orbit with entrances (box) und exits (cycle) in one *i*-block and directions of the *i*- and $(i\tilde{+}1)$-routing cycles

The *entrances* of an i-block are those nodes marked with a box in Fig. 1. The *exits* of an i-block are those nodes marked with a cycle in Fig. 1. Note that an i-block B has an entrance on each i-routing cycle crossing B and an exit on each $(i\tilde{+}1)$-routing cycle crossing B.

In [W95] these parking orbits are formally defined. They only use roughly one quarter of the edges of their i-blocks. This makes sure that, although an i-block can intersect with an $(i\tilde{+}1)$-block, the corresponding parking orbits are edge disjoint. Alltogether we obtain the following lemma.

Lemma 2. *All routing cycles and parking orbits are edge disjoint.* □

Let $V_0 := \left\{ \bar{v} \in [n]^d, \sum_{i=0}^{d-1} v_i \text{ is even} \right\}$ and $V_1 := \left\{ \bar{v} \in [n]^d, \sum_{i=0}^{d-1} v_i \text{ is odd} \right\}$. V_0 and V_1 partition the nodes of the (n, d)-torus into two equal sized sets.

Lemma 3. *Let P and P' be packets that start at the same time in a node from V_0 and V_1, resp.. P and P' never meet in a node during a hot-potato routing protocol.* □

If $d < 88$, $\left\lfloor \frac{\lfloor d/2 \rfloor}{44} n^d \right\rfloor$ packets are randomly distributed in the (n, d)-torus, at most one packet starts per node. If $d \geq 88$, $\left\lfloor \frac{\lfloor d/2 \rfloor}{44} \right\rfloor$ packets start in each node.

Now the routing protocol will be described from the point of view of a packet P that wants to travel from the source node $\bar{a}$ to the destination node $\bar{b}$. P will be enhanced by routing information, stored in two integers r and s.

Our routing protocol will try to send P from $\bar{a}$ to $\bar{b}$ along its *directed greedy path with starting dimension j*: This path in turn uses a part of a j-routing cycle, $(j\tilde{+}1)$-routing cycle ,..., $(j\tilde{+}(d-1))$-routing cycle. It bends from its i-routing cycle to its $(i\tilde{+}1)$-routing cycle in its *i-bending node* $(a_0, \ldots, a_{j-1}, b_j, \ldots, b_i, a_{i+1}, \ldots, a_{d-1})$, if $i \geq j$, and $(b_0, \ldots, b_i, a_{i+1}, \ldots, a_{j-1}, b_j, \ldots, b_{d-1})$ otherwise.

Bending at an i-bending node may cause a problem for P, because the first edge to be used on its $(i\tilde{+}1)$-routing cycle may be occupied by another packet. In order to ensure that a packet can finally bend, it will cycle along its i-parking orbit in its i-block, i.e. the i-parking orbit in the i-block where its i-bending node belongs to.

The main problem to be solved by our algorithm is to make sure that all packets can enter and exit their parking orbits fast, with high probability.

Let $c' > 0$ be an integer that only depends on d. We shall demand that P, in order to be successful at its i-parking orbit O, has to enter O during its first visit of its i-block and has to leave O within $c'\lceil \log n \rceil$ attempts.

Let $t := n + 4c + 16cc'\lceil \log n \rceil + 5 = n + O\left(d^2 \log n\right)$. Roughly speaking, t is an upper bound on the time needed by a successful packet for traveling on its i-routing cycle and for bending to its $(i\tilde{+}1)$-routing cycle.

Routing Protocol (for P, the k-th Paket of $\bar{a} \in V_l$)

1. Initially P is marked *successful*. If $d < 88$, set $j := 2r' + l$, with a randomly chosen $r' \in \left[\left\lfloor \frac{d}{2} \right\rfloor\right]$. Otherwise set $j := 88(k-1) + 2r' + l$, with a randomly chosen $r' \in [44]$. P now tries to use its directed greedy path with starting dimension j.

2. For $i := j, j\tilde{+}1, \ldots, j\tilde{+}(d-2)$, P runs through the following loop body in which it moves in an i-plane:
   - Set $s := 0$. P moves along its i-routing cycle until it has reached the entrance of its i-block B (compare Fig. 1). After each step, set $s := s+1$.
   - Routing in the i-block B (description see below). After this P is at the exit of B on its $(i\tilde{+}1)$-routing cycle (compare Fig. 1) and the edge to be used by P is free.
   - P makes $n - s$ steps along its $(i\tilde{+}1)$-routing cycle.

3. P moves along its $(j\tilde{+}(d-1))$-routing cycle, until it reaches $\bar{b}$. Here it is stored in the output buffer.

Observation:

a. All packets that have a source node from V_0 (V_1) have an even (odd) starting dimension.

Let P be a packet that moves in Fig. 2 from the entrance of its i-block B that is marked with a cross along the marked i-routing cycle through B, where it wants to use its i-parking orbit O. P *makes an attempt to enter O* at the nodes marked with a box in Fig. 2. P *has a possibility to enter O*, if it makes an attempt to enter O and the edge to be used by it is free. P *makes an attempt to leave O* at the node marked with a dashed box in Fig. 2 on its $(i\tilde{+}1)$-routing cycle. P *has a possibility to leave O*, if it makes an attempt to leave O and the edge to be used by it is free. If the $(i\tilde{+}1)$-routing cycle of P is the one marked in Fig. 2, P uses the exit of B marked with a cycle in Fig. 2.

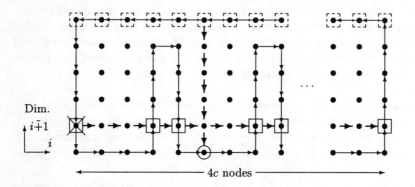

Fig. 2. i-parking orbit O in the i-block B and parts of an i- and $(i\tilde{+}1)$-routing cycle

Routing in the i-Block B (in the $(i+1)\tilde{-}j =: m$-th Pass of (2))

4. P randomly chooses $r \in [c]$. It moves along its i-routing cycle through its i-block B, until it has the $(r+1)$-th possibility to enter its i-parking orbit O. If it is not able to enter O during this first visit of B, it is marked *unsuccessful* (routing of unsuccessful packets see below).

5. P moves along O, without trying to leave O, until the end of step $(m-1)t + s + 4c$ is reached. (We shall see later that a successful P is at the end of its $(m-1)$-th pass of (2) before step $(m-1)t + 1$, so that it reaches (5) before step $(m-1)t + s + 4c + 1$.)

6. P moves along O, until it has the possibility to leave O. If it has no possibility to leave O before step $(m-1)t + s + 4c + 16cc'\lceil \log n \rceil + 1$, it is marked *unsuccessful*.

7. P moves to the exit of B on its $(i\tilde{+}1)$-routing cycle. I.e. either it is already there, or it makes 5 steps along its $(i\tilde{+}1)$-routing cycle (compare Fig. 2).

Observation:

b. Looking at Fig. 2, it can be seen that O consists of $16c$ edges. Thus P needs $16c$ steps to move once along O. Thus it can make $c'\lceil \log n \rceil$ attempts to leave O in the $16cc'\lceil \log n \rceil$ steps it is allowed to make in (6).

Routing of Unsuccessful Packets

8. Up to the end of step dt the unsuccessful packets are routed along arbitrary free edges. (We shall see later that, at this time, all successful packets are delivered.)
9. (Initially an Eulerian cycle E of the (n, d)-torus is fixed.) In step $dt + 1$ each unsuccessful packet chooses one free edge and then it is routed along E to its destination node. Here it is stored in the output buffer.

Observations:

c. An unsuccessful packet never disturbs any successful packet.
d. An unsuccessful packet makes in (9) at most $2dn^d$ steps before it reaches its destination, since the (n, d)-torus has $2dn^d$ edges.

3   Analysis of the Routing Protocol

Theorem 4. *Let $n \geq 2$, $d \geq 3$ and $\alpha \geq 1$, so that an $\epsilon \in \left(0, \frac{1}{2}\right)$ and an integer $\frac{2+\alpha}{1-\epsilon}d \leq c = O(d)$ exists, with $4c$ and 6 are divisors of n and $d \leq \frac{n^\epsilon}{38c(2+\alpha)\lceil \log n \rceil}$. The above protocol routes a random $f \in F\left(n^d, \frac{\lfloor d/2 \rfloor}{44}\right)$ in $dn + O\left(d^3 \log n\right)$ steps, with probability at least $1 - \left(n^d\right)^{-\alpha}$. In the worst case it needs $dn + O\left(d^3 \log n\right) + 2dn^d$ steps. In particular it is free of livelocks.*

Proof. We start with two lemmata exploring timing properties of our protocol.

Lemma 5. *For each successful packet the following is valid:*

1. *Its m-th pass of (2) is finished during the time interval $[(m-1)t + n, mt]$. In particular the packet is delivered after at most dt steps.*
2. *The packet spends at most $2(t - n)$ steps in its i-block, for each $i \in [d]$.*   □

Each packet reaches its destination in at most $dt + 2dn^d = dn + O\left(d^3 \log n\right) + 2dn^d$ steps, by the above lemma and observation (d). Thus the routing protocol is free of livelocks.

Lemma 6. *Successful packets with different starting dimensions never collide.*

Proof. Let P and P' successful packets with starting dimesion j and $j' \neq j$, resp.. Let $m \in \{1, \ldots, d-1\}$.
   By lemma 2, P can only collide in its m-th pass of (2) with P' in the following two cases:

- In a part of the time interval, in that P can enter its $\left(j \tilde{+} (m-1)\right)$-parking orbit O, P' moves along O.
- In a part of the time interval, in that P can exit O, P' moves along the $(j \tilde{+} m)$-routing cycle R of P.

First we investigate the time intervals in which P can enter or leave O:

- P can only enter O during the time interval $[\max(\{0, (m-2)t+n+1\}), mt]$, since it makes its m-th pass of (2) in this time interval, by lemma 5(1).
- P can only leave O during the time interval $[(m-1)t+1, mt]$, since it executes (6) in its m-th pass of (2) in this time intervall, by (5) and lemma 5(1).

Let $m' := (j \bar{+} m) \tilde{-} j'$.

- If $m' \geq 1$, P' can use O and R in its m'-th pass of (2).
- If $m' + 1 \leq d - 1$, P' can use R in its $(m' + 1)$-th pass of (2).
- If $m' + 1 = d$, P' can use R in (3).

Now we investigate the time intervals in which P' can use O or R:

- P' can only use O in the time interval $[\max(\{0, (m'-2)t+n+1\}), m't]$, since it makes its m'-th pass of (2) in this time interval (if $m' \geq 1$), by lemma 5(1).
- P' can only use R in the time interval $[\max(\{0, (m'-1)t+1\}), (m'+1)t]$, since it starts with (6) in its m'-th pass of (2) (if $m' \geq 1$), and finishes its $(m'+1)$-th pass of (2) (if $m'+1 \leq d-1$), or executes (3) (if $m'+1 = d$) in this time interval, by (5) and lemma 5(1).

If $j' \notin \{j \tilde{-} 1, j, j \bar{+} 1\}$, then $m' \notin \{m \tilde{-} 1, m, m \bar{+} 1\}$ and with the above observations can be concluded:

- If $1 \leq m' \leq m - 2$, P' can not use O after the step $m't \leq (m-2)t$ and R after the step $(m'+1)t \leq (m-1)t$. P can not enter O before the step $(m-2)t+n+1$ und leave O before the step $(m-1)t+1$.
- If $m+2 \leq m' \leq d-1$, P' can not use O before the step $(m'-2)t+n+1 \geq mt+n+1$ and R before the step $(m'-1)t+1 \geq (m+1)t+1$. P can not enter or leave O after the step mt.

Thus P and P' can never collide.

If $j' \in \{j \tilde{-} 1, j \bar{+} 1\}$, the source nodes from P and P' are from V_0 and V_1, resp., or vice versa, by observation (a). Then P and P' can never collide, by lemma 3. □

The above lemma implies that it suffies to analyse the protocol under the asumption that only packets with fixed starting dimension j travel. Without lost of generality, let $j := 0$. The rest of the proof is done under this asumption.

Lemma 7. *Fix $m \in [d-1]$ and $\bar{v} \in V_l$. There are at most n^m possibilities to choose a source node of a successful packet that is in $\bar{v}$ at the end of its m-th pass of (2). The source node is in V_l.*

Proof. The following algorithm describes the only way to choose a source node $\bar{v}'$ of a successful packet P, so that P is in $\bar{v}$ at the end of its m-th pass of (2).

- Set $\bar{v}' := \bar{v}$.

- For $i := m - 1, m - 2, \ldots, 0$ do the following:
  - Choose one i-block B through which the $(i+1)$-routing cycle R containing $\bar{v}'$ leads. Set $s :=$ number of steps P must make along R to get from the exit of B through which R leads (compare Fig. 1) to $\bar{v}'$.
  - Choose one entrance E of B (compare Fig. 1). Set $\bar{v}' :=$ the node from which P must make $n - s$ steps along an i-routing cycle to get to E.

Thus there are at most $\left(\frac{n}{6}6\right)^m$ possibilities to choose $\bar{v}'$.

$\bar{v}' \in V_l$ holds after each pass of the loop in the above algorithm, since all entrances and exits of an i-block are either from V_0 or from V_1 (compare Fig. 1), and since n must be divisible by 6 and hence is even. $\qquad \square$

Lemma 8. *A packet starts in a fixed node with probability at most $\frac{1}{44}$.* $\qquad \square$

Now we are ready to analyse our protocol, using the asumption that the routing function is randomly chosen from $F\left(n^d, \frac{\lfloor d/2 \rfloor}{44}\right)$.

Let $c' := (2 + \alpha)d = O(d)$. Recall that $\frac{2+\alpha}{1-\epsilon}d \leq c = O(d)$ and $t = n + 4c + 16cc'\lceil \log n \rceil + 5 = n + O\left(d^2 \log n\right)$. $4c$ and 6 are divisors of n, so that the partition of the (n, d)-torus can be done.

Now we have to bound the probability that a packet P becomes unsuccessful. Our routing protocol implies that P becomes unsuccessful in its m-th pass of (2), only if

- P fails to enter its $(m - 1)$-parking orbit in (4), or
- P fails to leave its $(m - 1)$-parking orbit in (6).

Lemma 9. *For each packet P and $m \in \{1, \ldots, d - 1\}$:*

$$Prob(P \text{ fails to enter its } (m-1)\text{-parking orbit in } (4)) \leq \left(n^d\right)^{-(2+\alpha)}.$$

Lemma 10. *For each packet P and $m \in \{1, \ldots, d - 1\}$:*

$$Prob(P \text{ fails to leave its } (m-1)\text{-parking orbit in } (6)) \leq \left(n^d\right)^{-(2+\alpha)}.$$

Using these lemmata it is easy to conclude the theorem. There are at most dn^d possibilities to choose a packet and there are $d - 1$ possibilities to choose m. Thus it can be concluded that there is an unsuccessful packet with probability at most

$$dn^d \cdot d \cdot 2 \cdot \left(n^d\right)^{-(2+\alpha)} = \frac{2d^2}{n^d}\left(n^d\right)^{-\alpha} \leq \left(n^d\right)^{-\alpha},$$

since $3 \leq d \leq \frac{n^\epsilon}{38c(2+\alpha)\lceil \log n \rceil} \leq n$ and $n \geq 2$.

The expected routing time is

$$dt + \left(n^d\right)^{-\alpha} \cdot 2dn^d \leq dt + 2d = dn + O\left(d^3 \log n\right),$$

by lemma 5(1) and observation (d) and since $\alpha \geq 1$.

Thus, with probability at least $1 - \left(n^d\right)^{-\alpha}$, all packets are successful. I.e., all packets reach their destinations in at most $dt = dn + O\left(d^3 \log n\right)$ steps, with probability at least $1 - \left(n^d\right)^{-\alpha}$. $\qquad \square$

Proof of lemma 9. Let P be a packet that does not manage in its m-th pass of (2) to enter its $(m-1)$-parking orbit O in (4).

Let T be the step in which P exits its $(m-1)$-block B, i.e. in which it is marked unsuccessful. Thus it was in B during the time interval $[T-4c+1,T]$ (compare Fig. 2). Let S_1 be the set of packets that are in O during a part of this time interval. As P has skipped at most $r \leq c-1$ many possibilities to enter O, at least $c+1$ possibilities have to have been blocked by packets from S_1 (compare Fig. 2). In Fig. 2 it can be seen that one packet from S_1 can only block one attempt of P to enter O. Thus $|S_1| \geq c$. Therefore the following proposition implies lemma 9.

Proposition 11. $Prob(|S_1| \geq c) \leq \left(n^d\right)^{-(2+\alpha)}.$

Proof. First the number of source nodes a packet $P' \in S_1$ can have is analysed.

- Let E be the entrance of B that lies on the $(m-1)$-routing cycle R' of P'. There are 6 possibilities to choose E.
  P' visits E within the time interval $[(T-4c+1)-(2(t-n)-1) =: T', T]$, because it is in B during a part of the time interval $[T-4c+1,T]$ and has left B after at most $2(t-n)$ steps, by lemma 5(2).
  - • If $m-1 \geq 1$, let $\bar{v}'$ be the node where P' is at the end of its $(m-1)$-th pass of (2). P' is in $\bar{v}'$ during the time interval $[(m-2)t+n,(m-1)t]$, by lemma 5(1).
  - • If $m-1 = 0$, let $\bar{v}'$ be the node where P' starts. P' is in $\bar{v}'$ in the step 0. Thus P' visits $\bar{v}'$ during the time interval $[\max(\{0,(m-2)t+n\}),(m-1)t]$. Then it can be concluded that P' makes at least $\max(\{0,T'-(m-1)t\})$ and at most $T-\max(\{0,(m-2)t+n\})$ steps along R' to get from $\bar{v}'$ to E, since it visits $\bar{v}'$ during the time interval $[\max(\{0,(m-2)t+n\}),(m-1)t]$ and E during the time interval $[T',T]$.
  Thus there are at most

$$(T-((m-2)t+n))-(T'-(m-1)t)+1 = T-T'+t-n+1$$
$$< 4c+3(4c+16cc'\lceil\log n\rceil+5)$$
$$< 48cc'\lceil\log n\rceil+18c$$
$$< 50cc'\lceil\log n\rceil$$

possibilities to choose $\bar{v}'$ on R', since $t = n+4c+16cc'\lceil\log n\rceil+5$, $c \geq \frac{2+\alpha}{1-\epsilon}d \geq 9$ and $c' = (2+\alpha)d \geq 9$.
- Let $l \in [2]$, so that $\bar{v}' \in V_l$. If $m-1 \geq 1$, the source nodes of successful packets being in $\bar{v}'$ at the end of their $(m-1)$-th pass of (2) are from V_l, by lemma 7. If $m-1 = 0$, $\bar{v}' \in V_l$ is the source node. Thus only one half of the possible positions of $\bar{v}'$ have to be considered, since only packets with source nodes from V_0 are in S_1, by observation (a).
- If $m-1 \geq 1$, there are at most n^{m-1} possibilities to choose a source node of a successful packet that is in $\bar{v}'$ at the end of its $(m-1)$-th pass of (2), by lemma 7. If $m-1 = 0$, $\bar{v}'$ is the source node.

Thus there are at most $\frac{6}{2}50cc'\lceil\log n\rceil n^{m-1}$ possibilities to choose a source node of a packet, so that it can be in S_1.

A packet starts in a fixed node with probability at most $\frac{1}{44}$, by lemma 8. A packet uses O with probability $\frac{4c}{n^m}$, since therefore its $(m-1)$-bending node must lie in B.

Thus it can be concluded

$$Prob(|S_1| \geq c) \leq \left(\frac{150cc'\lceil\log n\rceil n^{m-1}}{c}\right)\left(\frac{4c}{44n^m}\right)^c$$

$$\leq \left(\frac{600ecc'\lceil\log n\rceil}{44n}\right)^c$$

$$\leq \left(\frac{n^\epsilon}{n}\right)^{\frac{2+\alpha}{1-\epsilon}d} = \left(n^d\right)^{-(2+\alpha)},$$

since $\binom{a}{b} \leq \left(\frac{ea}{b}\right)^b$, $\frac{600e}{44} \leq 38$, $c' = (2+\alpha)d$, $d \leq \frac{n^\epsilon}{38c(2+\alpha)\lceil\log n\rceil}$ and $c \geq \frac{2+\alpha}{1-\epsilon}d$. $\qquad\square$

Proof of lemma 10. Let P be a packet that does not manage in its m-th pass of (2) to leave its $(m-1)$-parking orbit O in (6), i.e. it fails to leave O within $c'\lceil\log n\rceil$ attempts.

P needs $t' := (c'\lceil\log n\rceil - 1) \cdot 16c + 1$ steps from the first to the $(c'\lceil\log n\rceil)$-th attempt to leave O. It holds that $t' < 16cc'\lceil\log n\rceil < n^\epsilon < n$, since $d \leq \frac{n^\epsilon}{38c(2+\alpha)\lceil\log n\rceil}$ and $c' = (2+\alpha)d$.

Let S_2 be the set of packets that prevent P from leaving O in (6). Each packet from S_2 can prevent P only once from leaving O, because a packet needs $n > t'$ steps to cycle once around a routing cycle. Thus $|S_2| \geq c'\lceil\log n\rceil$. Therefore the following proposition implies lemma 10.

Proposition 12. $Prob(|S_2| \geq c'\lceil\log n\rceil) \leq \left(n^d\right)^{-(2+\alpha)}$.

Proof. Let $\bar{v}_P$ be the node where P wants to leave O, i.e. the node that is marked in Fig. 2 with a dashed box and that lies on the m-routing cycle R of P. Let T be the step in which P makes its $(c'\lceil\log n\rceil)$-th attempt to leave O. Thus all packets from S_2 visit $\bar{v}_P$ during the time interval $[T - t' + 1 =: T', T]$.

Now the number of source nodes a packet $P' \in S_2$ can have is analysed.

- Let $\bar{v}'$ be the node where P' is at the end of its m-th pass of (2). P' is in $\bar{v}'$ during the time interval $[(m-1)t + n, mt]$, by lemma 5(1).
  - If P' is at the end of its m-th pass of (2) before it collides in $\bar{v}_P$ with P, P' makes at least $\max(\{0, T' - mt\})$ and at most $T - ((m-1)t + n)$ steps along R to get from $\bar{v}'$ to $\bar{v}_P$.
  - Otherwise P' makes at least $\max(\{0, ((m-1)t + n) - T\})$ and at most $mt - T'$ steps along R to get from $\bar{v}_P$ to $\bar{v}'$.

Thus there are at most

$$T - T' + t - n + 1 = t' + t - n$$

$$= (16cc'\lceil\log n\rceil - 16c + 1) + (4c + 16cc'\lceil\log n\rceil + 5)$$

$$< 32cc'\lceil\log n\rceil$$

possibilities to choose $\bar{v}'$ on R, since $t = n + 4c + 16cc'\lceil \log n \rceil + 5$ and $c \geq \frac{2+\alpha}{1-\epsilon}d \geq 9$.

- Let $l \in [2]$, so that $\bar{v}' \in V_l$. The source nodes of successful packets being in $\bar{v}'$ at the end of their m-th pass of (2) are from V_l, by lemma 7. Thus only one half of the possible positions of $\bar{v}'$ have to be considered, since only packets with source nodes from V_0 are in S_2, by observation (a).

- There are at most n^m possibilities to choose a source node of a successful packet that is in $\bar{v}'$ at the end of its m-th pass of (2), by lemma 7.

Thus there are at most $\frac{32}{2}cc'\lceil \log n \rceil n^m$ possibilities to choose a source node of a packet, so that it can be in S_2.

A packet starts in a fixed node with probability at most $\frac{1}{44}$, by lemma 8. A packet moves along R with probability $\frac{1}{n^m}$, since therefore its $(m-1)$-bending node must lie on R.

Claim. *Let P' be a successful packet that is in $\bar{v}_P$ during the time interval $[T', T]$. Then $Prob(P' \in S_2) \leq \frac{1}{c}$.*

Proof. Let B' be the $(m-1)$-block, O' be the $(m-1)$-parking orbit of P'. Let $\bar{v}_{P'}$ be the node where P' wants to leave O', i.e. the node on R that is marked with a dashed box in Fig. 2.

P tries to leave O in the steps $T, T - 16c, \ldots, T - (c'\lfloor \log n \rfloor - 1) \cdot 16c$. Let u be the number of steps P' must make along R to get from $\bar{v}_{P'}$ to $\bar{v}_P$. Then P' must leave O' in a step $T - u - z \cdot 16c$, with $z \geq 0$, so that it can colide with P in $\bar{v}_P$. Thus P' must be in $\bar{v}_{P'}$ for the first time in a step $T - u - z \cdot 16c$, with $z \geq 0$, in order to be in S_2.

Let E' be the entrance of B' that lies on the $(m-1)$-routing cycle R' of P'. Let h_x be the number of steps P' needs to get from E' to $\bar{v}_{P'}$ for the first time, if P' enters O' in the $(x+1)$-th attempt, with $x \in [g]$. ($g \in \{2c, 4c\}$, depending where R' leads through B'.) In Fig. 2 it can be seen, that $|\{h_0, \ldots, h_{g-1}\}| = g$ and $\max(\{h_0, \ldots, h_{g-1}\}) - \min(\{h_0, \ldots, h_{g-1}\}) < 16c$.

That means that P' is permitted to enter O' only in one determined attempt, so that it can be in $\bar{v}_{P'}$ for the first time in a step $T - u - z \cdot 16c$, with $z \geq 0$, if this is possible at all. Then $Prob(P' \in S_2) \leq \frac{1}{c}$, since P' uses the $(r+1)$-th possibility to enter O' and $r \in [c]$ is randomly chosen. $\square$

Now we can conclude proposition 12:

$$Prob(|S_2| \geq c'\lceil \log n \rceil) \leq \binom{16cc'\lceil \log n \rceil n^m}{c'\lceil \log n \rceil} \left(\frac{1}{44n^m c}\right)^{c'\lceil \log n \rceil}$$

$$\leq \left(\frac{16e}{44}\right)^{(2+\alpha)d\lceil \log_{44/16e} n \rceil} = \left(n^d\right)^{-(2+\alpha)},$$

since $\binom{a}{b} \leq \left(\frac{ea}{b}\right)^b$ and $c' = (2 + \alpha)d$. $\square$

References

[AS92]   A. Acampora and S. Shah: Multihop lightwave networks: a comparison of store-and-forward and hot-potato routing. IEEE Transactions on Communications 40 (1992) 1082–1090

[FR92]   U. Feige and P. Raghavan: Exact analysis of hot-potato routing. Proceedings of the 33rd Symposium on Foundation of Computer Science (1992) 553–562

[GG93]   J. Goodman and A. Greenberg: Sharp approximate models of deflection routing in mesh networks. IEEE Transactions on Communications 41 (1993) 210–223

[GH92]   A. Greenberg and B. Hajek: Deflection routing in hypercube networks. IEEE Transaction on Communications 40 (1992) 1070–1081

[H91]   B. Hajek: Bounds on evacuation time for deflection routing. Distributed Computing 5 (1991) 1–6

[HK90]   B. Hajek and A. Krishna: Performance of shuffle-like switching networks with deflections. Proceedings of the IEEE INFOCOM (1990) 473–480

[KKR93]   C. Kaklamanis, D. Krizanc and S. Rao: Hot-potato routing on processor arrays. Proceedings of the 5th Symposium on Parallel Algorithms and Architectures (1993) 273–282

[M89]   N. Maxemchuk: Comparison of deflection and store-and-forward techniques in the manhattan street and shuffle-exchange networks. Proceedings of the IEEE INFOCOM (1989) 800–809

[S90]   T. Szymanski: An analysis of "hot-potato" routing in a fiber optic packet switched hypercube. Proceedings of the IEEE INFOCOM (1990) 918–925

[W95]   M. Westermann: Hot-potato routing. Diplomarbeit, University of Paderborn (1995)

On Devising Boolean Routing Schemes *

Michele Flammini[1,3], Giorgio Gambosi[2], Sandro Salomone[3]

[1] Dipartimento di Informatica e Sistemistica, University of Rome "La Sapienza", via Salaria 113, I-00184 Rome, Italy. E-mail: flammini@smaq20.univaq.it
[2] Dipartimento di Matematica, University of Rome "Tor Vergata", viale della Ricerca Scientifica, I-00133 Rome, Italy. E-mail: gambosi@mat.utovrm.it
[3] Dipartimento di Matematica Pura ed Applicata, University of L'Aquila, via Vetoio loc. Coppito, I-67010 l'Aquila, Italy. E-mail: salomone@smaq20.univaq.it

Abstract. In this paper, the problem of routing messages along shortest paths in a network of processors without using complete routing tables is considered.

The Boolean Routing model is considered and it is shown that it provides optimal representations of all shortest routes on some specific network topologies (such as paths, rings, trees, hypercubes, different types of d-dimensional grids, complete graphs and complete bipartite graphs). Moreover, it is also shown that the model deals efficiently with graphs resulting from applying some types of graph compositions, thus resulting in very efficient routing schemes for some classes of networks with regular topology. This is done by considering different significant cost measures of the space efficiency of the schemes considered.

Keywords: Parallel and distributed systems, compact routing tables, shortest paths representation.

1 Introduction

Routing messages between pairs of processors is a fundamental task in parallel and distributed computing systems. In such a framework, a network of processors is modeled by an (undirected) connected graph of n nodes each representing a processing element, where edges model communication links between processors. Assuming a cost function on the network edges, it is important to route each message along a shortest path from its source to the destination.

This can be trivially accomplished by referring at each node v_i to a complete routing table (with $\Theta(n)$ δ_i-bits entries, where δ_i is the node degree) that stores, for each destination v_j, the set of links incident to v_i which are on shortest paths between v_i and v_j.

Unfortunately, in the general case, such tables are too space consuming for large interconnection networks, requiring an overall $\Theta(n^3)$ bits of information.

* Work supported by the EEC ESPRIT II Basic Research Action Program under contract No.7141 "Algorithms and Complexity II", by the EEC "Human Capital and Mobility" MAP project and by the Italian MURST 40% project "Algoritmi, Modelli di Calcolo e Strutture Informative".

This introduces the need of devising routing schemes with smaller tables, accepting the possibility of broadcasting messages along paths which are not of minimal length.

Research activities focused on identifying classes of network topologies such that the shortest path information at each node can be stored succinctly, assuming that suitable "short" labels can be assigned to nodes and links at preprocessing time. Such labels are used to encode useful information about the network structure, with special regard to shortest paths.

In the ILS (Interval Labeling Scheme) routing scheme ([2], [3], [4]) node labels belong to the set $\{1, \ldots, n\}$, assumed cyclically ordered. Link labels are pairs of node labels representing disjoint intervals on $\{1, \ldots, n\}$. Note that such a scheme only represents one shortest path for any pair of nodes.

In this paper, we are interested in schemes which represent all shortest paths between pairs of nodes. This approach is particularly interesting since, by introducing more knowledge of the network topology at each node, it makes it possible to better deal also with problems of fault occurrence and efficient traffic distribution.

In particular, the *Boolean Routing* model presented in [1] is considered, and it is shown that it can be used to improve memory requirements with respect to Interval routing. According to the Boolean routing model, each node v has an assigned label which is a string of bits, and it is associated to a set of predicates, defined on node labels, where each predicate corresponds to a link incident to v. Given a destination node u, links to be chosen will be the ones whose corresponding predicates are verified when applied to the label assigned to u.

The approach is shown to obtain better performances in terms of both space complexity of the routing information required and number of shortest paths represented by the schemes. Moreover, it reduces, and in some cases even eliminates, the existing gap between the information required by ILS and the current lower bounds on the space complexity of routing schemes. Memory parameters of the Boolean Routing model are formalized and schemes for some classes of graphs obtained by composition operations over other graphs are presented. This shows that the model efficiently deals with highly modular graphs, since the optimality of the scheme applied to such a graph can be inferred by the optimality of the schemes of the basic graphs on which the composition is applied.

The paper is organized as follows: in section 2 we introduce the Boolean Routing model and all preliminary definitions. In section 3 we prove that this model can be efficiently applied to cases where network topologies are fixed or restricted. In sections 4 and 5 we show the modularity of the model by considering Boolean schemes for different types of graph products, under different cost measures.

Due to space constraints, many proofs are omitted. All of them will appear in the full paper.

2 The Boolean Routing Model

Let $G = (V, E)$ be the graph underlying the given network, with $n = |V|$. In the following, for any string σ, we will denote the length (in bits) of σ as $\lambda(\sigma)$.

Moreover, for any string of k bits $\sigma = b_k b_{k-1} \ldots b_1$, we will denote any substring $b_s b_{s-1} \ldots b_r$ ($1 \leq r \leq s \leq k$) as $\sigma[s, r]$. Given a partition of σ in t substrings, that is a $(t-1)$-tuple $(i_1, \ldots, i_{t-1})$ (with $i_1 > i_2 > \ldots > i_{t-1}$) such that $\sigma^1 = \sigma[k, i_1]$, $\sigma^j = \sigma[i_{j-1}, i_j]$, $\sigma^t = \sigma[i_{t-1}, 1]$, let $\lambda(\sigma^j) = i_{j-1} - i_j + 1$ be the length of σ^j.

In the Boolean Routing model each node v_i ($i = 1, \ldots, n$) is labeled by a string l_i with $\lambda(l_i) \leq c \log n$,[4] where c is a suitable constant whose value depends on the network topology. Let $L = \{l_i \mid i = 1, \ldots, n\}$.

Given a node v_k ($k = 1, \ldots, n$) with degree δ_k, let any arbitrary ordering $e_1^k, e_2^k, \ldots, e_{\delta_k}^k$ be defined among its incident links. We assume that a certain set $P_1^k, P_2^k, \ldots, P_{\delta_k}^k$ of *link predicates* defined on L is associated to v_k: given any node v_j $P_i^k(l_j)$ is true iff a shortest path from v_k to v_j includes link e_i^k. Given a message m at node v_k with destination v_j (we assume label l_j is specified in the header of m), m is routed on any link e_i^k such that $P_i^k(l_j)$ is verified.

As it can be realized, a fundamental design goal is to reduce the amount of space required by the scheme, i.e. to choose node labels so that link predicates can be efficiently coded.

In this paper we will use the following link predicates: given a node label l
1. ϵ_l is a link predicate such that, for any label l' with $\lambda(l') = \lambda(l)$, $\epsilon_l(l')$ is verified iff $l = l'$;
2. μ_l (μ_l^+) is a link predicate such that, for any label l' with $\lambda(l') = \lambda(l)$, $\mu_l(l')$ (μ_l^+) is verified iff $l \leq l'$ ($l \leq l' + 1$).

In general, link predicates need not be necessarily of the above kind, however if a node label l is partitioned in substrings $l^1, l^2, \ldots, l^h$, and $P : \{0, 1\}^{\lambda(l^j)} \to \{0, 1\}$ is any predicate, then a predicate $P^j : \{0, 1\}^{\lambda(l)} \to \{0, 1\}$ can be defined such that $P^j(l) = P(l^j)$ (i.e. P^j is P restricted over the j-th substring of l).

In general, the specification of a link predicate such as ϵ_l requires $c + \lambda(l)$ bits, where c bits are necessary to identify the predicate (among ϵ_l, μ_l and μ_l^+) and $\lambda(l)$ must be used to store label l.

However, if some predicates are defined at each node on the same label l, and in general if common substrings are used to encode link predicates, then it could be possible to store a common substring only once, referring to it by means of pointers. By the above observation we assume that each node v_i stores a code string $\mathcal{L}_i = \mathcal{L}_{i,1} \cdot \mathcal{L}_{i,2}$. If l_i is referred by some link predicate, then $\mathcal{L}_{i,1} = l_i$, otherwise $\mathcal{L}_{i,1}$ is the empty string. Moreover $\mathcal{L}_{i,2} = w_1 \cdot \ldots \cdot w_k$, where $w_1, \ldots, w_k$ is a set of strings used in predicates associated to links incident to v_i. We will assume that if $\mathcal{L}_{i,1} = l_i$ then l_i is stored only once, that is not twice as node label and as substring of $\mathcal{L}_i$.

Consequently, if we denote by $\mathcal{P}$ the set of all link predicates, the space required to encode link predicates is given by a function $\Lambda : \{0, 1\}^* \times \mathcal{P} \to N$

---

[4] All logarithms are assumed at the base 2

such that:

1. $\Lambda(\mathcal{L}, \epsilon_l), \Lambda(\mathcal{L}, \mu_l), \Lambda(\mathcal{L}, \mu_l^+) =$
   - c, for a suitable constant c, if $l = \mathcal{L}$;
   - $c + \lambda(l)$, for a suitable constant c, if l is stored outside $\mathcal{L}$;
   - $c + 2 \log \lambda(\mathcal{L})$, for a suitable constant c, if l is a substring of $\mathcal{L}$ ($2\lceil \log \lambda(\mathcal{L}) \rceil$ bits are necessary to identify l in $\mathcal{L}$);
2. $\mathcal{L}'$ is a substring of $\mathcal{L}$ then, for any predicate P, $\Lambda(\mathcal{L}', P) = \Lambda(\mathcal{L}, P) + 2\lceil \log \lambda(\mathcal{L}) \rceil$ ($2\lceil \log \lambda(\mathcal{L}) \rceil$ bits are necessary to identify $\mathcal{L}'$ in $\mathcal{L}$);
3. let each node label l be partitioned in substrings $l^1, l^2, \ldots, l^h$. For any $1 \leq j \leq h$, $\Lambda(\mathcal{L}, P^j) = \Lambda(\mathcal{L}, P) + 2\lceil \log \lambda(l) \rceil$ ($2\lceil \log \lambda(l) \rceil$ bits are necessary to specify the substring of the argument on which P has to be applied);
4. for any link predicate P, $\Lambda(\mathcal{L}, \neg P) = c + \Lambda(\mathcal{L}, P)$, for a suitable constant c;
5. for any pair of link predicates P_1, P_2, $\Lambda(\mathcal{L}, P_1 \wedge P_2) = c + \Lambda(\mathcal{L}, P_1) + \Lambda(\mathcal{L}, P_2)$, for a suitable constant c, and analogously $\Lambda(\mathcal{L}, P_1 \vee P_2) = c + \Lambda(\mathcal{L}, P_1) + \Lambda(\mathcal{L}, P_2)$.

Clearly, the total space required at each node is the sum of the space required by link predicates and the length of string $\mathcal{L}$, possibly plus $\lceil \log n \rceil$ if the node label is not contained in $\mathcal{L}$.

Occasionally, in a network $G = (V, E)$ it might be necessary to represent at each node only the set of shortest paths to a given subset $\overline{V} \subseteq V$ of nodes. Furthermore paths different from the shortest ones could be used for routing messages.

Given a graph $G = (V, E)$ and a scheme $\mathcal{S}_{\overline{V}}$ on G, which represents only (eventually not shortest) paths between pairs of nodes in $\overline{V}$, each node $v_i \in V$ requires $\Omega(\log \overline{n} + \overline{\delta}_i \log \overline{\delta}_i)$ bits of information, where $\overline{n} = |\overline{V}|$ and $\overline{\delta}_i \leq \delta_i$ is the number of links incident to v_i that belong to at least one path represented by the scheme. In fact, $\Omega(\log \overline{n})$ bits are necessary to specify the equality function needed to check, for each incoming message m, whether v_i is the destination node of m (only $\overline{n}$ nodes have to perform this test), while $\Omega(\overline{\delta}_i \log \overline{\delta}_i)$ are necessary to assign distinct link labels.

According to this observations, we have the following definitions:

Definition 1. For any $\overline{V} \subseteq V$, a scheme is k-balanced ($k \geq 1$) if at each node v_i it requires at most $k(\log \overline{n} + \overline{\delta}_i \log \overline{\delta}_i) + o(k(\log \overline{n} + \overline{\delta}_i \log \overline{\delta}_i))$ bits of information.

Definition 2. For any $\overline{V} \subseteq V$, a scheme is k-global ($k \geq 1$) if it requires at most $k(\overline{n} \log \overline{n} + \sum_{i=1}^{n} \overline{\delta}_i \log \overline{\delta}_i) + o(k(\overline{n} \log \overline{n} + \sum_{i=1}^{n} \overline{\delta}_i \log \overline{\delta}_i))$ overall bits of information.

It is immediate that a k-balanced scheme is also k-global, while the converse in general does not hold.

Notice that, in general, we may consider $f(n)$-balanced and $f(n)$-global schemes, where $f(n)$ is some function. In these cases, we are interested to evaluate the asymptotic behavior of $f(n)$, thus resulting in the definition of $O(f(n))$-balanced and $O(f(n))$-global schemes

So far, we have considered the overall amount of information necessary to store at each node its label and link predicates. However, since node labels require $\Theta(\log n)$ bits, it makes sense to take into account the space required to encode only link predicates (that is the length $\lambda(\mathcal{L})$ of the code string $\mathcal{L}$ plus, for each predicate P, $\Lambda(\mathcal{L}, P)$).

By the above observation we have the following definitions:

Definition 3. For any $\overline{V} \subseteq V$, a scheme is k-edge-balanced if at each node v_i at most $k\overline{\delta}_i \log \overline{\delta}_i + o(k\overline{\delta}_i \log \overline{\delta}_i)$ overall bits of information are required to encode link predicates.

Notice that in this definition some problems might arise if there is a node v_i with $\overline{\delta}_i = 1$. In fact in this case $\overline{\delta}_i \log \overline{\delta}_i = 0$ and even if the amount of space $\mathcal{O}_i$ used to represent the link predicate is constant, then there exists no $k > 0$ such that $\mathcal{O}_i \leq k\overline{\delta}_i \log \overline{\delta}_i$.

However, since the corresponding predicate (i.e. the inequality function of the node label) must be specified, we can compare the memory occupation simply with $\overline{\delta}_i$ instead of $\overline{\delta}_i \log \overline{\delta}_i$, that is , we implicitly assume that $\overline{\delta}_i \log \overline{\delta}_i = 1$ if $\overline{\delta}_i = 1$.

Definition 4. For any $\overline{V} \subseteq V$, a scheme is k-edge-global if it globally requires at most $k \sum_{i=1}^{n} \overline{\delta}_i \log \overline{\delta}_i + o(k \sum_{i=1}^{n} \overline{\delta}_i \log \overline{\delta}_i)$ overall bits of information to encode link predicates.

As in the previous case, if a scheme is k-edge-balanced then it is k-edge-global, while the converse in general does not hold. Furthermore, if a scheme is k-edge-balanced (k-edge-global) then it is k-balanced (k-global). Again, the converse in general does not hold.

Analogously to the observations above, $O(f(n))$-edge-balanced and $O(f(n))$-edge-global schemes can also be defined.

For the sake of simplicity and without loss of generality, when dealing with a scheme $\mathcal{S}_{\overline{V}}$ with $\overline{V} \subset V$, we assume that a unique label is assigned to all nodes in $V - \overline{V}$ (since it is not necessary to distinguish among the labels of such nodes). In particular, we assume that such a label has length $O(\log \overline{V})$ and its bits are identically equal to 0.

3 Schemes with Optimum Space Complexity

In this section we present boolean routing schemes for some common network types. All of them are k-balanced or k-global, for small constant values of k. Moreover their edge-balanced and edge-global occupation is considered.

3.1 Paths

In this case, the boolean routing scheme is very simple. Let us assign a label l of length $\lambda(l) = \lceil \log n \rceil$ to each node starting from one end and proceeding in increasing order ("forward") from one end to the other one, and let $\mathcal{L} = l$.

Moreover, let us introduce the following link predicates for a node whose label is l_i:
- Link (v_i, v_{i+1}): $P_f = \mu_{l_i}^+$; note that $\Lambda(\mathcal{L}, P_f) \leq c_1$ for a suitable constant c_1.
- Link (v_i, v_{i-1}): $P_b = \neg\mu_{l_i}$; note that $\Lambda(\mathcal{L}, P_b) \leq c_2$ for a suitable constant c_2.

So the scheme is 1-balanced (and thus 1-global). Furthermore, by definitions 3 and 4 the scheme is $(\log n)$-edge-balanced and $\frac{\log n}{2}$-edge-global. In fact all nodes require $\log n + c$ bits of information (included the code string) to represent link predicates, and their degree is 2, except the first and the last one.

3.2 Rings

In the case of rings, again we assign a label l of length $\lambda(l) = \lceil \log n \rceil$ to each node, starting from one node and proceeding in increasing order in a chosen *forward* direction.

Given a node v_i let v_{i_f} be the farthest node optimally reachable by following the link incident to v_i in the forward direction, and let v_{i_b} be the farthest node optimally reachable by following the link incident to v_i in the backward direction (note that it may result $v_{i_f} = v_{i_b}$). Let $\mathcal{L} = l_i \cdot l_{i_f}$. The following predicates can be defined for links incident to node v_i:
- Forward link: $R_f = \mu_{l_i}^+ \wedge \neg\mu_{l_{i_f}}^+$, with

$$\Lambda(\mathcal{L}, R_f) = \Lambda(l_i \cdot l_{i_f}, \mu_{l_i}^+ \wedge \neg\mu_{l_{i_f}}^+) = c + \Lambda(l_i \cdot l_{i_f}, \mu_{l_i}^+) + \Lambda(l_i \cdot l_{i_f}, \mu_{l_{i_f}}^+) \leq$$

$$c' + \Lambda(l_i, \mu_{l_i}^+) + 2\log\lambda(l_i \cdot l_{i_f}) + \Lambda(l_{i_f}, \mu_{l_{i_f}}^+) + 2\log\lambda(l_i \cdot l_{i_f}) \leq c'' + 4\log\log n,$$

for suitable constants c, c', c'';
- Backward link: $R_b = \neg\mu_{l_i} \wedge \mu_{l_{i_f}}^+$, if $l_{i_b} = l_{i_f}+1$, else if $l_{i_b} = l_{i_f}$ $R_b = \neg\mu_{l_i} \wedge \mu_{l_{i_f}}$, with $\Lambda(\mathcal{L}, R_b) \leq c + 4\log\log n$, for a suitable constant c.

Hence, the scheme is 2-balanced and $(\log n)$-edge-balanced. A more complex construction allows to obtain a 1-balanced scheme for unweighted rings (see [1]).

3.3 Trees

Choose any node r as the root. Label all nodes of the tree in preorder. Consider a node v with label l and let $lev(v)$ be the level of v in the rooted tree (with $lev(r) = 0$).

If v is a leaf, then the code string of v is $\mathcal{L} = l$, its link predicate is $f_{up} = \neg\epsilon_l$.

If v is not a leaf, let $d - 1$ be the number of links to nodes at level $lev(v) + 1$, $l_1, \ldots, l_{d-1}$ the labels of such nodes, and l_g be the greatest label in the subtree rooted at v. Notice that by the preorder labeling $l_1 = l + 1$, thus the code string of v is $\mathcal{L} = l \cdot l_2 \cdot \cdots \cdot l_{d-1} \cdot l_g$. The link predicates at v are:
- $f_{up} = \neg\mu_l \vee \mu_{l_g}$;
- $f_1 = \mu_l^+ \wedge \neg\mu_{l_2}$ $(f_1 = \mu_l^+ \wedge \neg\mu_{l_g}^+$ if $d = 2)$;
- $f_i = \mu_{l_i} \wedge \neg\mu_{l_{i+1}}$ $(i = 1, \ldots, d-2)$;
- $f_{d-1} = \mu_{l_{d-1}} \wedge \neg\mu_{l_g}^+$; clearly if $d = 2$ then $f_{d-1} = f_1$.

For what concerns space requirements, it is possible to prove the following theorem:

Theorem 5. *The scheme is* $O(max_{v \in V} \frac{d \log n}{\log n + d \log d})$-*balanced,* $(\log n)$-*edge-balanced and 2-global.*

Thus, the scheme has an optimal global occupation.

3.4 Complete Graphs

The scheme for complete graphs is very simple. All nodes are labeled using $\lceil \log n \rceil$ bits according to some order and let each code string $\mathcal{L}$ be equal to the node label. At node v_i, the link predicate for each edge (v_i, v_j) is ϵ_{l_j}.

Since at each node it is necessary to store its label and each link predicate requires at most $c + \log n$ bits, the scheme is 1-edge-balanced (and thus 1-balanced).

3.5 Complete Bipartite Graphs

Given a complete bipartite graph $G = (V_1 \cup V_2, V_1 \times V_2)$, label the nodes in V_1 in any order using $\lceil \log n \rceil - 1$ bits and at each node let each code string $\mathcal{L}$ be equal to the node label. Do the same for nodes in V_2. Place a 0-bit (1-bit) as the first bit in labels of nodes in V_1 (V_2).

In this case, labels are considered as composed of two substrings, the first one constituted by the first bit, corresponding to variable x_k (where $k = \lceil \log n \rceil$), and the second one by the remaining bits.

At node v_i, if link $e_{i,j}$ ($j = 1, \ldots, \delta_i$) reaches node v_h of the other partition, the corresponding predicate is $P_{i,j} = \neg \epsilon_{l_i} \wedge (\epsilon_{l_i}^1 \vee \epsilon_{l_h})$. Clearly $\Lambda(\mathcal{L}, P_{i,j}) \leq c + 4 \log \log n$, for a suitable constant c, thus the scheme is 1-edge-balanced.

4 Schemes for graph products

In the previous section we provided schemes for some simple graph topologies. Such schemes are basically obtained by simulating Interval Routing schemes. In this section, we show the added power of the Boolean model by presenting optimum schemes for some classes of graphs obtained by operations of composition of other graphs. This approach shows that the model allows an efficient treatment of highly modular graphs, since in such cases the optimality of schemes can be inferred by the optimality of the schemes of the basic graphs that are involved in the composition.

Let us consider a grid configuration of $n \times m$ nodes. Independently from the existence of wrap-arounds we can assign node labels according to the following strategy.

For each subgraph induced by nodes in a same row (row component), assign node labels $p_1, \ldots, p_m$ with a strategy depending from whether it is a path or a ring. Analogously, for each subgraph induced by a same column (column component), assign node labels $q_1, \ldots, q_n$ in the same way.

Let v_{ij} be a node on row i and column j, let p_j (q_i) the label assigned to v_{ij} as member of the row (column) component: we assign $q_i \cdot p_j$ (here $\cdot$ stands for the concatenation operator) as the label of v_{ij}.

Link predicates for arcs to nodes on row i are just the ones used for routing inside the row component, but restricted over the second part of the label, while link predicates for arcs to nodes on column j are the ones used for routing inside the column component, restricted over the first part of the label.

We can extend this strategy by letting the row and column components be any graph, thus considering the *cartesian product* of graphs. By allowing more than two dimensions, this operation can be defined as follows:

Definition 6. Given d graphs $G_1 = (V_1, E_1), \ldots, G_d = (V_d, E_d)$, $d > 0$, let us define the *product graph* $G_1 \times \ldots \times G_d$, as the graph $G = (V, E)$, where:

1. $V = V_1 \times \ldots \times V_d$; [5]
2. $E = \bigcup_{j=1}^{d} \{ (v_{i_1, \ldots, i_j, \ldots, i_d}, v_{i_1, \ldots, i'_j, \ldots, i_d}) \mid v_{i_1} \in V_1, \ldots, v_{i_{j-1}} \in V_{j-1}, v_{i_{j+1}} \in V_{j+1}, \ldots, v_{i_d} \in V_d, (v_{i_j}, v_{i'_j}) \in E_j \}$.

The definition states that each node $v_{i_1, \ldots, i_d} \in V$ (with $v_{i_j} \in V_j$) belongs to a d dimensional space. Moreover, the definition of the edge set E states that, for any $1 \leq k \leq d$, the subgraph induced by all nodes having the same $i_1, \ldots, i_{k-1}, i_{k+1}, \ldots i_d$ values is isomorphic to G_k.

This class of graphs includes the topologies of some interconnection networks commonly used in parallel architectures, such as hypercubes and d-dimensional grids.

In the following, we will use the notation below:
- $l_{i_1, \ldots, i_d}$ is the label associated to node $v_{i_1, \ldots, i_d} \in V$.
- $\mathcal{O}_{i_j}$ is the amount of information stored at node $v_{i_j} \in V_j$ (including the code string). In the case of edge–balanced and edge–global schemes this refers only to the information needed to encode link predicates (including the code string).
- $\mathcal{O}_{i_1, \ldots, i_d}$ is the overall space required at a node $v_{i_1, \ldots, i_d}$. Again, in the case of edge–balanced and edge–global schemes this refers only to encoding of link predicates.
- $\mathcal{O}$ is the overall space required in the whole network. Also in this case, if edge–balanced and edge–global schemes are considered, we refer only to the encoding of link predicates.
- δ_{i_j} is the degree of node $v_{i_j} \in V_j$.
- $\delta_{i_1, \ldots, i_d}$ is the degree of node $v_{i_1, \ldots, i_d} \in V$.

Notice that $\sum_{j=1}^{d} \delta_{i_j} = \delta_{i_1, \ldots, i_d}$; moreover, it is easy to see that, for all $k > 0$, $\sum_{j=1}^{d} k \delta_{i_j} \log \delta_{i_j} = O(k \delta_{i_1, \ldots, i_d} \log \delta_{i_1, \ldots, i_d})$ and $\sum_{j=1}^{d} k(\log n_j + \delta_{i_j} \log \delta_{i_j}) = O(k(\log n + \delta_{i_1, \ldots, i_d} \log \delta_{i_1, \ldots, i_d}))$, where $n = |V|$ and for each j $n_j = |V_j|$.

The boolean scheme for G is obtained as follows: for each node $v_{i_1, \ldots, i_d}$, label $l_{i_1, \ldots, i_d}$ derives from the concatenation of d substrings, i.e. $l_{i_1, \ldots, i_d} = l_{i_1} \cdot l_{i_2} \cdot \ldots \cdot l_{i_d}$ where, for $j = 1, \ldots, d$, l_{i_j} is the label of node v_{i_j} in the graph G_j. Moreover, at

---

[5] For the sake of simplicity we will denote any node $(v_{i_1}, \ldots, v_{i_d}) \in V$ by $v_{i_1, \ldots, i_d}$

each node $v_{i_1,...,i_d}$, the code string $\mathcal{L}_{i_1,...,i_d} = \mathcal{L}_{i_1} \cdot ... \cdot \mathcal{L}_{i_d}$, where $\mathcal{L}_{i_j}$ is the code string of node v_{i_j} in the graph G_j.

Notice that for any $v_{i_1,...,i_d} \in V$, $\lambda(l_{i_1,...,i_d}) = \sum_{j=1}^{d} \lambda(l_{i_j}) \le c \log n$ for a suitable constant c, as required by the definition of the model.

For what regards edges predicates, if $e = (v_{i_1,...,i_j,...,i_d}, v_{i_1,...,i'_j,...,i_d}) \in E$, the predicate associated to e at node $v_{i_1,...,i_j,...,i_d}$ is $P_e = P_{e_j}^j$, where P_{e_j} is the predicate associated to $e_j = (v_{i_j}, v_{i'_j}) \in E_j$ at node v_{i_j} in the scheme of the graph G_j.

It is then possible to prove the following lemma:

Lemma 7. *The scheme correctly routes messages along their shortest paths.*

In the following we will characterize the relationships between schemes defined on $G_1, ..., G_d$ and the scheme defined on G.

4.1 Balanced and global schemes

Theorem 8. *Let $G_1, ..., G_d$ be graphs such that each G_j has an $O(k)$-balanced scheme. Then the scheme derived for G is $O(k)$-balanced.*

Proof. For any $1 \le j \le d$, link predicates on the j-th component are derived as the predicates for graph G_j, evaluated to the j-th substring of a label. Thus, at each node $v_{i_1,...,i_d} \in V$ if $e = (v_{i_1,...,i_j,...,i_d}, v_{i_1,...,i'_j,...,i_d}) \in E$ is an incident link and the predicate associated to e is $P_e = P_{e_j}^j$ (where P_{e_j} is the predicate associated to $e_j = (v_{i_j}, v_{i'_j}) \in E_j$ at node v_{i_j} in the scheme of graph G_j), then
$$\Lambda(\mathcal{L}_{i_1,...,i_d}, P_e) = \Lambda(\mathcal{L}_{i_1,...,i_d}, P_{e_j}^j) = \Lambda(\mathcal{L}_{i_1,...,i_d}, P_{e_j}) + 2\lceil \log \lambda(l_{i_1,...,i_j,...,i_d})\rceil =$$
$$\Lambda(\mathcal{L}_{i_j}, P_{e_j}) + 2\lceil \log \lambda(l_{i_1,...,i_j,...,i_d})\rceil + 2\lceil \log \lambda(\mathcal{L}_{i_1,...,i_d})\rceil.$$
Since $\Lambda(\mathcal{L}_{i_j}, P_{e_j})$ is just the space required at node $v_{i_j} \in V_j$ to represent P_{e_j} in the scheme for V_j, it is:
$$\mathcal{O}_{i_1,...,i_d} = \sum_{j=1}^{d} \mathcal{O}_{i_j} + \sum_{j=1}^{d} 2\delta_{i_j}(\lceil \log \lambda(l_{i_1,...,i_j,...,i_d})\rceil + \lceil \log \lambda(\mathcal{L}_{i_1,...,i_d})\rceil) =$$
$$O(\sum_{j=1}^{d} k(\log n_j + \delta_{i_j} \log \delta_{i_j}) + \sum_{j=1}^{d} \delta_{i_j}(\log \lambda(l_{i_1,...,i_j,...,i_d}) + \log \lambda(\mathcal{L}_{i_1,...,i_d}))) =$$
$$O(k(\log n + \delta_{i_1,...,i_d} \log \delta_{i_1,...,i_d}) + \delta_{i_1,...,i_d} \log \lambda(l_{i_1,...,i_j,...,i_d}) +$$
$$\delta_{i_1,...,i_d} \log \lambda(\mathcal{L}_{i_1,...,i_d})). \text{ But,}$$
$$\delta_{i_1,...,i_d} \log \lambda(l_{i_1,...,i_j,...,i_d}) \le \delta_{i_1,...,i_d} \log(c \log n) = O(\log n + \delta_{i_1,...,i_d} \log \delta_{i_1,...,i_d}),$$
and as by hypothesis for each dimension j $\lambda(\mathcal{L}_{i_j}) = O(k(\log n_j + \delta_{i_j} \log \delta_{i_j}))$, it results $\delta_{i_1,...,i_d} \log \lambda(\mathcal{L}_{i_1,...,i_d}) =$
$$\delta_{i_1,...,i_d} \log(\sum_{j=1}^{d} \lambda(\mathcal{L}_{i_j})) = O(\delta_{i_1,...,i_d} \log(k(\log n + \delta_{i_1,...,i_d} \log \delta_{i_1,...,i_d}))) =$$
$$O(\delta_{i_1,...,i_d}(\log k + \log(\log n + \delta_{i_1,...,i_d} \log \delta_{i_1,...,i_d}))) =$$
$$O(k(\log n + \delta_{i_1,...,i_d} \log \delta_{i_1,...,i_d})).$$
Thus $\mathcal{O}_{i_1,...,i_d} = O(k(\log n + \delta_{i_1,...,i_d} \log \delta_{i_1,...,i_d}))$ and the theorem holds.

A similar theorem can be proved for the global case:

Theorem 9. *Let $G_1, ..., G_d$ be graphs such that each G_j admits an $O(k)$-global scheme. Then the scheme derived for G is $O(k)$-global.*

Proof. As for the balanced case, $\mathcal{O} = \sum_{v_{i_1,\ldots,i_d} \in V}(\mathcal{O}_{i_1} + \ldots + \mathcal{O}_{i_d}) +$
$\sum_{v_{i_1,\ldots,i_d} \in V} 2(\delta_{i_1} + \ldots + \delta_{i_d})(\lceil \log \lambda(l_{i_1,\ldots,i_d}) \rceil + \lceil \log \lambda(\mathcal{L}_{i_1,\ldots,i_d}) \rceil) =$
$\sum_{v_{i_1,\ldots,i_d} \in V}(\mathcal{O}_{i_1} + \ldots + \mathcal{O}_{i_d}) + \sum_{v_{i_1,\ldots,i_d} \in V} 2\delta_{i_1,\ldots,i_d} \lceil \log \lambda(l_{i_1,\ldots,i_d}) \rceil +$
$\sum_{v_{i_1,\ldots,i_d} \in V} 2\delta_{i_1,\ldots,i_d} \lceil \log \lambda(\mathcal{L}_{i_1,\ldots,i_d}) \rceil.$
But $\sum_{v_{i_1,\ldots,i_d} \in V}(\mathcal{O}_{i_1} + \ldots + \mathcal{O}_{i_d}) =$
$O(n_2 \cdot \ldots \cdot n_d k(n_1 \log n_1 + \sum_{v_{i_1} \in V_1} \delta_{i_1} \log \delta_{i_1}) + \ldots + n_1 \cdot \ldots \cdot n_{d-1} k(n_d \log n_d +$
$\sum_{v_{i_d} \in V_d} \delta_{i_d} \log \delta_{i_d})) =$
$O(k(n_1 \cdot \ldots \cdot n_d(\log n_1 + \ldots + \log n_d) + \sum_{v_{i_1,\ldots,i_d} \in V}(\delta_{i_1} \log \delta_{i_1} + \ldots + \delta_{i_d} \log \delta_{i_d}))) =$
$O(k(n \log n + \sum_{v_{i_1,\ldots,i_d} \in V} \delta_{i_1,\ldots,i_d} \log \delta_{i_1,\ldots,i_d})).$
Moreover,
$2\delta_{i_1,\ldots,i_d} \lceil \log \lambda(l_{i_1,\ldots,i_d}) \rceil \leq 2\delta_{i_1,\ldots,i_d} \lceil \log(c \log n) \rceil = O(\log n + \delta_{i_1,\ldots,i_d} \log \delta_{i_1,\ldots,i_d}),$
thus, $\sum_{v_{i_1,\ldots,i_d} \in V} 2\delta_{i_1,\ldots,i_d} \lceil \log \lambda(l_{i_1,\ldots,i_d}) \rceil =$
$O(n \log n + \sum_{v_{i_1,\ldots,i_d} \in V} \delta_{i_1,\ldots,i_d} \log \delta_{i_1,\ldots,i_d}).$
Finally, $\sum_{v_{i_1,\ldots,i_d} \in V} \delta_{i_1,\ldots,i_d} \log \lambda(\mathcal{L}_{i_1,\ldots,i_d}) =$
$\sum_{v_{i_1,\ldots,i_d} \in V} \delta_{i_1,\ldots,i_d}(\log \delta_{i_1,\ldots,i_d} + \log \frac{\lambda(\mathcal{L}_{i_1,\ldots,i_d})}{\delta_{i_1,\ldots,i_d}}) \leq$
$\sum_{v_{i_1,\ldots,i_d} \in V} \delta_{i_1,\ldots,i_d} \log \delta_{i_1,\ldots,i_d} + \sum_{v_{i_1,\ldots,i_d} \in V} \lambda(\mathcal{L}_{i_1,\ldots,i_d}) \leq$
$\sum_{v_{i_1,\ldots,i_d} \in V} \delta_{i_1,\ldots,i_d} \log \delta_{i_1,\ldots,i_d} + \sum_{v_{i_1,\ldots,i_d} \in V}(\mathcal{O}_{i_1} + \ldots + \mathcal{O}_{i_d}) =$
$O(k(n \log n + \sum_{v_{i_1,\ldots,i_d} \in V} \delta_{i_1,\ldots,i_d} \log \delta_{i_1,\ldots,i_d})).$
Thus, $\mathcal{O} = O(k(n \log n + \sum_{v_{i_1,\ldots,i_d} \in V} \delta_{i_1,\ldots,i_d} \log \delta_{i_1,\ldots,i_d}))$ and the theorem holds.

By the results concerning paths and rings we have the following corollary:

Corollary 10. *There exist $O(1)$-balanced boolean routing schemes for weighted d-dimensional grids (wrapped or unwrapped along to the various dimensions).*

4.2 Edge-balanced and edge-global schemes

Theorem 11. *Let $G_1, \ldots, G_d$ be graphs such that each G_j has an $O(k)$-edge-balanced scheme. Then the scheme derived for G is $O(max(k, \frac{\log \log n}{\log \delta_{min}}))$-edge-balanced, where δ_{min} is the minimum degree of a node in G.*

A similar theorem can be proved for the edge-global case:

Theorem 12. *Let $G_1, \ldots, G_d$ be graphs such that each G_j has an $O(k)$-edge-global scheme. Then the scheme derived for G is*
$$O(max(k, \frac{\sum_{v_{i_1,\ldots,i_d} \in V} \delta_{i_1,\ldots,i_d} \log \log n}{\sum_{v_{i_1,\ldots,i_d} \in V} \delta_{i_1,\ldots,i_d} \log \delta_{i_1,\ldots,i_d}}))) - edge - global.$$

Clearly theorem 11 still holds under the more restrictive condition that each G_j has exactly a k-edge-balanced scheme. Thus, it is possible to prove the following theorem

Theorem 13. *Let $G_1, \ldots, G_d$ be graphs such that for each $G_j \mid V_j \mid \leq c$ for a suitable constant G, then the scheme derived for G is $O(1)$-edge-balanced.*

Proof. For each $j = 1, \ldots, d$, if G_j has a constant number of nodes, then G_j has a k-edge-balanced scheme for a suitable constant k. This can be simply accomplished by associating to each node $v_{i_j} \in V_j$ a label of $\lceil \log c \rceil$ bits and an empty code string. Given any incident link e, the corresponding predicate is $P_e = \epsilon_{l_{s_1}} \wedge \ldots \wedge \epsilon_{l_{s_k}}$, where $v_{s_1}, \ldots, v_{s_k}$ are the nodes optimally reachable from v_{i_j} through link e.

By theorem 11 the scheme for G is $O(max(k, \frac{\log \log n}{\log \delta_{min}}))$-edge balanced, where δ_{min} is the minimum degree of a node in V. But $max(k, \frac{\log \log n}{\log \delta_{min}}) = O(1)$, since k is constant, $\delta_{min} \geq d$ and $n \leq c^d$, and the theorem holds.

Theorem 13 is very important, as it states that any product of graphs having a constant number of nodes has an optimal occupation even with respect to the edge-balanced case, that is the most restrictive case which implies all the others. In particular we have the following corollary:

Corollary 14. *There exist $O(1)$-edge-balanced boolean routing schemes for all d-dimensional hypercubes.*

4.3 Relations among different schemes

Theorem 15. *Let $G_1, \ldots, G_d$ be graphs such that each G_j has an $O(k)$-edge-balanced scheme. Then the scheme derived for G is*
$$O(max_{v_{i_1}, \ldots, i_d} \in V \frac{\log n + k\delta_{i_1, \ldots, i_d} \log \delta_{i_1, \ldots, i_d}}{\log n + \delta_{i_1, \ldots, i_d} \log \delta_{i_1, \ldots, i_d}}) - balanced.$$

A similar result can be proved for the global case:

Theorem 16. *Let $G_1, \ldots, G_d$ be graphs such that each G_j has an $O(k)$-edge-global scheme. Then the scheme for G is*
$$O(\frac{n \log n + k \sum_{v_{i_1}, \ldots, i_d} \in V \delta_{i_1, \ldots, i_d} \log \delta_{i_1, \ldots, i_d}}{n \log n + \sum_{v_{i_1}, \ldots, i_d} \in V \delta_{i_1, \ldots, i_d} \log \delta_{i_1, \ldots, i_d}}) - global.$$

The following corollaries can be directly derived from theorems 15 and 16:

Corollary 17. *Let $G_1, \ldots, G_d$ be graphs such that each G_j admits an $O(k)$-edge-balanced scheme. Then the scheme derived for G is $O(k)$-balanced. Moreover if for each $v_{i_1, \ldots, i_d} \in V$ $(\delta_{i_1, \ldots, i_d} \log \delta_{i_1, \ldots, i_d}) = O(\log n)$, then the scheme derived for G is $O(1)$-balanced.*

Corollary 18. *Let $G_1, \ldots, G_d$ be graphs such that each G_j admits an $O(k)$-edge-global scheme. Then the scheme derived for G is $O(k)$-global.*
Moreover, if $\sum_{v_{i_1, \ldots, i_d} \in V} \delta_{i_1, \ldots, i_d} \log \delta_{i_1, \ldots, i_d} = O(n \log n)$, then the scheme derived for G is $O(1)$-global.

Corollaries 17 and 18 are particularly interesting since in all reasonable networks, even if the number of processors is very large, the degree is usually small or constant. Finally, it is possible to prove the following theorems:

Theorem 19. *Let $G_1, \ldots, G_d$ be graphs such that each G_j has an $O(k)$-balanced scheme. Then the scheme derived for G is*
$$O(max(k, k\frac{\log n}{\delta_{min} \log \delta_{min}}))\text{-edge-balanced,}$$
where δ_{min} is the minimum degree of a node in G.

Proof. It follows directly from theorem 8 and from the definition of k-edge-balanced scheme.

Theorem 20. *Let $G_1, \ldots, G_d$ be graphs such that each G_j admits an $O(k)$-global scheme. Then the scheme derived for G is*
$$O(max(k, k\frac{n \log n}{\sum_{v_{i_1,\ldots,i_d} \in V} \delta_{i_1,\ldots,i_d} \log \delta_{i_1,\ldots,i_d}}))\text{-edge-global.}$$

Proof. It follows directly from theorem 9 and from the definition of k-edge-global scheme.

5 Schemes for partial graph products

Other classes of graphs have the same notion of dimension of the graph product, but not all nodes are projected in each dimension. This is the case of some common types of tori and meshes of graphs.

If for each graph $G_j = (V_j, E_j)$, we allow a subset of *border* nodes $\overline{V}_j \subseteq V_j$ to be projected in dimensions, we may define *partial products* of graphs:

Definition 21. Given $d \geq 1$ graphs $G_1 = (V_1, E_1), \ldots, G_d = (V_d, E_d)$ and d sets of border nodes $\overline{V}_i \subseteq V_i$, let us define the *partial graph product* $G = G_1 \times \ldots \times G_d$, as the graph $G = (V, E)$, where:
1. $V = \bigcup_{i=1}^{d} \overline{V}_1 \times \ldots \times \overline{V}_{i-1} \times V_i \times \overline{V}_{i+1} \times \ldots \times \overline{V}_d$;
2. $E = \bigcup_{j=1}^{d} \{(v_{i_1,\ldots,i_j,\ldots,i_d}, v_{i_1,\ldots,i'_j,\ldots,i_d}) \mid v_{i_1} \in \overline{V}_1, \ldots, v_{i_{j-1}} \in \overline{V}_{j-1}, v_{i_{j+1}} \in \overline{V}_{j+1}, \ldots, v_{i_d} \in \overline{V}_d, (v_{i_j}, v_{i'_j}) \in E_j\}$

In the following, we will refer to a *component* in dimension j $(j = 1, \ldots, d)$ as the subgraph of G induced by a maximal subset of nodes having the same $i_1 i_2 \ldots i_{j-1} i_{j+1} \ldots i_d$ value. Notice that, by definition, any component in dimension j is either isomorphic to G_j or one single node (we will refer to such a case as a *singleton component* in dimension j).

In order to represent all shortest paths for all pairs (u, v) of nodes in V, we consider three different cases:
1. $u = v_{i_1,\ldots,i_j,\ldots,i_d}$, $v = v_{i_1,\ldots,i'_j,\ldots,i_d}$, and $v_{i'_j} \in V_j - \overline{V}_j$. That is, u and v belong to the same component in dimension j and $v_{i'_j}$ is not a border node in G_j.
2. $u = v_{i_1,\ldots,i_j,\ldots,i_d}$, $v = v_{i'_1,\ldots,i'_j,\ldots,i'_d}$, $v_{i'_j} \in V_j - \overline{V}_j$, and there exists $k \neq j$ such that $i_k \neq i'_k$. That is, u and v belong to different components in dimension j and $v_{i'_j}$ is not a border node.
3. all the remaining cases.

With respect to the previous classification of node pairs, for any dimension j the following three schemes can be identified:

1. $\mathcal{S}_j^\alpha$: the scheme representing all shortest paths in G_j;

2. $\mathcal{S}_j^\beta$: the scheme that for any two nodes $u, v \in V_j$ represents all shortest paths between u and v which contain at least one border node $w \in \overline{V}_j$;

3. $\overline{\mathcal{S}}_j$: the scheme representing all shortest paths in G_j between pairs of nodes in $\overline{V}_j$ (recall that it is assumed that labels of nodes in $V_j - \overline{V}_j$ are equal, with all their bits set to 0).

Similarly to the previous section, we will denote by $l_{i_1,\ldots,i_d}$ the label associated to node $v_{i_1,\ldots,i_d} \in V$, and by $\delta_{i_1,\ldots,i_d}$ its degree.

The boolean scheme for G is obtained as follows: each label $l_{i_1,\ldots,i_d}$ is partitioned in $d+2$ substrings. Substrings $l_{i_1,\ldots,i_d}^1, \ldots, l_{i_1,\ldots,i_d}^d$ correspond to schemes $\mathcal{S}_{\overline{V}_j}$ $(j = 1,\ldots,d)$ by letting $l_{i_1,\ldots,i_d}^j = \overline{l}_{i_j}$, where $\overline{l}_{i_j}$ is the label of $v_{i_j} \in V_j$ in $\overline{\mathcal{S}}_j$.

The remaining two substrings correspond respectively to schemes $\mathcal{S}_j^\alpha$ and $\mathcal{S}_j^\beta$. In fact, given $v_{i_1,\ldots,i_d} \in V$, if there exists any j such that $v_{i_j} \in V_j - \overline{V}_j$ then $l_{i_1,\ldots,i_j,\ldots,i_d}^{d+1} = l_{i_j}^\alpha$ and $l_{i_1,\ldots,i_j,\ldots,i_d}^{d+2} = l_{i_j}^\beta$, where $l_{i_j}^\alpha$ and $l_{i_j}^\beta$ are respectively the labels of $v_{i_j} \in V_j$ in $\mathcal{S}_j^\alpha$ and $\mathcal{S}_j^\beta$. Otherwise, if for all j $v_{i_j} \in \overline{V}_j$, then all bits in the two substrings are set to 0.

Since labels $l_{i_j}^\alpha$ and $l_{i_j}^\beta$ may have different lengths with respect to the various dimensions, the $(d+1)$-th substring has length equal to that of labels $l_{i_k}^\alpha$ in $\mathcal{S}_k^\alpha$, where for each $j \neq k$ $\lambda(l_{i_j}^\alpha) \leq \lambda(l_{i_k}^\alpha)$, and the $(d+2)$-th substring has length equal to that of labels $l_{i_h}^\beta$ in $\mathcal{S}_h^\beta$, where for each $j \neq h$ $\lambda(l_{i_j}^\beta) \leq \lambda(l_{i_h}^\beta)$.

For dimensions such that labels $l_{i_j}^\alpha$ or $l_{i_j}^\beta$ are shorter than such maximum lengths, a suitable number of preceding 0-s is inserted.

Notice that the number of nodes of G is $n = \overline{n}_1 \cdot \ldots \cdot \overline{n}_d + (n_1 - \overline{n}_1)\overline{n}_2 \cdot \ldots \cdot \overline{n}_d + \ldots + (n_d - \overline{n}_d)\overline{n}_1 \cdot \ldots \cdot \overline{n}_{d-1}$, where for each j $\overline{n}_j = \overline{V}_j$ and $n_j = V_j$, thus $\sum_{j=1}^d \lambda(\overline{l}_{i_j}) \leq c_1 \log(\overline{n}_1 \ldots \overline{n}_d) \leq c_1 \log n$, $\lambda(l_{i_k}^\alpha) \leq c_2 \log n_k \leq c_2 \log n$, and $\lambda(l_{i_h}^\beta) \leq c_3 \log n_h \leq c_3 \log n$, for suitable constants c_1, c_2 and c_3, and it results $\lambda(l_{i_1,\ldots,i_j,\ldots,i_d}) = \sum_{j=1}^d \lambda(\overline{l}_{i_j}) + \lambda(l_{i_k}^\alpha) + \lambda(l_{i_h}^\beta) \leq c \log n$, for a suitable constant c, as required by the definition of the model.

Let us now consider the code string $\mathcal{L}_{i_1,\ldots,i_d}$ of $v_{i_1,\ldots,i_d}$. If for each j $v_{i_j} \in \overline{V}_j$, then $\mathcal{L}_{i_1,\ldots,i_d} = \overline{\mathcal{L}}_{i_1} \cdot \ldots \cdot \overline{\mathcal{L}}_{i_d} \cdot \mathcal{L}_{i_1}^\alpha \cdot \ldots \cdot \mathcal{L}_{i_d}^\alpha \cdot \mathcal{L}_{i_1}^\beta \cdot \ldots \cdot \mathcal{L}_{i_d}^\beta$, otherwise (if there exists j such that $v_{i_j} \in V_j - \overline{V}_j$) $\mathcal{L}_{i_1,\ldots,i_d} = \overline{\mathcal{L}}_{i_j} \cdot \mathcal{L}_{i_j}^\alpha \cdot \mathcal{L}_{i_j}^\beta$, where $\overline{\mathcal{L}}_{i_j}, \mathcal{L}_{i_j}^\alpha$ and $\mathcal{L}_{i_j}^\beta$ are respectively the code strings of $v_{i_j} \in V_j$ in $\overline{\mathcal{S}}_j$, $\mathcal{S}_j^\alpha$ and $\mathcal{S}_j^\beta$.

For any label $l = l_{i_1,\ldots,i_d}$ and $1 \leq j \leq d$, let $comp_j$ be the predicate verified by the labels of all nodes in the same component in dimension j containing node $v_{i_1,\ldots,i_d}$, i.e. $comp_j = \epsilon_{l_{i_1}}^1 \wedge \ldots \wedge \epsilon_{l_{i_{j-1}}}^{j-1} \wedge \epsilon_{l_{i_{j+1}}}^{j+1} \wedge \ldots \wedge \epsilon_{l_{i_d}}^d$.

For what regards link predicates, one of the three schemes is applied. In fact, for any link $e = (v_{i_1,\ldots,i_j,\ldots,i_d}, v_{i_1,\ldots,i_j',\ldots,i_d})$, the predicate P_e associated to e at node $v_{i_1,\ldots,i_j,\ldots,i_d}$ is:

$$(comp_j \wedge \epsilon_{0\ldots0}^j \wedge P_{e_j}^{\alpha^{d+1}}) \vee (\neg comp_j \wedge \epsilon_{0\ldots0}^j \wedge P_{e_j}^{\beta^{d+2}}) \vee (\neg \epsilon_{0\ldots0}^j \wedge \overline{P}_{e_j}^j),$$

where $P_{e_j}^\alpha$ is the predicate associated to e_j at node v_{i_j} in the scheme $\mathcal{S}_j^\alpha$, $P_{e_j}^\beta$ is the one associated to e_j at node v_{i_j} in $\mathcal{S}_j^\beta$, and $\overline{P}_{e_j}$ is the one associated to e_j at node v_{i_j} in $\overline{S}_j$. It is then possible to prove the following lemma:

Lemma 22. *The scheme correctly routes messages along their shortest paths.*

Proof. The proof will appear in the full paper.

In the following we will characterize the relationships between schemes defined on $G_1, \ldots, G_d$ and the scheme defined on G.

5.1 Balanced and global schemes

Theorem 23. *Let $G_1, \ldots, G_d$ be graphs such that schemes $\mathcal{S}_j^\alpha$, $\mathcal{S}_j^\beta$, and $\overline{S}_j$ are $O(k)$-balanced schemes. Then the scheme for G derived by the above construction is $O(k \frac{\log(n_1 \cdot \ldots \cdot n_d)}{\log n})$-balanced, where $n = |V|$ and $n_j = |V_j|$.*

Notice that theorem 23 is weaker than the corresponding one for normal product of graphs, but if n and $n_1 \cdot \ldots \cdot n_d$ are polynomially related, then the scheme for G is $O(k)$-balanced.

However, a better result can be found for the global case:

Theorem 24. *Let $G_1, \ldots, G_d$ be graphs such that schemes $\mathcal{S}_j^\alpha$, $\mathcal{S}_j^\beta$, and $\overline{S}_j$ are $O(k)$-global schemes. Then the scheme for G derived by the above construction is $O(k)$-global.*

5.2 Edge-balanced and edge-global schemes

Theorem 25. *Let $G_1, \ldots, G_d$ be graphs such that schemes $\mathcal{S}_j^\alpha$, $\mathcal{S}_j^\beta$, and $\overline{S}_j$ are $O(k)$-edge-balanced schemes. Then the scheme for G is $O(max(k, \frac{\log \log n}{\log \delta_{min}})))$-edge-balanced, where δ_{min} is the minimum degree of a node in G.*

A similar theorem can be proved for the edge-global case:

Theorem 26. *Let $G_1, \ldots, G_d$ be graphs such that schemes $\mathcal{S}_j^\alpha$, $\mathcal{S}_j^\beta$, and $\overline{S}_j$ are $O(k)$-edge-global schemes. Then the scheme for G is*
$$O\left(max\left(k, \frac{\sum_{v_{i_1}, \ldots, i_d} \in V \, \delta_{i_1, \ldots, i_d} \log \log n}{\sum_{v_{i_1}, \ldots, i_d} \in V \, \delta_{i_1, \ldots, i_d} \log \delta_{i_1, \ldots, i_d}}\right)\right)\text{-edge-global.}$$

Clearly theorem 25 still holds under the more restrictive condition that each G_j has exactly a k-edge-balanced scheme. Thus, it is possible to prove the following theorems

Theorem 27. *Let $G_1, \ldots, G_d$ be graphs such that for each $G_j \mid V_j \mid \leq c$ for a suitable constant G, then the scheme derived for G is $O(1)$-edge-balanced.*

5.3 Relations among different schemes

Theorem 28. *Let $G_1, \ldots, G_d$ be graphs such that schemes S_j^α, S_j^β, and $\overline{S}_j$ are $O(k)$-edge-balanced schemes. Then the scheme for G derived by the above construction is $O(max_{v_{i_1}, \ldots, i_d \in V} \frac{\log n + k\delta_{i_1, \ldots, i_d} \log \delta_{i_1, \ldots, i_d}}{\log n + \delta_{i_1, \ldots, i_d} \log \delta_{i_1, \ldots, i_d}})$-balanced.*

Theorem 29. *Let $G_1, \ldots, G_d$ be graphs such that schemes S_j^α, S_j^β, and $\overline{S}_j$ are $O(k)$-edge-global schemes. Then the scheme for G derived by the above construction is $O(\frac{n \log n + k \sum_{v_{i_1}, \ldots, i_d \in V} \delta_{i_1, \ldots, i_d} \log \delta_{i_1, \ldots, i_d}}{n \log n + \sum_{v_{i_1}, \ldots, i_d \in V} \delta_{i_1, \ldots, i_d} \log \delta_{i_1, \ldots, i_d}})$-global.*

Corollary 30. *Let $G_1, \ldots, G_d$ be graphs such that schemes S_j^α, S_j^β, and $\overline{S}_j$ are $O(k)$-edge-balanced schemes. Then the scheme derived for G is $O(k)$-balanced. Moreover if for each $v_{i_1, \ldots, i_d} \in V$ $(\delta_{i_1, \ldots, i_d} \log \delta_{i_1, \ldots, i_d}) = O(\log n)$, then the scheme is $O(1)$-balanced.*

Corollary 31. *Let $G_1, \ldots, G_d$ be graphs such that schemes S_j^α, S_j^β, and $\overline{S}_j$ are $O(k)$-edge-global schemes. Then the scheme derived for G is $O(k)$-global. Moreover if $\sum_{v_{i_1}, \ldots, i_d \in V} \delta_{i_1, \ldots, i_d} \log \delta_{i_1, \ldots, i_d} = O(n \log n)$, then the scheme derived for G is $O(1)$-global.*

Finally, it is possible to prove the following theorems:

Theorem 32. *Let $G_1, \ldots, G_d$ be graphs such that schemes S_j^α, S_j^β, and $\overline{S}_j$ are $O(k)$-balanced schemes. Then the scheme for G is $O(max(k, k\frac{\log n}{\delta_V \log \delta_V}, k\frac{\log(n_1 \cdot \ldots \cdot n_d)}{\delta_{\overline{V}} \log \delta_{\overline{V}}}))$-edge-balanced, where $n = |V|$ and $n_j = |V_j|$, $\delta_{\overline{V}}$ is the minimum degree of a node in $\overline{V} = \{v_{i_1, \ldots, i_d} \in V \mid \text{for each } j \, v_{i_j} \in \overline{V}_j\}$, and δ_V is the minimum degree of a node in $V - \overline{V}$.*

Theorem 33. *Let $G_1, \ldots, G_d$ be graphs such that schemes S_j^α, S_j^β, and $\overline{S}_j$ are $O(k)$-global schemes. Then the scheme derived for G is $O(max(k, k\frac{n \log n}{\sum_{v_{i_1}, \ldots, i_d \in V} \delta_{i_1, \ldots, i_d} \log \delta_{i_1, \ldots, i_d}}))$-global.*

References

1. M. Flammini, G. Gambosi, and S. Salomone. Boolean routing. In *Proceedings 7th International Workshop on Distributed Algorithms (WDAG)*, volume 725 of *Lecture Notes in Computer Science*, pages 219–233. Springer-Verlag, 1993.
2. N. Santoro and R. Khatib. Labelling and implicit routing in networks. *The Computer Journal*, 28:5–8, 1985.
3. J. van Leeuwen and R.B. Tan. Routing with compact routing tables. In G. Rozemberg and A. Salomaa, editors, *The book of L*, pages 259–273. Springer-Verlag, 1986.
4. J. van Leeuwen and R.B. Tan. Interval routing. *The Computer Journal*, 30:298–307, 1987.

Toward a General Theory of Unicast-Based Multicast Communication*

Barbara D. Birchler, Abdol-Hossein Esfahanian and Eric Torng

Department of Computer Science
Michigan State University
East Lansing, MI 48824-1027
{birchler,esfahani,torng}@cps.msu.edu

Abstract. Multicast, also known as one-to-many communication, is
the problem of sending a message from a single source node to several
destination nodes. In this paper, we study the problem of implementing multicasts via a technique called Unicast-Based Multicast (UBM).
In particular, we focus on the problem of finding optimal UBM calling schedules for *arbitrary* topologies using restricted *inclusive* routing
schemes (a natural class of restricted routing schemes which includes
most practically implemented restricted routing schemes) given *source
limited* routing information. Previous work has focused on finding optimal UBM calling schedules for *arbitrary* topologies using free routing
schemes given complete routing information and *specific* topologies using
specific restricted routing schemes given complete routing information.
We first show that implementing UBM in arbitrary topologies with restricted routing schemes is fundamentally different than implementing
UBM in arbitrary topologies with free routing schemes. We then develop
the Smart Centroid Algorithm, a polynomial time approximation algorithm which produces UBM calling schedules that are at most four times
as long as the optimal UBM calling schedule for arbitrary topologies
using restricted inclusive routing schemes given source limited routing
information. The question of determining the NP-hardness of produc-
ing optimal UBM calling schedules for arbitrary topologies using any
restricted routing scheme remains open.

1 Introduction

In this paper, we study communication paradigms for parallel systems which
utilize the popular *direct network* interconnection structure [1]. Systems based on
direct networks are typically organized as a collection of connected, homogeneous
nodes, where each node is a programmable computer with its own processor and
memory [2]. Because the nodes do not share a common memory, data sharing
is typically implemented via message-passing. A message-passing request is an
ordered pair $M = (S, D)$, where S is the set of source nodes, and D is the set of
destination nodes. Message passing paradigms are classified according to the sizes

---

* This work was supported in part by NSF grant MIP-9204066.

of the source and destination sets. We consider the one-to-many, or *multicast*, communication paradigm in which a single source sends the same message to multiple destinations. In this case, $|S| = 1$ and $|D| \geq 1$. Two specific cases of multicast are *unicast* and *broadcast*. Unicast, or one-to-one communication, involves a single source and a single destination, *i.e.*, $|S| = |D| = 1$. In broadcast, also called one-to-all communication, $D = V(G) - S$.

The particular implementation used to satisfy a multicast request depends on two physical characteristics of the system: the topology and the routing scheme. The underlying topology of a direct network is typically modeled as a graph or as a directed graph (digraph for short). We use a general graph model to introduce the main concepts and use directed graphs in special cases described later in the paper. We represent a direct network topology by a graph $G = (V, E)$, where V is the set of nodes in the network, and E is the set of the network's communication channels. If there is a physical communication link between node v_i and node v_j, then $e = v_i v_j \in E(G)$. We assume G is connected[2] and simple, *i.e.*, there are no loops or parallel edges.

The routing scheme of the network also influences the multicast implementation. The *routing scheme* of a network consists of the set of message passing primitives provided by the hardware. A message passing primitive P is a triple (S, D, e) where $S \subseteq V(G)$ is the set of source nodes, $D \subseteq V(G)$ is the set of destination nodes, and $e \subseteq E(G)$ is the set of edges such that the subgraph induced by e, $< e >$, is used by P to deliver the message. Most existing routing schemes only contain unicast primitives. Consequently, we redefine P to be the triple (s, d, e) where $s \in V(G)$ is the single source of the message, $d \in V(G)$ is the single destination of the message, and $< e >$ is a path[3] in G from s to d, denoted $p(s, d)$. If $e = \{(s, d)\}$, P is said to use a *local* unicast primitive.

In most routing schemes, at least one unicast primitive is provided for every pair of nodes $(s, d) \in V(G) \times V(G)$. A *restricted routing scheme* contains exactly one unicast primitive for every pair of nodes $(s, d) \in V(G) \times V(G)$. A *free routing scheme* contains a unicast primitive for each (s, d)-path in G for all pairs of nodes $(s, d) \in V(G) \times V(G)$.

Because existing systems only provide unicast primitives, multicast requests with more than one destination node can only be satisfied by using multiple unicast primitives. This software technique for implementing multicast requests is known as Unicast-Based Multicast (UBM). In particular, Farley introduced a *calling schedule* implementation for UBM (or, more accurately, for unicast-based broadcast) [3, 4]. A calling schedule consists of several time steps of unit length[4]. In each time step, one or more unicast primitives can be executed in parallel provided they meet certain constraints. In the *node-switching* model, all the unicast primitives during a single time step must be totally node disjoint. In the *line-switching* model, all the unicast primitives during a single time step must be pairwise edge disjoint. If the system uses only local unicast primitives,

---

[2] Strongly connected in a digraph model.
[3] A *path* is an alternating sequence of vertices and edges in which no vertex is repeated.
[4] The unit length assumption is realistic because of wormhole routing technology [1].

both of the above models have identical constraints, and we will refer to this as the *neighbor-switching* model.

In this paper, we study the problem of finding minimum length line-switching and node-switching calling schedules for implementing an arbitrary multicast request given an *arbitrary topology* and an *arbitrary routing scheme*. A lower bound on the length of a line-switching or node-switching calling schedule for implementing a multicast request with n destinations is $\lceil \lg(n + 1) \rceil$. The problem of finding $\lceil \lg(n + 1) \rceil$ length line-switching calling schedules for arbitrary multicasts with n destinations in systems which provide free routing schemes has been solved. Farley gave a polynomial-time algorithm for broadcasting in any topology, and McKinley *et al.* [2] extended this result to arbitrary multicast requests. Polynomial-time algorithms which find $\lceil \lg(n + 1) \rceil$ length line-switching calling schedules for multicast in specific topologies which use specific restricted routing schemes have also been developed [2, 5]. Most of these results were made possible by the development of natural graph theoretic models that captured both the topology and the routing schemes considered. We address the development of a similar graph theoretic model for restricted routing schemes.

The remainder of the paper is organized as follows. In Section 2, we describe the calling schedule implementation model for UBM algorithms and three classes of UBM communication. In Section 3, we describe some of the complexities involved in graphically modeling a system with an arbitrary topology and an arbitrary restricted routing scheme. Given these difficulties, we focus on *restricted inclusive routing schemes*, a natural class of routing schemes that includes the specific restricted routing schemes used in most current multicomputers. We further assume the routing table at each node only contains entries for the unicast primitives using itself as the source. We call such a routing table a *source limited* routing table, and we show that source limited routing tables for systems with restricted inclusive routing schemes can be represented by directed trees. In Section 4, we address the problem of finding algorithms for generating minimum length node-switching and line-switching calling schedules for implementing an arbitrary multicast in a system that uses a restricted inclusive routing scheme given a source limited routing table. We first prove non-trivial lower bounds which show that implementing UBM in systems with restricted routing schemes is fundamentally different than implementing UBM in systems with free routing schemes. That is, we show that line-switching schedules of length $\lceil \lg(n + 1) \rceil$ do not always exist in arbitrary topologies with arbitrary restricted routing schemes. We then present two approximation algorithms for this problem. The better (but slower) algorithm produces line-switching schedules for arbitrary multicast requests which are at most a factor of four times longer than necessary. Section 5 concludes the paper.

2 UBM Calling Schedule Models

Farley describes an implementation model, called a *calling schedule*, for several types of unicast-based broadcast in systems using free-routing.[5] We generalize Farley's calling schedule to accommodate UBM in restricted routing systems. A calling schedule C consists of a set of unicasts where each unicast is characterized by four parameters: source, destination, path, and time. We represent a unicast in a calling schedule as an ordered quadruple $(i, j, p(i, j), t)$ which is read as, "Node i sends a message to node j along path $p(i, j)$ during time step t." For a multicast request $M = (S, D)$, where $S = \{s\}$ and $D = \{d_1, d_2, \ldots, d_n\}$, we define the *informed set* of M at time step t with respect to a calling schedule C, denoted $I_t^C(M)$, to be the set of all nodes from $S \cup D$ that have received the message by the end of time step t under calling schedule C. Thus, $I_0^C(M) = \{s\}$ for any C and any M, *i.e.,* only s is informed before any calls are made.

Farley defines three classes of unicast-based broadcast which we generalize to unicast-based multicast. The distinguishing characteristic in the three variations is the requirements on the paths used during a given time step t of the calling schedule C. In the *line-switching*[6] model, all the paths must be edge disjoint. In the *node-switching*[7] model, all the paths must be totally node disjoint. If the system uses only local unicast primitives, it uses the *neighbor-switching*[8] model. A calling schedule C can be used to implement each of the above classes of UBM communication. There are several requirements for a calling schedule to be *legal*. We present below the requirements for a calling schedule C that implements a multicast request M under the line-switching model.[9] The requirements for legal calling schedules that implement node- and neighbor-switching are identical except for an appropriate modification to condition (vi).

(i) If $(i, j, p(i, j), t) \in C$ and $(m, n, p(m, n), t) \in C$ then $i \neq j$ and $m \neq n$.

(ii) For all $(i, j, p(i, j), t) \in C$, $i \in I_{t-1}^C(M)$.

(iii) For all $(i, j, p(i, j), t) \in C$, $j \notin I_{t-1}^C(M)$.

(iv) There exists a t such that $I_t^C(M) = S \cup D$. We call t the *length* of C.

(v) For all $(i, j, p(i, j), t) \in C$, $i, j \in S \cup D$.

(vi) If $(i, j, p(i, j), t) \in C$ and $(m, n, p(m, n), t) \in C$ are distinct unicasts, then $p(i, j)$ and $p(m, n)$ are edge-disjoint.

In studying UBM algorithms for each model, the objective is to find an algorithm that always produces legal calling schedules of minimum length. A lower bound on the minimum length legal calling schedule for any of the three models is $\lceil \lg(n + 1) \rceil$, where n denotes the size of the destination set D [3].

---

[5] Farley assumes a graph model. All concepts are easily extended to a digraph model by making edges directed.

[6] Farley calls this line-broadcasting.

[7] Farley calls this path-broadcasting.

[8] Farley calls this local-broadcasting.

[9] Farley's requirements for a corresponding legal line-broadcasting calling schedule include only conditions (ii) and (vi).

Farley presents an algorithm for generating a minimum length broadcast calling schedule for line-switching in the general graph model [4]. An algorithm for minimum length multicast calling schedules in the same model is given in [2]. The multicast algorithm is based on constructing a trail[10] that includes all of $S \cup D$. In each step the "middle" of the trail is informed, and the trail is broken into two equal sized subtrails. This process continues until all destinations are informed. Since each trail is reduced to half its size at each step, all destinations are informed after $\lceil \lg(n+1) \rceil$ steps, where n is the number of destinations.

3 Restricted Routing

We stated earlier that the problem of finding optimal line-switching schedules for arbitrary multicast requests in systems with arbitrary topologies using free routing schemes has been solved. One of the key tools in solving this problem was the development of a natural and useful graph-theoretic model for representing free routing schemes in an arbitrary topology. In this section, we address the problem of developing a useful graph-theoretic model for representing arbitrary restricted routing schemes in arbitrary topologies.

A restricted routing scheme R can be classified according to the properties of the dipaths used by the unicast primitives in R. The property we consider is whether or not a unicast dipath $p(i, j) \in R$ implies anything about other unicast dipaths in R. In the most general case, the existence of unicast dipath $p(i, j)$ gives no information about other unicast dipaths. Developing a natural and useful graph-theoretic model for such a restricted routing scheme in an arbitrary topology is a difficult problem. One may use a multigraph[11] in which each edge is labeled with the (source,destination)-pair that uses it. Unfortunately, multiple edges represent a single physical communication line, so the notion of physically edge-disjoint paths is not present in the multigraph model. Other models we have considered have similar drawbacks. As a result, we have been unable to use any graph-theoretic representations of restricted routing schemes in arbitrary topologies to develop reasonable algorithms for generating UBM calling schedules.

In order to develop a useful graph-theoretic model, we narrow our focus to a specific class of restricted routing schemes R in which the existence of unicast dipath $p(i, j) \in R$ does provide some information about other unicast dipaths in R.

Definition 3.1 *A routing scheme R is* inclusive *if the existence of unicast primitive $(v_i, v_j) \in R$ with dipath $p = v_1 v_2 \ldots v_{k-1} v_k$, where $v_1 = v_i$ and $v_k = v_j$, implies that there exists a unicast primitive $(v_m, v_n) \in R$ that uses the dipath $v_m v_{m+1} \ldots v_{n-1} v_n$, a subpath of p, for $1 \leq m < n \leq k$.*

Restricted inclusive routing schemes are extremely easy to implement in practice. For any message, the only routing information required in the message header

---

[10] A *trail* is an alternating sequence of vertices and edges in which no edge is repeated.
[11] In a *multigraph* there can be multiple edges between a pair of vertices.

is the destination of the message. Each intermediate node on the path from the source to the destination will use this destination information to determine which outgoing link to use when forwarding the message. Thus, routing tables at each node require only $n-1$ entries. In a general restricted routing scheme, the routing information in the message header must include both the source and the destination of the message. This implies routing tables at each node may require $\Omega(n^2)$ entries. Furthermore, in a free routing scheme, the only routing information required in the message header is the destination node. However, routing tables at each node need to store all possible outgoing links which can be used to reach each destination node. Finally, having to dynamically choose which outgoing link to use may significantly increase message handling time and/or significantly increase router complexity and cost. As a result, most common direct network implementations use restricted inclusive routing schemes. Examples include xy routing in a mesh and e-cube routing in a hypercube. Thus, we shall focus on restricted inclusive routing schemes in the rest of this paper.

We next address how much routing information is stored at each node for an algorithm to use in searching for a minimum length UBM calling schedule. In the ideal case, the algorithm has complete routing information. Unfortunately, we still do not have any natural and useful graph-theoretic models for representing arbitrary restricted inclusive routing schemes in arbitrary topologies. In the opposite extreme, it is easy to verify that having only the minimum amount of information (that is, each node is only aware of the outgoing link used for each destination node) means that calling schedules for worst-case multicasts with n destinations will have length n. As a result, we consider an intermediate option where a node x knows all the dipaths used by every unicast primitive with source node x in the routing scheme R. We call such a routing table a *source limited* routing table.

Source limited routing tables of restricted inclusive routing schemes are easy to work with because we can represent them by *directed trees* (or *ditrees*) as defined in [6][12]. For example, consider a direct network with eight nodes where the unicast primitives with source node v_1 use the dipaths shown in Table 1. We can represent all these dipaths using the ditree in Figure 1. It is not hard to verify that such a ditree can be constructed for any source limited routing table of any restricted inclusive routing scheme.

In the rest of this paper, we focus on the problem of constructing minimum length UBM calling schedules in systems with restricted inclusive routing schemes given access to source limited routing tables (*i.e.*, ditrees). We now define some notation specific to this problem.

Definition 3.2 *A vertex v in a rooted ditree T with root r is at* level i *if and only if the length of the dipath $p(r,v)$ in T is i.*

Definition 3.3 *The largest integer h for which there is a vertex at level h in a*

---

[12] A digraph $G(V,E)$ is said to have a *root* r if $r \in V$, and there is a (r,v)-dipath for all $v \in V$. A digraph is called a *directed tree*, or *ditree*, if it has a root, and its underlying graph is a tree.

Table 1. Routes from node v_1

Routes from node v_1		
source	destination	dipath
v_1	v_2	$v_1 v_2$
v_1	v_3	$v_1 v_3$
v_1	v_4	$v_1 v_4$
v_1	v_5	$v_1 v_2 v_5$
v_1	v_6	$v_1 v_3 v_6$
v_1	v_7	$v_1 v_3 v_7$
v_1	v_8	$v_1 v_3 v_6 v_8$

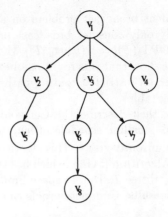

Fig. 1. Ditree with source v_1

rooted ditree is called its height. Let $H(T)$ denote the height of ditree T.

Definition 3.4 Let T be a directed tree. If directed edge $vu \in E(T)$, then u is called a child of v, and v is called the parent of u. Let $children_T(v)$ denote the set $\{u | u$ is a child of v in $T\}$, and $parent_T(u)$ denote node u's parent in T. If there is a dipath from v to u in T, then v is called an ancestor of u.

Definition 3.5 Let $|T|$ denote the number of nodes in T.

Definition 3.6 Let OPT_l, OPT_t, and OPT_n denote algorithms that only use dipaths available in the source limited routing table, and that always generate minimum length legal calling schedules under the line-switching, node-switching, and neighbor-switching models, respectively, in directed trees.

Definition 3.7 For any UBM algorithm A, let $A(M, T)$ denote the calling schedule produced by A for any multicast request M in any directed tree T, and let $|A(M, T)|$ denote the length of the calling schedule. When M is the broadcast problem, we only write $A(T)$.

4   Multicast in Directed Trees

In this section, we study UBM for arbitrary topologies that employ restricted inclusive routing and use only source limited routing tables. As was shown in the previous section, the routing table at the source node can be represented by a directed tree. Thus, this problem reduces to studying UBM in directed tree topologies.

We first show that for any multicast request M in any directed tree T, $|OPT_t(M, T)| \leq 2|OPT_l(M, T)|$. We subsequently restrict our attention to lower bounding and approximating OPT_t. We show that under the node-switching assumption, any multicast request in a directed tree T can be reduced to an equivalent broadcast problem on a reduced directed tree T'. Thus, we subsequently only consider broadcast problems. Within this context, we first show that $|\text{OPT}_t(T)| = \Omega(k \log_k |T|)$ when T is the complete, balanced, k-ary ditree, and we show that this result can be extended to apply to UBM algorithms which can use *all* restricted routing paths (*i.e.*, have access to a complete restricted routing table).

We then describe the Centroid Algorithm (CA) which generates a node-switching calling schedule of length at most $O(\log_{\frac{k+1}{k}} |T|)$ for any tree T with maximum outdegree k. This algorithm is modified to what we call the Smart Centroid Algorithm (SCA), which has the property that $|\text{SCA}(T)| \leq 2|\text{OPT}_t(M,T)|$ for any ditree T. Due to space limitations, we only present proof sketches of selected results. Complete proofs can be found in [7].

4.1 Relating Node-switching to Line-switching

An attractive feature of node-switching that makes analysis relatively easy is that each unicast decomposes the original multicast problem into two destination disjoint multicast subproblems. Specifically, suppose we want to find a calling schedule C to implement multicast request $M = (r, D)$, where D is the destination set in directed tree T. Suppose the call $(r, v, p(r, v), 1)$ is the single unicast in step 1 of C. Let T' be the subtree of T rooted at v. After the call is made, we can decompose M into two subproblems: implementing multicast request $M' = (r, D - V(T'))$ in directed tree $T - T'$, and implementing multicast request $M'' = (v, D - V(T - T'))$ in ditree T'.

The first question we study is: how does OPT_t compare to OPT_l?

Lemma 4.1 *There exists a ditree T such that $|OPT_t(T)| > |OPT_l(T)|$.*

Proof. Consider broadcast from the root of a tree T that has its root in level 0, one node at level 1, and two nodes at level 2. Clearly $|\text{OPT}_l(T)| = 2$ while $|\text{OPT}_t(T)| = 3$.

Lemma 4.2 *For all directed trees T and multicast requests M, $|OPT_t(M,T)| \leq 2|OPT_l(M,T)|$.*

Proof. We first state two facts about the directed paths used to perform unicasts in any time step of a legal line-switching calling schedule. First, for any uninformed node v, at most one unicast dipath will use that node. Second, for any informed node v, at most two unicast paths can use it. Now consider all unicasts occurring in time step t of a legal line-switching calling schedule. Let C_t be the set of all such unicasts. Because of the previous facts, the graph induced by the unicast dipaths in C_t is a *forest*, say F, of directed trees in which the maximum outdegree of any node is two. Furthermore, only leaves in F are destinations of unicast calls in step t. We will simulate the unicast calls of C_t using node disjoint paths. For each leaf ℓ in F, make the nearest informed ancestor responsible for informing ℓ. Since any such informed node can be responsible for at most two leaves, we can simulate the unicasts of C_t in at most two steps using node disjoint paths, and the lemma follows.

From Lemma 4.2, we know that a proof that $|OPT_t(M,T)| \geq c$ immediately translates into a proof that $|OPT_l(M,T)| \geq \frac{c}{2}$. Similarly, if we find an algorithm A such that $|A(M,T)| \leq c|OPT_t(M,T)|$ for any multicast request M on ditree T, then we know $|A(M,T)| \leq 2c|OPT_l(M,T)|$ for any multicast request M on ditree T. Therefore, we restrict our attention to lower bounding and approximating OPT_t in the remainder of this paper.

4.2 Reducing Multicast to Broadcast

Within the setting of node-switching algorithms, we show that we can reduce any multicast request M in ditree T to an equivalent broadcast request in ditree $f(T)$ where f is defined below.

Definition 4.1 *For any ditree T and multicast request $M = (r, D)$, define ditree $f(T)$ as follows. First, let $f(T) = T$. Suppose there are i levels in T. Consider each node v in level i sequentially. If $v \notin D$, then let $children_{f(T)}(parent_T(v))$ be $children_{f(T)}(parent_T(v)) \cup children_T(v) - \{v\}$ and remove v from T. Continue this process for levels $i-1, \ldots, 1$.*

Figure 2 gives an example of the mapping f. The multicast request $M = (r, \{v_2, v_3, v_4, v_6, v_7, v_8, v_{10}\})$ in directed tree T is transformed to its corresponding broadcast request in ditree $f(T)$. The dark nodes are destination nodes.

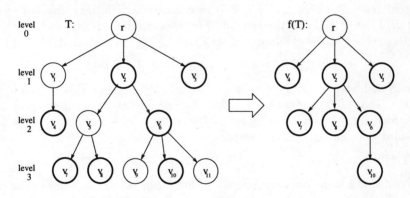

Fig. 2. Mapping multicast to broadcast.

Lemma 4.3 *For any multicast request M on any directed tree T, $|OPT_t(f(T))| = |OPT_t(M,T)|$.*

Proof. It is not hard to show that $|OPT_t(M,T)| \geq |OPT_t(f(T))|$ and $|OPT_t(f(T))| \geq |OPT_t(M,T)|$. Thus, $|OPT_t(f(T))| = |OPT_t(M,T)|$.

4.3 Lower Bound on Broadcast in k-ary Ditrees

In this section, we prove that performing broadcast in any complete, balanced k-ary directed tree T_k of order n requires $\Omega(k \log_k n)$ time steps under the line-switching model. First we give a lower bound on the length of an optimal neighbor-switching algorithm that implements broadcast.

Lemma 4.4 $|OPT_n(T_k)| = k\lfloor \log_k n \rfloor$, where T_k is the complete, balanced, k-ary ditree of order n.

Proof. Let T_k^ℓ be the complete, balanced k-ary ditree with ℓ levels. When the order of the ditree is specified, we simply write T_k since the number of levels depends on the number of nodes. Let $t(T_k^\ell)$ be the length of $OPT_n(T_k^\ell)$. It is not difficult to see that

$$t(T_k^\ell) = k + t(T_k^{\ell-1})$$
$$t(T_k^1) = k$$

Solving this recurrence relation gives the desired result that

$$t(T_k) = k\lfloor \log_k n \rfloor.$$

Now we show the relationship between OPT_n and OPT_t.

Lemma 4.5 *For all directed trees T, $|OPT_n(T)| \leq |OPT_t(T)| + H(T) - 1$.*

Proof. We prove this by induction on $H(T)$. The base case when $H(T) = 1$ is obvious. Now consider the inductive case. Assume that the lemma holds for all directed trees T' where $H(T') \leq k$. Now consider an arbitrary directed tree T with $H(T) = k + 1$. Let the root of T have m children. Name each of the m subtrees rooted at the children of T as T_i for $1 \leq i \leq m$ such that

$$|OPT_t(T_1)| \geq |OPT_t(T_2)| \geq \cdots \geq |OPT_t(T_m)|.$$

Note in each step of OPT_t, the root of T can only inform a single node in one subtree. Thus OPT_t cannot begin informing nodes in $m - 1$ of the subtrees until the second step. It cannot being informing nodes in $m - 2$ of the subtrees until the third step, and so on. This implies that

$$|OPT_t(T)| \geq \max\{|OPT_t(T_1)|, \ldots, |OPT_t(T_i)| + i - 1, \ldots, |OPT_t(T_m)| + m - 1\}.$$

We define a neighbor-switching algorithm for broadcast as follows. The root of T informs the root of T_1 in time step 1, the root of T_2 in time step 2, etc. In general, the root of T informs the root of T_i in time step i. The optimal neighbor-switching algorithm, OPT_n, clearly does at least as well as the algorithm we have defined, so

$$|OPT_n(T)| \leq \max\{|OPT_n(T_1)| + 1, \ldots, |OPT_n(T_i)| + i, \ldots, |OPT_n(T_m)| + m\}.$$

Let i be the index which maximizes $OPT_n(T_i) + i$ for $1 \leq i \leq m$. We know that $|OPT_n(T)| \leq |OPT_n(T_i)| + i$, $|OPT_t(T)| \geq |OPT_t(T_i)| + i - 1$, and $H(T_i) \leq H(T) - 1$. We apply our inductive hypothesis and these facts to show $|OPT_n(T)| \leq |OPT_t(T)| + H(T) - 1$.

Theorem 4.1 $|OPT_l(T_k)| \geq \frac{1+(k-1)\log_k n}{2} = \Omega(k \log_k n)$ *for any complete, balanced k-ary directed tree T_k of order n.*

Proof. The theorem follows directly from Lemmas 4.2, 4.5, and 4.4.

We now extend this lower bound example to show that even if an algorithm uses all available directed paths, it may still require at least $\Omega(k \log_k n)$ time to perform a broadcast in a specific network with n nodes and maximum indegree and outdegree of k. We construct a digraph D as follows. Let T_1 and T_2 be complete, balanced k-ary trees of order $\frac{n}{2}$, rooted at t_1 and t_2, respectively. Orient all edges of T_1 away from t_1, and orient all edge of T_2 toward t_2. We call t_1 the *source root* and t_2 the *sink root*. For each node v in T_1 with outdegree zero, make a directed edge vw, where w is the corresponding node with indegree zero in T_2. Finally, add directed edge t_2t_1. Figure 3 shows an example of digraph D when k is three. Suppose that our routing scheme R includes all the dipaths

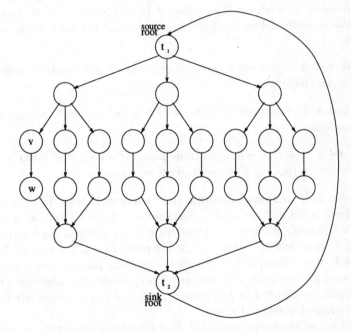

Fig. 3. Broadcast requires $\Omega(k \log_k n)$.

in D. We then have a routing scheme in which each node has at least one dipath to any other node in the network. Thus, if we can lower bound the time to perform broadcast under the line-switching model in this network, we also lower bound the time to perform broadcast under the line-switching assumption in this network with some restricted routing scheme. The key observation is that at most *one* dipath in any time step can use edge t_2t_1. Thus, performing broadcast in this line-switched network from the source root, t_1, can be done no faster than twice as fast as performing broadcast in the complete, balanced, k-ary tree. The number of nodes in the network modeled by D is twice the number of nodes in one of the complete, balanced, k-ary trees. Thus, the general lower bound result follows.

4.4   Centroid Algorithm

In Section 2, we described the technique used for generating a minimum-time line-switching calling schedule for direct networks that admit free routing. First, a trail containing the source and all destinations is constructed. Then, the trail is recursively broken into two trails of equal length during each time step. Breaking the problem in half at each step leads to $\lceil \lg(n+1) \rceil$ steps. In this section, we present the Centroid Algorithm (CA), which is built on a similar idea. At each step we attempt to break the directed tree representing the routing information into two *nearly equal* subtrees. We find an edge whose removal leaves two nearly equal sized trees. This edge will be adjacent to a *centroid* point of the underlying tree. We formally define the following terms that are used in the Centroid Algorithm.

Definition 4.2 *A* branch *at a vertex v of a tree T is a maximal subtree containing v as an endpoint.*

Definition 4.3 *The* weight *of a vertex v of T is the maximum number of edges in any branch at v.*

Definition 4.4 *A vertex c is a* centroid point *of a tree T if c has minimum weight, and the* centroid *of T consists of all such vertices.*

CA is a simple recursive algorithm that works as follows. In each step, we find a centroid point c of the tree T. Removal of c divides T into at most $k+1$ subtrees. We identify the largest one of these $k+1$ subtrees, and call it T_1. If the root of T_1 is a child of c, then inform the root of T_1. If the root of T_1 equals the root of T, then inform c. Now divide the problem into the two subproblems T_1 and $T - T_1$, and continue. CA is clearly a polynomial time algorithm because each call can be completed in polynomial time, and CA is recursively called at most a polynomial number of times.

Figure 4 shows the operation of CA on a broadcast request in an eight node ditree. The dark nodes are informed at the beginning of the current time step. In this case, CA produces an optimal calling schedule ($\lg(7+1) = 3$ time steps).

Theorem 4.2 $|CA(T)| = O(\log_{\frac{k+1}{k}} |T|)$ *for any directed tree T with maximum outdegree k.*

Proof. Let T' be the largest of the $k+1$ subtrees. Because c is a centroid point, $|T'| \leq \frac{|T|}{2}$ [8]. Because T' is the largest of the $k+1$ subtrees formed around c, $|T'| \geq \frac{|T|-1}{k+1}$. Clearly, the largest difference in size between T' and $T - T'$ occurs

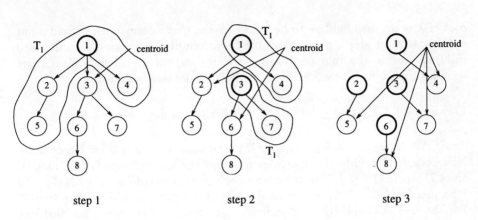

step 1   step 2   step 3

Fig. 4. Centroid Algorithm at work.

when $|T'| = \frac{|T|-1}{k+1}$, and $|T - T'| = |T| - \frac{|T|-1}{k+1} = \frac{k|T|+1}{k+1}$. Therefore, in the worst case, the size of the problem decreases by nearly a factor of $\frac{k}{k+1}$ in each time step and the result follows.

4.5 Smart Centroid Algorithm

While the Centroid Algorithm works well on many trees, particularly those which are relatively uniform in structure or those which have small maximum outdegree, it can perform quite poorly on others. The problem results from the fact that the algorithm always chooses to inform the subtree with the maximum number of nodes first. In some cases, this is a poor choice as another subtree with slightly fewer nodes may be the more time consuming tree in which to perform a broadcast. To illustrate this concept, consider the tree T which has nk nodes and is defined as follows. The root of T is its centroid point, and it has k children. The first $k - 1$ children all are the root of subtrees which are paths of length $n - 1$. The final child is the root of a tree which is structured identically to T but has only $n - 1$ nodes. The centroid algorithm will produce a calling schedule of length $\Omega(k \log_k |T|)$, whereas the optimal calling schedule will have length $O(k + \log_k |T|)$.

We present below a new algorithm called the Smart Centroid Algorithm (SCA). SCA is similar to CA in that for any unicast call $(v_1, v_2, p(v_1, v_2), t) \in CA$ there is a corresponding unicast call $(v_1, v_2, p(v_1, v_2), t') \in SCA$. The major difference is that t and t' may not be the same. SCA employs a better heuristic for determining which child of the centroid point to inform first. We prove that $|SCA(T)| \leq 4|OPT_l(T)|$ for all directed trees T.

The Smart Centroid Algorithm works as follows. After identifying a centroid point c which has outdegree $m - 1$, we decompose the problem into the subproblems $C_1, C_2, \ldots, C_m$ determined by the components of $T - c$. We compute $|SCA(C_i)|$ for each $i \leq m$. The root begins to inform these components in non-increasing order of calling schedule length. When the component containing the root of T is the next subtree to be informed, the root informs the centroid point c. In subsequent steps, c is responsible for informing its remaining uninformed children, freeing the root to inform its own component. While SCA is more complex than CA, it obviously is still a polynomial time algorithm.

Lemma 4.6 $|SCA(T)| \leq |OPT_t(T)| + \lg|T|$ *for all directed trees* T.

Proof. We prove this by induction on $|T|$. The base case when $|T| = 1$ is obvious. Now consider the inductive case. Assume that the lemma holds for all directed trees T' where $|T'| \leq j$. Now consider an arbitrary directed tree T with $|T| = 2j$ (the proof for $|T| = 2j + 1$ is analogous). Find the centroid of this tree T. For the sake of simplicity, assume it is the root (the other case is handled in a nearly identical fashion). Let the root have m children. Break the tree into the m distinct subtrees rooted at the children of the root. Name these m subtrees so that $|SCA(T_1)| \geq |SCA(T_2)| \geq \cdots \geq |SCA(T_m)|$. Note $|T_i| \leq \frac{|T|}{2} = j$ for $1 \leq i \leq m$, so the induction hypothesis applies to each of these subtrees. Note also

$$|SCA(T)| = \max\{|SCA(T_1)| + 1, \cdots, |SCA(T_i)| + i, \cdots, |SCA(T_m)| + m\}.$$

Now let $\sigma(i)$ be a permutation on $\{1, \cdots, m\}$ such that

$$|OPT_t(T_{\sigma(i)})| \geq |OPT_t(T_{\sigma(2)})| \geq \cdots \geq |OPT_t(T_{\sigma(m)})|.$$

Note, in any time step of OPT_t the root can only inform nodes in one of these m subtrees. This implies that $|OPT_t(T)| \geq \max_{1 \leq i \leq m}\{|OPT_t(T_{\sigma(i)})| + i - 1\}$. To relate these two sets of values, we know from our induction hypothesis that $|OPT_t(T_1)| \geq |SCA(T_1)| - \lg j$, so $|OPT_t(T_{\sigma(1)})| \geq |SCA(T_1)| - \lg j$. Similarly, $|OPT_t(T_2)| \geq |SCA(T_2)| - \lg j$ and $|OPT_t(T_1)| \geq |SCA(T_1)| - \lg j \geq |SCA(T_2)| - \lg j$, so $|OPT_t(T_{\sigma(2)})| \geq |SCA(T_2)| - \lg j$. In general, $|OPT_t(T_{\sigma(i)})| \geq |SCA(T_i)| - \lg j$.

Now let $1 \leq i \leq m$ be the value such that $|SCA(T)| = |SCA(T_i)| + i$. Combining this with the above fact gives $|OPT_t(T_{\sigma(i)})| + i + \lg j \geq |SCA(T)|$. We know that $|OPT_t(T)| \geq |OPT_t(T_{\sigma(i)})| + i - 1$, thus $|SCA(T)| \leq |OPT_t(T)| + \lg|T|$, and we are done.

Lemma 4.7 $|SCA(T)| \leq 2|OPT_t(T)|$ *for all directed trees* T.

Proof. This follows from Lemma 4.6 and the fact that $|OPT_t(T)| \geq \lg|T|$.

Theorem 4.3 $|SCA(T)| \leq 4|OPT_l(T)|$ *for all directed trees* T.

Proof. This follows from Lemma 4.7 and Lemma 4.2.

5 Conclusion

This paper represents a first step toward a general theory of unicast-based multicast in direct network systems. We consider only systems which use restricted inclusive routing schemes and source limited routing tables because we can represent these routing tables with directed trees. We show that multicast requests in such systems can be transformed into equivalent broadcast requests with only a constant factor difference in the length of the calling schedule. We then address broadcast in directed trees, showing that $\mathrm{OPT}_l(T) = \Omega(k\log_k n)$ for the complete, balanced k-ary directed tree T. Finally, we describe two algorithms, CA and SCA, for performing broadcast in directed trees and show that the length of the calling schedule produced by SCA is at most four times the length of an optimal line-switching calling schedule.

Several issues remain unresolved. First, the existence of useful graph-theoretic models for representing arbitrary restricted routing schemes and complete routing tables in arbitrary topologies is undetermined. Positive progress in this problem may lead to success in the next two problems. Second, the complexity of finding minimum length UBM calling schedules in systems with arbitrary restricted routing schemes or arbitrary restricted inclusive routing schemes given access to source limited or complete routing tables is unknown. Finally, assuming these problems are NP-hard, our approximation results only apply to systems with restricted inclusive routing schemes and source limited routing tables. The question of existence of similar approximation algorithms in systems with *general* restricted routing schemes and *complete* routing tables is undetermined.

References

1. L. M. Ni and P. K. McKinley, "A survey of wormhole routing techniques in direct networks," *IEEE Computer*, vol. 26, pp. 62–76, February 1993.
2. P. K. McKinley, H. Xu, A-H. Esfahanian, and L. M. Ni, "Unicast-based multicast communication in wormhole-routed networks," *IEEE Transactions on Parallel and Distributed Systems*, vol. 5, pp. 1252–1265, December 1994.
3. A. M. Farley, "Minimal broadcast networks," *Networks*, vol. 9, pp. 313–332, 1979.
4. A. M. Farley, "Minimum-time line broadcast networks," *Networks*, vol. 10, pp. 59–70, 1980.
5. D. F. Robinson, P. K. McKinley, and B. H. C. Cheng, "Optimal multicast communication in wormhole-routed torus networks," in *1994 International Conference on Parallel Processing*, pp. I–134–I–141, 1994.
6. S. Even, *Graph Algorithms*. Rockville, MD: Computer Science Press, Inc., 1979.
7. B. D. Birchler, A-H. Esfahanian, and E. Torng, "Toward a general theory of unicast-based multicast communication," Tech. Rep. CPS-95-3, Michigan State University, March 1995.
8. F. Harary, *Graph Theory*. Addison-Wesley Series In Mathematics, Reading, Massachusetts: Addison-Wesley Publishing Company, 1969.

Optimal Cutwidths and Bisection Widths of 2- and 3-Dimensional Meshes

José Rolim[1], Ondrej Sýkora[2,*], Imrich Vrt'o[2,**]

[1] Centre Universitaire d'Informatique, Université Genéve,
24 rue Général Dufour, CH 1211 Genéve, Switzerland
[2] Institute for Informatics, Slovak Academy of Sciences,
P.O.Box 56, 840 00 Bratislava, Slovak Republic

Abstract. We prove exact cutwidths and bisection widths of ordinary, cylindrical and toroidal meshes. This answers an open problem in [9]. We also give upper bounds for cutwidths and bisection widths of many dimensional meshes. Furthemore, we show the exact cyclic cutwidth of 2-dimensional toroidal meshes and prove optimal upper and lower bounds for other meshes.

1 Introduction

The cutwidth problem is to find a linear layout of vertices of a graph G so that the maximum number of cuts of a line separating consecutive vertices is minimized. The motivation comes from VLSI design [12]. The corresponding decision problem is $NP-$complete [6] but it is solvable in polynomial time for trees [15]. Very little is known about exact values of cutwidths for standard graphs. To our knowledge, known results include complete binary trees [10], hypercubes [3, 14] and some generalized hypercubes, i.e. products of equal paths of odd length [13] and products of complete graphs [11]. Some recent approximations of cutwidths of shuffle exchange graphs and related networks are in [2]. See also a survey [5]. The cutwidth problem is closely related to the so called isoperimetric problems, i.e. for a given m find a subset of m vertices of a graph with minimal edge boundary. A complete solution of this problem for finite graphs is known for hypercubes [7, 8], the product of complete graphs [11] and the product of two paths [1] only. For many dimensional equal sided ordinary and toroidal meshes Bollobás and Leader [4] solved the problem for reasonable large set of values of m.

In this paper, we prove exact cutwidths and bisection widths of ordinary, cylindrical and toroidal 2-dimensional meshes. E.g. the cutwidth of a cylindrical mesh (product of a path P_m with a cycle C_n) is $\min\{2m+1, n+2\}$. Further, we show cutwidths of 3-dimensional ordinary and toroidal meshes and extend the

---

* This research was partially supported by grant No. 2/1138/94 of Slovak Grant Agency and by EC Cooperative Action IC1000 "Algorithms for Future Technologies" (Project ALTEC)
** Partially supported by the Swiss National Science Foundation grant No. 20-40354-94

method to finding bisection widths of the above meshes. E.g. the bisection width of the mesh $P_{n_1} \times P_{n_2} \times P_{n_3}$ is $n_1 n_2 + n_1$, whenever n_3 is odd, n_2 is even and $n_1 \leq n_2 \leq n_3$. This partially answers the problem (R 1.277) of the Leighton's book [9] about the exact bisection width of many dimensional mesh. We also give some upper bounds on cutwidths and bisection widths of many dimensional meshes. Finally, we study the cyclic cutwidth problem, i.e. to find a circular layout of vertices of a graph so that the number of cuts of a line separating consecutive vertices is minimized. A partial result for the cyclic cutwidth of the hypercube is in [3]. For the cyclic cutwidth of a 2-dimensional toroidal mesh we give an exact result, while for other kinds of meshes we prove partially optimal upper and lower bounds.

2   Definitions and Notations

The cutwidth and the cyclic cutwidth are special cases of the so called congestion, the fundamental concept in the theory of interconnection networks. Therefore it is more convenient to define them through the congestion.
Let $G_1 = (V_1, E_1)$ and $G_2 = (V_2, E_2)$ be graphs such that $|V_1| = |V_2|$. An embedding of G_1 in G_2 is a couple of mappings (ϕ, ψ) satisfying

$$\phi : V_1 \rightarrow V_2 \quad \text{is a bijection}$$

$$\psi : E_1 \rightarrow \{ \text{ set of all paths in } G_2 \},$$

such that if $uv \in E_1$ then $\psi(uv)$ is a path between $\phi(u)$ and $\phi(v)$. Define the congestion of an edge $e \in E_2$ under the embedding ϕ, ψ of G_1 in G_2 as

$$cg(G_1, G_2, \phi, \psi, e) = |\{f \in E_1 \ : \ e \in \psi(f)\}|,$$

the congestion of G_1 in G_2 under ϕ, ψ as

$$cg(G_1, G_2, \phi, \psi) = \max_{e \in E_2}\{cg(G_1, G_2, \phi, \psi, e)\}$$

and the congestion of G_1 in G_2 as

$$cg(G_1, G_2) = \min_{\phi, \psi}\{cg(G_1, G_2, \phi, \psi)\}.$$

Let P_n and C_n denote the $n-$vertex path and cycle, respectively.
Let $V_{P_n} = \{1, 2, 3, ..., n\}$, $E_{P_n} = \{i(i+1) : i = 1, 2, ..., n-1\}$. Vertices and edges of C_n are defined similarly.
Let $G = (V, E)$ be an $n-$vertex graph. Define the cutwidth of G as

$$cw(G) = cg(G, P_n),$$

and the cyclic cutwidth of G as

$$ccw(G) = cg(G, C_n).$$

The bisection width of a graph G, denoted by $bw(G)$, is the minimum number of edges that have to be removed in order to partition the graph into two parts where the number of vertices differ by at most one.

For a set $A \subset V_G$ let $\partial_G(A)$ denote the set of all edges having one end in A and the second end in $V - A$. We call $\partial_G(A)$ the edge boundary of A. If it is clear from the context, which graph G is considered, then the subscript G is omitted. Hence $bw(G) = \min\{|\partial(A)| : A \subset V_G, |A| = \lfloor|V_G|/2\rfloor\}$.

Let $G_1 \times G_2$ denote the graph with the vertex set $V_1 \times V_2$ and in which $(i,j), (r,s)$ are adjacent iff either $i = r$ and $js \in E_2$ or $j = s$ and $ir \in E_1$.

Let $2 \le n_1 \le n_2 \le ... \le n_k$. Denote $M_k = \prod_{i=1}^{k} P_{n_i}$ and $TM_k = \prod_{i=1}^{k} C_{n_i}$. Fix a number i, $1 \le i \le k$, and a vertex $x = (x_1, x_2, ..., x_{i-1}, 1, x_{i+1}, ...x_k) \in V_{M_k}$. The i-section of M_k in the vertex x, denoted by $S_i(x)$ is defined as

$$S_i(x) = \{(x_1, x_2, ..., x_{i-1}, j, x_{i+1}, ..., x_k) : 1 \le j \le n_j\}.$$

For $A \subset V_{M_k}$, we define, $C_i(A_x)$, the i-compression of A according to the vertex x as

$$C_i(A_x) = \begin{cases} \emptyset, \text{if } A \cap S_i(x) = \emptyset, \\ \{(x_1, ..., x_{i-1}, l, x_{i+1}, ..., x_k) : 1 \le l \le |A \cap S_i(x)|\}, \text{otherwise.} \end{cases}$$

Then the i-compression of A, denoted by $C_i(A)$, is the union of all $C_i(A_x)$ over all $x = (x_1, x_2, ..., x_{i-1}, 1, x_{i+1}, ...x_k) \in V_{M_k}$. Finally, we say that A is compressed if it is i-compressed for all i. A similar definition can be introduced for TM_k.

3 Basic Lemmata

Our lower and upper bounds are essentially based on the following lemmata:

Lemma 3.1 Let $G = (V, E)$ be a graph. Then

$$cw(G) \ge \max_i \min_{\substack{A \subset V \\ |A|=i}} \{|\partial(A)|\},$$

$$ccw(G) \ge \max_i \min_{\substack{A \subset V \\ |A|=i}} \{|\partial(A)|\}/2.$$

Proof. We prove the first statement only. The proof of the second one is similar. Suppose G has n vertices. Let (ϕ, ψ) be the optimal (with respect to cutwidth) embedding of G in P_n. For $i = 1, 2, ..., n-1$ denote $B_i = \{1, 2, ..., i\} \subset V_{P_n}$. If $uv \in \partial(\phi^{-1}(B_i))$ then the edge $i(i+1)$ belongs to the path $\psi(uv)$. Hence

$$cw(G) \ge |\partial(\phi^{-1}(B_i))| \ge \min_{\substack{A \subset V \\ |A|=i}} \{|\partial(A)|\}.$$

As i was chosen arbitrarily, we get the claimed result. □

Lemma 3.2 For arbitrary graphs $G_1 = (V_1, E_1)$ and $G_2 = (V_2, E_2)$

$$cw(G_1 \times G_2) \le \min\{|V_2|cw(G_1) + cw(G_2), |V_1|cw(G_2) + cw(G_1)\}.$$

Proof. Consider an embedding $(\phi_1, \psi_1)((\phi_2, \psi_2))$ of $G_1(G_2)$ in $P_{|V_1|}(P_{|V_2|})$ with $cw(G_1)(cw(G_2))$. Define an embedding (ϕ, ψ) of $G_1 \times G_2$ in $P_{|V_1||V_2|}$ as follows

$$\phi(u, v) = \phi_1(u) + |V_1|(\phi_2(v) - 1),$$

$\psi((u_1, v_1), (u_2, v_2))$ is the unique path between $\phi(u_1, v_1)$ and $\phi(u_2, v_2)$. It is easy to see that

$$cw(G_1 \times G_2) \leq |V_1|cw(G_2) + cw(G_1).$$

The second embedding is obtained by changing indices 1 and 2 in the equation for ϕ. □

Lemma 3.3 *Let G_1, G_2 and G_3 be graphs such that $|V_{G_1}| = |V_{G_2}| = |V_{G_3}|$. Then*

$$cg(G_1, G_2)cg(G_2, G_3) \geq cg(G_1, G_3),$$

$$bw(G_2)cg(G_1, G_2) \geq bw(G_1).$$

Proof. The second claim was proved by Leighton [9]. Any embedding of G_1 into G_2 followed by any embedding of G_2 into G_3 defines an embedding (ϕ, ψ) of G_1 into G_3. It can be shown that

$$cg(G_1, G_3, \phi, \psi) \leq cg(G_1, G_2)cg(G_2, G_3).$$

□

Lemma 3.4 [1] *Let $A \subset M_2$.*

$$|\partial(A)| = \begin{cases} 2\left\lceil \sqrt{|A|} \right\rceil, & \text{if } |A| \leq \lfloor \frac{n_1^2}{4} \rfloor, \\ n_1 + \left\lceil \frac{|A| \bmod n_1}{n_1} \right\rceil, & \text{if } \lfloor \frac{n_1^2}{4} \rfloor \leq |A| \leq \frac{n_1 n_2}{2} \end{cases}$$

Lemma 3.5 [4] *For each subset A of M_k or TM_k and $i = 1, 2, ..., k$ it holds*

$$|\partial(A)| \geq \partial(C_i(A)).$$

Lemma 3.6 *Bisection widths of toroidal and ordinary meshes satisfy*

$$bw(TM_k) = 2bw(M_k).$$

Moreover if $cw(M_k) = \max_i \min\{|\partial(A)| : A \subset M_k, |A| = i\}$ then

$$cw(TM_k) = 2cw(M_k),$$

Proof. We sketch the proof of the second statement. The first one can be proved in a similar way. The inequality $cw(TM_k) \leq 2cw(M_k)$ follows from Lemma 3.3 by setting $G_1 = TM_k, G_2 = M_k$ and $G_3 = P_{n_1...n_k}$ and observing that $cg(TM_k, M_k) = 2$. Let A_0 be the subset of M_k for which

$$|\partial_{M_k}(A_0)| = \max_i \min\{|\partial(A)| : A \subset M_k, |A| = i\}.$$

Consider the set A_0 as a subset of TM_k. As we may assume that A_0 is compressed in M_k it is also compressed in TM_k. This implies

$$2cw(M_k) = 2|\partial_{M_k}(A_0)| = |\partial_{TM_k}(A_0)|.$$

We claim that

$$|\partial_{TM_k}(A_0)| \leq cw(TM_k).$$

Suppose by the contrary that

$$|\partial_{TM_k}(A_0)| > cw(TM_k).$$

Then

$$2cw(M_k) = |\partial_{TM_k}(A_0)| > cw(TM_k) \geq \max_i \min\{|\partial_{TM_k}(B)| : B \subset TM_k, |B| = i\}$$

$$= \max_i \min\{|\partial_{TM_k}(B)| : B \subset TM_k, |B| = i, B \text{ is compressed}\}$$

$$= \max_i \min\{2|\partial_{M_k}(B)| : B \subset M_k, |B| = i, B \text{ is compressed}\} = 2cw(M_k),$$

a contradiction. $\qquad\qquad\qquad\qquad\qquad\qquad\qquad\qquad\qquad\qquad\qquad\qquad\qquad\qquad\square$

4 Two-dimensional Meshes

In this section we give exact results on cutwidths and bisection widths of 2-dimensional meshes.

Theorem 4.1 For $m, n \geq 2$

$$cw(P_m \times P_n) = \begin{cases} 2, & \text{if } m = n = 2 \\ \min\{m+1, n+1\}, & \text{otherwise.} \end{cases}$$

For $m, n \geq 3$

$$cw(C_m \times C_n) = \min\{2m+2, 2n+2\}.$$

For $m \geq 2, n \geq 3$

$$cw(P_m \times C_n) = \begin{cases} 4, & \text{if } m = 2, n = 3 \\ \min\{2m+1, n+2\}, & \text{otherwise.} \end{cases}$$

Proof. The first result follows from Lemma 3.2 and Lemma 3.4, the second one from Lemmata 3.2, 3.6, and the first result. The results can also be proven by the method we use for the third result. Assume first that $m = 2$. If $n = 3$ then the upper bound is obtained by the following embedding ϕ of $P_2 \times C_3$ in P_6: $\phi(1, i) = i$ and $\phi(2, i) = 7 - i$, for $i = 1, 2, 3$. The lower bound follows from Lemma 3.1 showing that if $A \subset V_{P_2 \times C_3}, |A| = 4$ then $|\partial(A)| \geq 4$.
If $n > 3$ then the upper bound is obtained from Lemma 3.2. The lower bound follows from Lemma 3.1 showing that if $A \subset V_{P_2 \times C_3}, |A| = 5$ then $|\partial(A)| \geq 5$.

Assume $m > 2$. Then the upper bound follows from Lemma 3.2. Let A be an arbitrary subset of vertices of $P_m \times C_n$ satisfying

$$|A| = \left\lfloor \frac{mn}{2} \right\rfloor + 1.$$

To apply Lemma 3.1 we will estimate the cardinality of the edge boundary of A. Let us say, that the cylindrical mesh $P_m \times C_n$ has n rows and m columns. Let r and c denote the numbers of rows and columns containing at least one vertex from A, respectively. Clearly, $rc \geq |A|$. Let r^* and c^* denote the number of rows and columns containing only vertices from A, respectively. Distinguish 4 cases:

1. Let $r^* \neq 0, c^* \neq 0$. Then

$$|\partial(A)| \geq n - r^* + 2(m - c^*) \geq 2\sqrt{2(n - r^*)(m - c^*)}.$$

   Further

   $$mr^* + nc^* - r^*c^* \leq \left\lfloor \frac{mn}{2} \right\rfloor + 1,$$

   which implies

   $$(n - r^*)(m - c^*) \geq \left\lceil \frac{mn}{2} \right\rceil - 1.$$

   Hence

   $$|\partial(A)| \geq 2\sqrt{2\left\lceil \frac{mn}{2} \right\rceil - 2}.$$

   1.1. Suppose $2m > n + 1$.
        If $n = 3$ then $m \geq 3$ and $|\partial(A)| \geq 5$. If $n \geq 4$ then
        $$|\partial(A)| \geq 2\sqrt{2\left\lceil \frac{(n+2)n}{4} \right\rceil} - 2 \geq n + 2.$$
   1.2. Suppose $2m \leq n + 1$. Then for $m \geq 3$, $|\partial(A)| \geq 2\sqrt{2\left\lceil \frac{m(2m-1)}{2} \right\rceil} - 2 \geq 2m + 1$.
2. Assume $r^* \neq 0, c^* = 0$. Then $|\partial(A)| \geq r - r^* + 2m$. We claim that $r - r^* \geq 1$. Otherwise $r = r^*$, which implies that $\left\lfloor \frac{mn}{2} \right\rfloor + 1$ is divisible by m. But this is impossible when $m > 2$, a contradiction.
3. Assume $r^* = 0, c^* \neq 0$. Then $|\partial(A)| \geq n + 2(c - c^*)$. We claim that $c - c^* \geq 1$, otherwise $c = c^*$, which implies that $\left\lfloor \frac{mn}{2} \right\rfloor + 1$ is divisible by n, a contradiction.
4. Let $r^* = c^* = 0$. Then

$$|\partial(A)| \geq r + 2c \geq 2\sqrt{2rc} \geq 2\sqrt{2|A|} \geq 2\sqrt{2\left\lfloor \frac{mn}{2} \right\rfloor + 2} \geq \min\{2m+1, n+2\},$$

using the estimations from the case 1. $\qquad\square$

By using the same method we can prove:

Theorem 4.2 *For $2 \leq m \leq n$*

$$bw(P_m \times P_n) = \begin{cases} m, & \text{if } n \text{ is even} \\ m+1, & \text{if } n \text{ is odd.} \end{cases}$$

For $3 \leq m \leq n$

$$bw(C_m \times C_n) = 2bw(P_m \times P_n).$$

For $m \geq 2, n \geq 3$

$$bw(P_m \times C_n) = \begin{cases} \min\{2m, n\}, & \text{if } m, n \text{ are even} \\ \min\{2m, n+2\}, & \text{if } m \text{ is odd and } n \text{ is even} \\ \min\{2m+1, n\}, & \text{if } m \text{ is even and } n \text{ is odd} \\ \min\{2m+1, n+2\}, & \text{if } m, n \text{ are odd.} \end{cases}$$

5 Three and More Dimensional Meshes

Theorem 5.1 *For arbitrary $2 \leq n_1 \leq n_2 \leq n_3$ it holds*

$$cw(P_{n_1} \times P_{n_2} \times P_{n_3}) = \begin{cases} 5, & \text{if } 2 = n_1 = n_2 = n_3 \\ 6, & \text{if } 2 = n_1 = n_2 < n_3 \\ n_1 n_2 + n_1 + 1, & \text{otherwise.} \end{cases}$$

Proof. The first result, included for completeness, is the cutwidth of 3-dimensional hypercube, mentioned in the introduction. The second case follows from Theorem 4.1 as $P_2 \times P_2 \times P_{n_3} = C_4 \times P_{n_3}$. The upper bound for the third case follows by repeated application of Lemma 3.2. Now we give a sketch of proof of the lower bound for the third case. We use a similar approach as Bollobás and Leader [4] for estimating a topological boundaries of subsets of unit continuous cubes. Consider an arbitrary set $A \subset V_{M_3}, |A| = n_1 n_2 \lfloor n_3/2 \rfloor + n_1 \lfloor n_2/2 \rfloor + \lfloor n_1/2 \rfloor$. We will estimate the edge boundary of A. Therefore we may assume that A is a compressed set. For $t, 1 \leq t \leq n_3$, denote

$$A(t) = A \cap \{(x_1, x_2, t) : 1 \leq x_i \leq n_i, i = 1, 2.\}.$$

Then

$$n_1 n_2 \geq |A(1)| \geq |A(2)| \geq ... \geq |A(n_3)| \geq 0$$

and

$$|A| = \sum_{t=1}^{n_3} |A(t)|.$$

Let h be the maximum number s.t. $|A(h)| \neq 0$. Distinguish four cases

1. $|A(1)| = n_1 n_2, h = n_3$.
2. $|A(1)| = n_1 n_2, h < n_3$.
3. $|A(1)| < n_1 n_2, h = n_3$.
4. $|A(1)| < n_1 n_2, h < n_3$.

We will consider the fourth case only, other cases are similar. For each i, $1 \le i \le h$, there exists a real number $\alpha_i, 0 \le \alpha_i \le 1$, such that

$$|A(i)| = \alpha_i |A(1)| + (1 - \alpha_i)|A(h)|.$$

Then

$$|A| = \sum_{i=1}^{h} |A(i)| = |A(1)| \sum_{i=1}^{h} \alpha_i + |A(h)| \sum_{i=1}^{h} (1 - \alpha_i).$$

Denote $\sum_{i=1}^{h} \alpha_i = \alpha$. Then

$$|A| = \alpha |A(1)| + (h - \alpha)|A(h)|,$$

where $0 \le \alpha \le h$. Define a function $F : [0, n_1 n_2] \to \mathbf{R}$ as follows

$$F(x) = \begin{cases} 2\sqrt{x}, & \text{if } x \le \left(\frac{n_1}{2}\right)^2 \\ n_1, & \text{if } \left(\frac{n_1}{2}\right)^2 \le x \le n_1 n_2 - \left(\frac{n_1}{2}\right)^2, \\ 2\sqrt{n_1 n_2 - x}, & \text{if } x \ge n_1 n_2 - \left(\frac{n_1}{2}\right)^2. \end{cases}$$

According to Lemma 3.4, if $D \subset V_{M_2}$ then $|\partial(D)| \ge F(|D|)$. The function $F(x)$ is a concave function, hence

$$F(|A(i)|) = F(\alpha_i |A(1)| + (1 - \alpha_i)|A(h)|) \ge \alpha_i F(|A(1)|) + (1 - \alpha_i) F(|A(h)|).$$

Consequently

$$\sum_{i=1}^{h} F(|A(i)|) \ge \alpha F(|A(1)|) + (h - \alpha) F(|A(h)|).$$

Then counting the boundary edges of A according to dimensions we get

$$|\partial(A)| \ge |A(1)| + \sum_{i=1}^{h} |\partial(A(i))| \ge |A(1)| + \sum_{i=1}^{h} F(|A(i)|)$$
$$\ge |A(1)| + \alpha F(|A(1)|) + (h - \alpha) F(|A(h)|).$$

First suppose $|A(1)| = |A(h)|$. If

$$\left(\frac{n_1}{2}\right)^2 \le |A(1)| \le n_1 n_2 - \left(\frac{n_1}{2}\right)^2$$

then

$$|\partial(A)| \ge |A(1)| + hF(|A(1)|) \ge |A(1)| + \frac{|A|}{|A(1)|} n_1 \ge 2\sqrt{|A|n_1} \ge n_1 n_2 + n_1 + 1.$$

If

$$|A(1)| \ge n_1 n_2 - \left(\frac{n_1}{2}\right)^2$$

then

$$|\partial(A)| \ge |A(1)| + \frac{2|A|}{|A(1)|} \sqrt{n_1 n_2 - |A(1)|}.$$

The right hand side expression is a non-increasing function of the argument $|A(1)|$. Assume $|A(1)| = n_1 n_2 - 1$. Then it must hold $n_3 \geq n_1 + 1$, otherwise $|A|/(n_1 n_2 - 1)$ is not an integer. In this case

$$|\partial(A)| \geq n_1 n_2 - 1 + \frac{2|A|}{n_1 n_2 - 1} \geq n_1 n_2 + n_1 + 1.$$

Assume $|A(1)| \leq n_1 n_2 - 2$. Then similarly

$$|\partial(A)| \geq n_1 n_2 - 2 + \frac{2|A|\sqrt{2}}{n_1 n_2 - 2} \geq n_1 n_2 + n_1 + 1.$$

Second, suppose $|A(1)| > |A(h)|$. We distinguish four cases according to possible intervals for $|A(1)|, |A(h)|$, introduced in the definition of the function F. We describe the first case only. Other cases are similar. Let

$$\left(\frac{n_1}{2}\right)^2 \leq |A(h)| < |A(1)| \leq n_1 n_2 - \left(\frac{n_1}{2}\right)^2.$$

Then

$$|\partial(A)| \geq |A(1)| + \alpha n_1 + (h - \alpha) n_1 = |A(1)| + h n_1$$
$$\geq \frac{|A|}{h} + h n_1 \geq 2\sqrt{|A| n_1} \geq n_1 n_2 + n_1 + 1.$$

$\square$

Corollary 5.1 *The cutwidth of the 3-dimensional toroidal mesh satisfies*

$$cw(TM_3) = 2cw(M_3).$$

Proof. Let $A \subset V_{M_3}$ be the set of the first $n_1 n_2 \lfloor n_3/2 \rfloor + n_1 \lfloor n_2/2 \rfloor + \lfloor n_1/2 \rfloor$ vertices of M_3 in the lexicographic order, where the rightmost position is the most significant. Then it is easy to show that $|\partial(A)| = cw(M_3)$. $\square$

By a similar reasoning we are able to prove the following theorem on the bisection width of the 3-dimensinal meshes, which partially answers an open problem of Leighton [9].

Theorem 5.2 *For arbitrary $2 \leq n_1 \leq n_2 \leq n_3$ it holds*

$$bw(M_3) = \begin{cases} n_1 n_2, & \text{if } n_3 \text{ is even} \\ n_1 n_2 + n_1, & \text{if } n_2 \text{ is even and } n_3 \text{ is odd} \\ n_1 n_2 + n_1 + 1, & \text{otherwise.} \end{cases}$$

Corollary 5.2 *The bisection width of the 3-dimensional toroidal mesh satisfies*

$$bw(TM_3) = 2bw(M_3).$$

Cutwidths of many dimensional meshes have been determined in the following cases:

$$cw(M_k) = \left\lfloor \frac{2^{k+1}}{3} \right\rfloor, \text{if } n_1 = ... = n_k = 2, [3, 8, 14], \tag{1}$$

$$cw(M_k) = \frac{n_1^k - 1}{n_1 - 1}, \text{if } n_1 = ... = n_k = 2l + 1, [13].$$

We have the following upper bounds for cutwidths of many dimensional meshes in the remaining cases:

Theorem 5.3 *The cutwidth of k-dimensional ordinary mesh satisfies*

$$cw(M_k) \leq \begin{cases} n_1 n_2 ... n_{k-1} + ... + n_1 n_2 + n_1 + 1, \text{if } n_1 > 2 \\ 2^i(n_{i+1} ... n_{k-1} + ... + n_{i+1} n_{i+2} + n_{i+1} + 1) + \lfloor \frac{2^{i+1}}{3} \rfloor, \text{if } 2 = n_i < n_{i+1}, \\ \qquad \text{for some } i \text{ s.t. } 1 \leq i \leq k - 2. \\ \lfloor \frac{5}{3} 2^{k-1} \rfloor, \text{if } 2 = n_{k-1} < n_k \end{cases}$$

Proof. Using Lemma 3.2 and induction on k we first prove the case $n_1 > 2$. In general case we combine this result with the result (1) and using Lemma 3.2 again. □

Note that Theorem 5.3 gives immediately an upper bound for $cw(TM_k)$.

There are several partial answers to the problem of the bisection width of many dimensional mesh e.g. $bw(M_k) \geq n_1 n_2 ... n_{k-1}$ and equality holds if n_k is even, see [9]. We conjecture that the following upper bound is optimal:

Theorem 5.4 *The bisection width of the k-dimensional mesh satisfies*

$$cw(M_k) \leq \begin{cases} n_1 n_2 ... n_{k-1} + ... n_1 n_2 + n_1 + 1, \text{if } n_i \text{ is odd, for } i \geq 2 \\ n_1 n_2 ... n_{k-1} + ... + n_1 n_2 ... n_{i-1}, \text{if } i \geq 2 \text{ is maximal s.t. } n_i \text{ is even.} \end{cases}$$

Proof. It can be shown by induction on k, that the set of first $\lfloor n_1 n_2 ... n_k / 2 \rfloor$ vertices of M_k in lexicographic order, where the rightmost position is the most significant, has the boundary equal to the claimed upper bound. □

Note that Theorem 5.4 implies an upper bound for $bw(TM_k)$.

6 Cyclic Cutwidths

In this section we will study cyclic cutwidths of 2-dimensional meshes.

Theorem 6.1 *For $m, n \geq 2$ it holds*

1. $ccw(P_2 \times P_n) = 2$, if $n = 3, 4$.
2. $ccw(P_2 \times P_n) = 3$, if $n \geq 5$.
3. $ccw(P_n \times P_n) \leq n - 1$, if n is even.
4. $ccw(P_n \times P_i) \leq n$, if $i = n, n + 1$.
5. $\frac{\min\{m+1, n+1\}}{2} \leq ccw(P_m \times P_n) \leq \min\{m + 1, n + 1\}$.

Proof. The last result follows trivially from Theorem 4.1 and Lemma 3.1. We prove the third result only, other cases are similar.

Denote the vertices of $P_n \times P_n$ by tuples of indices $1 \leq i, j \leq n$ as (i, j). Let us divide $P_n \times P_n$ into four quadrants: $Q_{11} = \{(i, j) | 1 \leq i, j \leq n/2\}, Q_{12} = \{(i, j) | 1 \leq i \leq n/2, n/2 + 1 \leq j \leq n\}, Q_{21} = \{(i, j) | n/2 + 1 \leq i \leq n/2\}, Q_{22} = \{(i, j) | n/2 + 1 \leq i, j \leq n\}$. Let us divide C_{n^2} into four parts $A_{11} = \{1, 2, ..., n^2/4\}$, $A_{12} = \{n^2/4 + 1, ..., n^2/2\}, A_{22} = \{n^2/2 + 1, ..., 3n^2/4\}, A_{21} = \{3n^2/4 + 1, ..., n^2\}$. Now we map the quadrant Q_{ij} to A_{ij}. As the mapping of the quadrants to the parts of the cycle is symmetric, we need to describe only the mapping of the first quadrant. We do it in the following way. Let us divide the part A_{11} to the sectors which are assigned to the diagonals of the quadrant Q_{11}. As there are $n - 1$ diagonals, we have $n - 1$ sectors: $\bar{S}_{n/2}, ..., \bar{S}_2, S_1, S_2, ..., S_{n/2}$. Let us denote diagonals as follows:

$$D_1 = \{(1, 1), (2, 2), ..., (n/2, n/2)\}$$

$$D_i = \{(1, i), (2, i + 1), ..., (n/2 - i + 1, n/2)\}$$

$$\bar{D}_i = \{(i, 1), (i + 1, 2), ..., (n/2, n/2 - i + 1)\}, 2 \leq i \leq n/2.$$

We map the diagonals to the sectors in the natural manner: $\bar{D}_i \rightarrow \bar{S}_i, D_1 \rightarrow S_1, D_i \rightarrow S_i$.

First we show that the congestion of any cycle edge in S_1 is $n - 1$.

Each vertex (i, i) for $2 \leq i \leq n/2 - 1$ embedded in S_1 has two neighbors embedded in sector $\bar{S}_2$ and two neighbors embedded in sector S_2, the vertex $(1, 1)$ has one neighbor embedded in sector $\bar{S}_2$ and one neighbor embedded in sector S_2 and the vertex $(n/2, n/2)$ has one neighbor in sector $\bar{S}_2$, one neighbor in S_2, one neighbor in A_{21} and one in A_{12}. Let us take a cycle edge e in section S_1 and let there are x vertices of S_1 laying to the left of the edge e. Without loss of generality we can assume that the vertex $(1, 1)$ is to the left from the edge. Hence over the edge e there are embedded $(x - 1)2 + 1$ edges from S_1 to S_2 (possibly one edge even to A_{12} if the vertex $(n/2, n/2)$ is to the left from e) and $(n/2 - x)2$ edges from S_1 to $\bar{S}_2$ (possibly one edge even to A_{21} if the vertex $(n/2, n/2)$ is to the right from e. Therefore there are $(x - 1)2 + 1 + (n/2 - x)2 = n - 1$ edges of $P_n \times P_n$ which are embedded over the edge e. It means the congestion of any edge from S_1 is $n - 1$. Over the cycle edge between the sector S_1 and $\bar{S}_2$ there are embedded $2(n/2 - 1) + 1 = n - 1$ edges of $P_n \times P_n$ and it holds similarly for the cycle edge between S_1 and S_2.

Let us take S_i now.

It holds $|S_i| = |S_1| - i + 1$. Let us take a cycle edge e from the sector S_i and let us assume that there are x vertices to the left of the edge e in the S_i. Then there are $2x - 1$ edges from S_i to S_{i+1} and $(n/2 - i + 1 - x)2$ edges from S_i to S_{i-1} embedded over e. Then there are also $i - 1$ edges from $S_1, ..., S_{i-1}$ (one from each of $S_1, ..., S_{i-1}$) to Q_{12} which are embedded over e. Therefore the edge e has congestion: $(n/2 - i + 1 - x)2 + 2x - 1 + i - 1 = n - i$. Similarly the cycle edge between S_i and S_{i+1} has congestion $n - i$.

Similar claim could be shown for $\bar{S}_i$. □

Theorem 6.2 *For $m, n \geq 3$*

$$ccw(C_m \times C_n) = \min\{m + 2, n + 2\}.$$

Proof. Upper bounds follow easily by row by row or column by column placing of the vertices of the toroidal mesh on the cycle C_{mn} and considering the shortest paths between the images of adjacent vertices.

It is easy to see that if $m \geq n$ then $ccw(C_m \times C_n) \geq ccw(C_n \times C_n)$. Hence assume $m = n$. The Theorem 4.1, case cutwidth of toroidal meshes, implies that for each set $A \subset V_{C_n \times C_n}, |A| = \lfloor n^2/2 \rfloor + 1$ it holds $|\partial(A)| \geq 2n + 2$. Moreover, if $|\partial(A)| = 2n + 2$ then the set A has the only one possible form up to the symmetry and shifts. More precisely, let $\lfloor n^2/2 \rfloor + 1 = ln + r$, where $0 \leq r < n$. Then A is a subgraph of $C_n \times C_n$ induced by vertices $\{(i, j) : 1 \leq i \leq n, 1 \leq j \leq l\} \cup \{(i, l + 1), 1 \leq i \leq r\}$. If n is even then $l = n/2, r = 1$ and if n is odd then $l = r = (n - 1)/2$. Call the set A with this property extremal. For $i = 1, 2, ..., n^2$, let D_i denote the set $\{i, i + 1, i + 2, ...\}$ of $\lfloor n^2/2 \rfloor + 1$ consecutive vertices of C_{n^2}. If for some i, $\phi^{-1}(D_i)$ is not an extremal set then $|\partial(\phi^{-1}(D_i))| \geq 2n + 3$, which by Lemma 3.1 implies $ccw(C_n \times C_n) \geq n + 2$. Suppose that for all i, $\phi^{-1}(D_i)$ is an extremal set. Let n be odd. Take the sets D_1 and D_2. The sets $\phi^{-1}(D_1), \phi^{-1}(D_2)$ are extremal and the second one is obtained from the first one by deleting a vertex of degree 1 and inserting a new vertex of degree 1. Repeating this process for $D_2, D_3, ...$ it is easy to see that among the sets $\phi^{-1}(D_1), ..., \phi^{-1}(D_{2n+1})$ there are at lest two equal sets, a contradiction. The case n even is similar. □

Theorem 6.3 *For $m \geq 2, n \geq 3$*

$$\min\{m + 1, \frac{n}{2} + 1\} \leq ccw(P_m \times C_n) \leq \min\{m + 1, n + 2\}.$$

Proof. Upper bounds follow easily by row by row or column by column placing of the vertices of the meshes on the cycle C_{mn} and considering the shortest paths between the images of adjacent vertices. Lower bounds follow from Lemma 3.1 and Theorem 4.1, considering edge boundaries of sets of cardinality $\lfloor mn/2 \rfloor + 1$. □

Note, that if $m + 1 \leq n/2$ then the result is exact.

7 Conclusions

We proved exact cutwidths and bisection widths of ordinary, cylindrical and toroidal 2-dimensional meshes and cutwidths and bisection widths of 3-dimensional ordinary and toroidal meshes. We gave some upper bounds on cutwidths and bisection widths of many dimensional meshes. We also studied the cyclic cutwidth problem and for the cyclic cutwidth of a 2-dimensional toroidal mesh we gave an exact result, while for other kinds of meshes we proved partially optimal upper and lower bounds. We believe that our upper bounds are optimal.

References

1. Ahlswede, R., Bezrukov, S., Edge isoperimetric theorems for integer point arrays, *Technical Report No. 94-064*, University of Bielefeld, 1994.
2. Barth, D., Pellegrini, F., Raspaud, A., Roman, J., On bandwidth, cutwidth and quotient graphs, *RAIRO Informatique, Théorique et Applications*, to appear.
3. Bel Hala, A., Congestion optimale du plongement de l'ypercube $H(n)$ dans la chaine$P(2^n)$, *RAIRO Informatique, Théorique et Applications* **27** (1993), 1-17.
4. Bollobás, B., Leader, I., Edge-isoperimetric inequalities in the grid, *Combinatorica* **11** (1991), 299-314.
5. Chung, F. R. K., Labelings of graphs, in: L.Beineke and R.Wilson eds., *Selected Topics in Graph Theory 3*, Academic Press, New York, 1988, 151-168.
6. Garey, M.R., Johnson, D.S., Stockmayer, L., Some simplified $NP-$complete graph problems, *Theoretical Computer Science* **1** (1976), 237-267.
7. Harper, L. H., Optimal assignment of number to vertices, *J. Soc. Ind. Appl. Math.* **12** (1964) 131-135.
8. Hart, S., A note on the edges of the n-cube, *Discrete Mathematics* **14** (1976), 157-163.
9. Leighton, F.T., Introduction to Parallel Algorithms and Architectures, Morgan Kaufmann, 1992.
10. Lengauer, T., Upper and lower bounds for the min-cut linear arrangement of trees, *SIAM J. Algebraic and Discrete Methods* **3** (1982), 99-113.
11. Lindsey II, J.H., Assignment of numbers to vertices, *American Mathematical Monthly* **7** (1964), 508-516.
12. Lopez, A.D., Law, H.F.S., A dense gate matrix layout method for MOS VLSI, *IEEE Trans. Electron. Devices* **27** (1980), 1671-1675.
13. Nakano, K., Linear layout of generalized hypercubes, in: Proc. *19th Intl. Workshop on Graph-Theoretic Concepts in Computer Science*, Lecture Notes in Computer Science 790, Springer Verlag, Berlin, 1994, 364-375.
14. Nakano, K., Chen, W., Masuzawa, T., Hagihara, K., Tokura, N., Cutwidth and bisection width of hypercube graph, *IEICE Transactions* **J73-A** (1990), 856-862, (in Japanese).
15. Yannakakis, M., A polynomial algorithm for the Min Cut Linear Arrangement of trees, *J. ACM* **32** (1985), 950-988.

Searching for Faulty Leaves in Binary Trees

Peter Damaschke
FernUniversität, Theoretische Informatik II
58084 Hagen, Germany
peter.damaschke@fernuni-hagen.de

Abstract. Let us be given a complete binary tree where an unknown subset of the leaves is faulty. Suppose that we can test any node whether there is a faulty leaf in the subtree routed at that node. Our aim is to give a strategy for finding all faults that minimizes the worst-case search length for a previously estimated number of faults. This question is of immediate interest in VLSI circuit checking. In this paper we give an elementary proof of the optimality of a very simple test strategy.

1 Introduction

VLSI circuits have to be checked for faulty memory cells during production. In early stages of the production process many faults are expected, whereas before the final check one expects only a few or even no faults.

Consider a circuit with $n = 2^k$ cells. Each cell can be accessed by its address, that is a string of k bits. One can identify the faults by checking each cell separately; this requires n tests. But in a certain type of circuits, called parallel testable FAST-SRAMs [6] the faults can be located more efficiently, since a powerful kind of tests is available there: For any bit string $a_1 \ldots a_i$ ($0 \leq i \leq k$; $i = 0$ means the empty string) one can write in parallel a 0 into each cell with address $a_1 \ldots a_i * \ldots *$ (* means 0 or 1), and compute the logical OR of the contents of these 2^{k-i} cells in one step. Every faulty cell will change the stored 0 into 1, hence the result is 0 if all cells under consideration are intact, and 1 if at least one cell with address $a_1 \ldots a_i * \ldots *$ is faulty.

From the technical presumptions, it is appropriate to assume that the costs of a test (time, expenses) do not depend on the number of cells tested in parallel, hence the only interesting objective is to minimize the number of tests. Moreover, it is assumed that the faults are steady, that means, a faulty cell will always distort the stored 0 into 1 [7]. (There also exist more complicated types of faults.)

Obviously, our circuit checking problem can be formulated in an abstract way as follows. Let us be given a complete binary tree of depth k, where an unknown subset of the $n = 2^k$ leaves is faulty. An arbitrary node v is called faulty if the subtree with root v contains a faulty leaf. We can ask each node whether it is faulty. How many tests are necessary to find all faults surely?

Since there are 2^n possible subsets, we get the trivial information-theoretic lower bound of $\log_2 2^n = n$ tests, hence the capability of testing subtrees does not help in general. But, as mentioned above, one usually has an estimation of the number r of faulty leaves. One easily finds that for bounded r already $O(\log n)$ tests are sufficient. So it is natural to consider the following problem:

Given a depth k and an estimated number r of faulty leaves, what is a test strategy that minimizes the worst-case search length (i.e. the number of tests), provided that the actual number of faults is r?

Furthermore, the search length should not exceed the optimum very much if the estimated value r is approximately true – this is of course an important demand in view of the application.

Throughout this paper we use the term "fault" synonymously with "faulty leaf", rather than with "faulty inner node".

The introduced problem is related to group testing, a well-known topic in combinatorial search theory [1]. The question is there: Find r faulty elements in a set of size n, where any subset can be tested for containing a fault. This problem arises e.g. when a noxious substance must be detected in a set of samples. For fixed r there exist efficiently computable test strategies whose search lengths exceed the trivial lower bounds only by constant numbers [2] [8]. The case $r = 2$ which can be formulated as an edge search in graphs is particularly attractive [3] [1] [2] [4] [5].

An important difference to our present problem is the following. If at most r faulty samples are expected then we can apply a suboptimal strategy for this value r, and after that, the suspicious r samples can be checked by r additional tests for being really faulty, and one further group test of the remaining $n - r$ samples suffices to exclude further faults. In our tree problem the latter is not possible, since we cannot test arbitrary subsets of leaves in one go. So the considerations we need here are quite different from the "chemical" group testing problem.

In Section 2 we give a formal description of the search process in terms of labeled binary forests. In Section 3 we solve a case of our problem which appears rather special, but serves later as a component of the general solution. In Section 4 we calculate the search length of a certain strategy, called optimistic, which is suitable for few expected faults. Finally, in Section 5 we show that a simple generalization of this optimistic strategy for an arbitrary number of faults is worst-case optimal in complete binary trees. This settles the introduced problem completely.

2   Instantaneous Description

We generalize our search problem to arbitrary binary forests. Doing so, we can conveniently describe the instantaneous knowledge of the searcher after a sequence of tests.

A rooted tree is called a binary tree if every non-leaf has exactly two sons. A binary forest is a set of (non-empty) disjoint binary trees. We often omit the adjective "binary". A labeled tree is a tree B together with a label 1 or *. A labeled forest is a forest where each tree is labeled by 1 or *. B^1 symbolizes that the searcher already knows that the tree B contains at least one fault. Similarly, B^* means that the searcher does not know anything about faults in B.

The depth of a node of B is its distance to the root of B. In the complete binary tree C_k, all 2^k leaves have depth k. If v is a node of tree B, we denote the subtree with root v by B_v.

By $B - B_v$ we denote the binary tree obtained as follows.

- If v is the root then $B - B_v$ is empty.

- If v is the son of the root and u is the sibling of v then $B - B_v := B_u$.

- v has depth at least 2. Let u be the sibling of v, x the father of v, and y the father of x. Remove B_v and x from B and replace the edges yx and xu by the new edge yu. Let $B - B_v$ be the remaining binary tree.

Lemma 1 *The entire knowledge of the searcher at any time can be described by a labeled forest containing all faults.*

Proof (by induction on the number of tests).

Before the first test we have the tree C_k where nothing is known about faults. So the forest $\{C_k^*\}$ describes the situation.

In the induction step we first study what happens if a node v of a tree B is tested. There are four cases.

- v *is not faulty in* B^1. Then we can remove B_v without overlooking any fault. (Particularly, if v is the root then B can be completely removed.) Furthermore it is needless to test the father x of v later, instead we can test the sibling u of v just as well. Thus we can also remove x and, if x has a father y, also contract the path y–x–u to an edge yu. Finally note that there must be a fault among the remaining leaves, since we started with B^1. Altogether we obtain $(B - B_v)^1$.

- v *is not faulty in* B^*. Then we analogously get $(B - B_v)^*$.

- v *is faulty in* B^1. Then let $(v_0, v_1, \ldots, v_d)$ be the unique path from $v = v_0$ to the root v_d. The searcher knows that all v_i are faulty with v_0. Hence it is needless to test them later, and we can remove $v_1, \ldots, v_d$ together with their incident edges. Thus B decays into the trees B_{v_0} and $B_{v_{i+1}} - B_{v_i}$ $(0 \le i < d)$. The searcher knows that B_{v_0} contains a fault. Before the test he only knew that B contains at least one fault. So nothing can be said about faults in the other subtrees at this moment. Altogether we have to replace B^1 by $\{B_{v_0}^1, (B_{v_{i+1}} - B_{v_i})^* : 0 \le i < d\}$.

- v *is faulty in* B^*. Clearly we get the same result as in the previous case.

If a forest $\{B_1, \ldots, B_p\}$ (each B_i labeled by 1 or *) is given, some node of w.l.o.g. B_1 is tested, and $\{D_1, \ldots, D_q\}$ is the resulting labeled forest according to the above cases, then the knowledge after this test is completely described by $\{D_1, \ldots, D_q, B_2, \ldots, B_p\}$. (Note that a test in B_1 cannot provide any information about faults in the other trees.) This completes the induction step. $\square$

The search process ends when the instantaneous description is a forest of single-node trees, each with label 1.

Let $t_r(F)$ denote the number of tests which are necessary in the worst case for finding all faults in the labeled forest F, provided that r is the actual number of faults. (Our original question was to compute $t_r(C_k^*)$ and to give a corresponding strategy.) That means, the desired strategy should have minimal worst-case search length among all strategies if r faults are present, but it shall definitely work correctly also if this estimation r was wrong.

Let s be the number of labels 1 and l the total number of leaves in F. Clearly $t_r(F)$ is defined if and only if $s \le r \le l$, and from the viewpoint of the searcher every such r is possibly the true number of faults. Furthermore note that s can only increase and l can only decrease during the search process. An interesting case is reached when s becomes the actual r. The next section is concerned with this special case.

3 Some Partial Results

In this section we determine $t_r(F)$ for forests F with r labels 1.

Lemma 2

$$t_r(B_1^1, \ldots B_r^1, D_1^*, \ldots, D_p^*) = p + \sum_{i=1}^{r} t_1(B_i^1)$$

Proof In instances with actually r faults, each B_i contains exactly one fault, and each D_j is free of faults. The right side is a lower bound for t_r, since tests in one tree do not give information about faults in the other trees. The asserted number of tests is also sufficient: Apply an optimal strategy for one fault independently in each B_i^1, and test the roots of all D_j^* to exclude further faults. $\square$

It remains to compute $t_1(B^1)$ for trees B. The solution is given by the following recursion.

Lemma 3 *If B is a single node then $t_1(B^1) = 0$. Otherwise let u, v be the sons of the root such that w.l.o.g. $t_1(B_u^1) \leq t_1(B_v^1)$. Then we have*

$$t_1(B^1) = \max\{t_1(B_u^1) + 2, t_1(B_v^1) + 1\}.$$

Proof Consider the following strategy. Starting in $\{B^1\}$, first test u. If u is faulty then we get $\{B_u^1, B_v^*\}$, otherwise we get $\{(B - B_u)^1\} = \{B_v^1\}$ by Lemma 1. Hence, due to Lemma 2, $\max\{t_1(B_u^1) + 1, t_1(B_v^1)\}$ further tests are sufficient. This proves the upper bound.

We also show the lower bound. If the fault belongs to B_v then the searcher needs, no matter what his strategy is, at least one test to exclude faults in B_u. Since test results from B_u do not help to find the fault in B_v, in the worst case $t_1(B_v^1)$ further tests are required , even if the searcher knows that v is faulty. This shows $t_1(B^1) \geq t_1(B_v^1) + 1$.

Now, if $t_1(B_u^1) < t_1(B_v^1)$ then we also have $t_1(B^1) \geq t_1(B_u^1) + 2$. It remains to show $t_1(B^1) \geq t_1(B_u^1) + 2$ if $t_1(B_u^1) = t_1(B_v^1)$. Suppose w.l.o.g. that the searcher first tests some node w of B_u.

If $w = u$ and, unfortunately, w turns out to be faulty then we obtain $\{B_u^1, B_v^*\}$. By Lemma 2, $t_1(B_u^1) + 1$ further tests are necessary in the worst case.

If $w \neq u$ and, unfortunately, w is not faulty then the instantaneous description becomes a tree B'^1 with maximal subtrees $B_u - B_w$ and B_v. Trivially we have $t_1((B_u - B_w)^1) \leq t_1(B_u^1) = t_1(B_v^1)$. We have already proved above that $t_1(B'^1) \geq t_1(B_v^1) + 1$. Together this yields $t_1(B^1) \geq t_1(B_u^1) + 2$, since one test was already used for w. □

Note that the proof also yields an optimal strategy managing with $t_1(B^1)$ tests. Together with Lemma 2 we now easily conclude:

Theorem 4 *Let us be given a labeled forest F with r labels 1. Then the following strategy does not need more than $t_r(F)$ tests, provided that exactly r faults are present.*

(1) *Choose a tree B.*

(2) *If B has label * then test the root of B.*

(3) *If B has label 1 then test a son u of the root where $t_1(B_u^1)$ is minimal.*

(4) *Depending on the answer of the test, update F as in Lemma 1.*

(5) *Repeat these steps until a forest of single nodes, each with label 1, is obtained.*

Some remarks are advisable here.

Note that we test only roots or sons of roots. Therefore, each tree obtained in the search process is a subtree B_w of some original tree B from F. (In general this is not valid.) Consequently, the values needed in (3) can be efficiently computed in advance by the recursion of Lemma 3.

In (2) we always expect the answer "not faulty". If the result is "faulty" then we have $r+1$ labels 1 and continue the search with $r := r+1$. These cases are not relevant for determining $t_r(F)$.

The theorem completely solves our problem for forests F with $s = r$ labels 1. We finally conclude:

Corollary 5 *We have $t_0(C_k^*) = 1$, $t_1(C_k^1) = 2k$, $t_1(C_0^*) = 1$, and for $k \geq 1$ also $t_1(C_k^*) = 2k$.*

Proof The first and third assertion are trivial, the second follows immediately from Lemma 3. Furthermore we have trivially $t_1(C_k^*) \geq t_1(C_k^1) = 2k$, and in case $k \geq 1$, $2k$ tests are also sufficient: First test the sons u, v of the root. Provided that $r = 1$ is the actual number of faults, either u or v is faulty, hence we obtain $\{C_{k-1}^1\}$ which can be treated by $2k - 2$ further tests. $\square$

4   The Optimistic Strategy

The strategy of Theorem 4 could be called optimistic. At any time the searcher assumes that the true number of faults r is as small as possible, i.e. the number of labels 1. This hypothesis is kept until a fault is detected in a tree with label *, in this case the search is continued with $r := r+1$.

In view of our original problem we are particularly interested in the worst-case search length $t(k, r)$ when the optimistic strategy is applied to $\{C_k^*\}$ with really r faults in it. This should not be confused with $t_r(C_k^*)$, in fact note that $t_r(C_k^*) \leq t(k, r)$.

We derive a simple recursive formula for $t(k, r)$. Note that $t(0, 1) = t(k, 0) = 1$: First, 0 faults are expected, hence the searcher tests the root, and if $k = 0$ or $r = 0$ then the search ends after this test.

Now consider $B = C_k$ with $k, r \geq 1$ and let u, v be the sons of the root. Assume that B_u and B_v contains j and $r-j$ faults, respectively. The optimistic strategy first tests the root, the result is "faulty". Since $t_1(B_u^1) = t_1(B_v^1)$ we can w.l.o.g. assume that v is tested next. If $j < r$ then the answer is "yes", and we obtain $\{B_u^*, B_v^1\}$ where $t(k - 1, j) + t(k - 1, r - j) - 1$ further tests can be required. If $j = r$ then the answer is "no", and we get $\{B_u^1\}$ where $t(k-1, r)-1$ further tests can be required. The latter case is certainly not the worst case: If $j = 0$ then we need up to $1 + t(k - 1, r) - 1$ further tests which is more. Thus we have shown:

Lemma 6 $t(0,1)=t(k,0)=1$. *For $k, r \geq 1$ we have*

$$t(k,r) = \max_j t(k-1,j) + t(k-1,r-j) + 1$$

where j is drawn from the integers with $0 \leq j < r$, and $j, r-j \leq 2^{k-1}$.

In the following the term $\lceil \log_2 i \rceil$ occurs frequently, therefore we abbreviate it by $L(i)$.

Theorem 7 *For $k \geq 0$ and $0 \leq r \leq 2^k$ we have*

$$t(k,r) = 2kr + 1 - 2 \sum_{i=1}^{r} L(i).$$

Proof (by induction on k).

The formula yields correctly $t(0,0) = t(0,1) = 1$. (By the usual convention we have $\sum_{i=1}^{0} L(i) = 0$.)

Assume that the formula is valid for $k-1$. For $r = 0$ we correctly obtain $t(k,0) = 1$. We have to verify the formula for k and $r \geq 1$. Now Lemma 6 gives:

$$
\begin{aligned}
t(k,r) &= \max_j 2(k-1)j + 1 - 2\sum_{i=1}^{j} L(i) \\
&\quad + 2(k-1)(r-j) + 1 - 2\sum_{i=1}^{r-j} L(i) + 1 \\
&= \max_j 2(k-1)r + 3 - 2\left(\sum_{i=1}^{j} L(i) + \sum_{i=1}^{r-j} L(i) \right).
\end{aligned}
$$

Because of the monotonicity of $L(i)$ the sum in parantheses becomes minimal if $j = \lfloor r/2 \rfloor$. Thus we have

$$t(k,r) = 2kr + 1 - \left(2r - 2 + 4\sum_{i=1}^{\lfloor r/2 \rfloor} L(i) + 2\mu_r L(\lceil r/2 \rceil) \right)$$

where $\mu_r = 1/0$ if r is odd/even. It remains to show

$$\sum_{i=1}^{r} L(i) = r - 1 + 2\sum_{i=1}^{\lfloor r/2 \rfloor} L(i) + \mu_r L(\lceil r/2 \rceil).$$

For $r = 1$ this identity is valid. Observe that $L(2i) = L(i) + 1$ for $i \geq 1$, and $L(2i-1) = L(i) + 1$ for $i \geq 2$.

For even $r > 1$ this implies

$$r - 1 + 2\sum_{i=1}^{r/2} L(i) = \sum_{i=1}^{r/2} L(2i-1) + \sum_{i=1}^{r/2} L(2i) = \sum_{i=1}^{r} L(i).$$

For odd $r > 1$ we similarly get

$$r - 1 + 2\sum_{i=1}^{(r-1)/2} L(i) + L\left(\frac{r+1}{2}\right)$$

$$= \sum_{i=1}^{(r-1)/2} L(2i-1) + \sum_{i=1}^{(r-1)/2} L(2i) + 1 + L\left(\frac{r+1}{2}\right)$$

$$= \sum_{i=1}^{r-1} L(i) + L(r) = \sum_{i=1}^{r} L(i).$$

$\square$

5 A Simple Optimal Strategy for the Complete Binary Tree

Next we prove a lower bound for $t_r(C_k^*)$ $(r > 0)$. Define a subset A of the nodes of C_k as follows. Put $2^{L(r)} - r$ arbitrary nodes of depth $L(r) - 1$ into A. There remain $2^{L(r)-1} - (2^{L(r)} - r) = r - 2^{L(r)-1}$ nodes of depth $L(r) - 1$. Put their $2r - 2^{L(r)}$ sons of depth $L(r)$ into A. The so obtained set A of size r is a maximal antichain, i.e. no node of A is an ancestor of any other node of A, and every leaf of C_k is contained in exactly one subtree $C_{k,v}$ $(v \in A)$.

Lemma 8 *Let A be a maximal antichain in C_k with r nodes. Then we have* $t_r(C_k^*) \geq \sum_{v \in A} t_1(C_{k,v}^*)$.

Proof Here we argue informally, using again a kind of independence of disjoint subtrees. Consider only instances where every $C_{k,v}$ contains exactly one fault. We observe the sequence of instantaneous descriptions of a concrete search process starting with $\{C_k^*\}$. Simultaneously we observe the instantaneous descriptions of the same search process starting with $\{C_{k,v}^* : v \in A\}$ where tests of nodes above A are ignored. Verify that the searcher never has more knowledge about faults within the subtrees $C_{k,v}$ than documented in the second description. This is due to te fact that each test above A will answer "faulty". From this the assertion follows easily. $\square$

For our special choice of A we get by Corollary 5 for $r \leq 2^{k-1}$:

$$t_r(C_k^*) \geq (2^{L(r)} - r)2(k - L(r) + 1) + (2r - 2^{L(r)})2(k - L(r))$$
$$= 2r(k - L(r) - 1) + 2^{L(r)+1}.$$

For $r > 2^{k-1}$ we get similarly:

$$t_r(C_k^*) \geq (2^{L(r)} - r)2(k - L(r) + 1) + (2r - 2^{L(r)}) = (2^k - r)2 + (2r - 2^k) = 2^k,$$

that means $t_r(C_k^*) = 2^k$, since 2^k tests of the leaves are trivially sufficient.

We finally show that also in the case $r \leq 2^{k-1}$ a very simple strategy attains the lower bound.

Theorem 9 *The following strategy needs at most* $2r(k - L(r) - 1) + 2^{L(r)+1}$ *tests, and is therefore optimal, if* $r \leq 2^{k-1}$ *is the actual number of faults in* C_k: *Apply the optimistic strategy independently in all* $C_{k,v}$ *where* v *has depth* $L(r) + 1$.

Proof We claim that $\sum_v t(k - L(r) - 1, j_v)$ with $\sum_v j_v = r$ becomes maximal if $j_v \leq 1$ for all v. (The index v runs over all nodes of depth $L(r) + 1$.) This implies with help of Theorem 7:

$$
\begin{aligned}
t_r(C_k^*) &\leq r\, t(k - L(r) - 1, 1) + (2^{L(r)+1} - r)t(k - L(r) - 1, 0) \\
&= r(2(k - L(r) - 1) + 1) + (2^{L(r)+1} - r) \\
&= 2r(k - L(r) - 1) + 2^{L(r)+1}.
\end{aligned}
$$

Together with the lower bound this shows equality.

For proving the claim it suffices to consider two nodes u, v of depth $L(r) + 1$ such that $C_{k,u}$ and $C_{k,v}$ contains $p > 0$ faults and one fault, respectively. Let $h = k - L(r) - 1$ be the depth of these subtrees. Then Theorem 7 yields

$$t(h, p) + t(h, 1) = 2hp + 1 - 2\sum_{i=1}^{p} L(i) + 2h + 1 = 2h(p + 1) + 2 - 2\sum_{i=1}^{p} L(i)$$

and, on the other hand

$$t(h, p + 1) + t(h, 0) = 2h(p + 1) + 1 - 2\sum_{i=1}^{p+1} L(i) + 1$$

which is smaller. Hence the overall sum is maximized if the r faults spread over r subtrees $C_{k,v}$. $\square$

Since the strategy only refers to $L(r)$, it has even optimal worst-case behaviour if just $L(r)$ (rather than r) has been estimated correctly.

Acknowledgment

I wish to thank Christian Elm (Hagen) for drawing my attention to the subject.

References

[1] M.Aigner: *Combinatorial Search*, Wiley-Teubner 1988

[2] I.Althöfer, E.Triesch: Edge search in graphs and hypergraphs of bounded rank, *Discrete Math.* 115 (1993), 1-9

[3] G.J.Chang, F.K.Hwang, S.Lin: Group testing with two defectives, *Discrete Applied Math.* 4 (1982), 97-102

[4] P.Damaschke: A tight upper bound for group testing in graphs, *Discrete Applied Math.* 48 (1994), 101-109

[5] P.Damaschke: A parallel algorithm for nearly optimal edge search, *Information Processing Letters*, to appear

[6] C.Elm, D.Tavangarian: Fault detection and fault localization using I_{DDQ}-testing in parallel testable FAST-SRAMs, *12th IEEE VLSI Test Symposium*, Cherry Hill/NJ 1994

[7] A.J.van de Goor: *Testing Semiconductor Memories, Theory and Practice*, John Wiley & Sons, Chichester 1991

[8] E.Triesch: A group testing problem for hypergraphs of bounded rank, submitted to *Discrete Math.*

NC Algorithms for Partitioning Planar Graphs into Induced Forests and Approximating NP-Hard Problems

Zhi-Zhong Chen[1] and Xin He[2]

[1] Dept. of Mathematical Sciences, Tokyo Denki University, Hatoyama,
Saitama 350-03, Japan. E-mail: chen@r.dendai.ac.jp
[2] Dept. of Computer Science, State Univ. of New York at Buffalo,
Buffalo, NY 14260, U.S.A. E-mail: xinhe@cs.buffalo.edu

Abstract. It is well known that the vertex set of every planar graph can be partitioned into three subsets each of which induces a forest. Previously, there has been no NC algorithm for computing such a partition. In this paper, we design an optimal NC algorithm for computing such a partition for a given planar graph. It runs in $O(\log n \log^* n)$ time using $O(n/(\log n \log^* n))$ processors on an EREW PRAM. This algorithm implies *optimal NC* approximation algorithms for many NP-hard maximum induced subgraph problems on planar graphs with a performance ratio of 3. We also present optimal NC algorithms for partitioning the vertex set of a given K_4-free or $K_{2,3}$-free graph into two subsets each of which induces a forest. As consequences, we obtain *optimal NC* algorithms for 4-coloring K_4-free or $K_{2,3}$-free graphs which are previously unknown to our knowledge.

1 Introduction

The concept of *vertex-arboricity* is well known in graph theory [5]. For a (simple) graph G, the vertex-arboricity $a(G)$ of G is defined as the minimum number of subsets into which the vertex set of G can be partitioned so that each subset induces a forest [3, 4]. Chartrand *et al.* showed that $a(G) \le 3$ for every planar graph G [3]. This means that every planar graph G has a 3-*forest partition*, i.e., a partition of the vertex set of G into 3 subsets each of which induces a forest.

In this paper, we present the first NC algorithm for finding a 3-forest partition of a given planar graph. The algorithm runs in $O(\log n \log^* n)$ time using $O(n/(\log n \log^* n))$ processors on an EREW PRAM. Like most NC algorithms for coloring planar graphs, our algorithm uses the idea of repeatedly removing a large set X of vertices with degree ≤ 6 while introducing some changes to the remaining graph in order to make it possible to re-insert the vertices in X at small cost later. In the case of coloring, the color of each re-inserted vertex can be determined almost trivially from the colors of its neighbors. However, in our case, re-inserting the vertices is nontrivial since the re-inserted vertices may introduce cycles into the partitioned vertex subsets. This makes our algorithm more complicated than those for coloring planar graphs. Our parallel algorithms are based on lemmas on the structure of planar graphs. These lemmas assert that the neighborhood of each vertex of a planar graph displays certain nice *sparse* properties that are central to our parallel algorithms. These structural properties may find applications in designing parallel algorithms for solving other similar problems for planar graphs. To achieve optimality, we use a method of Hagerup *et al.* [9] which is originally based on the accelerating cascades technique of Cole and Vishikin [6].

We also give an interesting application of our above result. Let π be a property on planar graphs. π is *hereditary* if, whenever a planar graph G satisfies π, every induced subgraph of G also satisfies π. Throughout this paper, we consider only hereditary properties π. The *planar maximum subgraph problem* for π (PMSP(π)) is the following: Given a planar graph $G = (V, E)$, find a maximum subset U of V that induces a subgraph satisfying π. Yannakakis showed that most PMSP(π)'s are NP-hard [15]. On the other hand, Baker showed that some popular PMSP(π)'s have polynomial-time approximation schemes [2]. As have been observed in [12], Baker's approximation schemes can be simply parallelized using breadth-first-search and non-serial dynamic programming. Although the best known NC algorithm for breadth-first-search runs in poly-logarithmic time using a linear number of processors [12], the polynomial bounding its running time is of *very high* degree. Interestingly, using our 3-forest partition algorithm above, we immediately obtain an efficient NC approximation algorithm for PMSP(π). The approximation algorithm achieves a performance ratio of 3, i.e., always computes a subset satisfying π whose size is at least one-third of the maximum subset satisfying π. Moreover, it runs in $O(\log n \log^* n) + T_\pi(n)$ time using $O(n/(\log n \log^* n)) + P_\pi(n)$ processors on an EREW PRAM, where $T_\pi(n)$ is the time needed to solve PMSP(π) for n-vertex *trees* using $P_\pi(n)$ processors on an EREW PRAM. For many properties π of interest, PMSP(π) restricted to trees can be solved in $O(\log n \log^* n)$ time with $O(n/(\log n \log^* n))$ processors on an EREW PRAM using the well-known tree contraction technique [10, 13]. Therefore, for such properties π, we have an *optimal NC* approximation algorithm for PMSP(π) with a performance ratio of 3.

The result in [4] also shows that $a(G) \leq 2$ if G is a K_4-free graph or a $K_{2,3}$-free graph. (*Note*: By Kuratowski's Theorem, K_4-free graphs and $K_{2,3}$-free graphs are both planar.) This means that a K_4-free or $K_{2,3}$-free graph G has a *2-forest partition*, i.e., a partition of the vertex set of G into two subsets each of which induces a forest. We present two optimal NC algorithms for finding a 2-forest partition of a given K_4-free or $K_{2,3}$-free graph. The two algorithms resemble the one for finding a 3-forest partition of a given planar graph and both run in $O(\log n \log^* n)$ time using $O(n/(\log n \log^* n))$ processors on an EREW PRAM. From the two algorithms, we immediately get two *optimal NC* algorithms for 4-coloring K_4-free or $K_{2,3}$-free graphs. To our knowledge, no such algorithms are previously known.

The model of parallel computation we use is the *exclusive read exclusive write parallel random access machine* (EREW PRAM). The model consists of a number of identical processors and a common memory. The concurrent reads or concurrent writes of the same memory location by different processors are disallowed. (See [11] for a discussion of the PRAM models.)

2   3-Forest Partition of Planar Graphs

We first introduce some notations. Throughout this paper, a planar graph is always simple, that is, has neither parallel edges nor self-loops. Let $G = (V, E)$ be a planar graph. The *neighborhood* of a vertex v in G, denoted $N_G(v)$, is the set of vertices in G adjacent to v; $deg_G(v) = |N_G(v)|$ is the *degree* of v in G. For $U \subseteq V$, the *subgraph of G induced by U* is the graph (U, F) with $F = \{\{u, v\} \in E : u, v \in U\}$ and is denoted by $G[U]$. A *3-forest partition* of G is a partition of V into three subsets V_1, V_2, V_3 such that $G[V_1]$, $G[V_2]$, and $G[V_3]$ are all acyclic. An *independent set* of G is a subset U of V such that $G[U]$ contains no edge. A *maximal independent set* of G is an independent set that is not properly contained in some other independent set of G. For two nonadjacent vertices u and v in G, *merging u and v into a supervertex z* means identifying u and v with a new vertex z whose neighborhood is the union of the neighborhoods of u and v (resulting multiple edges are deleted). Note that merging two nonadjacent vertices in G yields a *simple* graph.

Our main result in this section is an optimal NC algorithm for finding a 3-forest partition of a given planar graph. Let us first give an outline of our algorithm. Given a planar graph $G = (V, E)$, the algorithm starts by finding a large independent set X of G in which all vertices have degree ≤ 6 and also have *certain* useful neighborhood properties. It then constructs a new (simple) planar graph G' from G by first merging two suitable (nonadjacent) neighbors of each vertex $x \in X$ with $deg_G(x) = 6$ into a supervertex and next deleting the vertices in X. The main point in the construction of X and G' is that every 3-forest partition of G' can be used to obtain a 3-forest partition of $G[V - X]$ which can be extended to a 3-forest partition of G. After constructing G', the algorithm recursively finds a 3-forest partition U_1, U_2, U_3 for G'. The large size of X guarantees that the depth of recursion is $O(\log n)$. Finally, the algorithm uses U_1, U_2, U_3 to obtain a 3-forest partition of $G[V - X]$ and then extend the partition to a 3-forest partition of G.

Before describing our algorithm precisely, we need to prove three lemmas. The first lemma is related to the construction of the independent set X mentioned above and has been shown in [9].

Lemma 1. [9]. Every planar graph $G = (V, E)$ contains an independent set X satisfying the following four conditions:
  (a) $|X| \geq c|V|$ for some $0 < c \leq 1$.
  (b) For all $x \in X$, $deg_G(x) \leq 6$.
  (c) For every two vertices x_1 and x_2 in X with $deg_G(x_1) = deg_G(x_2) = 6$, $N_G(x_1) \cap N_G(x_2)$ is empty and it is not the case that some $y_1 \in N_G(x_1)$ is adjacent to some $y_2 \in N_G(x_2)$ in G. (In other words, x_1 and x_2 are at least distance 4 apart in G.)

(d) For every vertex x in X with $deg_G(x) = 6$, each neighbor of x has degree at most 12 in G and $G[N_G(x)]$ has a Hamiltonian circuit.

The following two lemmas are related to the construction of the graph G' mentioned in the above outline of our algorithm.

Lemma 2. Let $G = (V, E)$ be a planar graph. Suppose that x is a vertex in G such that $deg_G(x) = 6$, $G[N_G(x)]$ has a Hamiltonian circuit H, and there is no $v \in V - (N_G(x) \cup \{x\})$ with $N_G(x) \subseteq N_G(v)$. Then, there is a pair $\langle y', y'' \rangle$ of two neighbors of x satisfying the following three conditions:

(i) $\{y', y''\} \notin E$.

(ii) For every $w \in V - (N_G(x) \cup \{x\})$, w is adjacent to at most one of y' and y''.

(iii) There is at most one $z \in N_G(x) - \{y', y''\}$ with $\{y', z\} \in E$ and $\{y'', z\} \in E$.

Proof. Let $u_1, u_2, \cdots, u_6$ be the neighbors of x in G and suppose that the Hamiltonian circuit H of $G[N_G(x)]$ is $u_1, u_2, \cdots, u_6, u_1$. Consider a planar embedding of G. Clearly, $u_1, u_2, \cdots, u_6$ must appear around x either clockwise or counterclockwise in this order in the embedding. W.l.o.g., we assume the former. For a pair $\langle u_i, u_j \rangle$ of two neighbors of x and a condition C of (i), (ii), and (iii) above, we say that $\langle u_i, u_j \rangle$ satisfies C if C is satisfied by setting $\langle y', y'' \rangle = \langle u_i, u_j \rangle$. Call $\langle u_i, u_j \rangle$ a *desired pair* if it satisfies all of the three conditions (i), (ii), and (iii). There are four cases that may occur.

Case 1: None of the three pairs $\langle u_1, u_4 \rangle$, $\langle u_2, u_5 \rangle$, and $\langle u_3, u_6 \rangle$ satisfies the condition (iii). Then, besides the six edges incident to x and the six edges on H, G must contain either the three edges $\{u_2, u_4\}$, $\{u_4, u_6\}$, $\{u_6, u_2\}$ or the three edges $\{u_1, u_3\}$, $\{u_3, u_5\}$, $\{u_5, u_1\}$. W.l.o.g., we may assume the former. Then, $\langle u_3, u_5 \rangle$ is clearly a desired pair.

Case 2: Exactly one pair of $\langle u_1, u_4 \rangle$, $\langle u_2, u_5 \rangle$, and $\langle u_3, u_6 \rangle$ satisfies the condition (iii). We consider only the case where $\langle u_1, u_4 \rangle$ is the pair; the other two cases are similar. Then, besides the six edges incident to x and the six edges on H, G must contain the two edges $\{u_3, u_5\}$ and $\{u_6, u_2\}$. Now, it is easy to see that $\langle u_1, u_4 \rangle$ is a desired pair.

Case 3: Exactly two pairs of $\langle u_1, u_4 \rangle$, $\langle u_2, u_5 \rangle$, and $\langle u_3, u_6 \rangle$ satisfy the condition (iii). We consider only the case where $\langle u_1, u_4 \rangle$ and $\langle u_2, u_5 \rangle$ are the two pairs; the other two cases are similar. Then, besides the six edges incident to x and the six edges on H, G must contain either the two edges $\{u_4, u_6\}$ and $\{u_6, u_2\}$ or the two edges $\{u_1, u_3\}$ and $\{u_3, u_5\}$. W.l.o.g., we may assume the former. Then, $\langle u_1, u_4 \rangle$ is obviously a desired pair.

Case 4: The three pairs $\langle u_1, u_4 \rangle$, $\langle u_2, u_5 \rangle$, and $\langle u_3, u_6 \rangle$ all satisfy the condition (iii). If the three pairs also satisfy the conditions (i) and (ii), then each of them is a desired pair. Otherwise, there are two cases that may occur.

Subcase 4.1: At least one pair of $\langle u_1, u_4 \rangle$, $\langle u_2, u_5 \rangle$, and $\langle u_3, u_6 \rangle$ does not satisfy the condition (i). We consider only the case where $\langle u_1, u_4 \rangle$ does not satisfy the condition (i); the other cases are similar. Then, G must contain the edge $\{u_1, u_4\}$. From this, it is easy to see that both $\langle u_2, u_5 \rangle$ and $\langle u_3, u_6 \rangle$ must satisfy the conditions (i) and (ii). Thus, both of them are desired pairs.

Subcase 4.2: At least one pair of $\langle u_1, u_4 \rangle$, $\langle u_2, u_5 \rangle$, and $\langle u_3, u_6 \rangle$ does not satisfy the condition (ii). We consider only the case where $\langle u_1, u_4 \rangle$ does not satisfy the condition (ii); the other cases are similar. Then, there is some $w \in V - (N_G(x) \cup \{x\})$ with $\{w, u_1\} \in E$ and $\{w, u_4\} \in E$. Taking the six edges incident to x into account, it is easy to see that both $\langle u_2, u_5 \rangle$ and $\langle u_3, u_6 \rangle$ satisfy the condition (i). We claim that at least one of $\langle u_2, u_5 \rangle$ and $\langle u_3, u_6 \rangle$ must satisfy the condition (ii). Assume, on the contrary, this is not the case. Then, there are w_1 and w_2 such that G contains the edges $\{u_2, w_1\}$, $\{u_5, w_1\}$, $\{u_3, w_2\}$, and $\{u_6, w_2\}$. By the planarity of G, both w_1 and w_2 must be w. However, this implies that w is a vertex in $V - (N_G(x) \cup \{x\})$ with $N_G(x) \subseteq N_G(w)$, contradicting the assumption of the lemma. Hence, at least one of $\langle u_2, u_5 \rangle$ and $\langle u_3, u_6 \rangle$ must be a desired pair. ∎

Lemma 3. Let $G = (V, E)$ be a planar graph, and let x be a vertex in G with $deg_G(x) = 6$. Then, there is at most one vertex $v \in V - (N_G(x) \cup \{x\})$ with $N_G(x) \subseteq N_G(v)$. Furthermore, if such a vertex v exists, then there are three neighbors y', y'', z of x satisfying the following conditions:

(1) $\{y', y''\} \notin E$.

(2) For every $w \in V - \{x, v, z\}$, w is adjacent to at most one of y' and y''.

(3) In the graph $G[V - \{x, v, y', y''\}]$, z is not reachable from any vertex in $N_G(x) - \{y', y'', z\}$.

Proof. The first assertion of the lemma is obvious. It remains to show the second assertion. Let u_1, u_2, $\cdots$, u_6 be the neighbors of x in G. Consider a planar embedding of G. W.l.o.g., we may assume that u_1, u_2, $\cdots$, u_6 clockwise appear around x in this order in the embedding. Now, it is easy to see that if we set $y' = u_1$, $y'' = u_3$ and $z = u_2$, then y', y'', and z must satisfy the conditions in the lemma. ∎

We are now ready to present our algorithm.

Algorithm 1

Input: A planar graph $G = (V, E)$ with n vertices.

Output: A 3-forest partition V_1, V_2, V_3 of G.

1. Find an independent set X of G satisfying the conditions (a), (b), (c), and (d) in Lemma 1.

2. If $V = X$, then set $V_1 = X$ and $V_2 = V_3 = \emptyset$ and halt.

3. Partition X into $X_{\leq 5}$ and $X_{=6}$, where $X_{\leq 5} = \{x \in X : deg_G(x) \leq 5\}$ and $X_{=6} = \{x \in X : deg_G(x) = 6\}$.

4. In parallel, for each $x \in X_{=6}$, perform the following steps:

   4.1. If there is no $v \in V - (N_G(x) \cup \{x\})$ with $N_G(x) \subseteq N_G(v)$, then first find a pair $\langle y', y'' \rangle$ of two neighbors of x satisfying the conditions (i), (ii), and (iii) in Lemma 2, and next set $pair(x) = \langle y', y'' \rangle$.

   4.2. If there is a (unique) $v \in V - (N_G(x) \cup \{x\})$ with $N_G(x) \subseteq N_G(v)$, then first set $rival(x) = v$, next find three neighbors y', y'', and z of x

satisfying the conditions (1), (2), and (3) in Lemma 3, and finally set $pair(x) = \langle y', y'' \rangle$ and $label(x) = z$.

5. From G, construct a new (simple) planar graph G' by first deleting all vertices of X and then for all $x \in X_{=6}$, merging the two vertices in $pair(x)$ into a supervertex $super(x)$.

6. Recursively call the algorithm for G' to obtain a 3-forest partition U_1, U_2, U_3 of G'.

7. For $1 \leq i \leq 3$, initialize V_i to be the set obtained from U_i by decomposing each supervertex $super(x)$ in U_i into the two vertices in $pair(x)$.

8. In parallel, for each $x \in X_{\leq 5}$, add x to a $V_i \in \{V_1, V_2, V_3\}$ such that x is adjacent to at most one vertex of V_i in the input graph G.

9. In parallel, for each $x \in X_{=6}$, perform the following steps:

   9.1. If $rival(x)$ is undefined (cf. step 4.1), then find a $U_i \in \{U_1, U_2, U_3\}$ containing at most one vertex in $(N_G(x) - \{y : y \text{ is in } pair(x)\}) \cup \{super(x)\}$ and add x into V_i.

   9.2. If $rival(x)$ is defined (cf. step 4.2), then perform the following steps:

      9.2.1. If there is some U_i with $1 \leq i \leq 3$ such that U_i contains at most one vertex in $N_G(x) - \{y : y \text{ is in } pair(x)\}$ but does not contain $super(x)$, then add x into V_i.

      9.2.2. If there is no U_i satisfying the condition in step 9.2.1 (i.e., one of U_1, U_2 and U_3 contains $super(x)$ and each of the other two contains two vertices in $N_G(x) - \{y : y \text{ is in } pair(x)\}$), then first find the unique U_i among U_1, U_2, and U_3 containing $super(x)$ and next add x into V_i if $rival(x) \notin V_i$ while add x into the unique V_j ($1 \leq j \neq i \leq 3$) with $label(x) \in V_j$ otherwise.

Lemma 4. Algorithm 1 correctly outputs a 3-forest partition for a planar graph G.

Proof. By induction on $d(G)$, the depth of recursion of Algorithm 1 on input G. In case $d(G) = 0$, clearly Algorithm 1 outputs a 3-forest partition for G. Assume that $d(G) > 0$ and Algorithm 1 correctly outputs a 3-forest partition for all planar graphs on which the depth of recursion is $< d(G)$. We want to establish that Algorithm 1 correctly outputs a 3-forest partition for G. To this end, we first prove a claim.

Claim 1 After step 7 and before step 8 are executed, $G[V_1]$, $G[V_2]$, and $G[V_3]$ are all acyclic.

Proof. We only show that $G[V_1]$ is acyclic; the proofs for $G[V_2]$ and $G[V_3]$ are similar. By the inductive hypothesis, $G'[U_1]$ is acyclic. Assume, on the contrary, that $G[V_1]$ contains a cycle C. Then, since $G'[U_1]$ is acyclic, C must contain at least two vertices y', y'' such that for some $x \in X_{=6}$, $pair(x) = \langle y', y'' \rangle$. We need to consider only the following three cases.

Case 1: The length of C is 3. Then, $\{y', y''\}$ must be an edge in G. However, this is impossible by step 4.

Case 2: The length of C is 4. Let v_1 and v_2 be the two vertices on C other than y' and y''. Then, since $\{y', y''\} \notin E$, both v_1 and v_2 are adjacent to y' and y'' in G. By step 4, x must satisfy the condition in step 4.2. That is, $rival(x)$ is defined and $N_G(x) \subseteq N_G(rival(x))$. Furthermore, one of v_1 and v_2 must be $rival(x)$ and the other is $label(x)$ by the condition (2) in Lemma 3. Note that $\{rival(x), label(x)\}$ $(= \{v_1, v_2\})$ must be an edge in G. Also, by step 1, neither $rival(x)$ nor $label(x)$ can be contained in some $pair(x')$ with $x' \in X_{=6}$, and thus both $rival(x)$ and $label(x)$ are vertices in G'. Now, v_1, v_2, and the supervertex $super(x)$ must induce a cycle in $G'[U_1]$, a contradiction.

Case 3: The length of C is at least 5. Let x'_1, x'_2, $\cdots$, x'_k be the vertices $x' \in X_{=6}$ such that the two vertices in $pair(x')$ both appear on C. W.l.o.g., we assume $x'_1 = x$. Let Q_1 be the (simple) graph obtained from C by merging the two vertices in $pair(x'_i)$ into a supervertex $super(x'_i)$ for all x'_i with $1 \leq i \leq k$. We want to show that Q_1 contains a cycle on which $super(x)$ $(= super(x'_1))$ appears. To this end, let Q_2 be the graph obtained from C by deleting y' and y'' together with their incident edges. (Recall that $pair(x) = \langle y', y'' \rangle$.) Since the edge $\{y', y''\}$ is not in C, Q_2 must consist of two connected components each of which is a simple path. Moreover, since the length of C is at least 5, one of the two paths in Q_2 contains at least one edge and hence has two distinct endpoints, say s and t. By step 1 of Algorithm 1, both s and t are vertices in Q_1. Noting that s is reachable from t in Q_2, it is easy to see that Q_1 contains a path from s to t on which $super(x)$ does not appear. Combining this path with the two edges $\{s, super(x)\}$ and $\{t, super(x)\}$, we get a cycle in Q_1 (on which $super(x)$ appears). However, corresponding to this cycle, there must exist a cycle in $G'[U_1]$ by step 7, a contradiction. ∎

By Claim 1 and step 8, we see that $G[V_1]$, $G[V_2]$, and $G[V_3]$ are all acyclic before step 9 is executed. The following claim shows that $G[V_1]$, $G[V_2]$, and $G[V_3]$ are also acyclic after step 9 is executed, establishing the lemma.

Claim 2 After step 9 is executed, $G[V_1]$, $G[V_2]$, and $G[V_3]$ are acyclic.

Proof. Assume, on the contrary, that some graph among $G[V_1]$, $G[V_2]$, and $G[V_3]$ contains a cycle C. We consider only the case where the graph is $G[V_1]$; the other two cases are similar. By Claim 1 and step 8, C must contain at least one vertex $x \in X_{=6}$. Let $x_1, \cdots, x_k$ be the vertices in both $X_{=6}$ and C. We need to consider only two cases.

Case 1: For all $1 \leq i \leq k$, the two neighbors of x_i in C are the two vertices contained in $pair(x_i)$. Then, for all $1 \leq i \leq k$, no neighbor of x_i other than the two in $pair(x_i)$ can appear on C, because otherwise U_2 or U_3 would contain at most one neighbor of some x_i and hence this x_i could have been added into V_2 or V_3 rather than into V_1 by steps 8 and 9. Moreover, for all $1 \leq i \leq k$, if $rival(x_i)$ is defined, then $rival(x_i)$ cannot be in V_1 (and hence cannot be in C) by step 9.2.2. Also, since the two vertices in each $pair(x_i)$ are not adjacent in G, the length of C is at least 4. We further distinguish two cases as follows.

Subcase 1.1: C has length 4. Then, $k = 1$ by step 1. Moreover, in the cycle C, the two vertices in $pair(x_1)$ must have a common neighbor (say, z) other than

x_1. If $rival(x_1)$ is undefined, then by the condition (ii) in Lemma 2, z must be in $N_G(x_1)$ which is impossible as argued above. Thus, $rival(x_1)$ must be defined. This implies that z cannot be $rival(x_1)$ as argued above. Now, z must be in $N_G(x_1)$ by the condition (2) in Lemma 3, still an impossibility as argued above.

Subcase 1.2: C has length at least 5. Let K be the (simple) graph obtained from C by first deleting x_i and next merging the two vertices in $pair(x_i)$ into a supervertex for all x_i with $1 \leq i \leq k$. Now, by step 1, K must be a simple cycle. However, corresponding to K, there must exist a cycle in $G'[U_1]$ by step 7, a contradiction.

Case 2: For some $1 \leq i \leq k$, at least one of the two neighbors of x_i in C is not contained in $pair(x_i)$. Then, by step 9, $super(x_i) \notin U_1$ and $rival(x_i)$ and $label(x_i)$ are defined. Since $super(x_i) \notin U_1$ and $x_i \in V_1$, $rival(x_i)$ is not in V_1 (and hence not in C) by step 9.2.2. Moreover, by step 9.2.2, one of the two neighbors of x_i in C must be $label(x_i)$. Let u be the neighbor of x_i in C other than $label(x_i)$. Then, u must be in $N_G(x_i) - (\{label(x_i)\} \cup \{y : y \text{ is in } pair(x_i)\})$. Furthermore, by step 9.2.2, among the neighbors of x_i in C, only $label(x_i)$ and u are in C. Now, deleting x_i from C gives us a path from $label(x_i)$ to u in the subgraph of G induced by $V - (\{x_i, rival(x_i)\} \cup \{y : y \text{ is in } pair(x_i)\})$, which is a contradiction by step 4.2. ∎

Now, the lemma immediately follows from Claim 2. ∎

Lemma 5. Algorithm 1 can be implemented in $O(\log n \log^* n)$ time using n processors.

Proof. We assume that the input graph G to Algorithm 1 is given in the form of a set of doubly linked adjacency lists. Vertices in G are encoded by $\lceil \log n \rceil$-bit integers. The adjacency list of each vertex v contains exactly one entry for each of its neighbors u in G. This entry contains the encoding of u and in addition contains a pointer to the entry for v in the adjacency list of u. To each vertex, there is an associated processor.

Step 1 can be done in $O(\log^* n)$ time [9]. Clearly, steps 2 and 3 can be done in constant time. Since the neighbors of each vertex x in $X_{=6}$ have degree at most 12, the processor associated with x can decide in $O(1)$ time whether $N_G(x) \subseteq N_G(v)$ for some $v \in V - (N_G(x) \cup \{x\})$ or not. Moreover, the processor associated with each vertex $x \in X_{=6}$ can find a Hamiltonian circuit of the graph $G[N_G(x)]$ in $O(1)$ time. Recall that this Hamiltonian circuit tells us the order of the neighbors of x in *every* planar embedding of G. Using these facts and the proofs of Lemma 2 and Lemma 3, we know that step 4 can be done in $O(1)$ time. Obviously, steps 5, 7, 8, and 9 can also be done in $O(1)$ time. Since the independent set X of G computed at step 1 contains a constant fraction of vertices of G, the recursion depth of Algorithm 1 is $O(\log n)$. Thus, Algorithm 1 runs in $O(\log n \log^* n)$ time using n processors. ∎

To see that Algorithm 1 has an optimal implementation, we use the following observation due to Hagerup, Chrobak, and Diks [9].

Observation [9]. Any algorithm can be implemented in $O(\log n \log^* n)$ time with $O(n/(\log n \log^* n))$ processors if it has an n-processor implementation consisting of $O(\log n)$ stages with the following characteristics:

(1) Each stage consists of some constant-time computation plus a constant number of computations of maximal independent sets in simple cycles or simple paths by the Cole&Vishkin method [6].

(2) The number of active processors in each stage decreases geometrically. Once a processor has become inactive, it remains so.

Let us now see why the above observation is applicable to Algorithm 1. By the proof of Lemma 5 and Hagerup *et al.*'s implementation of step 1 of Algorithm 1 [9], it is easy to see that the implementation of Algorithm 1 described in the proof of Lemma 5 indeed has the characteristic (1) above. Moreover, this implementation of Algorithm 1 clearly has the characteristic (2) above. Thus, by Lemma 4 and Lemma 5 together with the above observation, we have the following theorem:

Theorem 6. A 3-forest partition of a given n-vertex planar graph can be found in $O(\log n \log^* n)$ time with $O(n/(\log n \log^* n))$ processors.

Theorem 7. Let π be a hereditary property on graphs such that PMSP(π) restricted to trees can be solved in $T_\pi(n)$ time with $P_\pi(n)$ processors. Then, there is an NC approximation algorithm for PMSP(π) that runs in $O(\log n \log^* n) + T_\pi(n)$ time with $O(n/(\log n \log^* n)) + P_\pi(n)$ processors and achieves a performance ratio of 3.

Proof. Consider the following simple algorithm:

Input: A planar graph $G = (V, E)$ with n vertices.
Output: A subset U of V such that $G[U]$ satisfies π.
1. Use Algorithm 1 to find a 3-partition V_1, V_2, V_3 of V.
2. For each $1 \leq i \leq 3$, find a maximum subset U_i of V_i such that $G[U_i]$ satisfies π.
3. Output the subset U_k such that $|U_k| \geq |U_i|$ for $1 \leq k \neq i \leq 3$.

The above algorithm is clearly correct. To estimate the size of its output U_k, let U_{max} be a maximum subset of V such that $G[U_{max}]$ satisfies π. Let l be an integer such that $|U_{max} \cap V_l| \geq |U_{max} \cap V_i|$ for all $1 \leq l \neq i \leq 3$. Then, $|U_{max} \cap V_l| \geq |U_{max}|/3$. On the other hand, $G[U_{max} \cap V_l]$ satisfies π due to the hereditariness of π. Thus, $|U_{max} \cap V_l| \leq |U_k|$ by the choice of U_k. Therefore, $|U_k| \geq |U_{max}|/3$. This implies that the above approximation algorithm achieves a performance ratio of 3.

We next analyze the complexity of the above algorithm. By Theorem 6, step 1 can be done in $O(\log n \log^* n)$ time with $O(n/(\log n \log^* n))$ processors. Since each $G[V_i]$ ($1 \leq i \leq 3$) is a forest, the connected component in $G[V_i]$ can be computed in $O(\log n \log^* n)$ time with $O(n/(\log n \log^* n))$ processors [7]. Thus, step 2 can be done in $O(\log n \log^* n) + T_\pi(n)$ time with $O(n/(\log n \log^* n)) + P_\pi(n)$ processors. Step 3 can easily be done in $O(\log n \log^* n)$ time with $O(n/(\log n \log^* n))$

processors. Therefore, the above algorithm runs in $O(\log n \log^* n) + T_\pi(n)$ time with $O(n/(\log n \log^* n)) + P_\pi(n)$ processors. ∎

Corollary 8. There is an NC approximation algorithm for PMSP(π) that runs in $O(\log n \log^* n) + T_\pi(n)$ time with $O(n/(\log n \log^* n))$ processors and achieves a performance ratio of 3 if π is one of the following graph properties: 1-colorable, acyclic, outerplanar, series-parallel, bipartite, strongly chordal, chordal, comparability, perfect.

Proof. According to [10], a maximum 1-colorable set in a tree can be computed in $O(\log n \log^* n)$ time with $O(n/(\log n \log^* n))$ processors. Since a tree satisfies each of the rest properties, the corollary follows. ∎

3 2-Forest Partition of Special Planar Graphs

We first give several definitions. Let G be a planar graph. If $e = \{u, v\}$ is an edge in G and w is not a vertex in G, then e is *subdivided* when it is replaced by the edges $\{u, w\}$ and $\{w, v\}$. G is said to be *homeomorphic* to another graph H if G can be obtained from H by a sequence of edge subdivisions. For another graph H, G is said to be H-*free* if G contains no subgraph homeomorphic to H. Let K_4 be the complete graph with 4 vertices, and let $K_{2,3}$ be the complete bipartite graph with bipartition (X, Y) in which $|X| = 2$ and $|Y| = 3$. By Kuratowski's Theorem, K_4-free graphs and $K_{2,3}$-free graphs are both planar. A *2-forest partition* of G is a partition of V into two subsets V_1, V_2 such that both $G[V_1]$ and $G[V_2]$ are acyclic.

In this section, we present two *optimal NC* algorithms for finding a 2-forest partition of a given K_4-free or $K_{2,3}$-free graph. The algorithms resemble Algorithm 1. Before describing the algorithms, we need to prove several lemmas.

Lemma 9. Suppose G is an n-vertex graph having at most $2n$ edges. For $0 \le i \le n - 1$, let n_i be the number of vertices with degree i in G. Let n_4' be the number of vertices v in G such that $deg_G(v) = 4$ and v has no neighbor of degree ≥ 24 in G. Then, $\Sigma_{i=0}^{3} n_i + n_4' \ge n/30$.

Proof. Let m be the number of edges in G. Since $m \le 2n$, $\Sigma_{i=0}^{n-1} i n_i \le 4\Sigma_{i=0}^{n-1} n_i$ or equivalently, $\Sigma_{i=24}^{n-1} i n_i \le 4\Sigma_{i=0}^{n-1} n_i - \Sigma_{i=0}^{23} i n_i$. On the other hand, $n_4 - n_4' \le \Sigma_{i=24}^{n-1} i n_i$ by the definition of n_4'. Thus, we have

$$n_4' \ge n_4 - \Sigma_{i=24}^{n-1} i n_i$$
$$\ge n_4 - 4\Sigma_{i=0}^{n-1} n_i + \Sigma_{i=0}^{23} i n_i$$
$$= -4n_0 - 3n_1 - 2n_2 - n_3 + n_4 + \Sigma_{i=5}^{23}(i-4)n_i - 4\Sigma_{i=24}^{n-1} n_i$$
$$\ge -5n_0 - 4n_1 - 3n_2 - 2n_3 + \Sigma_{i=0}^{23} n_i - 4\Sigma_{i=24}^{n-1} n_i.$$

Therefore, $5(\Sigma_{i=0}^{3} n_i + n_4') \ge 5n_0 + 4n_1 + 3n_2 + 2n_3 + n_4' \ge \Sigma_{i=0}^{23} n_i - 4\Sigma_{i=24}^{n-1} n_i$. Note that $\Sigma_{i=24}^{n-1} n_i \le n/6$ because $m \le 2n$. This implies that $\Sigma_{i=0}^{23} n_i \ge 5n/6$. Hence, $\Sigma_{i=0}^{3} n_i + n_4' \ge (\Sigma_{i=0}^{23} n_i - 4\Sigma_{i=24}^{n-1} n_i)/5 \ge (5n/6 - 4n/6)/5 = n/30$. ∎

It is well known that a K_4-free or $K_{2,3}$-free graph with n vertices has at most $2n - 2$ edges [1]. Thus, by Lemma 9, we have $\sum_{i=0}^{3} n_i + n_4' \geq n/30$ for K_4-free or $K_{2,3}$-free graphs G. Using this result together with the ideas in the proof of Lemma 5 in [9], we can show the following lemma (which resembles Lemma 1):

Lemma 10. Suppose $G = (V, E)$ is an n-vertex K_4-free or $K_{2,3}$-free graph. Then, an independent set X of G satisfying the following three conditions can be found in $O(\log^* n)$ time using n processors:
  (a) $|X| \geq cn$ for some $0 < c \leq 1$.
  (b) For all $x \in X$, $deg_G(x) \leq 4$.
  (c) For all $x \in X$ with $deg_G(x) = 4$, $deg_G(y) \leq 23$ for all $y \in N_G(x)$.
  (d) For every two vertices x_1 and x_2 in X with $deg_G(x_1) = deg_G(x_2) = 4$, $N_G(x_1) \cap N_G(x_2)$ is empty and it is not the case that some $y_1 \in N_G(x_1)$ is adjacent to some $y_2 \in N_G(x_2)$ in G. (In other words, x_1 and x_2 are at least distance 4 apart in G.)

The following two lemmas are similar to Lemma 2.

Lemma 11. Suppose $G = (V, E)$ is a K_4-free graph. Let x be a vertex in G such that $deg_G(x) = 4$ and there is no $v \in V - (N_G(x) \cup \{x\})$ with $N_G(x) \subseteq N_G(v)$. Then, there is a pair $\langle y', y'' \rangle$ of two neighbors of x satisfying the following three conditions:
  (i) $\{y', y''\} \notin E$.
  (ii) For every $w \in V - (N_G(x) \cup \{x\})$, w is adjacent to at most one of y' and y''.
  (iii) There is at most one $z \in N_G(x) - \{y', y''\}$ with $\{y', z\} \in E$ and $\{y'', z\} \in E$.
Moreover, if $deg_G(y) \leq 23$ for all $y \in N_G(x)$, then $\langle y', y'' \rangle$ can be found in $O(1)$ time using a single processor.

Proof. Let u_1, u_2, u_3, u_4 be the neighbors of x in G. Consider a planar embedding of G. W.l.o.g., we may assume that u_1, u_2, u_3, u_4 clockwise appear around x in this order in the embedding. Since G is K_4-free, every pair of two neighbors of x in G must satisfy the condition (iii) in the lemma. There are two cases that may occur.
  Case 1: $\{u_1, u_3\} \in E$ or there is some $w \in V - (N_G(x) \cup \{x\})$ with $\{w, u_1\} \in E$ and $\{w, u_3\} \in E$. Then, since G is planar and there is no $v \in V - (N_G(x) \cup \{x\})$ with $N_G(x) \subseteq N_G(v)$, neither the edge $\{u_2, u_4\}$ is in G nor there is some $w' \in V - (N_G(x) \cup \{x\})$ such that w' is adjacent to both u_2 and u_4. By setting $y' = u_2$ and $y'' = u_4$, the three conditions in this lemma are clearly satisfied.
  Case 2: Case 1 does not occur. Clearly, we may set $y' = u_1$ and $y'' = u_3$. ∎

Lemma 12. Suppose $G = (V, E)$ is a $K_{2,3}$-free graph. Let x be a vertex of degree 4 in G. Then, there is a pair $\langle y', y'' \rangle$ of two neighbors of x satisfying the following three conditions:
  (i) $\{y', y''\} \notin E$.

(ii) For every $w \in V - (N_G(x) \cup \{x\})$, w is adjacent to at most one of y' and y''.

(iii) There is at most one $z \in N_G(x) - \{y', y''\}$ with $\{y', z\} \in E$ and $\{y'', z\} \in E$.

Moreover, if $deg_G(y) \leq 23$ for all $y \in N_G(x)$, then $\langle y', y'' \rangle$ can be found in $O(1)$ time using a single processor.

Proof. Let us first make two observations. Firstly, there is no $w \in V - (N_G(x) \cup \{x\})$ with $N_G(x) \subseteq N_G(w)$; otherwise G would have a subgraph homeomorphic to $K_{2,3}$. Secondly, if u_1, u_2, u_3, u_4 are the neighbors of x in G and they clockwise appear around x in this order in a planar embedding of G, then at least one of the four edges $\{u_1, u_2\}$, $\{u_2, u_3\}$, $\{u_3, u_4\}$, $\{u_4, u_1\}$ is not contained in G because G is $K_{2,3}$-free. Using the two observations, we can simply modify the proof of Lemma 11 to prove this lemma. ∎

The following lemma resembles Lemma 3.

Lemma 13. Suppose $G = (V, E)$ is a K_4-free graph. Let x be a vertex of degree 4 in G. Then, there is at most one $v \in V - (N_G(x) \cup \{x\})$ with $N_G(x) \subseteq N_G(v)$. Moreover, if such a v exists, then *every* pair $\langle y', y'' \rangle$ of two neighbors of x in G must satisfy the following three conditions:

(1) $\{y', y''\} \notin E$.

(2) For every $w \in V - \{x, v\}$, w is adjacent to at most one of y' and y''.

(3) The two vertices in $N_G(x) - \{y', y''\}$ are not reachable from each other in the graph $G[V - \{x, v, y', y''\}]$.

Proof. Easy and thus omitted. ∎

Lemma 14. Let H be a graph in which each vertex has degree at most 3. Suppose G is an H-free graph. Let x be a vertex in G, and let y', y'' be two neighbors of x in G. Let Q be the (simple) graph obtained from G by first deleting x and then merging y' and y'' into a supervertex $super(x)$. Then, Q is also H-free.

Proof. Assume, on the contrary, that Q has a subgraph homeomorphic to H. Let Q' be a minimal subgraph of Q homeomorphic to H. Since G is H-free, Q' must contain $super(x)$. By the minimality of Q' and the condition on the maximum degree of H, $super(x)$ has degree at most 3 in Q'. W.l.o.g., we may assume that $super(x)$ has degree exactly 3 in Q'; the other cases are simpler. Let v_1, v_2, and v_3 be the three neighbors of $super(x)$ in Q'. Then, in the graph G, each v_j, $1 \leq j \leq 3$, must be adjacent to y' or y''. By symmetry, we need to consider only the following two cases:

Case 1: In G, all of v_1, v_2, and v_3 are adjacent to y'. Clearly, replacing $super(x)$ with y' yields a subgraph of G homeomorphic to H, a contradiction.

Case 2: In G, v_1 and v_2 are adjacent to y' and v_3 is adjacent to y''. Let G' be the subgraph of G obtained from Q' by first removing $super(x)$ together with the three edges incident to it and then adding the three vertices y', x, y'' and the five edges $\{y', v_1\}$, $\{y', v_2\}$, $\{y', x\}$, $\{x, y''\}$, $\{y'', v_3\}$. Then, it is easy to see that G' is homeomorphic to H, a contradiction. ∎

We are now ready to present an algorithm for finding a 2-forest partition of a given K_4-free graph. The algorithm is very similar to Algorithm 1.

Algorithm 2
Input: A K_4-free graph $G = (V, E)$ with n vertices.
Output: A 2-forest partition V_1, V_2 of G.
1. Find an independent set X of G satisfying the conditions (a), (b), (c), and (d) in Lemma 10.
2. If $V = X$, then set $V_1 = X$ and $V_2 = \emptyset$ and halt.
3. Partition X into $X_{\leq 3}$ and $X_{=4}$, where $X_{\leq 3} = \{x \in X : deg_G(x) \leq 3\}$ and $X_{=4} = \{x \in X : deg_G(x) = 4\}$.
4. In parallel, for each $x \in X_{=4}$, perform the following steps:
   4.1. If there is no $v \in V - (N_G(x) \cup \{x\})$ with $N_G(x) \subseteq N_G(v)$, then first find a pair $\langle y', y'' \rangle$ of two neighbors of x satisfying the conditions (i), (ii), and (iii) in Lemma 11, and next set $pair(x) = \langle y', y'' \rangle$.
   4.2. If there is a (unique) $v \in V - (N_G(x) \cup \{x\})$ with $N_G(x) \subseteq N_G(v)$, then set $rival(x) = v$ and $pair(x) = \langle y', y'' \rangle$, where y' and y'' are two neighbors of x satisfying the conditions (1), (2), and (3) in Lemma 13.
5. From G, construct a new (simple) K_4-free graph G' by first deleting all vertices of X and then for all $x \in X_{=4}$, merging the two vertices in $pair(x)$ into a supervertex $super(x)$. (Comment: Using Lemma 14 repeatedly, we can show that G' is indeed K_4-free.)
6. Recursively call the algorithm for G' to obtain a 2-forest partition U_1, U_2 of G'.
7. For $1 \leq i \leq 2$, initialize V_i to be the set obtained from U_i by decomposing each supervertex $super(x)$ in U_i into the two vertices in $pair(x)$.
8. In parallel, for each $x \in X_{\leq 3}$, add x to a $V_i \in \{V_1, V_2\}$ such that x is adjacent to at most one vertex of V_i in the input graph G.
9. In parallel, for each $x \in X_{=4}$, perform the following steps:
   9.1. If $rival(x)$ is undefined (cf. step 4.1), then find the $U_i \in \{U_1, U_2\}$ containing at most one vertex in $(N_G(x) - \{y : y \text{ is in } pair(x)\}) \cup \{super(x)\}$ and add x into V_i.
   9.2. If $rival(x)$ is defined (cf. step 4.2), then perform the following steps:
      9.2.1. If there is some U_i with $1 \leq i \leq 2$ such that U_i contains at most one vertex in $N_G(x) - \{y : y \text{ is in } pair(x)\}$ but does not contain $super(x)$, then add x into V_i.
      9.2.2. If there is no U_i satisfying the condition in step 9.2.1 (i.e., one of U_1 and U_2 contains $super(x)$ and the other contains the two vertices in $N_G(x) - \{y : y \text{ is in } pair(x)\}$), then first find the $U_i \in \{U_1, U_2\}$ containing $super(x)$ and next add x into V_i if $rival(x) \notin V_i$ while add x into V_j $(1 \leq j \neq i \leq 2)$ otherwise.

Lemma 15. Algorithm 2 correctly outputs a 2-forest partition for G.

Proof. Using Lemma 10, 11, 13, and 14, we can show the lemma in almost the same way as we did in the proof of Lemma 4. ∎

Theorem 16. A 2-forest partition of a given n-vertex K_4-free graph can be found in $O(\log n \log^* n)$ time with $O(n/(\log n \log^* n))$ processors.

Proof. By Lemma 15, it suffices to show that Algorithm 2 runs in $O(\log n \log^* n)$ time with $O(n/(\log n \log^* n))$ processors. To this end, using a proof similar to that of Lemma 5, we can show that Algorithm 2 can be implemented in $O(\log n \log^* n)$ time with n processors. Moreover, this implementation has the two characteristics in the observation stated in the last section. This completes the proof. ∎

Theorem 17. A 2-forest partition of a given n-vertex $K_{2,3}$-free graph can be found in $O(\log n \log^* n)$ time with $O(n/(\log n \log^* n))$ processors.

Proof. Let Algorithm 3 be the algorithm obtained from Algorithm 2 by first modifying the input to be a $K_{2,3}$-free graph G, next replacing Lemma 11 in step 4.1 with Lemma 12, and finally removing steps 4.2 and 9.2. Then, it is easy to see that Algorithm 3 correctly finds a 2-forest partition of G and runs in $O(\log n \log^* n)$ time with $O(n/(\log n \log^* n))$ processors. ∎

Corollary 18. Given an n-vertex K_4-free or $K_{2,3}$-free graph G, a 4-coloring of G can be computed in $O(\log n \log^* n)$ time with $O(n/(\log n \log^* n))$ processors.

Proof. Using Algorithm 2 or 3, we get a 2-forest partition V_1, V_2 of G. It is easy to see that Algorithm 2 and Algorithm 3 can be modified to compute the connected components of $G[V_1]$ and $G[V_2]$ without increasing their complexity. Since each connected component T of $G[V_1]$ is a tree, we can use the Euler tour technique to find a 2-coloring of T (and hence of $G[V_1]$) in $O(\log n \log^* n)$ time with $O(n/(\log n \log^* n))$ processors. Similarly, a 2-coloring of $G[V_2]$ can be found within the same bounds. Combining the 2-coloring of $G[V_1]$ and that of $G[V_2]$, we get a 4-coloring of G. ∎

4 Discussion

We presented the first NC algorithm for finding a 3-forest partition of a given planar graph. It is worth mentioning that the vertex set of every planar graph can be partitioned into three subsets each of which induces a forest in which every component is a path [8, 14]. The proof for this fact given by Poh also exhibits that partitioning the vertex set of a given planar graph into such three subsets can be done in polynomial time [14]. It is very natural to ask whether this can be done in NC or not. However, we have not been able to settle this question.

References

1. T. Asano, An approach to the subgraph homeomorphism problem, *Theoret. Comput. Sci.* 38, 249-267 (1985).

2. B.S. Baker, Approximation algorithms for NP-complete problems on planar graphs, *J. ACM* 41, 153-180 (1994).

3. G. Chartrand, H.V. Kronk, and C.E. Wall, The point-arboricity of a graph, *Israel J. Math.* 6, 169-175 (1968).

4. G. Chartrand and H.V. Kronk, The point-arboricity of planar graphs, *J. London Math. Soc.* 44, 612-616 (1969).

5. G. Chartrand and L. Lesniak, *Graphs and Digraphs* (2nd ed.), (Wadsworth, Belmont, 1986).

6. R. Cole and U. Vishikin, Deterministic coin tossing with applications to optimal parallel list ranking, *Inform. and Control* 70, 32-53 (1986).

7. H. Gazit, Optimal EREW parallel algorithms for connectivity, ear Decomposition and st-Numbering of planar graphs, *in* "Proceedings, 5th IEEE International Parallel Processing Symposium 1991," pp. 84-91.

8. W. Goddard, Acyclic colorings of graphs, *Discrete Math.* 91, 91-93 (1991).

9. T. Hagerup, M. Chrobak, and K. Diks, Optimal parallel 5-colouring of planar graphs, *SIAM J. Comput.* 18, 288-300 (1989).

10. X. He, Binary tree algebraic computation and parallel algorithms for simple graphs, *J. Algorithms* 9, 92-113 (1988).

11. R. M. Karp and V. Ramachandran, "A survey of parallel algorithms for shared-memory machines," *in* "Handbook of Theoretical Computer Science, Volume A: Algorithms and Complexity," The MIT Press/Elsevier, 1990, 869-941.

12. P.N. Klein and S. Subramaniam, A linear-processor polylog-time algorithm for shortest paths in planar graphs, in: Proceedings of 31st IEEE Symposium on Foundations of Computer Science (IEEE, 1993) 173-182.

13. G.L. Miller and J.H. Reif, Parallel tree contraction and its applications, in: Proceedings of 26th IEEE Symposium on Foundations of Computer Science (IEEE, 1985) 478-489.

14. K.S. Poh, On the linear vertex-arboricity of a planar graph, *J. Graph Theory* 14, 73-75 (1990).

15. M. Yannakakis, Node- and edge-deletion NP-complete problems, *in* "Proceedings, 10th ACM Sympos. on Theory of Comput. 1978," pp. 253-264.

Efficient Parallel Modular Decomposition
(Extended Abstract)

Elias Dahlhaus

Basser Dept. of Computer Science,
University of Sydney,
NSW 2006, Australia,
e-mail: dahlhaus@cs.su.oz.au and dahlhaus@cs.uni-bonn.de

Abstract. Modular decomposition plays an important role in the recognition of comparability graphs and permutation graphs [12]. We prove that modular decomposition can be done in in polylogarithmic time with a linear processor bound.

Keywords. Parallel algorithms, graph theory, graph algorithms

1 Introduction

By a *module* of a graph $G = (V, E)$, we define a subset V' of the vertex set V such that all vertices in V' have the same neighbors outside V'. The decomposition of a graph into modules has many applications. First, modular decomposition allows a more compact representation of a graph. Modular decomposition plays also an important role in the recognition of certain graph classes like interval graphs, comparability graphs, and permutation graphs (see for example [12]). One only has to check that the 'module components' are permutation, interval, or comparability graphs respectively. Modular decomposition is also helpful to solve classical graph problems like graph coloring or maximum clique. The best known sequential algorithms for modular decomposition have a time bound of $O(n + m)$ [9, 4]. The only problem of these optimal algorithms is that a big mathematical effort is necessary to understand them. Previous efficient (not linear time) algorithms are due to [10] and [14]. Parallel algorithms with a non linear processor bound are due to M. Morvan and L. Viennot and to M. Novick (both unpublished).

Very related to the problem of modular decomposition is the problem to recognize cographs [3]. In some sense, cographs are those graphs which are totally decomposable into modules, i.e. each induced subgraph with at least four vertices has a module of size at least two.

The goal of this paper is to develop a parallel algorithm for modular decomposition. The processor number is linear, and the time is polylogarithmic. Compared to the algorithms of [9, 4], the algorithm presented here is quite simple. We combine some ideas of the parallel cograph recognition algorithm of [5] and the sequential divide and conquer modular decomposition method of [6].

As in [5], we distinguish between low degree, middle degree, and high degree vertices. In case of the existence of a middle degree vertex, we consider its neighbors, non neighbors, do modular decompositions separately, and combine the modular decomposition to a big one. If there are only low degree vertices then we compute a spanning tree and compute a first approximation of an elimination ordering as done by P. Klein [8]. Up to modules containing the root of the spanning tree, all modules appear only in one level. We make a modular decomposition in each level and combine these. In case that all vertices have high degree we only have to consider the complement.

The only situation that is left is that we have low degree vertices and high degree vertices but no middle degree vertices. There we distinguish between small size and large size modules. We shall find out that first contain only low degree vertices or only high degree vertices. Large size modules have the property that the neighborhood consists only of large degree vertices and the non neighborhood consists only of low degree vertices.

Section 2 introduces the notations and basic notions that are of relevance in the whole paper. Section 3 discusses describes the algorithm. Section 4 presents a complexity analysis of the algorithm.

2   Notation

A *graph* $G = (V, E)$ consists of a *vertex set* V and an *edge set* E. Multiple edges and loops are not allowed. The edge joining x and y is denoted by xy.

We say that x is a *neighbor* of y iff $xy \in E$. The *neighborhood* of x is the set $\{y : xy \in E\}$ consisting of all neighbors of x and is denoted by $N(x)$. The neighborhood of a set of vertices V' is the set $N(V') = \{y | \exists x \in V', yx \in E\}$ of all neighbors of some vertex in V'.

A *subgraph* of (V, E) is a graph (V', E') such that $V' \subset V$, $E' \subset E$. An *induced subgraph* is an edge-preserving subgraph, i.e. (V', E') is an induced subgraph of (V, E) iff $V' \subset V$ and $E' = \{xy \in E : x, y \in V'\}$.

By a *module* of $G = (V, E)$, we mean a subset V' of V such that with $y \in V \setminus V'$ and $u, v \in V'$, both uy and vy are in E or none of uy and vy is in E.

We compute not all modules but those modules X which do not *overlap* with other modules, i.e. there is no module Y that has a nonempty intersection with X and neither $X \subset Y$ nor $Y \subset X$. We call such modules also *overlap free*. Note that the overlap-free modules of any graph form a tree with respect to set inclusion, i.e. two overlap free modules are disjoint or comparable with respect to set inclusion. The notions of parent and child modules can be defined in an obvious way. The parent $P(X)$ of an overlap-free module X is the unique smallest overlap free module that is a proper superset of X. Y is a child module of X if and only if Y is an inclusion maximal proper overlap free submodule of X. The algorithm we present later computes, for a given graph G, a root directed tree T_G representing the overlap free modules in their tree like ordering with respect to set inclusion.

There are the following types of overlap free modules:

Prime Modules First define a graph $G = (V, E)$ with more than two vertices as *prime* if all the only modules are V and the subsets of V containing exactly one vertex. An overlap free module $X \subseteq V$ is called *prime* if the graph that results from $G[X]$ after identification of all vertices in the same child module of X is prime.

Degenerated Modules An overlap free module X is *degenerated* if the graph that results from $G[X]$ after identification of all vertices appearing in the same child module is complete or forms an independent set. X is called *positively degenerated* or also called a 1-module if the resulting graph is complete, i.e. all vertices are pairwise adjacent. Otherwise, if the resulting graph forms an independent set, the degenerated submodule X is called *negatively degenerated module* or a 0-module.

Note that any module, overlap free or not, is a prime module or a subset of a degenerated module X with the property that each child module is a subset of X or disjoint with X (compare [9]).

The goal of this algorithm is, given a graph (V, E), to compute a rooted tree T_G such that the leaves are the vertices of G and each inner node t represents an overlap free module, i.e. the leaves that are descendents of t form an overlap free module of G. We call this tree T_G also the *modular tree* of G.

The computation model is the CRCW-PRAM. We measure each primitive operation by one time unit.

3 The Algorithm

The algorithm is a divide and conquer strategy. We reduce the problem to subgraphs with at most $3/4|V|$ vertices to get a logarithmic recursion depth. We distinguish between vertices of low degree, middle degree, and high degree. A vertex v of $G = (V, E)$ is of *low degree* if it is of degree at most $1/4|V|$. If the degree is at least $3/4|V|$ then the vertex is called a *high degree* vertex. If a vertex is neither a high degree nor a low degree vertex then it is called a *middle degree* vertex.

3.1 First Case: There is a Middle Degree Vertex

Suppose there is a middle degree vertex u. Clearly each module X that does not contain u has the property that all vertices of X are in the neighborhood of u or none of them. Therefore every module not containing u is a prime module of the neighborhood or the non neighborhood of u or a subset of a degenerate module of the neighborhood or of the non neighborhood of u.

We first find out the modules that do not contain u. Let L_1 be the set of neighbors of u and L_2 be the set of non neighbors of u. Note that each module that does not contain u is a subset of L_1 or a subset of L_2. Moreover, each module in L_1 or L_2 has the property that all vertices have the same vertices outside L_1 or L_2 respectively. A prime module of the whole graph that is in

L_1 or L_2 is also a prime module of G restricted to L_1 or L_2 respectively. A degenerated module of the whole graph that appears as a subset of L_1 or L_2 is a subset of a degenerated module of G restricted to L_1 or L_2 respectively (not necessarily the whole degenerated module).

We assume that modular trees T_1 and T_2 of G restricted to L_1 and G restricted to L_2 respectively are computed recursively.

We first find out those overlap-free modules of G restricted to L_1 and L_2 that remain modules of the whole graph. We call a module of L_1 or L_2 *inhomogeneous* if it has descendents in the modular tree with different neighborhoods outside L_1 or L_2 respectively. We find out inhomogeneous modules, i.e. the tree nodes that represent inhomogeneous modules as follows. We assume that the trees T_1 and T_2 are embedded into the plane and the leaves of T_1 and T_2, i.e. the vertices of L_1 and L_2 are enumerated from left to right, say $L_1 = \{x_1, \ldots x_k\}$ and $L_2 = \{y_1, \ldots y_l\}$. Such enumerations can be computed in $O(\log n)$ time with $O(n/\log n)$ processors using Eulerian cycle techniques on trees (see for example [15, 2]). We assign those (x_i, x_{i+1}) and (y_i, y_{i+1}) that distinguish in the neighborhood in L_2 and L_1 respectively. We determine the least common ancestors s_i and t_i of all assigned pairs (x_i, x_{i+1}) and (y_i, y_{i+1}) in $O(\log n)$ time with $O(n/\log n)$ processors [11]. We assign all tree nodes as inhomogeneous that are ancestors of at least one assigned tree node s_i or t_i in $O(\log n)$ time with $O(n/\log n)$ processors [15].

The second step is to find out new degenerate modules, i.e. homogeneous subsets of inhomogeneous degenerated modules of L_1 and of L_2. We consider vertices of L_1 or L_2 as *equivalent* if they have the same neighbors outside L_1 or L_2 respectively. One can find the equivalence classes by lexicographic sorting in $O(\log n)$ time using $O(n + m)$ processors (to compare the neighborhoods of two vertices with degrees d_1 and d_2, one needs $d_1 + d_2$ processors and constant time on a CRCW-PRAM. To sort n elements, one needs $O(n)$ processors and $O(\log n)$ time [1]). Let A be an equivalence class and X be an inhomogeneous degenerated module and $X_1, \ldots, X_k$ be the child modules of X that are subsets of A. If X contains more than one such child module in A then we create a new module X_A with $X_1, \ldots, X_k$ as its children, i.e. suppose t has been assigned as inhomogeneous and represents a degenerate module. Suppose t has more than one child that are not assigned as inhomogeneous. For each non inhomogeneous child t' of t, we pick a vertex $v_{t'}$ of G that is a descendent of t' ($v_{t'}$ is a leaf). This can be done in $O(\log n)$ time with $O(n/\log n)$ processors by list ranking along the Eulerian cycle along the double edged tree. Consider children t' and t'' of t as equivalent if $v_{t'}$ and $v_{t''}$ are equivalent. We determine the equivalence classes by sorting (in $O(\log n)$ time with $O(n)$ processors for all t simultaneously). For each equivalence class A, we create a new node with the children of t in A as children.

It remains to find out the modules that contain u. Let X be a module containing u. Note that every neighbor of X that is not in X is in the neighborhood of u. Let L_1 be the set of neighbors of u and L_2 be the set of non neighbors of u. Since all neighbors of X not being in X are in L_1, $X \cap L_2$ is a union of connected components of G restricted to L_2. Note that each module of G is also a module of the complement of G. Therefore for symmetry reasons, $X \cap L_1$ is the union of connected components of the complement of G restricted to L_2.

Shortly we call connected components of G restricted to L_2 *2-components* and connected components of the complement of G restricted to L_1 *1-components*. Note that all 1-components and 2-components are known recursively as modules of G restricted to L_1 and L_2 respectively.

We consider the situation that D_1 is a 2-component not in the module X and D_2 is a 2-component in X. Let C_1 be a 1-component, such that there is an edge joining some vertex in $x \in D_1$ with some vertex $y \in C_1$. Then also C_1 is not in X. Since C_1 is a subset of the neighborhood of u, and therefore in the neighborhood of X, all vertices in X are adjacent to all vertices in C_1. Therefore also all vertices in D_2 are adjacent to all vertices in C_1.

A 1-component C is called *strongly adjacent* to a 2-component D if each vertex of C is joined by an edge with each vertex of D. We call C also a *strong neighbor* of C and vice versa. A 1-component C is called *weakly* adjacent to a 2-component D if there is an edge joining some vertex of C with some vertex of D. We call D also a *weak neighbor* of C and vice versa.

We can reformulate above observation as follows.

Let X be a module containing u, D_1 be a 2-component not in X, and D_2 be a 2-component in X. Then all weak neighbors of D_1 are strong neighbors of D_2.

Next observe that any 1-component C with the property that some 2-component in X is not a strong neighbor of C must be in X and any 1-component that is a weak neighbor of some 2-component not in X must not be in X.

Theorem 1. *Let $u \in X \subseteq V$.*
Then X is a module if and only if

1. *For all 1- and 2-components D, $D \subseteq X$ or $D \cap X = \emptyset$*
2. *For all 2-components $D_1 \not\subseteq X$, $D_2 \subseteq X$, all weak neighbors of D_1 are strong neighbors of D_2,*
3. *all 1-components not in X are strong neighbors of all 2-components in X, and*
4. *all 1-components being weak neighbors of some 2-component not in X are not in X.*

Proof. The second statement follows from the first by previous discussions.

Suppose now, the second condition is satisfied. First we show that all neighbors of X not in X are in L_1. Since $L_1 \cap X$ is a union of connected components of G restricted to L_2, no edge appears between a vertex of $L_2 \cap X$ and a vertex of $L_2 \setminus X$. Since all 1-components being weak neighbors of some 2-component not in X are not in X, there is also no edge between a vertex in $L_2 \setminus X$ and a vertex in $L_1 \cap X$. Therefore no neighbor of X in L_2 exists.

Since every connected component of the complement of G restricted to L_1 is either completely in X or disjoint with X, every vertex in $X \cap L_1$ has all vertices in $L_1 \setminus X$ as a neighbors. Therefore all vertices in $X \cap L_1$ have all the same neighbors outside X.

Since all 1-components not in X are strong neighbors of all 2-components in X, $L_1 \setminus X$ is also the neighborhood outside X of every vertex in $L_2 \cap X$. Q.E.D.

We continue with some remarks on the structure of positively and negatively degenerated modules containing u.

First we consider the case that X is a negatively degenerated overlap free module. Let Y be the child module of X containing u. Then all child modules $Y' \neq Y$ of X are 2-components. The weak neighborhood of any such $Y' \neq Y$ is also its strong neighborhood, and all these child modules $Y' \neq Y$ of X have the same 1-components as weak and strong neighbors. Since X is a negatively degenerated module, there is no edge joining any vertex in a child module $Y' \neq Y$ with some vertex in $X \setminus Y'$. Therefore these 1-components are not in X and therefore not in Y.

Note that Y is not a negatively degenerated module and therefore connected. Therefore each 2-component in Y has at least one 1-component in Y as a neighbor and the number of weak neighbors of any 2-component in Y is larger than the number of weak neighbors of any child module $Y' \neq Y$ of X.

Now let D' be any 2-component not in X. Note that every weak neighbor of D' is a strong neighbor of any 2-component D in X. This is particularly true for the child modules $Y' \neq Y$ of X. Therefore D' has at most as many weak neighbors as any child module $Y' \neq Y$ of X. We assume that the number of weak neighbors of D and of Y' is equal. Then Y' and D have the same weak neighbors, because the weak neighborhood of D' is a subset of the strong neighborhood of Y'. The strong neighborhood of D' must be a proper subset of the strong neighborhood of Y' (which is also the weak neighborhood of Y'). Otherwise $D' \cup X \setminus Y$ and X are overlapping modules. This is a contradiction.

Negatively degenerated overlap free modules X with Y as child module containing u have therefore the following characteristic structure.

1. All 2-components in $X \setminus Y$ have the same 1-components as weak neighbors.
2. the 2-components in $X \setminus Y$ are the child modules of X that are not identical to Y,
3. the weak neighbors of any 2-component in $X \setminus Y$ are also strong neighbors,
4. the weak (and therefore strong) neighbors of any 2-component in $X \setminus Y$ are all not in X
5. the weak (and therefore strong) neighborhood of any 2-component in $X \setminus Y$ is a proper subset of the weak neighborhood of any 2-component in Y
6. for any 2-component D' not in X and any 2-component D in $X \setminus Y$, the weak neighborhood of D' is a subset of the weak (and therefore strong) neighborhood of D and the strong neighborhood of D' is a proper subset of the strong neighborhood of D.

Now let X be a positively degenerated overlap free module and Y be the child module of X containing u.

Note that all child modules $Y' \neq Y$ of X are 1-components. All these child modules $Y' \neq Y$ are strong neighbors of each 2-component in X and have no

2-components not in X as weak neighbors. Moreover, each 1-component C in Y is a non strong neighbor (not even necessarily a weak neighbor) of some 2-component in X. Otherwise $(X \setminus Y) \cup C$ and X would be overlapping module (contradiction). Observe that each 1-component C with no 2-component outside X as weak neighbor and all 2-components in X as strong neighbors is a child module of X. Otherwise X and $(X \cup C) \setminus Y$ would be overlapping modules (contradiction).

Therefore positively degenerated overlap free modules can be characterized as follows.

1. Y and X have the same 2-components
2. each 1-component in Y is a non strong neighbor of some 2-component in Y,
3. the 1-components in $X \setminus Y$ are the child modules of X not identical to Y,
4. all 1-components $X \setminus Y$ are strong neighbors of any 2-component in X and not weak neighbors of any 2-component not in X,
5. each 1-component not in X is a weak neighbor of some 2-component not in X.

We assume that the modules of G restricted to L_1 and G restricted to L_2 are known. Then we also know the 1-components (as positively degenerated modules of L_1 that are maximal submodules of L_1) and the 2-components (as negatively degenerated modules of L_2 that are maximal submodules of L_2).

We get the overlap free modules containing u in several steps. We first sort the 2-components with respect to the number of weak and strong neighbors and we determine the final segments of the sorted sequence "representing" modules, and finally we determine the modular tree.

In detail we proceed as follows.

1. We sort all 2-components with respect to the number of weak neighbors to a sequence $(D_1, \ldots, D_l)$. If the number of weak neighbors of two 2-components is equal then the 2-component with the larger number of strong neighbors is considered as the larger 2-component. This can be done in $O(\log n)$ time with $O(n+m)$ processors (we compute, for each 2-component, the number of its weak and strong neighbors in $O(\log n)$ time with $O(n+m)/\log n$ processors and sort in $O(\log n)$ time with $O(n)$ processors [1]).
2. For each 1-component C, we determine the minimum index $i = Max(C)$ such that for all $j > i$, D_j is a strong neighbor of C, (i.e. the maximum index of a 2-component that is not a strong neighbor of C) and the minimum $j = Min(C)$ such that D_j is a weak neighbor of C. This can be done, for all C simultaneously, in $O(\log n)$ time with $O(n+m)/\log n$ processors (standard minimum computation).
3. Let $I_i := \bigcup_{j \geq i} D_j$. We determine all I_i with the property that there is a module M_i with $I_i = M_i \cap L_2$. In that case, we say that i represents a module. Note that i represents a module iff there is no interval $[Min(C), Max(C)]$ with $Min(C) < i$ and $Max(C) \geq i$. We check whether i represents a module as follows. For each j, we determine the maximum $j' = m(j)$, such that there

is a C with $Min(C) = j$ and $Max(C) = j'$. Let $ma(i)$ be the maximum $Max(C)$ with $Min(C) < i$. Then i represents a module iff $ma(i) < i$. Note that $ma(i) = \max_{j<i} m(j)$. $m(j)$ can be computed in $O(\log n)$ time with $O(n + m)/\log n$ processors (standard minimum computation). $ma(i)$ can be computed in $O(\log n)$ time with $O(n/\log n)$ processors using paràllel prefix computation.

4. We determine the negatively degenerated modules as follows. We compute the sets of maximal intervals I such that
   - each $i \in I$ represents a module and
   - all D_i with $i \in I$ have only strong neighbors as weak neighbors and the same strong neighbors. It is sufficient to check that the number of strong and weak neighbors coincide and that numbers of strong neighbors are equal.
   - $k_I = max\, I + 1$ represents a module (this has has to be tested only in case that I has only one element).

If these conditions are satisfied then we say that I represents a negatively degenerate module.

The negatively degenerated module module X_I represented by I consists of all D_i, $i \in I$ and

$$\bigcup_{j \geq k_I} D_j \cup \{u\} \cup$$

$$\{x \in L_1 | x \text{ is in the neighborhood of some } D_j, \, j \geq k_I$$

$$\text{and not in the neighborhood of some } D_j, \, j < k_I\}$$

as submodules.

To find out such intervals I, we first find out (in $O(\log n)$ time with $O(n + m)/\log n$ processors) those i representing a module such that D_i has only strong neighbors as weak neighbors. Then we find out those i, such that i and $i+1$ represent modules, D_i and D_{i+1} have only strong neighbors as weak neighbors, and D_i and D_{i+1} have the same number of strong neighbors. This is done in constant time with $O(n)$ processors. This enables us to find the left and right borders of such intervals I in constant time with $O(n)$ processors. To get the intervals, we compute, for each i the maximum left border of such an interval, using parallel prefix computation ($O(\log n)$ time and $O(n/\log n)$ processors). Especially we have, for each right border, its corresponding left border and therefore all the intervals I representing a negatively degenerated module.

5. We compute the positively degenerated modules containing u as follows. For each i representing a module, we determine the set $P_i := \{C : C$ strong neighbor of all D_j, $j \geq i$ and C not neighbor of some D_j, $j < i\} = \{C | Min(C) = i$ and $Max(C) = i - 1\}$. This can be done in $O(\log n)$ time with $O(n/\log n)$ processors. If $P_i \neq \emptyset$ then there is a positively degenerated module PD_i consisting of all $C \in P_i$ and a module U_i consisting of u, all vertices in some D_j, $j \geq i$, and all vertices in some C with $Max(C) > i$ as child modules. If P_i is not empty then we say that i *represents a positively degenerate module.*

(Note that it can be checked in constant CRCW time with a linear number of processors whether P_i is empty or not). If there are 1-components that are even not weak neighbors of any 2-components then they form together with u a positively degenerate module. In that case, $k+1$ represents a positively degenerate module, where k is the largest index of a 2-component.

6. The prime modules are determined as follows. A prime module is necessarily a module U_i as defined in the last step and is a prime module if it does not appear in some interval I as defined in the step that determines the negatively degenerated modules.

We say that i *represents a prime module* if i represents a module and does not appear in an interval representing a negatively degenerated module. We can check whether i appears in an interval representing a prime module in logarithmic time, and for all i simultaneously, in the same time bound with a linear number of processors.

We have to complete the modular tree.

We get the following tree nodes that are created by the indices representing modules.

1. The tree nodes s_i, such that i represents a prime module containing u,
2. the tree nodes t_i, such that i represents a positively degenerate module containing u, and
3. the tree nodes u_I, such that I represents a negatively degenerated module containing u.

We determine the parent function *Parent* of the modular tree as follows.

First we consider $Parent(u)$. If $k+1$ represents a positively degenerated module then $Parent(u)$ is set to be t_{k+1}. Otherwise let k' be the largest index that represents a module. We can determine k' by list ranking in $O(\log n)$ time with $O(n)$ processors [2]. If k' represents a prime module then $Parent(u)$ is $s_{k'}$. Otherwise $Parent(u)$ is the u_I such that I contains k'.

Next we consider the parents of s_i, t_i, and u_I. For each i representing a module, we determine the largest $i' < i$, say $P(i)$, that represents a module, by list ranking [2] in $O(\log n)$ time with $O(n/\log n)$ processors.

Suppose i represents a prime module or is the smallest index of an interval I representing a negatively degenerated module. If i represents also a positively degenerated module then $Parent(s_i) := t_i$ or $Parent(u_I) := t_i$ respectively. If i does not represent a positively degenerate module then there are two possible cases.

- $P(i)$ (the next smaller i' representing a module) represents a prime module. Then $Parent(s_i) := s_{P(i)}$ or $Parent(u_I) := s_{P(i)}$.
- $P(i) = i - 1$ and i is the largest index of an interval I' representing a negatively degenerate module. Then $Parent(s_i)$ and $Parent(u_I)$ are set to be $u_{I'}$ respectively.

If i represents a positively degenerated module then there are the same two possible cases as mentioned above and $Parent(t_i) = s_{P(i)}$ if $P(i)$ represents a prime module and $Parent(t_i) := u_{I'}$ if $P(i)$ appears in the interval I' representing a negatively degenerate module.

It remains to determine the parents of tree nodes representing modules that are subsets of L_1 or L_2.

We have to determine the parents of homogeneous nodes with an inhomogeneous parent that have not a new node as new parent and the parents of new nodes.

Let t such a node. We first determine (by list ranking in $O(\log n)$ time with $O(n/\log n)$ processors) a leaf descendent v_t of t. Note that v_t is a vertex of L_1 or L_2. We determine the 1- or 2-component C_t or D_t v_t belongs to (in constant time with $O(n)$ processors).

Suppose t represents a module in L_2 and let $D_t = D_i$. Then i_t is the largest index $\leq i$ representing a module . Note that $i_t = P(i+1)$ and that i_t therefore can be determined in constant time. If i_t represents a prime module then $Parent(t)$ is set to be the node s_i representing this prime module. If i_t is in an interval I representing a negatively degenerate module (this I can be determined in constant time) then $Parent(t)$ is set to be u_I if t does not represent a negatively degenerate module. Otherwise t and u_I are contracted to one node. Note that in that case, $i = i_t$.

Suppose t represents a module in L_1. Note that all vertices of L_2 that are leaf descendents of t have the same neighbors in L_2.

First case: $Min(C_t) = Max(C_t) + 1$ and $i = Min(C_t)$ represents a module. Then i also represents a positively degenerated module and C_t belongs to the positively degenerated module t_i. If t itself represents a positively degenerate module then t and t_i are contracted to one node. If t does not represent a positively degenerate (this is only possible if t represents exactly the module C_t) then $Parent(t)$ is set to be t_i. This part can be done in constant time with a linear processor number)

Second case: $Min(C_t) \leq Max(C_t)$ or $Min(C_t)$ does not represent a module. Let i_t be the largest $j \leq Min(C_t)$ representing a module (in constant time by knowledge of $P(i_t + 1)$). $Parent(t)$ is just the prime module s_{i_t}. Note that i_t cannot appear in an interval representing by a negatively degenerate module.

Therefore:

Lemma 2. *If the conditions of the first case are satisfied then one recursion step can be done in logarithmic time with a linear processor bound on a CRCW-PRAM.*

3.2 Second Case: All Vertices are Low Degree Vertices

Here we proceed in a similar way as in the low degree refinement of the perfect elimination ordering algorithm of Klein [8]. We compute a spanning forest F of G (in $O(\log n)$ time with $O(n+m)$ processors [13]). The connected components of G, i.e. the trees of F are modules. If G is not connected then the whole vertex set V forms a negatively degenerated module. For each tree T of F with a root u, we compute a preorder $\{u_1, \ldots, u_n\}$ of T. We set $L_1 := \{u\}$ and

$$L_{i+1} := \{x | u_i \text{ is the neighbor of } x$$

with minimum index $\}$.

By a similar argument as in the first case, each module not containing u is a subset of some L_i. To compute the modules in L_i, we compute, as in the first case, the modules of G restricted to L_i and compute, in the same way as in the first case, the modules of G in L_i by comparison of the neighborhood outside L_i.

The modules containing u can be found out in the same way as the modules containing u in the first case.

The processor and time bounds of one recursion step are therefore as in the first case.

3.3 Third Case: All Vertices are of Low or High Degree

We distinguish between *small* modules that contain only $1/4|V|$ vertices and *large* modules that contain more than $1/4|V|$ vertices.

Lemma 3. *If X is a small module then all vertices of X are of low degree or all vertices of X are of high degree.*

Proof. Suppose $|X| < 1/4|V|$, $x, y \in X$, $degree(x) < 1/4|V|$, and $degree(y) > 3/4|V|$. Then x has at most $1/4|V|$ neighbors outside X and y has at least $1/2|V|$ neighbors outside X. Q.E.D.

Lemma 4. *If X is a large module then all neighbors of X that are not in X are of high degree.*

Therefore

Lemma 5. *If X is a large module containing low degree and high degree vertices then the set X' of low degree vertices of X is a union of connected components of G restricted to the set of low degree vertices.*

Corollary 6. *If X is a large module containing low degree and high degree vertices then the set X'' of high degree vertices of X is a union of connected components of the complement of G restricted to the set of high degree vertices.*

First, we compute all modules of the set L of low degree vertices and of the set H of high degree vertices. The modules of L are computed as in the second case. To compute the modules of H, we first compute the complement of G restricted to H, and afterwards, we proceed as in the second case. Note that the number of edges in G restricted to H is smaller than the number of edges adjacent to some vertex in H and therefore the number of edges of the complement of G restricted to H is $O(m)$. By comparison of neighbors outside L or H, we get all modules in L and H that are also modules of G. By previous lemma and corollary, modules having high degree and low degree vertices satisfy the same conditions as modules containing the set u in the first case. In so far, we can determine them in the same way as in the first case.

Again, as in the second and first case, one recursion step can be done in $O(\log n)$ time with $O(n + m)$ processors.

4 Complexity Analysis

In all three cases, the problem of modular decomposition is reduced to components of size at $3/4$ of the number of vertices. Therefore the recursion depth is $O(\log n)$. In all three cases, one recursion can be done in $O(\log n)$ time by a CRCW-PRAM with $O(n + m)$ processors. Therefore the overall complexity on a CRCW-PRAM is of $O(n + m)$ processors and $O(\log^2 n)$ time.

Theorem 7. *Modular decomposition can be done by a CRCW-PRAM in $O(\log^2 n)$ time with $O(n + m)$ processors.*

5 Conclusions

The linear time modular decomposition algorithms of [9, 4] is quite complicated. It should not be too difficult to transform the parallel modular decomposition algorithm as presented here to a simpler linear time algorithm. It is almost obvious how to turn the algorithm as presented here into an $O(n + m) \log n$ algorithm. One might also ask whether the techniques can be applied to other structures like directed graphs or graphs with colored edges. We believe that this is possible with not too much effort.

Very related to the problem of modular decomposition is also the problem of transitive orientation of an undirected graph [12]. Further research should be done to extend the algorithm presented in this paper to an efficient parallel algorithm for transitive orientation and permutation graph recognition. Note that permutation graphs are exactly those graphs which are transitively orientable and where the complement is transitive orientable.

The parallel modular decomposition algorithm induces also a parallel cograph recognition algorithm and therefore generalizes a result of X. He [7]. An improved cograph algorithm working in $O(\log^2 n)$ time on a CREW-PRAM with a linear processor number is developed in [5].

References

1. R. Cole, Parallel Merge Sort, *27. IEEE-FOCS* (1986), pp. 511-516.
2. R. Cole, U. Vishkin, Deterministic Coin Tossing with Applications to Optimal Parallel List Ranking, *Information and Control* 70 (1986), pp. 32-53.

3. D. Corneil, H. Lerchs, L. Burlingham, Complement Reducible Graphs, *Discrete Applied Mathematics* 3 (1981), pp. 163-174.

4. A. Cournier, M. Habib, A New Linear Time Algorithm for Modular Decomposition, *Trees in Algebra and Programming*, LNCS 787 (1994), pp. 68-84.

5. E. Dahlhaus, Efficient Parallel Recognition Algorithms of Cographs and Distance Hereditary Graphs, *Discrete Applied Mathematics* 57 (1995), pp. 29-44.

6. A. Ehrenfeucht, H. Gabow, R. McConnell, S. Sullyvan, An $O(n^2)$ Divide-and-Conquer Algorithms for the Prime Tree Decomposition of Two-Structures and Modular Decomposition of Graphs, *Journal of Algorithms*, 16 (1994), pp. 283-294.

7. X. He, Parallel Algorithms for Cograph Recognition with Applications, *Journal of Algorithms* 15 (1993), pp. 284-313.

8. P. Klein, Efficient Parallel Algorithms for Chordal Graphs, 29th IEEE-FOCS (1988), pp. 150-161.

9. R. McConnell, J. Spinrad, Linear-Time Modular Decomposition and Efficient Transitive Orientation of Comparability Graphs, *Fifth Annual ACM-SIAM Symposium of Discrete Algorithms* (1994), pp. 536-545.

10. J.H. Muller, J. Spinrad, Incremental Modular Decomposition, *Journal of the ACM*, 36 (1989), pp. 1-19.

11. B. Schieber, U. Vishkin, On Finding Lowest Common Ancestors: Simplification and Parallelization, *SIAM Journal on Computing* 17 (1988), pp. 1253-1262.

12. J. Spinrad, On Comparability and Permutation Graphs, *SIAM-Journal on Computing*, 14 (1985), pp. 658-670.

13. Y. Shiloach, U. Vishkin, An O(log n) Parallel Connectivity Algorithm, *Journal of Algorithms*, 3 (1982), pp. 57-67.

14. J. Spinrad, P4-Trees and Substitution Decomposition, *Discrete Applied Mathematics*, 39 (1992), pp. 263-291.

15. R. Tarjan, U. Vishkin, Finding biconnected components and computing tree functions in logarithmic parallel time, *SIAM Journal on Computing* 14 (1985), pp. 862-874.

Modular Decomposition of Hypergraphs

P. Bonizzoni and G. Della Vedova

Dipartimento di Scienze della Informazione*
Università Degli Studi di Milano
via Comelico 39, 20135 Milano – ITALY
e-mail bonizzon@ghost.dsi.unimi.it

Abstract. We propose an $O(n^4)$ algorithm to build the modular decomposition tree of hypergraphs of dimension 3 and show how this algorithm can be generalized to compute efficiently the decomposition of hypergraphs of fixed dimension k.

1 Introduction

The *modular decomposition* of graphs may solve efficiently a large number of combinatorial problems on graphs and partial orders, as well as some scheduling problems [8].

The importance of this form of representation for graphs has been recently pointed out by the first linear-time algorithms [7, 1] for computing the modular decomposition which have improved the time bounds for solving problems on *comparability graphs* [6], i.e graphs obtained by eliminating the orientation in digraphs representing partial orders.

The modular decomposition of a graph gives a tree representation (or *partitive tree* in [10]) of the graph by means of *modules*, subsets of vertices which are adjacent to the same vertices outside the module. An important advantage of such a tree is that it represents in $O(n)$ space all modules of a graph, even though they can be in exponential number w.r.t. n.

This form of tree representation turns out to be of interest for other relational structures besides graphs. In fact, a generalization of it has been to *k-ary relations* in [10] and to *2-structures* in [3], where a 2-structure on a domain D is a labeling of all antireflexive pairs on D. This second generalization of the decomposition reveals to be useful in solving other different problems in Computer Science [5]. Another generalization has been proposed with the notion of a *k-structure* [4], which is a labeling of all antireflexive k-tuples on a domain D, that is of the form $\{x_1, \cdots, x_k\}$ with $x_j \neq x_i$ for some $1 \leq i, j \leq k$. A first polynomial algorithm for computing the modular decomposition tree for k-ary relations has been shown in [9].

In this paper, we propose an efficient algorithm to build the decomposition tree for hypergraphs, seen as a natural generalization of the modular decomposition for graphs.

---
* This work has been supported by MURST 40% *Algoritmi e strutture di calcolo* and ASMICS 2 $N°$ 6317.

We first propose an $O(n^4)$ algorithm for the modular decomposition of hypergraphs of dimension 3. Then we show how our algorithm generalizes to higher dimensions, by using the decomposition tree of hypergraphs of dimension $k - 1$ to compute that of hypergraphs of dimension k. This general algorithm requires $O(n^{3k-5})$ time.

2 Preliminaries

Let A, B be two sets of elements. Then A, B overlap iff $A - B$, $B - A$ and $A \cap B$ are nonempty sets.

A *hypergraph* is a pair $H = (V, E)$, where V is a finite set of elements, called *vertices* and $E \subset 2^V$ is such that if $e \in E$ then $|e| \geq 2$. The elements of E are called the *hyperedges* of H. A hyperedge e is an *i-hyperedge* iff $|e| = i$.

We will say that two vertices v and w are *i-adjacent* if there exists a hyperedge consisting of exactly i vertices and containing both v and w.

A *hypergraph of dimension k*, is a hypergraph with hyperedges which have at most k elements, that is if $e \in E$ then $|e| \leq k$.

Let $H = (V, E)$ be a hypergraph and assume $X \subseteq V$. The *subhypergraph induced by X*, denoted as $H|X$ is the hypergraph $H' = (X, E \cap 2^X)$. The subhypergraph H' is *proper* if X is a proper subset of V.

A family F of sets on a domain D is a *partitive set family* [10] iff $D \in F$, the singleton subsets of D are each members of F, and whenever X and Y are overlapping members of F, then $X \cap Y$, $X \cup Y$, $X - Y$, $Y - X$ are also members of F.

A *partitive set family* has a tree representation, the *partitive tree* [10]: the members F' of the family that overlap no other are nodes of the tree and the containment relation on these sets gives the parent relation in the tree. Each internal node X of the partitive tree is labeled as:

- *q-complete* if the union of any subset of its children is a member of F,
- *q-primitive* if each of its children is a member of F, while no nontrivial union of its children is a member of F,
- *q-linear* if there is a linear order of children of X such that the union U of a subset of its children is a member of F iff the elements in U are consecutive in such a linear order.

It has been proved [10] that a set $X \subseteq D$ is a member of F iff it is a node in the partitive tree or a union of children of an internal node of such a tree.

The family of modules of some relational structures, as graphs, k-ary relations [10] and k-structures [4], forms a partitive family; the partitive tree of these structures is their *modular decomposition tree*.

Let $G = (V, E)$ be a graph, we will say that $M \subseteq V$ is a *module* if there does not exist a vertex $v \notin M$ such that v is adjacent to at least one vertex in M, but not to all vertices in M. If there exists such a vertex v then we will say that v *distinguishes* the set M (in this case M cannot be a module).

We now generalize the notion of a *module* to hypergraphs. We will define this notion by using the following relation on sets.

Definition 2.1 (C-relation) *Let $H = (V, E)$ be a hypergraph and let A, B and C be subsets of V. The two sets A, B are in C-relation, which is denoted as $A =_C B$, iff $A - C \neq \emptyset$, $A \cap C \neq \emptyset$, $B - C \neq \emptyset$, $B \cap C \neq \emptyset$ and $A - C = B - C$.*

Two sets are in C-relation when they are identical outside C and the intersection with C of each of these sets is nonempty.

Definition 2.2 (module) *Let $H = (V, E)$ be a hypergraph. A set $M \subseteq V$ is a module of H iff for $A, B \subseteq V$, $A =_M B$ implies that $A \in E$ iff $B \in E$.*

Informally, a module is a subset of vertices which cannot be distinguished from outside. This concept is more clear if we give the following definition of a set X which *distinguishes* a set Y: X is a set such that for some $A, B \subseteq V$, where $A =_Y B$ and $A - Y = B - Y = X$ it is $A \in E$ and $B \notin E$. Similarly, we can define the notion of a vertex v which distinguishes a set of vertices.

Clearly, by the previous Definition 2.2, V, $\emptyset$ and the singleton subsets of V are modules of H: these are the *trivial* modules of a hypergraph.

The following lemma has been proved in [10] for arbitrary set system.

Lemma 2.1 *The family of nonempty modules of a hypergraph is a partitive set family.*

The *modular decomposition of a hypergraph H* is the partitive tree associated to its partitive set family.

Let P be a partition of vertices of H. A *system of representatives* of P is a set containing a vertex from each set in P. If P is a partition of V such that each member of P is a module, then P is called a *congruence partition*; each system of representatives of such a partition induces an identical subhypergraph, which is the *quotient hypergraph H/P*.

The results in [4] on k-structures suggest interesting facts about the modular decomposition of hypergraphs. Similarly as for k-structures it can be easily proved that the partitive tree of a hypergraph has no internal nodes which are labeled *q-linear*.

Moreover, as for graphs and k-structures, a *modular restriction rule* for hypergraphs holds: given a module M of a hypergraph $H = (V, E)$ and $Y \subseteq V$, then $M \cap Y$ is a module of the induced hypergraph $H|Y$.

3 An Algorithm for the Modular Decomposition

In this section we propose an algorithm for the modular decomposition of hypergraphs of dimension 3 on which is based the general algorithm for the decomposition of hypergraphs of fixed dimension k, $k > 3$, illustrated at the end of the paper.

Observe that the notion of a module given in the Definition 2.2 is still valid for hypergraphs of fixed dimension if we pose a restriction to the subsets of V which can be in a C-relation, because in this case we must consider hyperedges of cardinality at most k. More precisely, if H is a hypergraph of dimension k, given $A, B, C \subseteq V$, if A and B are in C-relation then it must be $|A| \leq k$ and $|B| \leq k$; this condition must be added to Definition 2.1.

Similarly as for arbitrary hypergraphs we can show that the nonempty modules of a hypergraph of fixed dimension form a partitive family, from which we can derive its tree representation.

We now give some preliminary definitions and properties used in the algorithm of decomposition for hypergraphs. In the following by a hypergraph we mean a hypergraph of fixed dimension.

Definition 3.1 (clique) *Let $H = (V, E)$ be a hypergraph of dimension k and let $X \subseteq V$. Then X is a* clique *of H iff every subset of X with at most k elements is a hyperedge of H.*

Definition 3.2 (maximal module excluding v) *Let X be a module of a hypergraph H. Then X is a* maximal module excluding v *iff $v \notin X$ and for any other module Y of H such that $X \subset Y$, it is $v \in Y$.*

By $M(H, v)$ we denote the set of all maximal modules of H excluding v.

Lemma 3.1 *Let $H = (V, E)$ be a hypergraph and $v \in V$. Then $\{\{v\}\} \cup M(H, v)$ is a partition of the vertex set V.*

Proof. Assume that an arbitrary vertex w, with $w \neq v$ is not in any member of $M(H, v)$. This is not possible, as $\{w\}$ is a module of H excluding v (in fact, $\{w\}$ is a trivial module of H). Assume now that there are two members of $M(H, v)$ that contain the vertex w. Since these modules overlap, their union is a module containing w and excluding v, but this contradicts the fact that $M(H, v)$ contains maximal modules excluding v. $\qquad\square$

Definition 3.3 (underlying graph) *Let $H = (V, E)$ be a hypergraph of dimension 3, and let P be a partition of V. Let v be a vertex in the set C in P. The* underlying graph *w.r.t v and P, denoted as $G_{v,P}$, is the graph with vertex set $V - \{v\}$ and set of edges $E_1 = \{(v_1, v_2) : \{v_1, v_2, v\} \in E, \text{ where } v_1 \notin C \text{ or } v_2 \notin C\}$.*

The algorithm for the construction of the decomposition tree of a hypergraph H of dimension 3 consists of three basic procedures:

- **Partition(v,P)** constructs a refinement of the partition P, analyzing the 2-adjacency and the 3-adjacency to v, checking if v can *distinguish* some sets of vertices in P (see definition in the preliminaries section).
- **Maxmodule(H,v)** computes the maximal modules of H which exclude the vertex v,
- **Construct-tree(H)** computes for each vertex $v \in V$ all maximal modules excluding v and uses these modules in the procedure **Create-node** to construct the partitive tree of H,

Let us describe in detail each step of the algorithm.

The procedure $Partition(v, P)$ has a vertex v and a partition P of V as arguments; initially it refines P into vertices 2-adjacent and not 2-adjacent to v, and then w.r.t. the 3-adjacency to v, by computing the underlying graph $G_{v,P}$ and looking for some modules of it that satisfy certain properties.

Partition(v, P)

Input:

$v \in V$, a partition P of V, where $v \in C_1$, with $C_1 \in P$.

Each set $Y \in P$, with $Y \neq C_1$ is partitioned into the set Y_a and Y_{na} of vertices, where $Y_a = \{y \in Y : (y, v) \in E\}$ and $Y_{na} = \{y \in Y : (y, v) \notin E\}$. The elements in Y_a are labeled *adjacent* while those in Y_{na} are labeled *non-adjacent*. We obtain a first refinement P_1 of P.

Construct the *underlying graph* $G_{v,P}$ w.r.t. v and P.

Compute the decomposition tree of $G_{v,P}$.

Then a new refinement P_2 of the partition P_1 is obtained by constructing maximal sets which satisfy the following properties:

each set $A \in P_2$, $A \neq C_1$ must be a module of $G_{v,P}$ and either

- A is a clique of $G_{v,P}$ if A is contained in a set of P_1 which contains only vertices labeled adjacent, or
- A induces in $G_{v,P}$ a graph without edges if A is contained in a set of P_1 which contains only vertices labeled non-adjacent.

The partition P_2 is given as output of the procedure.

Observe that the procedure *Partition* uses the modular decomposition of the underlying graphs in order to analyze the 3-adjacency in a hypergraph. This is done by studying the 2-adjacency in the underlying graph.

Maxmodule(H,v)

The vertex v is labeled *new*, while all other vertices are labeled *old*.

Pose $P = \{\{v\}, V - \{v\}\}$.

Repeat

let v_1 be a *new* vertex, $P_1 := Partition(v_1, P)$;

v_1 is labeled *old* and all vertices of a set of P which is not in P_1 are labeled *new*;

$P := P_1$;

until all vertices are labeled *old*.

Then $P - \{\{v\}\}$ is the output of the procedure.

The properties described in Lemmas 6.2, 6.3, 6.4, given in the Appendix, show that $Maxmodule(H, v)$ computes the maximal modules excluding v, which form a partition of the vertex set V as proved in Lemma 3.1.

The procedure *Construct-tree*, as said before, constructs the decomposition tree of a hypergraph H, given all maximal modules excluding each vertex in V. The following lemma which generalizes to hypergraphs a result shown in [2] describes how the ancestor of a given vertex v in the decomposition tree is related to the maximal modules excluding v.

Lemma 3.2 *Let $H = (V, E)$ be a hypergraph of fixed dimension k, and let $v \in V$. Let U be an ancestor of $\{v\}$ in the decomposition tree of H, and let Z be the child of U in the tree containing v. Then*

- *1. If U is q-complete, then the union of all children of U, except Z, is a maximal module of H excluding v.*
- *2. If U is q-primitive, then each child X of U, with $X \neq Z$ is a maximal module of H excluding v.*

The set of all maximal modules of H excluding v is obtained by applying statement 1. and 2. to all ancestors of $\{v\}$ in H.

Proof. Let U be the lowest common ancestor of vertices v and w in the decomposition tree, and let Z and W be the children of U containing v and w, respectively. By a property of a partitive family (see the preliminaries section), each module of H is obtained as a union of some children of a module of the partitive tree of H. Then the maximal module of H excluding v and containing w is a union of children of U, such that this union contains W and excludes Z. But, if U is q-primitive, by the definition of q-primitive it follows that W is the biggest union that gives the maximal module excluding v and containing w. Similarly, if U is q-complete, then the union of all children of U, except Z is the maximal module excluding v and including w. □

By Lemma 3.2, it should be easy for the reader to understand how to construct a common ancestor for descendent nodes of the partitive tree, once all maximal modules excluding each vertex in V are known. In order to make clear this final step of the algorithm, we want to illustrate in detail the procedure *Construct-tree*.

Given $v, w \in V$, by $E\text{-}I(v, w)$ we denote the maximal module of H which excludes v and includes the vertex w. Moreover, the set $Inc(w)$ will denote the set $\{E\text{-}I(v, w) : v \in V - \{w\}\}$ of all maximal modules which include w but exclude a vertex in $V - \{w\}$. Note that since there can be different vertices v_1, v_2 such that $E\text{-}I(v_1, w) = E\text{-}I(v_2, w)$, then by definition, $Inc(w)$ can contain copies of the same set.

The set $Minc(w)$ will consist of all sets in $Inc(w)$ of minimum cardinality.

The *Construct-tree* procedure determines, starting from the leaves of the tree, the common internal node which is parent of modules in the decomposition tree and the type of such a node, that is whether it is *q-complete* or *q-primitive*. This step is realized by the *Create-node(v)* procedure, in which all vertices of a module are represented by a single vertex, as we choose to describe modules by means of a *system of representatives* (see the preliminaries section).

The *Create-node(v)* procedure verifies that a given vertex w satisfies one of the following conditions 1, 2; these two conditions correspond to the two properties stated in Lemma 3.2.

1. All the sets in $Minc(w)$ are distinct, and their union is the set $\{w\} \cup \{v' \in V : E\text{-}I(v', w) \in Minc(w)\}$.

This condition corresponds to the case that the parent of w is q-complete. In fact, let U be the parent of the set $\{v\}$. If U is q-complete, as stated in Lemma 3.2, the union of all children of U except the module $\{v\}$ gives a maximal module X excluding v. This means that the set of maximal modules of minimum cardinality containing a vertex w and excluding any other vertex v', i.e. the sets in $Minc(w)$ are all obtained by removing a given vertex v' from the same set X, where v' is such that $E\text{-}I(v', w) \in Minc(w)$.

2. The sets in $Minc(w)$ are equal to $\{w\}$.

This condition corresponds to the case that the parent of w is q-primitive. In fact, if U is the parent of the set $\{w\}$ and U is q-primitive, by statement 2 of Lemma 3.2 all maximal modules excluding a vertex $v \in U$ are the descendants of U which do not contain v. This implies that all maximal modules of minimum cardinality excluding a given vertex v and including w consist exactly of the singleton $\{w\}$.

The procedure *Create-node* associates recursively to a vertex $v \in V$ an internal node of the tree which is an ancestor of v, and such that the associated module is denoted as $N(v)$ (see the example in Sect. 7).

Now we can describe the procedure Construct-tree.

Construct-tree(H)

Initially $N(v) := \{v\}$, for each $v \in V$ and $V' = V$.

Compute $Maxmodule(H, v)$, for each $v \in V$ and the set $M(V') = \{E\text{-}I(v, w) : v, w \in V'\}$.

Create-node(M(V'))

Follows the description of the subprocedure Create-node.

Create-node(M(V'))

For all vertices $v \in V'$ compute $Inc(v), Minc(v)$.

If there is a vertex v which satisfies the condition 1,

begin

$\quad F := \{v\} \cup \{v' \in V' : E\text{-}I(v', v) \in Minc(v)\}$,

$\quad X := \cup_{v \in F} N(v)$.

$\quad$ Then X is the nearest ancestor q-complete of each node $N(v')$ such that $\quad\quad v' \in F$.

end

If there is no vertex v which satisfies the condition 1, then determine each vertex w which verifies the condition 2.

begin

$\quad$ For each of such vertex w, compute the set $F(w) := \{w\} \cup \{v' \in V' : E\text{-}$ $\quad I(v', w) = \{w\}\}$.

$\quad$ All sets $F(w)$ are ordered by their cardinality.

$\quad$ Choose a vertex v such that $F(v)$ is of minimum cardinality and pose $\quad\quad F := F(v)$,

$\quad X := \cup_{v \in F} N(v)$.

$\quad$ Then X is the nearest ancestor q-primitive of each node $N(v')$ such that $\quad\quad v' \in F$.

end

Then $N(v) := X$, that is X is the ancestor associated to the vertex v.
$V' := (V' - F) \cup \{v\}$.
We remove from every set $E\text{-}I(w_1, w_2)$ in $M(V')$ all vertices in $F - \{v\}$, that
is we leave v in each set $E\text{-}I(w_1, w_2)$ as a representative of the module X.
This can be done since X is a module, that is all vertices in X cannot be
distinguished from outside of X, and v is chosen as a representative of X.
(Note that, by Lemma 3.2 a vertex in $F - \{v\}$ is in $E\text{-}I(w_1, w_2)$ for $w_1 \notin F$
or $w_2 \notin F$ iff all vertices in $F - \{v\}$ are in $E\text{-}I(w_1, w_2)$).
If $X \neq V$, that is the root of the tree has not yet been constructed, then
Create-node(M(V')).

The correctness of the whole procedure *Construct-tree* should be clear from
the previous observations on Lemma 3.2.

4   On the Complexity

The complexity of the procedure $Partition(v, P)$ is mainly determined by the
construction of the underlying graph w.r.t. v and P, i.e. $G_{v,P}$. We assume that
$\mu(v, P)$ is the number of 3-hyperedges examined to build $G_{v,P}$. Then its decomposition tree will be computed in $O(n + \mu(v, P))$ time, if we use the $O(n + m)$
algorithm for graphs by R. McConnell and J. Spinrad.

The procedure $Maxmodule(H, v)$ may call $Partition(v_i, P)$ for every vertex
$v_i \in V$. Since each vertex v_i can be labeled *new* each time the set of the partition containing v is refined, there will be at most n calls of the procedure
$Partition(v_i, P)$ (clearly a subset of vertices can be split at most n times, since
$|V| = n$).

We observe that after the first call of $Partition(v_i, P)$, the successive calls of
the procedure $Partition(v_i, P)$ require the construction of the underlying graph
w.r.t. the set C containing v, where C is in the partition given as an argument
in the previous call to $Partition(v_i, P)$. In other words, it is not necessary to
examine all the underlying graph w.r.t. v_i and the partition P. It follows that
each call examines distinct 3-hyperedges containing v. Thus the total number
of 3-hyperedges examined in all calls of $Partition(v_i, P)$ for a given vertex v_i, is
$O(n^2)$, i.e. the number of distinct hyperedges containing v_i: this will be the sum
of all $\mu(v_i$, of each call of *Partition* with the vertex v_i. Since, as observed above,
there can be calls to *Partition* for every vertex in V, we obtain the $O(n^3)$ time
complexity of *Maxmodule*.

Finally we can conclude that, since *Construct-tree* calls n times *Maxmodule*,
the time complexity of the algorithm is $O(n^4)$.

5   The Algorithm for Hypergraphs of Dimension k

In the previous section we proposed an $O(n^4)$ algorithm for the modular decomposition of hypergraphs of dimension 3. In this algorithm the procedure *Partition*

decomposes the underlying hypergraphs, in order to partition the vertices which are not distinguished w.r.t. 3-adjacency.

Following this approach, we generalize the procedure *Partition* to determine the maximal modules excluding a vertex in a hypergraph $H = (V, E)$ of dimension $k \geq 3$.

Thus we define the *underlying hypergraph of H w.r.t. to a vertex v and a partition P of V*, which is the hypergraph $H_{v,P} = (V', E')$, where $V' = V - \{v\}, v \in C, C \in P$, and $X \in E'$ iff $\{v\} \cup X \in E$ and there is a vertex $v' \in X$ such that $v' \notin C$. Clearly, H_v is a hypergraph of dimension $k - 1$.

Then the general procedure $Partition(v, P_1)$ works similarly as for the case of hypergraphs of dimension 3.

Initially, the sets of P_1 are partitioned w.r.t. to the fact that they can be distinguished w.r.t. 2-adjacency in the hypergraph H, thus computing a refinement P_2 of the partition P_1, based on the distinction between *adjacent* and *non-adjacent* vertices. Then the procedure computes the underlying hypergraph H_v of H w.r.t. the vertex v and the partition P_2. Given the decomposition tree of H_v, a new refinement of P_2 is computed by analyzing the maximal sets which are modules and are cliques of the hypergraph if containing vertices labeled adjacent, or hypergraphs with no edges if containing vertices labeled non-adjacent. The underlying hypergraph is just used to partition the sets of vertices in P_2 w.r.t. to the fact that they can be distinguished w.r.t. i-adjacency, for $3 \leq i \leq k - 1$.

Once, for any vertex $v \in V$, all maximal modules excluding the vertex v are determined with the $Maxmodule(H, v)$ procedure, the construction of the decomposition tree for hypergraphs of dimension k is naturally defined as for the case of $k = 3$.

Since decomposing a hypergraph of dimension k can require to decompose at most n^3 hypergraphs of dimension $k - 1$ we get that the time complexity of the algorithm is $O(n^{3k-5})$, which is obtained solving a simple recurrence equation.

References

1. Alain Cournier and Michel Habib. A new linear algorithm for modular decomposition. In Sophie Tison, editor, *Trees in algebra and programming, CAAP '94: 19th international colloquium*, LNCS 787, pages 68–84, 1994.
2. A. Ehrenfeucht, H.N. Gabow, R.M. McConnell, and S.J. Sullivan. An $O(n^2)$ divide and conquer algorithm for the prime tree decomposition of 2-structures. *Journal of Algorithms*, 16:283–294, 1994.
3. A. Ehrenfeucht and G. Rozenberg. Theory of 2-structures, part 2: representations through labeled tree families. *Theoretical Computer Science*, 70:305–342, 1990.
4. Andrzej Ehrenfeucht and Ross M. McConnell. A k-structure generalization of the theory of 2-structures. *Theoretical Computer Science*, 132:209–227, 1994.
5. A. Engelfriet, T. Harjan, A. Proskurowsky, and G. Rozenberg. Survey on graphs as 2-structures and parallel complexity of decomposition. Technical Report TR93-06, University of Leiden, Department of Computer Science, 1993.
6. M.C. Golumbic. *Algorithmic Graph Theory and Perfect Graphs*. Academic Press, New York, 1980.

7. Ross M. McConnell and Jeremy P. Spinrad. Linear-time modular decomposition of undirected graphs and efficient transitive orientation of comparability graphs. In *5th ACM-SIAM Symposium on Discrete Algorithms*, pages 536–545, 1994.

8. R.H. Möhring. Algorithmic aspects of comparability graphs and interval graphs. In I. Rival, editor, *Graphs and Orders*, pages 41–101. D. Reidel, Boston, 1985.

9. R.H. Möhring. Algorithmic aspects of the substitution decomposition in optimization over relations, set systems and boolean functions. *Annals of Operations Research*, 4:195–225, 1985/6.

10. R.H. Möhring and F.J. Radermacher. Substitution decomposition for discrete structures and connections with combinatorial optimization. *Annals of Discrete Mathematics*, 19:257–356, 1984.

6 Appendix

We can easily prove the following lemma.

Lemma 6.1 *Let $H = (V, E)$ be a hypergraph of dimension 3, let M be a module excluding v, and let P be a partition of V such that v and M are not contained in the same set in P. Then for every $v_1 \in V - M$, M is a module of the underlying graph of H w.r.t. v_1 and P.*

We recall that by Lemma 3.1, the set consisting of $\{v\}$ and all maximal modules of H excluding v is a partition of V.

Lemma 6.2 *Let $H = (V, E)$ be a hypergraph of dimension 3. Let $v_1, v_2 \in V - \{v\}$ be two vertices contained in two distinct sets of the partition P_v computed by Maxmodule(H, v), where $v \in V$. Then v_1 and v_2 are in two distinct modules of H excluding a vertex v' which separated v_1 from v_2 in Partition(v', P).*

Proof. Let P be the last partition computed by *Maxmodule* in which v_1, v_2 where contained in the same set $C_{1,2}$ of P. Since v_1 and v_2 are in two distinct sets of the partition P_v, they were separated by a call of the procedure *Partition(v', P)*, where v' is a vertex which cannot be in the same set $C_{1,2}$, by definition of *Partition*.

These are the possible cases which determined the separation of the vertices v_1, v_2:

i) $(v_1, v') \in E$ and $(v_2, v') \notin E$, which implies that v' distinguishes the set $\{v_1, v_2\}$, and this proves the lemma.

ii) $(v_1, v') \in E$ and $(v_2, v') \in E$, $\{v_1, v_2\}$ is not contained in a module M of the underlying graph $G_{v', P}$, where M is contained in $C_{1,2}$. This implies that $\{v_1, v_2\}$ is distinguished by a vertex in $G_{v', P}$. By Lemma 6.1, v_1 and v_2 are not contained in the same module of H excluding v'.

iii) $(v_1, v') \in E$, $(v_2, v') \in E$ and $\{v_1, v_2\}$ is contained in a module M of the underlying graph $G_{v', P}$, where M is contained in $C_{1,2}$ whose vertices are labeled adjacent, but M is not a clique in $G_{v', P}$. We assume that M is the smallest module of the underlying graph which contains $\{v_1, v_2\}$. Since M is not a clique in $G_{v, P}$, there are two vertices v_3, v_4 of M which are not adjacent in $G_{v', P}$, that is $\{v', v_3, v_4\}$ is not a 3-hyperedge of H while $\{v', v_1\}$ and $\{v', v_2\}$ are 2-hyperedges: thus M is not a module of H which excludes v'. Assume that M' is a module of H which excludes v' and contains $\{v_1, v_2\}$. By Lemma 6.1, M' must be a module of $G_{v',P}$ which contains $\{v_1, v_2\}$ and $M' \supseteq M$. Similarly as for M we have that M' must not be a module of H excluding v'. It follows that $\{v_1, v_2\}$ are in two distinct modules of H excluding v'.

The case that M is contained in a set of P whose vertices are labeled non-adjacent is similar to the former ones. $\qquad\Box$

Lemma 6.3 *If the two vertices v_1 and v_2 are not in the same set of the partition constructed by* Maxmodule(H, v), *then v_1 and v_2 are not in the same module of the hypergraph H excluding v.*

Proof. Let P be the biggest partition computed in *Maxmodule(H, v)* such that v_1 and v_2 are in the same set of P, $C_{1,2}$. By previous Lemma 6.2, v_1 and v_2 are in two distinct modules of H excluding a vertex $v_3 \notin C_{1,2}$ such that *Partition(v_3, P)* separated v_1 and v_2. This implies that H has no module containing $\{v_1, v_2\}$ and excluding v_3. If $v_3 = v$, the lemma follows, otherwise consider the partition P' in which $\{v_1, v_2, v_3\}$ are in the same set $C_{1,2,3}$. Since v_3 separated v_1 and v_2, then there is a vertex v_4 such that *Partition(v_4, P')* separated v_3 from v_1, v_2. By Lemma 6.2 there are two distinct modules of H excluding v_4 and containing v_3 and $\{v_1, v_2\}$. But there will be a call to *Partition* with v_3 which will separate v_1, v_2 consequently H has no module containing $\{v_1, v_2\}$ and excluding v_4. If $v \neq v_4$, iteratively we construct a sequence of sets $C_{1,2,\dots,i}$, with $1 \leq i \leq |V|$, such that *Partition(v_i, P_i)* separated v_{i-1} from v_{i-2} and by transitivity v_{i-1} is separated from $\{v_1, v_2\}$, that is it follows that H has no module containing $\{v_1, v_2\}$ and excluding $\{v_i\}$.

Since $1 \leq i \leq |V|$, it follows that for some i it must be $v_i = v$. $\qquad\Box$

Lemma 6.4 *Let P be the partition of $V - \{v\}$ constructed by* Maxmodule(H, v). *Then each element of P is a module of H excluding v.*

Proof. Clearly, if $P = V - \{v\}$ then the lemma follows. Assume that there is a set $X \in P$, $X \subset V - \{v\}$ such that is not a module of H. This means that there is a vertex v_1 or a couple of vertices v_1, v_2 that distinguishes X in H; the following two cases are possible.

i) The vertex v_1 is adjacent to some vertices of X, but not all vertices, or v_1 with a pair of elements in X, but not with all pairs, forms a 3-hyperedge of H. By definition of *Partition* and *Maxmodule*, then it is easy to verify that v_1 is labeled *new* at least once. It follows that there is a call of *Partition* with the vertex v_1 which divides X into two sets. But, we obtain a contradiction, as X must not be in P.

ii) There is a vertex $v_2 \notin X$ such that $\{v_1, v_2\}$ distinguishes X. This means that given the underlying graph $G_{v_1,P}$ (or $G_{v_2,P}$) X is not a module of such a graph, that is X must be divided by a call of *Partition* with v_1 or v_2. But, it is easy to verify that one of such a call is possible. As above, we obtain a contradiction. $\qquad\Box$

Lemmas 6.3 and 6.4 prove the following fact.

Corollary 6.1 *The procedure* Maxmodule(H, v) *constructs the set of all maximal modules excluding* v.

7 An Example

Let $H = (V, E)$ be a hypergraph of dimension 3, so that $V = \{v_1, \ldots, v_6\}$ and $E = \{\{v_1, v_2, v_6\}; \{v_1, v_3, v_6\}; \{v_1, v_4, v_6\}; \{v_1, v_5, v_6\}; \{v_1, v_2\}; \{v_1, v_6\}; \{v_2, v_6\}; \{v_1, v_3\}; \{v_3, v_6\}; \{v_1, v_4\}; \{v_4, v_6\}; \{v_1, v_5\}; \{v_5, v_6\}; \{v_2, v_3\}; \{v_3, v_4, v_5\}; \{v_1, v_2, v_3\}; \{v_1, v_2, v_4\}; \{v_1, v_3, v_4\}; \{v_2, v_3, v_6\}; \{v_2, v_4, v_6\}; \{v_3, v_4, v_6\}; \{v_1, v_2, v_5\}; \{v_1, v_3, v_5\}; \{v_1, v_4, v_5\}; \{v_2, v_5, v_6\}; \{v_3, v_5, v_6\}; \{v_4, v_5, v_6\}\}$.

Initially, Construct-tree(H) computes Maxmodule(H, v) for all $v \in V$. We can note that two vertices w_1, w_2 can behave similarly w.r.t. the adjacency relation, in the sense that given $X \subseteq V - \{w_1, w_2\}$ it holds that $\{w_1\} \cup X$ is a hyperedge iff $\{w_2\} \cup X$ is a hyperedge: we will say that w_1 and w_2 are symmetric. This is the case of vertices v_1 and v_6 and of vertices v_4 and v_5.

Let us see the main steps of the execution of Construct-tree(H).

- Maxmodule(H, v_1)

$P_1 := \{\{v_1\}, V - \{v_1\}\}$

We now compute Partition(v_1, P_1).

The partition of P_1 w.r.t. 2-adjacency to v_1 gives the partition P_2.

$P_2 := \{\{v_1\}, \{v_2, v_3, v_4, v_5, v_6\}\}$

Let us compute the underlying graph G_{v_1, P_2} to compute the partition P_3 of P_2 w.r.t. the 3-adjacency: note that the set of vertices $\{v_2, v_3, v_4, v_5, v_6\}$ induces a complete graph in G_{v_1, P_2}, so it has a simple modular decomposition tree represented in Fig. 1.

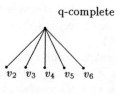

q-complete

Fig. 1.

Visiting the modular decomposition tree of G_{v_1, P_2} we get that $P_3 = \{\{v_1\}, \{v_2, v_3, v_4, v_5, v_6\}\}$ since $\{v_2, v_3, v_4, v_5, v_6\}$ induces a clique in G_{v_1, P_2}, it is a (trivial) module of G_{v_1, P_2} and its vertices are all 2-adjacent to v_1.

Since the call to Partition(v_1, P_1) does not modify the partition P_1 , all vertices in P_1 will be labeled *old*; then Maxmodule(H, v_1) returns $\{v_2, v_3, v_4,$

v_5, v_6}. Vertices v_1 and v_6 are symmetric so we get that Maxmodule(H, v_6) returns $\{v_1, v_2, v_3, v_4, v_5\}$.

- Maxmodule(H, v_2)

$P_1 := \{\{v_2\}, V - \{v_2\}\}$

We now compute Partition(v_2, P_1)

Let us partition P_1 w.r.t. 2-adjacency to v_2:

$P_1 := \{\{v_2\}, \{v_1, v_3, v_6\}, \{v_4, v_5\}\}$

Let us compute the underlying graph G_{v_2, P_1} and its modular decomposition tree: they are represented in Fig. 2.

Fig. 2.

By analyzing the cliques and the edgeless graphs in the modular decomposition tree of G'_{v_2, P_1} we get that P_1 is partitioned into $P_2 = \{\{v_1, v_6\}, \{v_3\}, \{v_4, v_5\}, \{v_2\}\}$. For example we note that $\{v_1, v_6\}$ induces a clique in G_{v_2, P_1}, it is a module of G_{v_2, P_1} and v_1, v_6 are both 2-adjacent to v_2. A call to Partition(v_3, P_2) will split v_4 and v_5, so Maxmodule(H, v_2) returns $\{\{v_1, v_6\}, \{v_3\}, \{v_4\}, \{v_5\}\}$. This call to Partition(v_3, P_2) has not been examined closely since it is similar to the one that will be examined in Maxmodule(H, v_3) that we are going to show immediately.

- Maxmodule(H, v_3)

$P_1 := \{\{v_3\}, V - \{v_3\}\}$.

We now compute Partition(v_3, P_1)

Let us partition P_1 w.r.t. 2-adjacency to v_3.

$P_1 := \{\{v_3\}, Y_a, Y_{na}\} = \{\{v_3\}, \{v_1, v_2, v_6\}, \{v_4, v_5\}\}$

Let us compute the underlying graph G_{v_3, P_1} and its modular decomposition tree, both ones are represented in Fig. 3.

By analyzing this modular decomposition we get the partition $\{\{v_1, v_6\}, \{v_2\}, \{v_3\}, \{v_4\}, \{v_5\}\}$.

Note that v_4 and v_5 are split because they are 2-adjacent, but not 3-adjacent to v_3 (they form a clique in the modular decomposition tree).

- Maxmodule(H, v_4)

This call returns $\{\{v_2\}; \{v_3\}; \{v_5\}; \{v_1, v_6\}\}$ and, since v_4 and v_5 are symmetric Maxmodule(H, v_5) returns $\{\{v_2\}; \{v_3\}; \{v_4\}; \{v_1, v_6\}\}$.

Fig. 3.

Now we have to apply the Create-Node procedure. Table 1 shows all the sets $E\text{-}I(v,w)$.

v \ w	v_1	v_2	v_3	v_4	v_5	v_6
v_1		v_2, v_3, v_4 v_5, v_6	v_2, v_3, v_4 v_5, v_6	v_2, v_3, v_4 v_5, v_6	v_2, v_3, v_4 v_5, v_6	v_2, v_3, v_4 v_5, v_6
v_2	v_1, v_6		v_3	v_4	v_5	v_1, v_6
v_3	v_1, v_6	v_2		v_4	v_5	v_1, v_6
v_4	v_1, v_6	v_2	v_3		v_5	v_1, v_6
v_5	v_1, v_6	v_2	v_3	v_4		v_1, v_6
v_6	v_1, v_2, v_3 v_4, v_5	v_1, v_2, v_3 v_4, v_5	v_1, v_2, v_3 v_4, v_5	v_1, v_2, v_3 v_4, v_5	v_1, v_2, v_3 v_4, v_5	

Table 1.

We analyze the sets in $Minc(v_2)$: we have that $E\text{-}I(v_3, v_2)$, $E\text{-}I(v_4, v_2)$ and $E\text{-}I(v_5, v_2)$ satisfy condition 2, so $\{v_2, v_3, v_4, v_5\}$ is the q-primitive parent of $\{v_2\}$, $\{v_3\}$, $\{v_4\}$, $\{v_5\}$ in the modular decomposition tree. We will say that v_2 represents the internal node we have just determined. We drop all the sets $E\text{-}I(v,w)$ we are not going to use any more, and we will change the remaining ones by removing v_3, v_4, v_5.

In Table 2 are shown all the remaining sets $E\text{-}I(v,w)$.

If we look at the sets in $Minc(v_1)$ we have that $E\text{-}I(v_2, v_1)$ and $E\text{-}I(v_6, v_1)$ satisfy condition 1, so $\{v_1, v_2, v_3, v_4, v_5, v_6\}$ is the q-complete parent of $\{v_1\}$, $\{v_2, v_3, v_4, v_5\}$, $\{v_6\}$.

$v \diagdown w$	v_1	v_2	v_6
v_1		v_2, v_6	v_2, v_6
v_2	v_1, v_6		v_1, v_6
v_6	v_1, v_2	v_1, v_2	

Table 2.

Note that the halting condition of the Construct-tree procedure is now satisfied, and we have built the modular decomposition tree that is represented in Fig. 2.

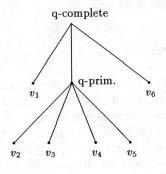

Fig. 4.

Partition Coefficients of Acyclic Graphs *

John L. Pfaltz

Dept. of Computer Science
University of Virginia
Charlottesville, VA 22903
jlp@cs.virginia.edu

Abstract. We develop the concept of a "closure space" which appears with different names in many aspects of graph theory. We show that acyclic graphs can be almost characterized by the partition coefficients of their associated closure spaces. The resulting nearly total ordering of all acyclic graphs (or partial orders) provides an effective isomorphism filter and the basis for efficient retrieval in secondary storage. Closure spaces and their partition coefficients provide the theoretical basis for a new computer system being developed to investigate the properties of arbitrary acyclic graphs and partial orders.

1 Binary Partitions

In this paper we combine two mathematical threads and apply them in a graph-theoretic context. The first thread of binary partitions was studied by Euler as early as 1750. The second thread involving closure spaces is of more recent origin. A binary partition of a positive integer N is its expression as a sum of powers of 2. Mahler [16], and Churchhouse [3] [4] have studied binary partitions from a number theoretic point of view. Because our intention is to connect these partitions with closure spaces, we will confine our attention to the special case where N is also a power of 2.

By a *binary partition* of 2^n we mean a sequence of non-negative integers $< \cdots, a_k \cdots >, 0 \leq k \leq n$ such that

$$a_n \cdot 2^n + a_{n-1} \cdot 2^{n-1} + a_{n-2} \cdot 2^{n-2} + \cdots + a_1 \cdot 2^1 + a_0 \cdot 2^0 = 2^n \qquad (1)$$

or $\sum_{k=0}^{n} a_k \cdot 2^k = 2^n$. The set of all such partitions we denote by $\mathcal{P}^n$. (From now on we frequently omit the adjective "binary".)

Several characteristics of (1) are readily apparent. First, $a_n \neq 0$ if and only if $a_k = 0$ for all $0 \leq k < n$. Second, since the right hand side is even and all terms $a_k \cdot 2^k$, $k > 0$ must be even, the coefficient a_0 must be even. Third, if $< \cdots, a_k, a_{k-1}, \cdots >$ is a partition, then $< \cdots, a_k - 1, a_{k-1} + 2, \cdots >$ must be as well. And fourth, if $< a_n, \cdots, a_k, \cdots, a_o >$ is a partition of 2^n then $< a_n, \cdots, a_k, \cdots, a_0, 0 >$ is a partition of 2^{n+1}.

With these observations, it is not difficult to write a process which generates all partitions in lexicographic order. Doing so, and displaying each partition,

```
       n = 3                    n = 4
   1  0  0  0          1  0  0  0  0       0  0  2  1  6
   0  2  0  0          0  2  0  0  0       0  0  2  0  8
   0  1  2  0          0  1  2  0  0       0  0  1  6  0
   0  1  1  2          0  1  1  2  0       0  0  1  5  2
   0  1  0  4          0  1  1  1  2       0  0  1  4  4
   0  0  4  0          0  1  1  0  4       0  0  1  3  6
   0  0  3  2          0  1  0  4  0       0  0  1  2  8
   0  0  2  4          0  1  0  3  2       0  0  1  1 10
   0  0  1  6          0  1  0  2  4       0  0  1  0 12
   0  0  0  8          0  1  0  1  6       0  0  0  8  0
                       0  1  0  0  8       0  0  0  7  2
                       0  0  4  0  0       0  0  0  6  4
                       0  0  3  2  0       0  0  0  5  6
                       0  0  3  1  2       0  0  0  4  8
                       0  0  3  0  4       0  0  0  3 10
                       0  0  2  4  0       0  0  0  2 12
                       0  0  2  3  2       0  0  0  1 14
                       0  0  2  2  4       0  0  0  0 16
```

Fig. 1. $\mathcal{P}^3$ and $\mathcal{P}^4$

generates the following enumerations of $\mathcal{P}^3$ and $\mathcal{P}^4$. It is quite easy to verify by inspection that each sequence is a partition of 2^n. And because they are in lexicographic order, one can verify that all possible partitions have been generated.

If one were to run the same program with $n = 5$ there would be 202 generated partitions which are are impractical to display in a paper of this length.

2   Closure Spaces

The preceding discussion of binary partitions will take on additional interest if we introduce the concept of a closure space. We let **U** denote some *universe* of elements of interest. Lower case letters $a, b, \cdots, x, y, z$ will denote individual elements of **U**, and upper case letters will denote subsets. A set, **U**, and a closure operator, φ, satisfying the following three closure axioms[2]

$$X \subseteq X.\varphi$$
$$X \subseteq Y \text{ implies } X.\varphi \subseteq Y.\varphi \qquad (2)$$
$$X.\varphi.\varphi = X.\varphi^2 = X.\varphi$$

* Research supported in part by DOE grant DE-FG05-95ER25254.

[2] We will write these expressions using the mixed infix/suffix form more common in algebra. That is, binary set operators will be written using infix and unary transformations will be written using suffix notation, as in $(X \cap Y).f$ to denote the image of $X \cap Y$ under f. This notation greatly simplifies expressions involving transformations of closure spaces; and the redundant dot delimiter is of great value when using computer parsing techniques.

are said to be a *closure space* $(\mathbf{U}, \varphi)$, as in [12]. X is said to be *closed* [3] if $X.\varphi = X$. A closure operator, φ, is said to be *uniquely generated* if it also satisfies the following fourth axiom, which serves to distinguish it from a topological closure,

$$X.\varphi = Y.\varphi \quad \text{implies} \quad (X \cap Y).\varphi = X.\varphi = Y.\varphi \qquad (3)$$

Closure operators satisfying (3) above are uniquely generated in the sense that for any set Z, there exists a unique minimal set $X \subseteq Z$, called its *generator* and denoted $Z.gen$, such that $X.\varphi = Z.\varphi$.[4] The importance of uniquely generated closure spaces lies in the fact that in discrete systems they play a role that is in many respects analogous to the vector spaces of classical mathematics. We establish this parallel in the next paragraph.

A closure operator σ, satisfying the three closure axioms of (2), together with the Steinitz-MacLane *exchange* property

$$\text{if } y \notin X.\sigma \text{ then } y \in (X \cup \{x\}).\sigma \text{ implies } x \in (X \cup \{y\}).\sigma \qquad (4)$$

can be shown to be the closure operator of a matroid [25] [26]. Recall that a *matroid* is a set system which generalizes the independent sets of a linear algebra, and a *vector space* is the closure, usually called the *spanning operator*, of one or more of these independent sets. Now (4) has the familiar interpretation: if y is not in the vector subspace spanned by X, but is in the vector space formed by adjoining x as a basis vector, then x must be in the vector space spanned when y is adjoined to X.

Similarly, a closure φ satisfying the three closure axioms and the *anti-exchange* property

$$\text{if } x, y \notin X.\varphi \text{ then } y \in (X \cup \{x\}).\varphi \text{ implies } x \notin (X \cup \{y\}).\varphi \qquad (5)$$

is the closure operator of an anti-matroid [7] [15]. In [21] it is shown that

Theorem 1. *A closure operator is uniquely generated if and only if it satisfies the anti-exchange property (5).*

Therefore, uniquely generated closure spaces are precisely the analogs of vector spaces, but with respect to anti-matroids. From now on, we will simply call them *closure spaces*. Because they are uniquely generated, any closure space is completely characterized by enumerating its closed sets and their generators, that is by enumerating $[X.\varphi, X.gen], \forall X \subseteq \mathbf{U}$.

Closure spaces are fairly common in computer science and its applications, although they frequently have other names. Transitive closure, for example of the set of edges in an acyclic graph or of functional dependencies in an acyclic database schema, gives rise to a closure space. The term "convexity" is often

---

[3] The family $\mathcal{C}$ of closed sets is closed under intersection, and this characterization is equivalent to (2), *c.f.* [9].

[4] Readily, if X_1 and X_2 were distinct minimal generators of $Z.\varphi$, then because $X_1.\varphi = X_2.\varphi = Z.\varphi$, we must have, by (3), $(X_1 \cap X_2).\varphi = Z.\varphi$ contradicting minimality.

applied to closure concepts, and many examples of convexity concepts occurring in graphs can be found in [8] [11] and [14]. Convexity in discrete geometries also yields a number of intuitively satisfying closure spaces. The convex hull operator is the closure operator. See [10] for an excellent treatment of *convex geometries*. Finally, numerous examples of anti-matroids, whose closure will yield a closure space, can be found in the survey of anti-matroids [7] or the text on *greedoids* [15] which generalize an important class of computer algorithms.

We have found that *ideal* and *interval* operators in partially ordered sets, or acyclic graphs provide an abundance of easily accessible examples. It is not hard to show that the path structure of an acyclic graph is uniquely generated [21]. That is, there is a unique, minimal representation[5] of any acyclic graph which we call a *basic graph* [18]. These are commonly used in the implementation of acyclic data structures and processes.

One can organize the *closed sets* of a closure space in many ways. The most natural is to partially order them by inclusion, in which case it can be shown that the partial order will be a lower semi-modular (or meet-distributive) lattice [17] [9]. A more interesting partial order, $\leq_\varphi$, of *all subsets* is given by

$$X \leq_\varphi Y \qquad \text{if and only if} \qquad Y \cap X.\varphi \subseteq X \subseteq Y.\varphi \qquad \forall X, Y \subseteq \mathbf{U}. \qquad (6)$$

which is described in [21]. The closure space with this partial order can be shown to be a lattice, $\mathcal{L}_{(U,\varphi)}$, called the *closure lattice* of $(\mathbf{U}, \varphi)$. Figure 2 illustrates the structure of a small 7 point closure space. The closed sets of $(\mathbf{U}, \varphi)$ are set in bold face, and connected by solid lines. These closed sets form a sublattice whose partial order is by inclusion. It can be instructive to diagram the points and their set membership of this space. Since $\{g\}$ is the generator of $\mathbf{U} = \{abcdefg\}$, the closure of $\{g\}$, or any set containing the point g, is the whole space. The generator of $\{abce\}$ are the points $\{ae\}$, and so forth. There are 64 subsets whose closure is $\{abcdefg\}$; they constitute the lattice interval $[abcdefg, g]$. To avoid clutter, we simply denote all of them by a single dotted ellipse. Only one of its elements $\{efg\}$ is indicated. (From now on, we also ignore $\{\cdots\}$ delimiting sets of enumerated points.)

Closure lattices such as this have a number of unique properties which are explored in [20]. Central to this development is

Theorem 2. *The poset* $\{Y_i | Y.\varphi \leq_\varphi Y_i \leq_\varphi Y.gen\}$, *is a boolean algebra on* n *elements, where* $n = |Y.\varphi| - |Y.gen|$.

These boolean algebras, $[Y.\varphi, Y.gen]$ are denoted by dotted ellipses in Figure 2. It is not hard to see that each lattice interval, $[Y.\varphi, Y.gen] = \{Y_i | Y.gen \subseteq Y_i \subseteq Y.\varphi\}$, and that $|[Y.\varphi, Y.gen]| = 2^n$. Since every subset $Y \subseteq \mathbf{U}$ is an element of some closure/generator interval, the decomposition of $2^{\mathbf{U}}$ into these intervals is a binary partition of $2^{|U|}$, which we call the *partition coefficients* of $(\mathbf{U}, \varphi)$. For

---

[5] Minimal in the sense that removal of any edge yields a graph with a different path structure, transitive closure, or partial order. We usually illustrate acyclic relationships with basic representations; they are far less cluttered.

2. Richard A. Brualdi, Hyung Chan Jung, and William T. Trotter, Jr. On the poset of all posets on n elements. *Discrete Applied Mathematics*, 1994. To appear.

3. R.F. Churchhouse. Congruence properties of the binary partition function. *Proc. Cambridge Phil. Soc.*, 66(2):371–376, 1969.

4. R.F. Churchhouse. Binary partitions. In A.O.L. Atkin and B.J. Birch, editors, *Computers in Number Theory*, pages 397–400. Academic Press, 1971.

5. Joseph C. Culberson and Gregory J. E. Rawlins. New results from an algorithm for counting posets. *Order*, 7:361–374, 1991.

6. Dragos Cvetkovic, Peter Rowlinson, and Slobodan Simic. A study of eigenspaces of graphs. *Linear Algebra and Its Applic.*, 182:45–66, Mar. 1993.

7. Brenda L. Dietrich. Matroids and antimatroids — a survey. *Discrete Mathematics*, 78:223–237, 1989.

8. Feodor F. Dragan, Falk Nicolai, and Andreas Brandstadt. Convexisty and hhd-free graphs. Technical Report SM-DU-290, Gerhard-Mercator Univ., Duisburg, Germany, May 1995.

9. Paul H. Edelman. Meet-distributive lattices and the anti-exchange closure. *Algebra Universalis*, 10(3):290–299, 1980.

10. Paul H. Edelman and Robert E. Jamison. The theory of convex geometries. *Geometriae Dedicata*, 19(3):247–270, Dec. 1985.

11. Martin Farber and Robert E. Jamison. Convexity in graphs and hypergraphs. *SIAM J. Algebra and Discrete Methods*, 7(3):433–444, July 1986.

12. George Gratzer. *General Lattice Theory*. Academic Press, 1978.

13. Frank Harary. *Graph Theory*. Addison-Wesley, 1969.

14. Robert E. Jamison-Waldner. Partition numbers for trees and ordered sets. *Pacific J. of Math.*, 96(1):115–140, Sept. 1981.

15. Bernhard Korte, Laszlo Lovasz, and Rainer Schrader. *Greedoids*. Springer-Verlag, Berlin, 1991.

16. K. Mahler. On a special functional equation. *J. London Math. Soc.*, 15(58):115–123, Apr. 1940.

17. John L. Pfaltz. Convexity in directed graphs. *J. of Comb. Theory*, 10(2):143–162, Apr. 1971.

18. John L. Pfaltz. *Computer Data Structures*. McGraw-Hill, Feb. 1977.

19. John L. Pfaltz. Partitions of 2^n. Technical Report TR CS-94-22, University of Virginia, June 1994.

20. John L. Pfaltz. Closure lattices. *Discrete Mathematics*, 1995. (to appear), preprint available as Tech. Rpt. CS-94-02 through home page http://uvacs.cs.virginia.edu/.

21. John L. Pfaltz. Partially ordering the subsets of a closure space. *ORDER*, 1995. (submitted).

22. A. J. Schwenk. Computing the characteristic polynomial of a graph. In R. Bari and F. Harary, editors, *Graphs and Combinatorics*, pages 153–172. Springer Verlag, 1974.

23. N. J. A. Sloane. *A Handbook of Integer Sequences*. Academic Press, 1973. On-line version at 'sequences@research.att.com'.

24. Richard P. Stanley. *Enumerative Combinatorics, Vol 1*. Wadsworth & Brooks/Cole, 1986.

25. W. T. Tutte. *Introduction to the Theory of Matroids*. Amer. Elsevier, 1971.

26. D.J.A. Welsh. *Matroid Theory*. Academic Press, 1976.

The closure space of Figure 2 is normal.

These partition coefficients constitute an invariant of the closure space that is independent of representation or isomorphic mappings. It is evident from Theorem 2 and the preceding discussion that for every closure space there is a corresponding binary partition of $2^{|U|}$. It can also be shown that for every binary partition of 2^n there exists a closure space on n elements with that [closed_set, generator] structure.

3 Partition Coefficients of Acyclic Graphs

In this section, we apply the concept of closure spaces and their binary partition coefficients to the study of acyclic graphs and partially ordered sets.

With any graph one can postulate a number of invariants. They may be any of a variety of scalar quantities, such as covering or independence numbers [13] or various polynomial expressions, e.g. chromatic polynomials [1]. It is desirable if the invariant conveys information about the graph. A fairly popular invariant of G is its *characteristic polynomial* [22]. In fact this terminology is slightly misleading. One is really associating the graph G with a linear transformation τ, for which the adjacency matrix of G is a representation. Now, the characteristic polynomial, eigenvalues, and eigenspaces of τ can be regarded as invariants of G [6].

We now do much the same. Given a poset, or acyclic graph $G = (P, E)$, one can use the path relation ρ to induce a partial order on the point set, P. Now we set $\mathbf{U} = P$, and let

$$
\begin{aligned}
Y.\varphi_L &= \{x | (x, y) \in \rho, \ y \in Y\}, \\
Y.\varphi_R &= \{z | (y, z) \in \rho, \ y \in Y\}, \quad \text{or} \\
Y.\varphi_C &= \{x | (y_1, x) \in \rho, (x, y_2) \in \rho, \ y_1, y_2 \in Y\}.
\end{aligned}
\tag{7}
$$

The first two closures are *ideal* operators on $\mathbf{U}$, and the last is an *interval* operator.[7]

For any acyclic graph G and uniquely generated closure φ, such φ_L, φ_R or φ_C above, we have an induced closure space. In Figure 3, we illustrate the three different closure spaces obtained by applying φ_L, φ_R, and φ_C to a single 5 point graph. Again, the sub-lattice of closed sets is denoted by solid lines. And, as usual, we will denote the [closed_set, generator] intervals by dashed ellipses. The partition coefficients of these three closure spaces are $< 0\ 0\ 1\ 3\ 3\ 2 >$,

---

[7] In [11], φ_L is called *downset alignment* and φ_C is called *order convexity*, but just plain *convexity* in [17]. There are many conventions for drawing partially ordered sets. In an effort to distinguish between the underlying acyclic graph and its closure space, the author prefers to orient the former horizontally and the latter vertically. Because we illustrate with a left to right horizontal orientation, we use the subscripts, L(eft) and R(ight), to distinguish the ideal operators. The terms *upper/lower ideal* and $\downarrow$ operators are also encountered.

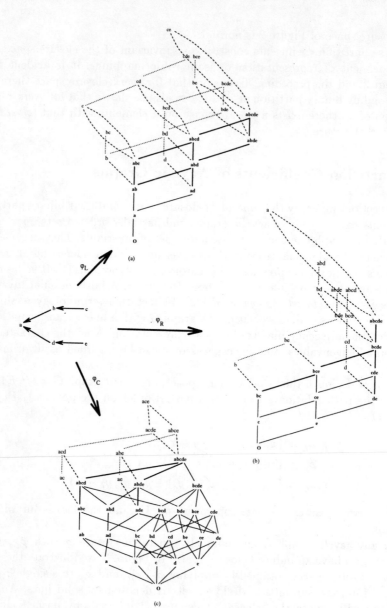

Fig. 3. Different closure spaces arising from the closure operators, φ_L (a), φ_R (b), and φ_C (c).

$< 0\ 1\ 0\ 1\ 4\ 4 >$, and $< 0\ 0\ 0\ 1\ 4\ 20 >$ respectively. Readily, different closure operators give rise to different partition coefficients.

We now treat the partition coefficients of this closure space as invariants of G. As observed above, this invariant depends on the closure operator. For the rest of this paper, we use only the ideal closure φ_L of (7). In Figure 4 we show $\mathcal{G}^4$, that is the collection of all basic, acyclic graphs on 4 points, together with

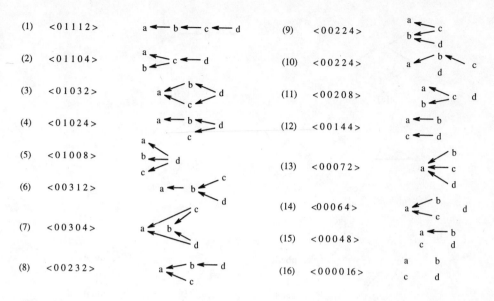

Fig. 4. All basic, acyclic 4 point graphs, $\mathcal{G}^4$ and their partition coefficients (w.r.t. φ_L)

the partition coefficients of their closure spaces. Because φ_L is path derived, any graph with additional edges, but the same transitive closure, must have the same associated closure space.

The graphs of $\mathcal{G}^4$ are not uniquely characterized by their coefficients; consider graphs (9) and (10) which both have $< 0\,0\,2\,2\,4 >$ as partition coefficients. But (9) is connected whereas (10) is disconnected. Unfortunately, this distinction is of little value. The connected, non-isomorphic graphs of Figure 5 both have partition coefficients $< 0\,1\,0\,2\,2\,4 >$ with respect to φ_L. We would note that while the partition coefficients of Figures 5(a) and (b) are the same, their corresponding closure spaces, as illustrated by the lattices are distinct. This follows from

Theorem 3. Fundamental Theorem of Distributive Lattices *If* $(\mathbf{U}, \varphi)$ *is a finite closure space in which* $\mathbf{U}$ *is partially ordered and* φ *is an ideal operator, then the set of closed sets, partially ordered by inclusion, is a distributed lattice. Moreover, there is a one-to-one correspondence between the set of all distributive lattices and such closure spaces.*

Proof. See theorem 3.4.1 of [24]   □

Distinct, non-isomorphic, graphs must have distinct closure spaces, but distinct closure spaces may have the same partition coefficients, just as two distinct linear transformations may have the same characteristic polynomial. Consequently, acyclic graphs cannot be completely characterized by their partition

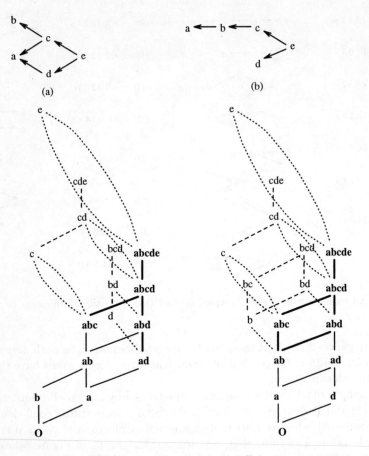

Fig. 5. Two graphs (a) and (b) having the partition coefficients < 0 1 0 2 2 4 > together with their corresponding closure spaces

coefficients. Nevertheless, these coefficients convey significant information about the graphs and can be quite useful when manipulating them in computer systems.

The author has created one such computer system, capable of representing arbitrary graphs, whose primary purpose is the study of properties of graph transformations. For many of the studies of interest to us, we must generate all, or a large sample of, non-isomorphic graph on n points. Comparing binary partition coefficients is a useful filter for eliminating obviously non-isomorphic pairs. In Table 1 we display the expected number of acyclic graphs on n points that have the same identical binary partition coefficients, $exp(|G|\ per\ bp)$. For $n = 8$, there exist 16,999 distinct, non-isomorphic, acyclic graphs,[8] having 5,187

---

[8] The number of distinct n point acyclic graphs, or posets, grows exponentially. It is known that $|\mathcal{G}^n|$ is: 183,231 ($n = 9$), 2,567,284 ($n = 10$) and 46,794,427 ($n = 11$) [5]. No general enumeration formula is known.

$n = \|P\|$	3	4	5	6	7	8
$\|\mathcal{G}^n\|$	5	16	63	318	2,405	16,999
$exp(\|G\|\ per\ bp)$	1.00	1.07	1.21	1.53	2.13	3.28
$exp(\|G\|\ per\ bp\ and\ \|E\|)$	1.00	1.00	1.03	1.12	1.30	1.66

Table 1. Densities of acyclic graphs on n points when partitioned w.r.t. binary partition (bp) coefficients and w.r.t number of *edges*.

distinct partition coefficient sequences; so that an expected 3.277 have the same binary partition coefficients. But two graphs with the same partition coefficients need not have the same number of edges. They frequently do not. As shown on the next line of Table 1, the expected number of graph with identical partition coefficients and the same number of edges, $exp(\|G\|\ per\ bp\ and\ \|E\|)$, drops to 1.656. In practice, these expectations transalate into a effective filter. In a recent application that involved testing 1,034 M random pairs of 8-point graphs for isomorphism (equality), we first applied the edge cardinality filter; 193 M pairs passed this filter. Of these, only 148,762 had identical partition coefficients, and of these 87,710 were actually isomorphic. The probability of being isomorphic, given equal partition coefficients and numbers of edges was 1.69, compared to 1.66 as predicted by the table.

A quick measure of the effectiveness of invariant partition coefficients as an isomorphism filter can be attained by comparing it with other common filters. In Table 2, we count the number of equivalence classes generated in the family $\mathcal{G}^n$ of all n point acyclic graphs, assuming (a) partition coefficients alone, (b) partition coefficients plus equal edge cardinalities, (c) equal in (left) and out (right) degrees, (d) equal in (left) and out (right) ideals, and (e) equal ideals plus equal edge cardinalities. Readily, the expected number of graphs passing

		nbr of equivalence classes				
		(a)	(b)	(c)	(d)	(e)
n	$\|\mathcal{G}^n\|$	coeff	$+\|E\|$	degree	ideal	$+\|E\|$
4	16	15	16	16	15	16
5	63	52	61	63	52	61
6	318	208	285	125	208	284
7	2,045	962	1,570	432	951	1,551
8	16,999	5,187	10,263	1,588	4,932	9,863

Table 2. Comparison of isomorphism filters on graphs with n points

any filter, as in Table 1, is the expected number of graphs per equivalence class. The similarity of (a) and (b) with (d) and (e) is striking. This should not be too surprising, since φ_L is an ideal operator. But, it is a one-sided ideal operator, whereas (d) and (e) in Table 2 are based on two-sided ideals. Moreover, storage of filter (e) requires $2 \cdot n + 1$ integers whereas filter (b) consists of just $n + 1$ integers. In terms of information content, the partition coefficients are nearly twice as efficient. There may be more effective isomorphism filters, but we know of none with as dense information content.

A lexicographic ordering of the partition coefficients is an invariant, *nearly total* ordering of all acyclic graphs on n points. This can be of considerable value. In particular, we can use binary search to quickly obtain the neighborhood of any desired graph. The 4 point graphs of Figure 4 have been displayed in this order.

Another use of our graph manipulation system has been to gather various counts regarding basic, acyclic graphs on $n = |P|$ points with $e = |E|$ edges. Some of these results are summarized in Table 3. The numbers of trees on n

$|P| =$	3		4		5		6		7		8	
$|E|$	nc	c	nc	c	nc	c	nc	c	nc	c	nc	c
0	1		1		1		1		1		1	
1	1		1		1		1		1		1	
2		3	4		4		4		4		4	
3				8	11		12		12		12	
4				2	2	27	43		46		47	
5						12	14	91	156		170	
6						5	5	87	110	350	670	
7								45	50	532	721	1,376
8								12	12	475	550	3,272
9								3	3	201	216	4,298
10										71	74	3,197
11										14	14	1,565
12										7	7	554
13												186
14												44
15												16
16												4
Totals	2	3	6	10	19	44	80	238	395	1,650	2,487	14,512
$|\mathcal{G}^n|$	5		16		63		318		2,045		16,999	

Table 3. Numbers of disconnected (nc) and connected (c) acyclic graphs on $|P|$ points with $|E|$ edges

points, connected graphs with $n - 1$ edges, is evident. We would observe that the counts are quite different from the similar table of [5] which has graphs with many more edges. They enumerate the *transitively closed* graphs (or partial orders) with e edges, whereas we enumerate the basic (or minimal) graphs with that order. Using the terminology of this paper, they count the edges in the closure of a partial order, while we count the edges in its generator.

The partition coefficients appear to encode a considerable amount of additional graph specific information. For example, it is not difficult to prove that:

Theorem 4. *If the closure operator is φ_L, then a_0 must be a power of two, whose exponent denotes the number of minimal (leftmost) elements.*

It also appears that partition coefficients encode a measure of connectivity information. After a tedious sequence of minor lemmas such as

Lemma 5. *Let φ be a path based closure and let $G^{(n)} = (P, E)$ on n points have the closure coefficients $< a_n, a_{n-1}, \cdots, a_0 >$. Then, $G^{(n+1)} = (P \cup \{x\}, E)$ has the closure coefficients $< 2 \cdot a_n, 2 \cdot a_{n-1}, \cdots, 2 \cdot a_0 >$.*

Lemma 6. *Let φ be the left (right) ideal closure. $G^{(n)} = (P, E)$ has a greatest (least) point if and only if the partition coefficient $a_{n-1} = 1$.*

one finally derives a curious result,

Theorem 7. *Let φ be an ideal closure. If all the binary partition coefficients of a graph, $< a_n, a_{n-1}, \cdots, a_1, a_0 >$ (w.r.t an ideal closure) are even, then the graph is disconnected or else there exists a disconnected graph with these binary partition coefficients.*

Suggested by this result, but not stated is the fact that if all the binary partition coefficients associated with a graph are even, then, with very high probability, the graph is disconnected. On the other hand, if even one coefficient is odd, the graph is probably connected.

Of major concern with the use of closure spaces and their binary partition coefficients as tools for the analysis and filtering of acyclic graphs, is the expected cost of generating them. The straightforward approach of generating all 2^n subsets and calculating their closures is clearly impractical for even moderate sized graphs. Fortunately, this is unnecessary. Given any closed set in a closure space, and its generator, one can easily determine all closed sets that it covers because,

Theorem 8. *If φ is uniquely generated, and if $X \neq \emptyset$ is closed, then $p \in X.gen$ if and only if $Z - \{p\}$ is closed.*

Proof. See Lemma 3.1, [20].

This theorem is treated as the defining property of *extreme points*, which are the generators of convex sets in [10]. It appears in one form or another in many efficient graph algorithms. For example, this property is used in [5] [2] to generate partial orders, where the universe **U** is the edge set; their closure is transitive closure; and a "cover" is a minimal edge set that generates the transitive closure. In [11] and [8] it is exploited to characterize properties of undirected graphs, and efficient algorithms to recognize them.

In our case, we use Theorem 8, to determine the closure space of a graph and its partition coefficients by first putting the entire point set P, which must be closed, in a queue. We then successively remove closed sets Y from the queue, verify that we have not already processed it,[9] then apply $generator(Y)$. We increment a_k where $k = |Y| - |Y.gen|$. For each $y \in Y.gen$, we add $Y - \{y\}$ to the queue. The cost of generating the closure space is approximately $|closedsets| \cdot cost_{generator(Y)}$. Assuming $cost_{generator(Y)}$ is nearly constant[10] given G, the cost of generating partition coefficients will be clearly dominated by the number of closed sets to be processed.

So, the key question becomes "what is the expected number of closed sets in an acyclic graph on n points?" The number of closed sets in any particular G is given by the sum of its partition coefficients, $\sum_{i=0}^{n} a_i$. Table 4 enumerates these expected values for $3 \le n \le 8$, first for those graphs with precisely $|E|$ edges, and then for all graphs in $\mathcal{G}^n$. Readily, the worst case behavior is $O(2^n)$; but

	$|P|$					
$|E|$	3	4	5	6	7	8
0	8.0	16.0	32.0	64.0	128.0	256.0
1	6.0	12.0	24.0	48.0	96.0	192.0
2	4.7	9.2	18.5	37.0	74.0	148.0
3		7.1	14.2	28.2	56.5	113.0
4		6.5	10.9	21.8	43.5	86.9
5			9.5	16.6	33.1	66.0
6			9.6	14.5	25.6	50.7
7				13.7	21.7	39.0
8				13.9	20.2	33.0
9				13.0	19.5	29.7
10					19.2	28.1
11					18.3	27.1
12					18.8	26.5
13						26.1
14						26.0
15						26.8
16						25.2
all graphs	5.6	8.4	12.2	17.1	23.6	32.3

Table 4. Expected numbers of closed sets in graphs with $|P|$ points and $|E|$ edges

this occurs only if $|E| = 0$. As $|E|$ increases the number of closed sets decreases towards an asymptote. If G must be connected, so that $|E| \ge n - 1$, the number of number of closed sets is close to the asymptote itself.

---

[9] Because we use a queue, this is a level by level processing of the closed sets. It is possible to reach the same closed set twice. *C.f.* figure 2.

[10] With these small, finite graphs there is a hard upper bound for any n.

4 Counting Binary Partitions

We close by once again considering binary partitions. The space of acyclic graphs grows exponentially. Is it reasonable to expect to characterize them by partition coefficients as n becomes large? How many binary partitions are there on n points? It is customary to let $b(n)$, called the binary partition function, denote the *number* of binary partitions of n. As before, our interest is the number of binary partitions of 2^n, that is $b(2^n)$. In [3], it is shown that

$$b(2^n) = c_{n,0} + c_{n,1} \tag{8}$$

where $c_{1,0} = c_{1,1} = 1$, $c_{n+1,0} = c_{n,0} = 1$, and $c_{n+1,i} = \sum_{k=0}^{2i} c_{n,k}$. This particularly simple formulation was executed by Churchhouse on an Atlas computer in 1968 to obtain initial values of the binary partition function. A more complex, but somewhat faster code is given in [19]. With this code one can generate the following Table 5 of partitions of 2^n. The second column counts the number of

n	$b(2^n) = |\mathcal{P}^n|$	$|normal|$	$|\mathcal{G}^n|$
1	2	1	1
2	4	2	2
3	10	6	5
4	36	26	16
5	202	166	63
6	1,828	1,626	318
7	27,338	25,510	2,045
8	692,004	664,666	16,999
9	30,251,722	29,559,718	183,231
10	2,320,518,948	2,290,267,226	2,567,284

Table 5. Number of partitions of 2^n, of normal partitions, of acyclic graphs

normal partitions in which $a_0 \neq 0$, which in accordance with observation four in Section 1, is always $|\mathcal{P}^n| - |\mathcal{P}^{n-1}|$. Closure spaces associated with acyclic graphs must be normal. The third column counts the number of such non-isomorphic graphs on n points. It is easy to verify all sequences in Sloane's Handbook of Integer Sequences [23]. The point of this table is to illustrate that while the diversity of acyclic graphs on n points has exponential growth, the variety of closure spaces has super exponential growth; specifically $b(2^n) \sim (2^n)^{n/2}$. The concept of uniquely generated closure spaces is clearly rich enough to be embraced as a tool in the study of acyclic graphs and partially ordered spaces.

References

1. G. D. Birkhoff and D. Lewis. Chromatic polynomials. *Trans. Amer. Math. Soc.*, 60:355–451, 1946.

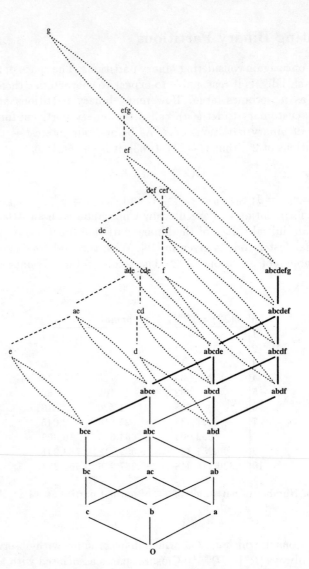

Fig. 2. The closure lattice $\mathcal{L}_{(U,\varphi)}$ of a small 7 point closure space.

example, the binary partition corresponding to the closure space $(\mathbf{U}, \varphi)$ of Figure 2 is $< 0\ 1\ 0\ 1\ 3\ 4\ 0\ 8 >$ (where $a^n = a^7 = 0$ is the leading coefficient). There is a single interval, $[abcdefg, g]$, of size 2^6; so $a_6 = 1$. The three intervals of size 2^3, $[abcde, de]$, $[abcdf, cf]$, and $[abdf, f]$, imply that $a_3 = 3$. There are eight singleton elements, abc, ab, ac, bc, a, b, c and $\emptyset$, where $X.\varphi = X.gen$. Consequently, $a_0 = 8$. Those partitions for which $a_0 \neq 0$ we call *normal*. Customarily the closure of $\emptyset$ is empty[6], even though it is not required in a the general theory of closure spaces.

---

[6] That the convex closure of the empty set should be empty is so reasonable, it is taken to be an axiom in [10] and [11].

Sub-Cubic Cost Algorithms for the All Pairs Shortest Path Problem[*]

Tadao TAKAOKA

Department of Computer Science
Ibaraki University, Hitachi, Ibaraki 316, JAPAN
E-mail: takaoka@cis.ibaraki.ac.jp

Abstract. In this paper we give three sub-cubic cost algorithms for the all pairs shortest distance (APSD) and path (APSP) problems. The first is a parallel algorithm that solves the APSD problem for a directed graph with unit edge costs in $O(\log^2 n)$ time with $O(n^\mu/\sqrt{\log n})$ processors where $\mu = 2.688$ on an EREW-PRAM. The second parallel algorithm solves the APSP, and consequently APSD, problem for a directed graph with non-negative general costs (real numbers) in $O(\log^2 n)$ time with $o(n^3)$ subcubic cost. Previously this cost was greater than $O(n^3)$. Finally we improve with respect to M the complexity $O((Mn)^\mu)$ of a sequential algorithm for a graph with edge costs up to M into $O(M^{1/3} n^{(6+\omega)/3} (\log n)^{2/3} (\log \log n)^{1/3})$ in the APSD problem.

1 Introduction

The all pairs shortest path (APSP) problem is to compute shortest paths between all pairs of vertices of a directed graph with non-negative real numbers as edge costs. The all pairs shortest distance problem (APSD) is defined similarly, the word "paths" replaced by distances. Traditionally the latter was called the APSP problem. Alon, Galil and Margalit [1] gave a sub-cubic algorithm for the APSD problem for a graph with small integer edge costs. Recently Alon, Galil, Margalit and Naor [2] distinguished these two problems and faced a higher complexity for the APSP problem with the same class of graphs. The best time complexities for the APSD and APSP problems for undirected graphs with unit edge costs are given by Seidel [10]. On the other hand, the complexity for the all pairs shortest distance (APSD) problem with general edge costs was slightly improved by Takaoka [12], from that by Fredman [5], which is n^3 divided by polylog of n, very close to n^3. With this algorithm [12], we can solve the APSP problem at the same time, and hence no need to distinguish between APSD and APSP.

The technique for obtaining paths in [2] is based on the concept of witnesses for Boolean matrix multiplication. They compute the witnesses of Boolean matrix multiplication for (n, n) matrices in $O(n^\omega \log^c n)$ where $c < 5$ and $\omega = 2.376$. Using this result they solve other shortest path problems, attaching $\log^c n$ as factors to the complexities of corresponding distance problems.

[*] Partially supported by Grant-in-aid No. 07680332 by Monbusho Scientific Research Program and a research grant from Hitachi Engineering Co., Ltd.

In this paper we design a parallel algorithm for the APSD problem for a directed graph with unit edge costs with $O(\log^2 n)$ time (worst case) and $O(n^{(3+\omega)/2}/\sqrt{\log n})$ processors. This result is compared with a parallel algorithm for a directed graph with general edge costs in $O(\log \log n)$ expected time and $O(n^{2.5+\varepsilon})$ processors in [13] where ε is a small positive real number.

Next we improve the parallel algorithm for the APSP problem with general edge costs in [12] whose cost (= number of processors × time) is slightly above $O(n^3)$. The cost of $O(n^3 \log n)$ in Dekel, Nassimi and Sahni [4] was also improved by Han, Pan and Reif [8] into that still avove $O(n^3)$. The cost of our new algorithm is slightly below $O(n^3)$ and the time is polylog of n, that is, in NC.

Finally we present a sequential algorithm for a graph with edge cost up to M whose complexity is sub-cubic when $M = O(n^{0.624})$.

2 Basic definitions

A directed graph is given by $G = (V, E)$, where $V = \{0, \cdots, n-1\}$ and E is a subset of $V \times V$. The edge cost of $(i, j) \in E$ is denoted by d_{ij}. The (n, n) matrix D is one whose (i, j) element is d_{ij}. We assume that $d_{ij} \geq 0$ and $d_{ii} = 0$ for all i, j. If there is no edge from i to j, we let $d_{ij} = \infty$. The cost, or distance, of a path is the sum of costs of the edges in the path. The length of a path is the number of edges in the path. The shortest distance from vertex i to vertex j is the minimum cost over all paths from i to j, denoted by d_{ij}^*. Let $D^* = \{d_{ij}^*\}$. We call n the size of the matrices.

Let A and B are (n, n)-matrices. The three products are defined using the elements of A and B as follows:

(1) Ordinary multiplication over a ring $\quad C = AB$

$$c_{ij} = \sum_{k=0}^{n-1} a_{ik} b_{kj},$$

(2) Boolean matrix multiplication $\quad C = A \cdot B$

$$c_{ij} = \bigvee_{k=0}^{n-1} a_{ik} \wedge b_{kj},$$

(3) Distance matrix multiplication $\quad C = A \times B$

$$c_{ij} = \min_{0 \leq k \leq n-1} \{a_{ik} + b_{kj}\}.$$

The best algorithm [3] computes (1) in $O(n^\omega)$ time. To compute (2), we can regard Boolean values 0 and 1 in A and B as integers and use the algorithm for (1), and convert non-zero elements in the resulting matrix to 1. Therefore this complexity is $O(n^\omega)$. If we have an algorithm for (3) with $T_D(n)$ $(T_P(n))$ time we can solve the APSD (APSP) problem in $O(T_D(n)\log n)$ $(O(T_P(n)\log n))$ time by the repeated squaring method, described as the repeated use of $A \leftarrow A \times A$. In this description, $T_P(n)$ is the time for the algorithm to compute (3) with witnesses, that is, giving k that gives the minimum for c_{ij}. The witnesses of (2) is the matrix $W = \{w_{ij}\}$ where $w_{ij} = k$ for some k such that $a_{ik} \wedge b_{kj} = 1$. If there is no such k, $w_{ij} = 0$.

The definition of computing shortest paths is to give a path matrix of size n by which we can give a shortest path from i to j in $O(\ell)$ time where ℓ is the length of the path. More specifically, if $w_{ij} = k$ in the path (or witness) matrix $W = \{w_{ij}\}$, it means that the path from i to j goes through k. Therefore a recursive function $path(i,j)$ is defined by $(path(i,k), k, path(k,j))$ if $path(i,j) = k > 0$ and nil if $path(i,j) = 0$, where a path is defined by a list of vertices excluding endpoints. In the following sections, we record k in w_{ij} whenever we can find k such that a path from i to j is modified or newly set up by paths from i to k and from k to j.

3 All pairs shortest paths

We review the algorithm in [1] in this section. Let the costs of edges of the given graph be ones. Let $D^{(\ell)}$ be the ℓ-th approximate matrix for D^* defined by $d_{ij}^{(\ell)} = d_{ij}^*$ if $d_{ij}^* \leq \ell$, and $d_{ij}^{(\ell)} = \infty$ otherwise. Then we can compute $D^{(r)}$ by the following algorithm.

Algorithm 1   Shortest distances by Boolean matrix multiplication

1   $A := \{a_{ij}\}$ where $a_{ij} = 1$ if $d_{ij} \leq 1$, and 0 otherwise;
2   $B := I;$ {Boolean identity matrix}
3   **for** $\ell := 2$ **to** r **do begin**
4      $B := A \cdot B;$ {Boolean matrix multiplication}
5   **for** $i := 0$ **to** $n-1$ **do for** $j := 0$ **to** $n-1$ **do**
6   **if** $b_{ij} = 1$ **then** $d_{ij}^{(\ell)} := \ell$ **else** $d_{ij}^{(\ell)} = \infty;$
7   **if** $d_{ij}^{(\ell-1)} \leq \ell$ **then** $d_{ij}^{(\ell)} := d_{ij}^{(\ell-1)}$
8   **end**.

In this algorithm, $D^{(\ell)}$ is computed in increasing order of ℓ. Note that $D^{(1)} = D$. Since we can compute line 4 in $O(n^\omega)$ time, the computing time of this algorithm is $O(rn^\omega)$.

The following algorithm in [1] for the APSD problem uses Algorithm 1 as an "accelerating phase" and repeated squaring as a "cruising phase."

Algorithm 2   Solving APSD

   {Accelerating phase}
1 **for** $\ell := 2$ **to** r **do** compute $D^{(\ell)}$ using Algorithm 1;
   {Cruising phase}
2 $\ell := r;$

3 **for** $s := 1$ **to** $\lceil \log_{3/2} n/r \rceil$ **do begin**

4   **for** $i := 0$ **to** $n - 1$ **do**

5   Scan the i-th row of $D^{(\ell)}$ and find the smallest set of equal $d_{ij}^{(\ell)}$'s such that $\lceil \ell/2 \rceil \leq d_{ij}^{(\ell)} \leq \ell$ and let the set of corresponding indices j be S_i;

6     $\ell_1 := \lceil 3\ell/2 \rceil$;

7   **for** $i := 0$ **to** $n - 1$ **do for** $j := 0$ **to** $n - 1$ **do begin**

8   **if** $S_i \neq \emptyset$ **do** $m_{ij} := \min_{k \in S_i} \{d_{ik}^{(\ell)} + d_{kj}^{(\ell)}\}$ **else** $m_{ij} := \infty$

9   **if** $d_{ij}^{(\ell)} \leq \ell$ **then** $d_{ij}^{(\ell_1)} := d_{ij}^{(\ell)}$

10   **else if** $m_{ij} \leq \ell_1$ **then** $d_{ij}^{(\ell_1)} := m_{ij}$

11   **else** $\{m_{ij} > \ell_1\}$ $d_{ij}^{(\ell_1)} := \infty$

12   **end**;

13     $\ell := \ell_1$

14 **end**

Algorithm 2 computes $D^{(\ell)}$ from $\ell = 2$ to r in the accelerating phase spending $O(rn^3)$ time and computes $D^{(\ell)}$ for $\ell = r$, $\lceil \frac{3}{2}r \rceil$, $\left\lceil \frac{3}{2}\lceil \frac{3}{2}r \rceil \right\rceil$, $\cdots$, n' by repeated squaring in the cruising phase, where n' is the smallest integer in this series of ℓ such that $\ell \geq n$. The key observation in the cruising phase is that we only need to check S_i at line 8 whose size is not larger than $2n/\ell$. Hence the computing time of one iteration beginning at line 3 is $O(n^3/\ell)$. Thus the time of the cruising phase is given with $N = \lceil \log_{3/2} n/r \rceil$ by

$$O\left(\sum_{s=1}^{N} n^3/((3/2)^s r)\right) = O(n^3/r).$$

Balancing the two phases with $rn^\omega = n^3/r$ yields the time of Algorithm of $O(n^{(\omega+3)/2})$ with $r = O(n^{(3-\omega)/2})$.

When we have a directed graph G whose edge costs are between 0 and M where M is a positive integer, we can convert the graph G to $G' = (V', E')$ by adding auxiliary vertices $v_1, \cdots, v_{M-1}$ for $v \in V$. The edge set is also modified to E'. If $c(v, w) = \ell$, w is connected from $v_{\ell-1}$ in E' where $v_0 = v$. Obviously we can solve the problem for G by applying Algorithm 2 to G', which takes $O\left((Mn)^{(\omega+3)/2}\right)$ time.

Now witnesses can be kept at lines 4 – 7 of Algorithm 1 in $O(n^\omega \log^c n)$ time. At line 8 of Algorithm 2 witnesses are obtained by storing k in the witness matrix. This does not increase the order of complexity of the cruising phase. Thus we have the complexity of the APSP of $O\left(n^{(\omega+3)/2}\sqrt{(\log n)^c}\right)$. For the graph with edge costs of M or less, the complexity becomes $O\left((Mn)^{(\omega+3)/2}\sqrt{(\log(Mn))^c}\right)$.

4   Parallelization for graphs with unit costs

We design a parallel algorithm on an EREW-PRAM for a directed graph with unit edge costs. In this section and the next section, we mainly describe our algorithm using a CREW-PRAM for simplicity. The overhead time to copy data in $O(\log n)$ time with a certain number of processors depending on each phase is absorbed in the dominant complexities. Let A be the adjacency matrix used in Algorithm 1. That is, $a_{ij} = 1$ if there is an edge from i to j and 0 otherwise. All diagonal elements are 1. There is a path from i to j of length $\leq \ell$ if and only if the (i, j) element A^ℓ is 1, where A^ℓ is the ℓ-th power of A by Boolean matrix multiplication. By repeated squaring, we can get A^ℓ $(\ell = 1, 2, 4, \cdots, n')$ with $\lceil \log_2 n \rceil$ Boolean matrix multiplications, where n' is the smallest integer in this series of ℓ such that $\ell \geq n$. These matrices give a kind of approximate estimation on the path lengths. That is, if the (i, j) element of A^{2^r} becomes 1 for the first time, we can say that the shortest path length from i to j is between $2^{r-1}+1$ and 2^r for $r \geq 1$. As r gets large, the gap between $2^{r-1}+1$ and 2^r gets large. As Gazit and Miller [6] observe we can fill the gap in increasing order of r in the following way. Let shortest paths up to 2^{r-1} be computed already. Then a shortest path from i to j, whose length is between $2^{r-1} + 1$ and 2^r consists of a shortest path from i to k whose length is up to 2^{r-1} and a path from k to j of length 2^{r-1}. Formally we have the following algorithm, in which the approximation phase is shown for explanation purposes.

Algorithm 3   Shortest distances up to 2^R
  {Approximation phase}
  1 $A^{(1)} := A$; $\ell := 1$;
  2 **for** $s := 1$ **to** R **do begin**
  3    $A^{(2\ell)} := A^{(\ell)} \cdot A^{(\ell)}$; {Boolean matrix multiplication}
  4    $\ell := 2\ell$
  5 **end**;
  {Gap filling phase}
  6 $D^{(1)} := D$;
  7 **for** $s := 1$ **to** R **do begin**
  8   **for** $\ell := 2^{s-1} + 1$ **to** 2^s **do begin**
  9      $A^{(\ell)} := A^{(\ell - 2^{s-1})} \cdot A^{(2^{s-1})}$; {Boolean matrix multiplication}
  10   **for** $i := 0$ **to** $n - 1$ **do for** $j := 0$ **to** $n - 1$ **do**
  11   **if** $a_{ij}^{(\ell)} = 1$ **then** $b_{ij}^{(\ell)} := \ell$ **else** $b_{ij}^{(\ell)} := \infty$
  12   **end**;
  13   **for** $i := 0$ **to** $n - 1$ **do for** $j := 0$ **to** $n - 1$ **do**
  14      $d_{ij}^{(2^s)} := \min_{2^{s-1} < \ell \leq 2^s} \{b_{ij}^{(\ell)}\}$
  15 **end**.

In this algorithm arrays $B^{(\ell)} = \{b_{ij}^{(\ell)}\}$ are used as working space to compute $d_{ij}^{(2^s)}$ at line 14. We fill the gaps for $A^{(\ell)}$, but do not do so for $D^{(\ell)}$, since we are only interested in $D^{(2^R)}$. The computing time of this algorithm is $O(2^R n^\omega)$. If we let $R = \lceil \log_2 n \rceil$, we can solve the APSD, but then the time becomes $O(n^{\omega+1})$, which is not efficient. If we substitute Algorithm 3 for Algorithm 1 in Algorithm 2 and let $2^R = n^{(3-\omega)/2}$, however, we can solve the APSD problem in $O(n^{(\omega+3)/2})$ time, which is on a par with Algorithm 2.

The merit of Algorithm 3 is that it is easy to parallelize. It is well-known that we can multiply two matrices over a ring in $O(\log n)$ time with $O(n^\omega)$ processors in parallel. We substitute Algorithm 3 for Algorithm 1 in Algorithm 2 and call the resulting algorithm Algorithm 2'. We can perform 2^{s-1} multiplications in parallel at line 9. Also we can compute 2^{s-1} matrices $B^{(\ell)}$ at lines 10 and 11 in parallel. At line 14, we can find the minimum in $O(s)$ time with $O(2^s/s)$ processors.

Turning our attention to the cruising phase of Algorithm 2' (or 2), we can find the minimum at line 8 in $O(\log n)$ time with $O(n/(\ell \log n))$ processors. The rest is absorbed in these complexities. Now we summarize the complexities for the parallel algorithm. T is for time and P is for the number of processors.

Accelerating phase $T = O(R \log n), P = O(n^\omega \cdot 2^R)$
Cruising phase $T = O(\log(n/2^R) \cdot \log n), P = O(n^3/(2^R \log n))$

If we let $2^R = n^{(3-\omega)/2}/\sqrt{\log n}$, we have the overall complexity as follows:

$$T = O(\log^2 n), P = n^{(3+\omega)/2}/\sqrt{\log n}.$$

If we have a graph with edges costs up to M we can replace n by Mn in the above complexities.

5   Parallelization for graphs with general costs

If edges costs are non-negative real numbers, we can not apply techniques in the previous sections. Even in the previous section, if M, the magnitude of edge costs, is $O(n)$, the efficiencies of both sequential and parallel algorithms get much worse than primitive methods.

Fredman [5] first gave an algorithm for the APSD problem in $o(n^3)$, that is, $O(n^3 (\log \log n/ \log n)^{1/3})$ time, by showing that distance matrix multiplication can be solved in this complexity. Takaoka [12] improved this to $O(n^3 (\log \log n / \log n)^{1/2})$ and pointed out that the APSP problem can be solved in the same complexity. This algorithm was also parallelized in [12]. The parallel version takes the repeated squaring approach. The parallel algorithm for distance matrix multiplication has complexities of $T = O(\log n)$ and $P = O(n^3 (\log \log n)^{1/2}/(\log n)^{3/2})$ on an EREW-PRAM. Therefore the APSP problem can be solved with $T = O(\log^2 n)$ and $P = O(n^3 (\log \log n)^{1/2}/(\log n)^{3/2})$. In this algorithm, we can keep track of witnesses easily and thus the APSP problem can be solved in the same complexities. The cost $PT = O(n^3 (\log n \log \log n)^{1/2})$ is slightly above $O(n^3)$. Since then, it has been a major open problem whether there is an NC algorithm whose cost is $O(n^3)$.

In this section, we show a stronger result that there exists an NC algorithm for the APSP problem, whose cost is $o(n^3)$.

Definition 1. Let a parallel algorithm have time complexity T and use $P(t)$ processors at the t-th step. Then the cost complexity C is defined by

$$C = \sum_{t=1}^{T(n)} P(t).$$

A time interval I_i over which the number of processors is fixed is called a processor phase. That is, $P(t)$ are equal to P_i for all $t \in I_i$ and interval $[0..T]$ is divided as $[0..T] = I_0 \cup \cdots \cup I_{k-1}$ where k is the number of processor phases. Then we have

$$C = \sum_{i=0}^{k-1} P_i T_i,$$

where T_i is the size of interval I_i.

Brent's theorem[7] states that other processor phases can be simulated by a smaller number of processors at the expense of increasing computing time, without mentioning the overhead time for rescheduling processors. We suggest that the number of processor phases be finite so that the rescheduling does not cause too much overhead time. In the following parallel algorithm, the number of processor phases is two.

The engine, so to speak, in the acceleration phase in Algorithm 2 was a fast algorithm for Boolean matrix multiplication. We use the fast distance matrix multiplication algorithm in [12] as the engine and modify the cruising phase slightly to fit our parallel algorithm. In Algorithm 2, there is no difference between distances and lengths of paths since the edge costs are ones. In line 5 of Algorithm 2, we choose set S_i based on the distances $d_{ij}^{(\ell)}$ ($j = 0, \cdots, n-1$) satisfying $\lceil \ell/2 \rceil \leq d_{ij}^{(\ell)} \leq \ell$ to guarantee the correct computation of distances between ℓ and $\lceil 3\ell/2 \rceil$. We observe that the computation of S_i is essentially based on path lengths, not distances. If we keep track of path lengths, therefore, we can adapt Algorithm 2 to our problem. The definition of $d_{ij}^{(\ell)}$ here is that it gives the cost of the shortest path whose length is not greater than ℓ. The algorithm follows with a new data structure, array $Q^{(\ell)}$, such that $q_{ij}^{(\ell)}$ is the length of a path that gives $d_{ij}^{(\ell)}$.

Algorithm 4
{Accelerating phase}

1 $\ell := 1;\ D^{(1)} := D;\ Q^{(1)} := \{q_{ij}^{(1)}\}$, where $q_{ij}^{(1)} = \begin{cases} 0, & \text{if } i = j \\ 1, & \text{if } i \neq j \text{ and } (i,j) \in E \\ \infty, & \text{otherwise}; \end{cases}$

2 **for** $s := 1$ **to** $\lceil \log_2 r \rceil$ **do begin**

3    $D^{(2\ell)} := D^{(\ell)} \times D^{(\ell)}$;   {distance matrix multiplication in [12]}

4    $Q^{(2\ell)} := \{q_{ij}^{(2\ell)}\}$ where $q_{ij}^{(2\ell)} = \begin{cases} q_{ik}^{(\ell)} + q_{kj}^{(\ell)}, & \text{if } d_{ij}^{(2\ell)} \text{ is updated by} \\ & \qquad d_{ik}^{(\ell)} + d_{kj}^{(\ell)} \\ q_{ij}^{(\ell)}, & \text{otherwise}; \end{cases}$

5    $\ell := 2\ell$

6 **end**;

  {Cruising phase}

7 **for** $s := 1$ **to** $\lceil \log_{3/2} n/r \rceil$ **do begin**

8   **for** $i := 0$ **to** $n - 1$ **do**

9   Scan the i-th row of $Q^{(\ell)}$ and find the smallest set of equal $q_{ij}^{(\ell)}$'s such

       that $\lceil \ell/2 \rceil \leq q_{ij}^{(\ell)} \leq \ell$ and let the set of corresponding indices j be S_i;

10   $\ell_1 := \lceil 3\ell/2 \rceil$;

11   **for** $i := 0$ **to** $n - 1$ **do for** $j := 0$ **to** $n - 1$ **do begin**

12   **if** $S_i \neq \emptyset$ **then begin**

13        $m_{ij} := \min_{k \in S_i} \{d_{ik}^{(\ell)} + d_{kj}^{(\ell)}\}$;

14        $k :=$ one that gives the above minimum and satisfies that $q_{ik}^{(\ell)} + q_{kj}^{(\ell)}$

          is minimum among such k;

15        $L := q_{ik}^{(\ell)} + q_{kj}^{(\ell)}$;

16   **end else** $\{S_i = \emptyset\}$ $L := \infty$;

17   **if** $m_{ij} < d_{ij}^{(\ell)}$ **then begin** $d_{ij}^{(\ell_1)} := m_{ij}$; $q_{ij}^{(\ell_1)} := L$ **end**

18   **else begin** $d_{ij}^{(\ell_1)} := d_{ij}^{(\ell)}$; $q_{ij}^{(\ell_1)} := q_{ij}^{(\ell)}$ **end**;

19     $\ell := \ell_1$

20   **end**

21 **end**.

As described in [12], we can parallelize the distance matrix multiplication at line 3 on an EREW-PRAM. We index time T and the number of processors P in the accelerating phase and cruising phase by 1 and 2. Then we have

$$T_1 = O(\log r \log n), \quad P_1 = O(n^3 (\log \log n)^{1/2} / (\log n)^{3/2}).$$

The computation of all S_i can be done in $O(\log n)$ time with $O(n^2)$ processors. The dominant complexity in the cruising phase is at line 13. This part can be computed in $O(\log(n/r))$ time with $O((n/r)/\log(n/r))$ processors. Thus we have

$$T_2 = O(\log n \log(n/r)), \quad P_2 = O(n^2(n/r)/\log(n/r)).$$

Letting $r = (\log n / \log \log n)^{3/2}$ yields

$$T_1 = O(\log n \log \log n), \quad P_1 = O(n^3 (\log \log n)^{1/2} / (\log n)^{3/2})$$
$$T_2 = O(\log^2 n), \qquad\qquad P_2 = O(n^3 (\log \log n)^{3/2} / (\log n)^{5/2}).$$

Thus the cost is given by

$$P_1 T_1 + P_2 T_2 = O(n^3 (\log \log n)^{3/2} / (\log n)^{1/2}) + O(n^3 (\log \log n)^{3/2} / (\log n)^{1/2})$$
$$= O(n^3 (\log \log n)^{3/2} / (\log n)^{1/2})$$
$$= o(n^3).$$

We note that we can solve the APSP problem with the same order of cost by this algorithm. We only need to keep track of witnesses at distance matrix multiplication and the minimum operation at line 13.

If we perform the accelerating phase with P_2 processors, the time for this phase will become $O(\log^2 n)$, and the cost will be the same as above for the whole computation. That is, we can keep the number of processors uniform and claim that the algorithm has the above complexities under the traditional definition of cost by $C = PT$.

6 An algorithm for graphs with small edge costs

When the edge costs are bounded by a positive integer M, we can do better than we saw in previous sections. We briefly review Romani's algorithm [9] for distance matrix multiplication.

Let A and B be distance matrices whose elements are bounded by M or infinite. Let the diagonal elements be 0. Then we convert A and B into A' and B' where

$$a'_{ij} = \begin{cases} n^{M-a_{ij}}, & \text{if } a_{ij} \neq \infty \\ 0 & \cdot, & \text{if } a_{ij} = \infty \end{cases}$$

$$b'_{ij} = \begin{cases} n^{M-b_{ij}}, & \text{if } b_{ij} \neq \infty \\ 0 & , & \text{if } b_{ij} = \infty. \end{cases}$$

Let $C' = A'B'$ be the product by ordinary matrix multiplication and $C = A \times B$ be that by distance matrix multiplication. Then we have

$$c'_{ij} = \sum_{k=0}^{n-1} n^{2M-(a_{ik}+b_{kj})}, \quad c_{ij} = 2M - \lfloor \log_n c'_{ij} \rfloor.$$

Thus we can compute C with $O(n^{\omega})$ arithmetic operations on integers up to n^M. Since these values can be expressed by $O(M \log n)$ bits and Schönhage and Strassen's algorithm [11] for multiplying k-bit numbers takes $O(k \log k \log \log k)$ bit operations, we can compute C in $O(n^{\omega} M \log n \log(M \log n) \log \log(M \log n))$ time.

We replace the accelerating phase of Algorithm 4 by the following and call the resulting algorithm Algorithm 5.

Algorithm 5

{Another accelerating phase}

1 $\ell := 1$; $D^{(1)} := D$; $Q^{(1)} := \{q_{ij}^{(1)}\}$; where $q_{ij}^{(1)} = \begin{cases} 0, & \text{if } i = j \\ 1, & \text{if } i \neq j \text{ and } (i,j) \in E \\ \infty, & \text{otherwise} \end{cases}$

2 **for** $s := 1$ **to** r **do begin**

3    $D^{(\ell+1)} := D^{(\ell)} \times D$; {distance matrix multiplication by Romani in [7]}

4    $Q^{(\ell+1)} := \{q_{ij}^{(\ell+1)}\}$; where $q_{ij}^{(\ell+1)} = \begin{cases} q_{ij}^{(\ell)} + 1, & \text{if } d_{ij}^{(\ell+1)} \text{ is updated} \\ q_{ij}^{(\ell)}, & \text{otherwise} \end{cases}$

5    $\ell := \ell + 1$

6 **end;**

{Cruising phase} same as that in Algorithm 4.

Note that the bound M is replaced by ℓM in the distance matrix multiplication. The time for the accelerating phase is given by

$$O(n^\omega r^2 M \log n \log(rM \log n) \log\log(rM \log n)).$$

We assume that M is $O(n^k)$ for some constant k. Balancing this complexity with that of cruising phase, $O(n^3/r)$, yields the total computing time of

$$O(n^{(6+\omega)/3}(M \log n \log(nM \log n) \log\log(nM \log n))^{1/3})$$

with the choice of

$$r = O(n^{(3-\omega)/3}(M \log n \log(nM \log n) \log\log(nM \log n))^{-1/3}).$$

This complexity is simplified into

$$O(M^{1/3} n^{(6+\omega)/3}(\log n)^{2/3}(\log\log n)^{1/3}).$$

The value of M can be almost $O(n^{0.624})$ to keep the complexity within sub-cubic. This bound on M is a considerable improvement over $O(n^{0.116})$ given in [1].

In the above we solved only the APSD problem. In the Romani algorithm, we can not keep track of witnesses. If we could, we would be able to replace the accelerating phase by that based on repeated squaring and would have a better complexity for both the APSD and APSP problem with small edge costs.

7 Concluding remarks

The balancing parameters between the accelerating and cruising phases change depending on what engine we use in the accelerating phase. We may find more results if we use other algorithms for the engine in the accelerating phase.

Acknowledgment.

A comment on Definition 1 by Zhi-Zhong Chen is greatly appreciated.

References

1. N. Alon, Z. Galil and O. Margalit, On the exponent of the all pairs shortest path problem, Proc. 32th IEEE FOCS (1991), pp 569–575.
2. N. Alon, Z. Galil, and O. Margalit and M. Naor, Witnesses for Boolean matrix multiplication and for shortest paths, Proc. 33th IEEE FOCS (1992), pp. 417–426.
3. D. Coppersmith and S. Winograd, Matrix multiplication via arithmetic progressions, Journal of Symbolic Computation 9 (1990), pp. 251–280.

4. E. Dekel, D. Nassimi and S. Sahni, Parallel matix and graph algorithms, SIAM Jour. on Comp. 10 (1981), pp. 657–675.

5. M. L. Fredman, New bounds on the complexity of the shortest path problem, SIAM Jour. Comput. 5 (1976), pp. 49–60.

6. H. Gazit and G. Miller, An improved parallel algorithm that computes the bfs numbering of a directed graph, Info. Proc. Lett. 28 (1988) pp. 61–65.

7. A. Gibbons and W. Rytter, Efficient Parallel Algorithms, Cambridge Univ. Press (1988).

8. Y. Han, V. Pan and J. Reif, Efficient parallel algorithms for computing all pairs shortest paths in directed graphs, Proc. 4th ACM SPAA (1992), pp. 353–362.

9. F. Romani, Shortest-path problem is not harder than matrix multiplications, Info. Proc. Lett. 11 (1980) pp.134–136.

10. R. Seidel, On the all-pairs-shortest-path problem, Proc. 24th ACM STOC (1990), pp. 213–223.

11. A. Schönhage and V. Strassen, Schnelle Multiplikation Großer Zahlen, Computing 7 (1971) pp. 281–292.

12. T. Takaoka, A new upperbound on the complexity of the all pairs shortest path problem, Info. Proc. Lett. 43 (1992), pp. 195–199.

13. T. Takaoka, An efficient parallel algorithm for the all pairs shortest path problem, WG 88, Lecture Notes in Computer Science 344 (Springer, Berlin, 1988) pp. 276–287.

Diametral Path Graphs

J.S. Deogun[1] *   and   D. Kratsch[2] **

[1] Department of Computer Science and Engineering
University of Nebraska – Lincoln
Lincoln, NE 68588-0115, USA
[2] Fakultät für Mathematik und Informatik
Friedrich-Schiller-Universität
Universitätshochhaus
07740 Jena, Germany

Abstract. We introduce a new class of graphs called *diametral path graphs* that properly contains the class of asteroidal triple-free graphs and the class of dominating pair graphs. We characterize the trees as well as the chordal graphs that are diametral path graphs. We present an $O(n^3 m)$ algorithm deciding whether a given graph has a dominating diametral path. Finally, we study the structure of minimum connected dominating sets in diametral path graphs.

1 Introduction

Distance is one of the most fundamental concepts in the study of graphs and networks. Distance problems in graphs and related parameters have been of interest to researchers for quite some time. Ore's characterization of diameter maximal graphs, i.e., graphs to which the addition of any edge decreases the diameter, may be considered as a significant milestone in this direction [16]. Several studies investigating distance properties of graphs have appeared in the literature [1, 3, 4]. Recently, a number of researchers have shown considerable interest in problems involving distance and its derivative parameters [1, 7, 15].

Traditionally, graph theoretic parameters such as diameter, the distance between two nodes, vertex-connectivity, and edge-connectivity have played an important role in the design and analysis of networks. Recent investigations of vertex-integrity, edge-integrity and change in the diameter of a graph as a result of node or edge failures have yielded a wealth of new information and have resulted in new questions and avenues for further research [1, 2]. For many of these questions the basic underlying concept of distance turns out to be important and significant. Distance plays an important role in the design and analysis of networks in a variety of networking environments like communication networks, electric power grids, and transportation networks.

* This research was supported in part by the Office of Naval Research under Grant No. N0014-91-J-1693.
** Part of this research was done while the second author was at IRISA Rennes (France) supported by CHM Fellowship. Email: kratsch@minet.uni-jena.de

Recently, we discovered the class of *diametral path graphs* while investigating models for the design and analysis of networks in an adversarial environment [11]. The richness of the class of diametral path graphs can not be overestimated. The class of diametral path graphs turns out to have several nice structural properties. Moreover, the class of diametral path graphs properly contains several well-known and well researched classes of graphs. The class of diametral path graphs is expected to be important from the point of view of designing networks. The diametral path graphs are naturally suited as a topology for networks. The basic idea is that by increasing the reliability of the nodes and the edges on the diametral path, the integrity of the graph can be increased considerably, i.e., by assuming that the nodes and arcs which are on the diametral path are hardened and can not be destroyed easily, the network becomes less susceptible to disruption in an adversarial environment. Moreover, as the number of nodes on the diametral path will be much less in comparison to the total number of nodes in the graph, the cost of increasing the reliability of the network will not be high.

Our paper is organized as follows. Diametral path graphs are introduced in Section 2. The relation between diametral path graphs, asteroidal triple-free graphs and dominating pair graphs is considered in Section 3. An $O(n^3 m)$ algorithm deciding whether a given graph has a dominating diametral path is given in Section 4. In Section 5 a partial list of minimal forbidden subgraphs for the class of diametral path graphs is given. Furthermore we characterize those diametral path graphs that are trees and chordal graphs, respectively. Finally in Section 6 we show that any diametral path graph of diameter at least five has a minimum connected dominating set that induces a path.

2 Preliminaries

We present some specific definitions, concepts and properties related to diametral path graphs. For standard graph theory notations not given here we refer the reader to [5].

We denote by $G = (V, E)$ a finite, undirected and simple graph, n denotes the number of vertices and m denotes the number of edges. $G[W]$ denotes the subgraph of the graph $G = (V, E)$ induced by the vertex set $W \subseteq V$. For any $x \in V$ let $N(x) = \{y : \{x, y\} \in E\}$ and $N[x] = N(x) \cup \{x\}$. Furthermore, $N[A] = \bigcup_{a \in A} N[a]$ for any $A \subseteq V$.

A set of vertices $D \subseteq V$ is called a *dominating set* of the graph G if every vertex in $V \setminus D$ has a neighbour in D. A dominating set D of G is called a *connected dominating set* if $G[D]$ is connected. The *connected domination number* of a connected graph $G = (V, E)$ is defined as the cardinality of a *minimum connected dominating set* (MCDS), and is denoted by $\gamma_{conn}(G)$. A vertex $y \in V$ is a *private neighbour* of the vertex $x \in D$ with respect to $D \subseteq V$ if x is the only neighbour of y in D, i.e., $N[y] \cap D = \{x\}$. Accordingly, we define $Pr(x, D) := N(x) \setminus N[D \setminus \{x\}]$.

A simple path $P = (u = x_0, x_1, \ldots, x_k = v)$ of G is an *induced (or chordless) path* if the induced subgraph $G[V(P)]$ is a path itself. The *distance* between two vertices u and v in G, denoted by $d_G(u,v)$, is the shortest length of a path between u and v. The *diameter* of a graph G is defined as $diam(G) = \max\{d_G(u,v) : u, v \in V\}$.

An induced path P of a graph G is a *dominating (induced) path* if $V(P)$ is a dominating set of G. An x,y-path P of a graph G is a *dominating shortest path* of G if P is a dominating path and the length of P is $d_G(x,y)$. A pair of vertices $u, v \in V$ with $d_G(u,v) = diam(G)$ is called a *diametral pair* and a shortest path between the vertices of a diametral pair is called a *diametral path*. In other words a diametral path in a graph is a shortest path between two vertices that are at maximum distance in G. The following lemma gives a simple and useful observation.

Lemma 1. *Let* $P = (u = x_0, x_1, \ldots, x_k = v)$, $k \geq 0$, *be a shortest* x,y-*path in a graph* $G = (V, E)$. *Then for any vertex* $z \in V \setminus V(P)$ *the neighbourhood of* z *in the path* P, *i.e., the set* $N(z) \cap V(P)$, *is exactly one of the following sets:* $\emptyset$, $\{x_i\}$ *with* $0 \leq i \leq k$, $\{x_i, x_{i+1}\}$ *with* $0 \leq i \leq k - 1$, $\{x_i, x_{i+2}\}$ *with* $0 \leq i \leq k - 2$, $\{x_i, x_{i+1}, x_{i+2}\}$ *with* $0 \leq i \leq k - 2$.

Definition 2. A path $P = (x = x_0, x_1, \ldots, x_k = y)$ of a graph $G = (V, E)$ is a *dominating diametral path* (DDP) if P is a diametral path and the vertex set of P is a dominating set of G.

Clearly any complete graph has a DDP. However, there are graphs G with $diam(G) = 2$ that do not have a DDP, as e.g. the Petersen graph.

Obviously every DDP of a graph G is also a dominating shortest path of G. The following sufficient condition for the existence of a DDP shows that in a weaker sense even the reverse holds.

Proposition 3. *If a graph* $G = (V, E)$ *has a dominating shortest path then it also has a dominating diametral path.*

Proof. Let $P = (x = x_0, x_1, \ldots, x_k = y)$ be a dominating shortest path of G, hence $k = d_G(x, y)$. If $d_G(x, y) = diam(G)$ then P is a DDP of G. Suppose $d_G(x, y) < diam(G)$. Let (u, v) be a diametral pair of G. P is a dominating path, thus both u and v have neighbours in $V(P)$. Thus there is a u, v-path of the form $u, x_i, x_{i+1}, \ldots, x_j, v$. Such a path must have length at least $diam(G)$, hence $k \geq diam(G) - 2$. Moreover, w.l.o.g. we may choose the diametral pair (u, v) such that either (a) $u = x$ and $v \in Pr(y, V(P))$, or (b) $u \in Pr(x, V(P))$ and $v \in Pr(y, V(P))$. (All other cases are similar.) In both cases there is a DDP P' of G that extends P. If $diam(G) = k + 1$ we could choose $P' = (x = x_0, x_1, \ldots, x_k = y, v)$ in both cases. If $diam(G) = k + 2$ then only case (b) is possible and we choose $P' = (u, x = x_0, x_1, \ldots, x_k = y, v)$. P is a dominating shortest path, hence P' is a DDP by construction.

Definition 4. A graph $G = (V, E)$ is called a *diametral path graph* if every connected induced subgraph H of G has a dominating diametral path.

Proposition 3 implies immediately the following characterization of diametral path graphs.

Theorem 5. *A graph G is a diametral path graph if and only if every connected induced subgraph H of G has a dominating shortest path.*

The class of diametral path graphs includes several well-known classes of graphs. Figure 1 shows the relationship between diametral path graphs and some classes of perfect graphs.

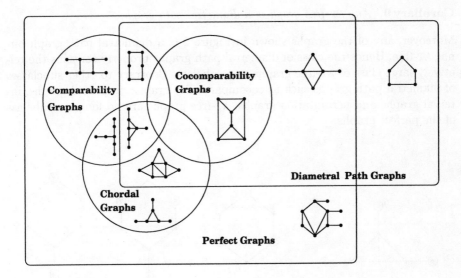

Fig. 1. Diametral path graphs and classes of perfect graphs.

For more informations on structural properties of classes of perfect graphs we refer to [6, 12].

3 AT-Free Graphs and Dominating Pair Graphs

Diametral path graphs not only include several classes of perfect graphs. They also include some other important graph classes, as e.g. the class of asteroidal triple-free graphs which has been investigated extensively by Corneil, Olariu and Stewart [7, 8].

Definition 6. A set of three independent vertices x, y, z of a graph G is called an *asteroidal triple* (AT) if for any two of these vertices there exists a path joining them that avoids the (closed) neighborhood of the third. A graph G is called an *asteroidal triple-free* (AT-free) graph if G does not contain an asteroidal triple.

Notice that the class of AT-free graphs is an *hereditary* graph class. One of the major structural properties of AT-free graphs is given in [7].

Theorem 7. *Any connected* AT-*free graph contains a* dominating pair, *i.e., a pair* (x, y) *of vertices of* G *such that* $V(P)$ *is dominating for every path* P *between* x *and* y.

This has been strengthened in [8].

Theorem 8. *Any connected* AT–*free graph has a dominating pair that is also a diametral pair.*

Corollary 9. *Any* AT-*free graph is a diametral path graph.*

Moreover, any of the graphs shown in Figure 2 is a diametral path graph but not AT-free. Hence the class of diametral path graphs properly contains the relatively large class of AT-free graphs as a subclass. Contrary to other subclasses of diametral path graphs such as cocomparability graphs, trapezoid graphs, interval graphs and permutation graphs, AT-free graphs are no longer a subclass of the perfect graphs.

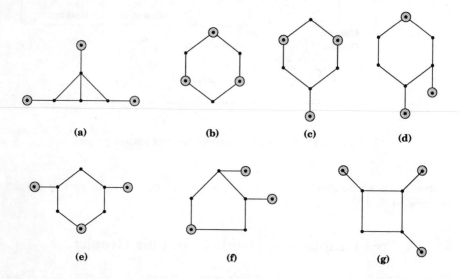

(a) (b) (c) (d)

(e) (f) (g)

Fig. 2. Diametral path graphs having an asteroidal triple.

We introduce another graph class that is of interest in relation to AT-free graphs and diametral path graphs.

Definition 10. A graph $G = (V, E)$ is called a *dominating pair graph* if every induced connected subgraph H of G has a dominating pair.

By Theorem 7, any AT-free graph is a dominating pair graph. We explore the relationship between dominating pair graphs and diametral path graphs.

Lemma 11. *Any connected graph having a dominating pair also has a* DDP.

Proof. Let (x, y) be a dominating pair of G and let P be a shortest x, y-path in G. Then P is a dominating shortest path of G, hence G has a DDP by Proposition 3.

Corollary 12. *Any dominating pair graph is a diametral path graph.*

Notice that the diametral path graphs properly contain the dominating pair graphs. An example is the graph shown in Fig. 2 (g). Furthermore, the dominating pair graphs properly contain the AT-free graphs. An example is the graph C_6 shown in Fig. 2 (b).

4   Finding a Dominating Diametral Path

We present an $O(n^3 m)$ algorithm for checking whether a given graph $G = (V, E)$ has a dominating diametral path. This algorithm, called $\mathtt{ddp}(G)$, computes for every $x \in V$ the BREADTH-FIRST-SEARCH-tree of G with start vertex x and checkes whether the BFS-tree 'contains' a DDP of G with end vertex x. If this check fails for all $x \in V$ then G has no DDP.

The main idea of checking for a DDP with a fixed end vertex x is to proceed through the levels $H_0 = \{x\}, H_1 = N(x), \ldots, H_i = \{w \in V : d_G(x, w) = i\}, \ldots, H_k = \{w \in V : d_G(x, w) = k\}$ of the BFS-tree of x by Dynamic Programming. This relies on the fact that $\{u, v\} \in E$, $u \in H_i$ and $v \in H_j$ implies $|i - j| \leq 1$. A state (u, v) of the Dynamic Programming corresponds to an edge $\{u, v\} \in E$ with $u \in H_{i-1}$, $v \in H_i$ for some $i \in \{1, 2, \ldots, k\}$. The aim of the algorithm is to compute queues A_i, $i \in \{1, 2, \ldots, k\}$, such that A_i contains exactly the succesful states (u, v) with $u \in H_{i-1}$, $v \in H_i$. Thereby a state (u, v) is *succesful* if there is a path $P = (x, \ldots, u, v)$ in G such that P contains exactly one vertex from each of the levels $H_0, H_1, \ldots, H_i$ and $V(P)$ dominates $\bigcup_{j=0}^{i-1} H_j$. Notice that due to the structure of the BFS-tree only paths P with these properties can possibly be extended to a dominating diametral path with end vertex x.

Algorithm.
```
algorithm ddp(G)

Input:   A graph G = (V, E).
Output:  'YES' if G has a DDP and 'NO' otherwise.
FOR every x ∈ V DO
  BEGIN
  Label every edge of G as 'free';
  Compute the BFS-tree with start vertex x:
```

$H_0 = \{x\}, H_1 = N(x), \ldots, H_k = \{w \in V : d_G(x,w) = k\}$;
IF $k = diam(G)$ **THEN**
  BEGIN
  $i := 1$;
  Initialize the queue A_1 that contains the pairs (x,z) for
  all $z \in N(x)$;
  WHILE $A_i \neq \emptyset$ **AND** $i < k$ **DO**
    BEGIN
    i:=i+1;
    FOR all states (u,v) in the queue A_{i-1} **DO**
      BEGIN
      $D_{(u,v)} := H_{i-1} \setminus (N[u] \cup N[v])$;
      [* Comm.: $D_{(u,v)}$ denotes the set of vertices of H_{i-1} that are
      not dominated by any path P corresponding to state (u,v). *]
      $C_{(u,v)} := \{w \in H_i : w$ dominates all vertices in $D_{(u,v)}\}$;
      FOR every $w \in C_{(u,v)} \cap N(v)$ with $\{v,w\}$ labelled 'free'
      insert (v,w) in the queue A_i and label $\{v,w\}$ 'used';
      END;
    END;
  END;
  IF there is a state (u,v) in the queue A_k with
  $H_k \subseteq (N[u] \cup N[v])$ **THEN** output 'YES' and **STOP**.
  END;
END;
output 'NO' and **STOP**.

Theorem 13. *Algorithm* ddp(G) *decides in time* $O(n^3 m)$ *whether a given graph G has a* DDP.

Proof. Since for any BFS-tree of the input graph G adjacent vertices appear either in the same level or in consecutive levels, any diametral path P of G with end vertex x must contain exactly one vertex from each level of the BFS-tree with start vertex x. (In this proof all paths are supposed to be of this type.) Moreover, x is in a diametral pair if and only if $k = diam(G)$.

For concluding the correctness proof we claim that after termination of the algorithm any queue A_i, $i \in \{1, 2, \ldots, k\}$, contains exactly the succesful states (u,v) with $u \in H_{i-1}$ and $v \in H_i$. We are going to show this by induction on i. Notice that A_1 contains all states (x,z) with $x \in H_0$ and $z \in H_1$ and that all of them are succesful. Assume that the queue A_i, $i \in \{1, 2, \ldots, k-1\}$, contains exactly the states (u,v) with $u \in H_{i-1}$ and $v \in H_i$ that are succesful, i.e., those for which a path $Q = (x = a_0, a_1, \ldots, a_{i-1} = u, a_i = v)$ with $\bigcup_{j=0}^{i-1} H_j \subseteq N[V(Q)]$ exists.

Let (v,w) be a succesful state with $v \in H_i$ and $w \in H_{i+1}$. Thus there is a path $P' = (x = a_0, a_1, \ldots, a_{i-1}, a_i = v, a_{i+1} = w)$ with $\bigcup_{j=0}^{i} H_j \subseteq N[V(P')]$. Then $P = (x = a_0, a_1, \ldots, a_{i-1}, a_i = v)$ is a path with $\bigcup_{j=0}^{i-1} H_j \subseteq N[V(P)]$

since $N(w) \cap \bigcup_{j=0}^{i-1} H_j = \emptyset$. Hence (a_{i-1}, v) is a succesful state with $a_{i-1} \in H_{i-1}$ and $v \in H_i$. Hence by the assumption of the induction (a_{i-1}, v) is a state in A_i. Since $\{v, w\} \in E$ and w dominates all vertices of H_i that are not dominated by $V(P)$, i.e., $D_{a_{i-1},v} \subseteq N[w]$, the algorithm will either store (v, w) in A_{i+1} when processing the state (a_{i-1}, v) or the edge $\{a_{i-1}, v\}$ has already been labelled 'used' before processing (a_{i-1}, v). In both cases (v, w) is in A_{i+1}.

On the other hand, let (v, w) be a state in A_{i+1}. Suppose (v, w) has been stored in A_{i+1} when processing a state (u, v) of A_i. Hence $\{v, w\} \in E$ and $D_{u,v} \subseteq N[w]$, i.e., w dominates all vertices of H_i that are adjacent neither to u nor to v. By the assumption of the induction the state (u, v) is succesful, hence there is a path $Q = (x = a_0, a_1, \ldots, a_{i-1} = u, a_i = v)$ with $\bigcup_{j=0}^{i-1} H_j \subseteq N[V(Q)]$. Then $Q' = (x = a_0, a_1, \ldots, a_{i-1} = u, a_i = v, a_{i+1} = w)$ is a path with $\bigcup_{j=0}^{i} H_j \subseteq N[V(Q')]$. Consequently (v, w) is a succesful state. This proves the claim.

Therefore a state (u, v) is in A_k if and only if there is a diametral path $R = (x = a_0, a_1, \ldots, a_{k-1} = u, a_k = v)$ with $\bigcup_{j=0}^{k-1} H_j \subseteq N[V(R)]$. Thus (u, v) in A_k corresponds to a DDP of G if and only if $H_k \subseteq N[u] \cup N[v]$.

Now let us consider the running time of the algorithm during the performance of the outer **for** loop for a fixed vertex $x \in V$. The BFS-tree with start vertex x can be computed in time $O(n + m)$ (see e.g. [12]). There are at most m states that are contained in any of the queues since each state corresponds to an edge of the graph. Moreover each state occurs in at most one queue. Since the execution of the **for** loop for a fixed state (u, v) requires $O(n^2)$ time we obtain that the time for performing the outer **for** loop for a vertex $x \in V$ is $O(n^2 m)$. Consequently the overall running time of the algorithm is $O(n^3 m)$.

Remark. The algorithm ddp(G) can be modified such that it outputs a dominating diametral path P of the input graph G, if G has a DDP. Simply add a pointer from each state (v, w) inserted in queue A_i, $i \in \{2, 3, \ldots, k\}$, to the state (u, v) of queue A_{i-1} that has been processed when (v, w) got inserted. For a state (a_{k-1}, a_k) with $N(a_{k-1}) \cup N(a_k) \supseteq H_k$ follow the pointers to find a DDP P.

5 Diametral Path Graphs and Classes of Perfect Graphs

It is a matter of routine to verify that all graphs shown in Figure 3 are *minimal forbidden subgraphs* for the class of diametral path graphs, i.e., each of the graphs in Figure 3 is not a diametral path graph but any of its proper induced subgraphs is a diametral path graph. Hence, any minimal forbidden graph can not be contained in a diametral path graph as induced subgraph. The list helps in understanding the structure of diametral path graphs, although the list is not complete.

Theorem 14. *A tree is a diametral path graph if and only if it does not contain the graph AT-1, shown in Figure 3(a), as induced subgraph.*

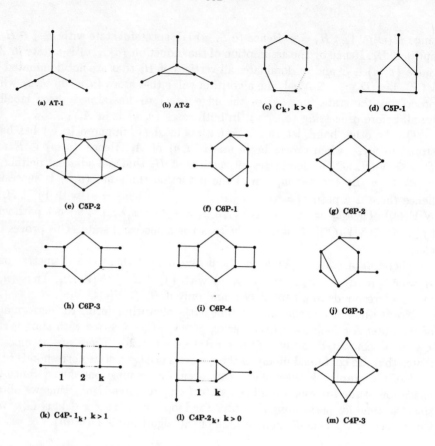

Fig. 3. Minimal forbidden subgraphs for the diametral path graphs.

Proof. A graph G that is a tree and a diametral path graph does clearly not contain the graph AT-1. On the other hand, a graph G that is a tree and does not contain the graph AT-1 is a caterpillar and every induced connected subgraph of G is also a caterpillar. Since a caterpillar is a tree consisting of a (induced) path, the body, and vertices of degree one only adjacent to one vertex of the body, any caterpillar has a DDP. Hence G is a diametral path graph.

The next theorem gives a characterization of chordal diametral path graphs.

Theorem 15. *A chordal graph G is a diametral path graph if and only if it does not contain the graphs* AT-1 *and* AT-2, *shown in Figure 3(a) and (b), as induced subgraph.*

Proof. Clearly a chordal graph that is also a diametral path graph can not contain the graphs AT-1 and AT-2 as induced subgraphs.

Suppose the other direction of the theorem would not be true and let G be a counterexample with smallest number of vertices, thus G is a chordal graph that does not contain the graphs AT-1 and AT-2 as induced subgraphs but G is not a diametral path graph, while any proper induced subgraph of G is a diametral path graph. Hence, G has no DDP. We distinguish two cases.

Case 1: $diam(G) \leq 3$.

Since $diam(G) \leq 3$, the chordal graph G has a dominating clique D (see [13]). Let C be a maximal clique containing D. Since G is chordal and does not contain the graph AT-2, there cannot be three different vertices $x, y, z \in C$ such that $Pr(x, C) \neq \emptyset$, $Pr(y, C) \neq \emptyset$ and $Pr(z, C) \neq \emptyset$. Consequently, any minimal dominating clique of G has size at most two. Hence there is a dominating clique $C' \subseteq C$ of G with $|C'| \leq 2$.

Hence G either has a dominating vertex u or a dominating clique $\{v, w\}$. Thus G has a dominating shortest path, namely either $(x = u = y)$ or $(x = v, w = y)$. Consequently by Proposition 3, G has a DDP, a contradiction.

Case 2: $diam(G) \geq 4$.

Since G is a minimal counterexample, no diametral path of G is dominating. We choose a diametral path $P = (x = x_0, x_1, \ldots, x_k = y)$ of G such that $|N[V(P)]|$ is maximum. Notice that $N[V(P)] \neq V$ since P is not a dominating path of G.

Let z be an arbitrary vertex that has no neighbour in $V(P)$ and that fulfils $d_G(z, V(P)) = \min\{d_G(w, z) : w \in V(P)\} = 2$. Let b be any neighbour of z that also belongs to $N[V(P)]$. By Lemma 1 and since G is chordal the set $N(b) \cap N[V(P)]$ is equal to $\{x_i\}$ with $0 \leq i \leq k$, $\{x_i, x_{i+1}\}$ with $0 \leq i \leq k-1$ or $\{x_i, x_{i+1}, x_{i+2}\}$ with $0 \leq i \leq k-2$. The following considerations will show that any shortest path from z to $V(P)$ meets the path P in a vertex x_i with $i \in \{0, 1, k-1, k\}$.

Case 2.1: $N(b) \cap N[V(P)] = \{x_i\}$ with $0 \leq i \leq k$.

The vertices $x_{i-2}, x_{i-1}, x_i, x_{i+1}, x_{i+2}, b, z$ induce the graph AT-1 if $2 \leq i \leq k-2$. Since AT-1 is not contained in G, we get $i \in \{0, 1, k-1, k\}$.

Case 2.2: $N(b) \cap N[V(P)] = \{x_i, x_{i+1}\}$ with $0 \leq i \leq k-1$.

The vertices $x_{i-1}, x_i, x_{i+1}, x_{i+2}, b, z$ induce the graph AT-2 if $1 \leq i \leq k-1$. Since AT-2 is not contained in G, we get $i \in \{0, k\}$.

Case 2.3: $N(b) \cap N[V(P)] = \{x_i, x_{i+1}, x_{i+2}\}$ with $0 \leq i \leq k-2$.

By the choice of P there must be a vertex $w \in Pr(x_{i+1}, V(P) \cup \{b\})$ since otherwise $|N[V(P')]| > |N[V(P)]|$ for the path $P' = (x = x_0, x_1, \ldots, x_i, b, x_{i+2}, \ldots, x_k = y)$. Since G is chordal we have $\{w, z\} \notin E$. Then the vertices $x_{i-1}, x_i, x_{i+1}, w, b, z$ induce the graph AT-2 if $i > 0$ and the vertices $x_{i+1}, x_{i+2}, x_{i+3}, w, b, z$ induce the graph AT-2 if $i < k$. Consequently this case is impossible for G since $k \geq 4$.

Hence we have shown that any shortest path from z to $V(P)$ meets P in a vertex x_i with $i \in \{0, 1, k-1, k\}$. W.l.o.g. we may assume that there is a shortest path $Q = (z, b, x_i)$ with $i \in \{0, 1\}$ where x_{i+1} is not adjacent to b. Since $d_G(y, z) \leq d_G(x, y)$ there must be another shortest path $Q' = (z = z_0, z_1, \ldots, z_{t-1} = b', z_t = x_j)$, $t \geq 2$, from z to $V(P)$ meeting P in a vertex x_j with $j \geq 2$. Hence $j \in \{k-1, k\}$ and we may assume that x_{j-1} is not adjacent to b'. Notice that $b \notin V(Q')$ since $N(b) \cap N[V(P)] \subseteq \{x_0, x_1\}$ and $k \geq 4$. Hence $C = (z, b, x_i, x_{i+1}, \ldots, x_j = z_t, z_{t-1}, \ldots, z_0 = z)$ is a (simple) cycle in the chordal graph G. Consider the edge $\{b, x_i\}$ of the cycle C. By the choice of x_i the only vertices of C that could be adjacent to x_i are vertices of P and the vertices b, b'. However $j \geq k - 1$ and $k \geq 4$ imply $\{b', x_i\} \notin E$. Finally the only vertex of $V(P) \cap V(C)$ adjacent to b is x_i. Hence, b and x_i have no common neighbour in the cycle C. This implies that there is a chordless cycle of length at least four in $G[C]$, contradicting the assumption that G is chordal.

6 Minimum Connected Dominating Sets

We are going to investigate a structural property which is already known to be true for some well-known subclasses of diametral path graphs, namely that each connected graph G in the class has a PATH-MCDS, i.e., a minimum connected dominating set that induces a path in G. This is known to be true for permutation graphs [10], for cocomparability graphs [14] and it has been shown recently for AT-free graphs [8]. We show that every connected diametral path graph G of diameter at least five has a PATH-MCDS. We start with two simple observations.

Lemma 16. *Let G be a connected graph. Then $\gamma_{conn}(G) \geq diam(G) - 1$.*

Proof. Let C be an MCDS of G. Then $d_G(a, b) \leq diam(G[C]) + 2$ for all $a, b \in V$. Consequently, $diam(G) \leq diam(G[C]) + 2 \leq |C| + 1 = \gamma_{conn}(G) + 1$.

Lemma 17. *Let $P = (x = x_0, x_1, \ldots, x_k = y)$ be a DDP of a graph G. Then the following implications hold:*

1. *$\gamma_{conn}(G) \neq k$ implies that G has a PATH-MCDS.*
2. *If $\gamma_{conn}(G) = k$ and D is an MCDS of G such that $G[D]$ is not a path then $x, y \notin D$, and $d_{G[D]}(a, b) = k - 2$ for all $a \in N(x) \cap D$ and $b \in N(y) \cap D$.*

Proof. Let $P = (x = x_0, x_1, \ldots, x_k = y)$ be a DDP of G. Hence $k = diam(G) = d_G(x, y)$ and $\gamma_{conn}(G) \leq k + 1$. Let D be an MCDS of G. Then $k - 1 \leq |D| = \gamma_{conn}(G) \leq k + 1$, by Lemma 16. Let $a \in D \cap N[x]$ and $b \in D \cap N[y]$.

If $\gamma_{conn}(G) = k + 1$ then P is a PATH-MCDS. If $\gamma_{conn}(G) = |D| = k - 1$ then D is a PATH-MCDS, since otherwise $d_{G[D]}(a, b) < k - 2$ implies $d_G(x, y) < k$, a contradiction.

Finally, suppose $\gamma_{conn}(G) = |D| = k$ and $G[D]$ is not a path. Let $a \in N[x] \cap D$ and $b \in N[y] \cap D$. Clearly $d_{G[D]}(a, b) \leq k - 2$. As above $d_{G[D]}(a, b) < k - 2$ would lead to a contradiction. Hence, $d_{G[D]}(a, b) = k - 2$ and $x, y \notin D$.

Theorem 18. *Any diametral path graph G with $diam(G) > 4$ has a PATH-MCDS.*

Proof. Let $P = (x = x_0, x_1, \ldots, x_k = y)$ be a DDP of G and D an MCDS of G. Suppose G does not have a PATH-MCDS. Then $G[D]$ is not a path and by Lemma 17 we may assume $|D| = \gamma_{conn}(G) = k$, $x, y \notin D$ and $d_{G[D]}(a, b) = k-2$ for all $a \in N(x) \cap D$ and $b \in N(y) \cap D$.

Among all shortest a, b-paths of $G[D]$ between a vertex $a \in N(x) \cap D$ and a vertex $b \in N(y) \cap D$ we choose a path $A = (a = a_1, a_2, \ldots, a_{k-1} = b)$ such that $|N[V(A) \cup \{x, y\}]|$ is maximum. Then $d_G(x, y) = k$ implies that $\{x_i, a_j\} \in E$ requires $j \in \{i-2, i-1, i\}$ and that $x_i = a_j$ requires $i = j$. Consequently, $A' = (x = a_0, a = a_1, a_2, \ldots, a_{k-1} = b, a_k = y)$ is a shortest x, y-path of G. Let u be the only vertex of D not belonging to the path A. Since $G[D]$ is connected u has at least one neighbour in $V(A)$ and since $G[D]$ is not a path it is not possible that u has exactly one neighbour in A that is either a or b.

Case 1: *A' is a dominating diametral path of G.*

If any of the two subpaths $x = a_0, a = a_1, a_2, \ldots, a_{k-1} = b$ and $a = a_1, a_2, \ldots, a_{k-1} = b, a_k = y$ of A' is dominating then it is a PATH-MCDS of G. Hence we assume that this is not the case. Therefore, there is a vertex $p_x \in Pr(x, V(A'))$ and a vertex $p_y \in Pr(y, V(A'))$. Clearly both are neighbours of u. Hence x, p_x, u, p_y, y is a path in G which is a contradiction to our assumption $d_G(x, y) = diam(G) > 4$.

Case 2: *A' is not a dominating diametral path of G.*

Consequently, there is a vertex $w \in Pr(u, D \cup \{x, y\})$. We consider the neighbourhood of u in A'. By Lemma 1 the set $N(u) \cap V(A')$ is equal to $\{a_i\}$ with $0 \le i \le k$, $\{a_i, a_{i+1}\}$ with $0 \le i \le k - 1$, $\{a_i, a_{i+2}\}$ with $0 \le i \le k - 2$, or $\{a_i, a_{i+1}, a_{i+2}\}$ with $0 \le i \le k - 2$. The following case analysis is somewhat similar to that in the proof of Theorem 15.

Case 2.1: $N(u) \cap V(A') = \{a_i\}$ *with* $0 \le i \le k$.

Then $a_{i-2}, a_{i-1}, a_i, a_{i+1}, a_{i+2}, u, w$ induce an AT-1 in G if $2 \le i \le k-2$. Furthermore $i \notin \{0, k\}$ since u must have a neighbour in $V(A)$ and $i \notin \{1, k-1\}$ since otherwise $G[D]$ would be a path.

Case 2.2: $N(u) \cap V(A') = \{a_i, a_{i+1}\}$ *with* $0 \le i \le k - 1$.

Then $a_{i-1}, a_i, a_{i+1}, a_{i+2}, u, w$ induce an AT-2 in G if $1 \le i \le k - 2$. Hence, $i \in \{0, k-1\}$ since G does not contain AT-2. Furthermore, $i \in \{0, k-1\}$ would imply that u has exactly one neighbour in $V(A) = D \setminus \{u\}$, namely either a or b, implying that $G[D]$ is a path, a contradiction.

Case 2.3: $N(u) \cap V(A') = \{a_i, a_{i+2}\}$ *with* $0 \le i \le k - 2$.

By the choice of the path A the set $Pr(a_{i+1}, V(A') \cup \{u\})$ is nonempty since otherwise $|N[V(Q')]| > |N[V(A')]|$ for the path $Q' = (x = a_0, a = a_1, \ldots, a_i, u, a_{i+2}, \ldots, a_{k-1} = b, a_k = y)$. Suppose there would be a vertex $z \in Pr(a_{i+1}, V(A') \cup \{u\})$ that is not adjacent to w. Then $a_{i-1}, a_i, a_{i+1}, a_{i+2}, a_{i+3}, u, w, z$ induce an C4P-1_2 in G if $1 \le i \le k - 3$. Moreover, if $2 \le i \le k-2$ then $a_{i-2}, a_{i-1}, a_i, a_{i+1}, u, w, z$ induce an AT-1 and if $0 \le i \le k-4$ then $a_i, a_{i+1}, a_{i+2}, a_{i+3}, a_{i+4}, u, w, z$ induce an AT-1. Since $k \ge 5$ and since G contains neither C4P-1_2 nor AT-1 any vertex $z \in Pr(a_{i+1}, V(A') \cup \{u\})$ is adjacent to w. Furthermore, $\{w, z\} \in E$

and $z \in Pr(a_{i+1}, V(A') \cup \{u\})$ imply that $a_{i-1}, a_i, a_{i+1}, a_{i+2}, a_{i+3}, w, z$ induce an AT-1 if $1 \leq i \leq k - 3$. Consequently, we have $i \in \{0, k - 2\}$ such that $N(u) \cap V(A') = \{x, a_1\}$ or $N(u) \cap V(A') = \{a_{k-2}, y\}$.

Case 2.4: $N(u) \cap V(A') = \{a_i, a_{i+1}, a_{i+2}\}$ with $0 \leq i \leq k - 2$.

By the choice of the path A the set $Pr(a_{i+1}, V(A') \cup \{u\})$ is nonempty since otherwise $|N[V(Q')]| > |N[V(A')]|$ for the path $Q' = (x = a_0, a = a_1, \ldots, a_i, u, a_{i+2}, \ldots, a_{k-1} = b, a_k = y)$. Suppose there would be a vertex $z \in Pr(a_{i+1}, V(A') \cup \{u\})$ that is not adjacent to w. Then $a_{i-1}, a_i, a_{i+1}, u, w, z$ induce an AT-2 in G if $1 \leq i \leq k-2$ and $a_{i+1}, a_{i+2}, a_{i+3}, u, w, z$ induce an AT-2 in G if $0 \leq i \leq k - 3$. Since $k \geq 5$ and since G does not contain AT-2 any vertex $z \in Pr(a_{i+1}, V(A') \cup \{u\})$ is adjacent to w. Furthermore, $\{w, z\} \in E$ and $z \in Pr(a_{i+1}, V(A') \cup \{u\})$ imply that $a_{i-1}, a_i, a_{i+1}, a_{i+2}, a_{i+3}, w, z$ induce an AT-1 if $1 \leq i \leq k - 3$. Consequently, we have $i \in \{0, k-2\}$ such that $N(u) \cap V(A') = \{x, a, a_1\}$ or $N(u) \cap V(A') = \{a_{k-2}, b, y\}$.

By symmetry the final case, we have to analyze, is as follows. Every vertex $z \in Pr(a, V(A') \cup \{u\})$ is adjacent to the vertex $w \in Pr(u, D \cup \{x, y\})$ and either $N(u) \cap V(A') = \{x, a_2\}$ or $N(u) \cap V(A') = \{x, a, a_2\}$. We consider an arbitrary vertex $v \in Pr(a, D)$. v can not be adjacent to y since the path x, a, v, y would imply $d(x, y) \leq 3$. If $v \in Pr(a, D \cup \{y\})$ and v adjacent to x then v must be adjacent to w since otherwise x, a, a_2, a_3, u, w, z induce an AT-2 in G. Thus $v \in Pr(a, V(A') \cup \{u\}) \subseteq N[w]$. Consequently, G has a PATH-MCDS, namely the set $(D \cup \{w\}) \setminus \{a\}$.

There is a linear time algorithm computing a dominating path of a given AT-free graph G. Moreover, the dominating path computed by this algorithm contains at most two vertices more than an MCDS of G [9]. By Lemma 16 the algorithm ddp(G) of Section 4 can be used to obtain a similar approximation algorithm on diametral path graphs.

Proposition 19. *If the $O(n^3 m)$ algorithm* ddp(G) *outputs a dominating diametral path P of the input graph G then $V(P)$ is a connected dominating set of G with $|V(P)| \leq \gamma_{conn}(G) + 2$.*

7 Open Problems

There are many open questions concerning diametral path graphs. In our opinion finding a polynomial time recognition algorithm as well as finding a complete list of minimal forbidden subgraphs are the most interesting open problems. Moreover, we conjecture that any connected diametral path graph has a PATH-MCDS.

References

1. K.S. Bagga, L.W. Beineke, W.D. Goddard, M.J. Lipman, and R.E. Pippert, A survey of integrity, to appear in *Discrete Applied Mathematics*.

2. C.A. Barefoot, R. Entringer, and H. Swart, Vulnerability in graphs – A comparative survey, *J. Combin. Math. Combin. Comput.* **1** (1987), 12–22.

3. J.C. Bermond, and B. Bollobás, The diameter of graphs – a survey, *Congressus Numerantium* **81** (1981), 3–27.

4. J.C. Bermond, J. Bond, M. Paoli, and C. Peyrat, Graphs and interconnection networks: diameter and vulnerability, *Surveys in Combinatorics*, London Math. Soc. Lecture Notes **82** (1983), 1–30.

5. J.A. Bondy, U.S.R. Murty, *Graph theory with applications*, Macmilliam Press Ltd., 1976.

6. A. Brandstädt, Special graph classes – a survey, Schriftenreihe des Fachbereichs Mathematik, SM-DU-199, Universität Duisburg, 1991.

7. D.G. Corneil, S. Olariu, and L. Stewart, Asteroidal triple-free graphs, *Proceedings of the 19th International Workshop on Graph-Theoretic Concepts in Computer Science WG'93*, Lecture Notes in Computer Science, Vol. 790, Springer-Verlag, 1994, pp. 211–224.

8. D.G. Corneil, S. Olariu, and L. Stewart, Asteroidal triple-free graphs, Manuscript, 1994, (revised version of [7]).

9. D.G. Corneil, S. Olariu, and L. Stewart, A linear time algorithm to compute a dominating path in an AT-free graph, *Information Processing Letters*, **54** (1995), 253–258.

10. C.J. Colbourn, and L.K. Stewart, Permutation graphs : connected domination and steiner trees, *Discrete Mathematics* **86** (1991), 179–190.

11. J.S. Deogun, Diametral path graphs, presented at the *Second International Conference On Graph Theory and Algorithms*, San Francisco, August 1989.

12. M.C. Golumbic, *Algorithmic graph theory and perfect graphs*, Academic Press, New York, 1980.

13. D. Kratsch, P. Damaschke, and A. Lubiw, Dominating cliques in chordal graphs, *Discrete Mathematics* **128** (1994), 269–276.

14. D. Kratsch, L. Stewart, Domination on cocomparability graphs, *SIAM Journal on Discrete Mathematics* **6** (1993), 400–417.

15. R. Ogier, and N. Shacham, A distributed algorithm for finding shortest pairs of disjoint paths, *IEEE INFOCOM'89* **1** (1989), 173-182.

16. O. Ore, Diameters in graphs, *J. Combin. Theory* **5** (1968), 75-81.

Chordal graphs and their clique graphs

Philippe Galinier, Michel Habib and Christophe Paul

LIRMM
UMR 9928 UNIVERSITE MONTPELLIER II/CNRS
161, Rue Ada
34392 Montpellier cedex 5 France
email: paul@lirmm.fr

Abstract. In the first part of this paper, a new structure for chordal graph is introduced, namely the clique graph. This structure is shown to be optimal with regard to the set of clique trees. The greedy aspect of the recognition algorithms of chordal graphs is studied. A new greedy algorithm that generalizes both Maximal cardinality Search (MCS) and Lexicographic Breadth first search is presented. The trace of an execution of MCS is defined and used in two linear time and space algorithms: one builds a clique tree of a chordal graph and the other is a simple recognition procedure of chordal graphs.

Introduction

Since chordal graphs have no chordless cycle of length more than 3, they can be considered as a generalization of trees. In [9], chordal graphs have been considered as the intersection graphs of subtrees of a tree. Chordal graphs are often represented by a clique tree (see [9, 19]). This is a structure which translates most of the information contained in a chordal graph. The structure of clique tree does not only appeared in the graph theory litterature, but in the context of the acyclic database schemes [6, 1] and in the context of sparse matrix computations too [3, 12, 14]. This is probably why some results have been rediscovered independently several times. Some results of this paper are contained in [3], but we present here a unifying graph theoretical point of view.

Chordal graphs can also be characterized using perfect elimination orderings (PEO) [16]. A vertex is simplicial if and only if its neighbourhood is a complete subgraph. An elimination ordering $x_1, x_2, ..., x_n$ is perfect if and only if each x_i is simplicial in the subgraph induced by $x_i, ..., x_n$.

Several greedy recognition algorithms of chordal graphs are known. The most famous are the Lexicographic Breadth First Search (BFS for short) [17] and the Maximum Cardinality Search (MCS for short) [20]. Chordal graphs can be represented with clique-trees (see for example [19]). The linear recognition of chordal graphs involves two distinct phases: the execution of MCS or BFS in order to compute an elimination ordering and a checking procedure to decide whether this elimination ordering is perfect (PEO).

In the first part of this paper, a new structure namely the clique graph is introduced. Some graph properties of this structure are studied with regard to clique trees. And the clique graph is justified as being the optimal structure containing all clique trees of a chordal graph. In the second section, an explanation of the greedy aspect of the recognition algorithms, MCS and BFS, is given. The strong properties of the clique graph allow a simple algorithm which computes maximum weighted spanning tree. It is shown that such an algorithm generalizes both MCS or BFS. In fact, in [3] MCS has yet been compared to Prim's algortihm [15]. This correspondence gives a better knowledge of the greediness of chordal graph recognitions and implies new proof for these algorithms. In the last part of this paper, we present two linear algorithms: one for building a clique-tree, and the other for the recognition of PEO. We define the trace of an elimination ordering as the mark level of vertices when they are numbered by MCS. Both algorithms are based on the study of the increasing sequences of the trace induced by the elimination ordering . This study produces a new and simplier understanding of the recognition of chordal graphs. These two algorithms can also be adapted for BFS.

1 The clique graph of a chordal graph

One of the most widely used representations of chordal graphs is the clique tree defined above (see [9]). In this section, we will introduce and study a new structure called the *clique graph* of a chordal graph. We will show the ties between clique graphs and the clique trees. In [19], the clique tree is studied with regard to the clique intersection graph, which can be seen as the cliques hypergraph [2]. The clique graph defined here is a subgraph of the clique intersection graph. We will prove that a clique graph can be seen as the minimal graph containing all clique trees. All graphs considered here are supposed to be connected, if not each connected component has to be considered separately.

Definition 1. Given an undirected graph $G = (V, E)$, and two non-adjacent vertices a and b, a subset $S \subset V$ is an a, b-separator if the removal of S separates a and b in distinct connected components. If no proper subset of S is an a, b-separator then S is a minimal a, b-separator. A (minimal) separator is a set of vertices S for which there exist non adjacent vertices a and b such that S is a (minimal) a, b-separator.

It is well known [8], that the minimal separators of a chordal graphs are complete subgraphs. Here is the definition of the clique graph of a chordal graph.

Definition 2. Let $G = (V, E)$ be a chordal graph. The clique-graph of G, denoted by $C(G) = (V_c, E_c, \mu)$, with $\mu : E_c \to N$, is defined as follows :

1. The vertex set V_c, is the set of maximal cliques of G;
2. The edge (C_1, C_2) belongs to E_c if and only if the intersection $C_1 \cap C_2$ is a minimal a, b-separator for each $a \in (C_1 \setminus C_2)$ and each $b \in (C_2 \setminus C_1)$;
3. The edges of $(C_i, C_j) \in V_c$ are weighted by the cardinality of the corresponding minimal separator $S_{ij} : \mu(C_i, C_j) = |S_{ij}|$.

Let C_i and C_j be two maximal cliques of a chordal graph. Hereafter, we will note $S_{ij} = C_i \cap C_j$ if and only if S_{ij} is a minimal a, b-separator for each $a \in (C_i \backslash C_j)$ and each $b \in (C_j \backslash C_i)$. Let us now prove several structure properties of the clique graph.

Triangle Lemma Let (C_1, C_2, C_3) be a 3-cycle in $C(G)$ and let S_{12}, S_{13}, S_{23} be the associated minimal separators of G. Then two of these three minimal separators are equal and included in the third.

Proof: Assume that two minimal separators among S_{12}, S_{13}, S_{23} are incomparable for the inclusion order. Let S_{12} and S_{13} be these minimal separators. Then there exist two vertices x and y such that $x \in (S_{12} \backslash S_{13})$ and $y \in (S_{13} \backslash S_{12})$. Since C_1, C_2, C_3 are distinct maximal cliques, $C_2 \backslash C_3$ and $C_3 \backslash C_2$ are not empty. The vertices x and y do not belong to $C_2 \cap C_3$. For each $a \in (C_2 \backslash C_3)$ and each $b \in (C_3 \backslash C_2)$, the path a, x, y, b exists and is not cut by $C_2 \cap C_3$. A contradiction, therefore $C_2 \cap C_3$ is not a a, b-separator and the edge between C_2 and C_3 does not exist.

Therefore if there exists a 3-cycle in $C(G)$, then the three minimal separators on the edges can be linearly ordered by inclusion. Without loss of generality, assume that $S_{12} \subset S_{13} \subseteq S_{23}$. Therefore $S_{13} \subset C_1$ and $S_{13} \subset C_2$. And so $S_{13} \subset (C_1 \cap C_2)$. This leads to a contradiction : $S_{13} \subset S_{12}$. We have proved that $S_{12} = S_{13} \subseteq S_{23}$. □

Note that the converse is false. Let C_1, C_2, C_3 be three maximal cliques such that $(C_1, C_2) \in E_c$ and $(C_1, C_3) \in E_c$, then $S_{12} = S_{13}$ does not imply that the edge $(C_2, C_3) \in E_c$. But the following property stands.

Lemma 3. *Let $C(G)$ be the clique graph of the chordal graph G. Let C_1, C_2, C_3 be three maximal cliques such that $(C_1, C_2) \in E_c$ and $(C_1, C_3) \in E_c$, then $S_{12} \subset S_{13} \Rightarrow (C_2, C_3) \in E_c$.*

Proof: By the triangle lemma, $S_{12} \subset S_{13}$ (strict inclusion) implies that $C_2 \cap C_3 = S_{12}$, otherwise there is no edge between C_1 and C_2. By definition S_{12} is a minimal separator for all $a \in C_2 \backslash C_1$ and $b \in C_1 \backslash C_2$. Since $S_{12} \subset S_{13}$, every vertex $c \in C_3 \backslash C_2$ is in the same connected component of $G[V - S_{12}]$ than b. And so $C_2 \cap C_3$ is a minimal separator for every vertices like a and c. □

The clique graph of a chordal graph is not always chordal. Let us now examine the structure of clique trees in full details. After recalling the definition, we will show that a kind of chordality property of the clique tree stands.

Definition 4. Let $G = (V, E)$ be a chordal graph. A clique tree of G is a tree $T_C = (C, F)$ such that C is the set of maximal cliques of G and for each vertex $x \in E$, the set of maximal cliques containing x induces a subtree of T_C.

Lemma 5. *Let T be a clique tree of the chordal graph G and let C_1 and C_2 be two adjacent maximal cliques. Then $C_1 \cap C_2$ is a minimal separator for all $a \in C_1 \backslash C_2$ and $b \in C_2 \backslash C_1$.*

Proof: If $C_1 \cap C_2$ is not an a, b-separator, then there exists a path between a and b which avoids $C_1 \cap C_2$. Every edge of this path belongs to a maximal clique. Therefore a and b are contained in some of these maximal cliques. But T is a tree, hence the set of maximal cliques containing a or b can not induce a subtree of T. So $C_1 \cap C_2$ is an a, b-separator and is necessarily minimal otherwise C_1 or C_2 should not be a clique. $\qquad\square$

Weak Triangulation Lemma Let $P_{ik} = [C_1, ..., C_k]$, $k \geq 4$, be a path in a clique tree T of a chordal graph G. If (C_1, C_k) is an edge of $C(G)$, then either (C_2, C_k) or (C_1, C_{k-1}) is an edge of $C(G)$.
Proof:

For each $1 \leq i \leq k-1$, $S_{i,i+1}$ is a a, b-minimal separator for each $a \in C_i \backslash C_{i+1}$ and each $b \in C_{i+1} \backslash C_i$. Then by lemma 5 each $S_{i,i+1}$ is a x, y-separator for $x \in C_1 \backslash C_2$ and $y \in C_k \backslash C_{k-1}$, but not necessarily minimal. Since $(C_1, C_k) \in V_c$, S_{1k} is a minimal x, y-separator. Hence S_{1k} must be included in all $S_{i,i+1}$. The triangle lemma proves the existence of a chord in the cycle $C_1, ..., C_k$ in $C(G)$. And so the triangle lemma applied iteratively, proves that either (C_2, C_k) or (C_1, C_{k-1}) is an edge of $C(G)$. $\qquad\square$

In other words, the above lemma shows that any path of a clique tree induces a chordal subgraph of $C(G)$. Next theorem shows ties between clique trees and maximum weighted spanning trees of $C(G)$.

Theorem 6. *Let $G = (V, E)$ be a chordal graph and $C(G)$ its clique graph. Let $T = (V_c, F)$ be a spanning tree of $C(G)$. Then T is a clique tree of G if and only if T is a maximum weighted spanning tree of $C(G)$.*

Proof: Since the set of nodes of T and $C(G)$ are the same, we just need to prove that the set of nodes containing a vertex $x \in E$ induces a subtree of T if and only if T is a maximum weighted spanning tree of $C(G)$.

$\Rightarrow$ Assume that T is not a maximum weighted spanning-tree of $C(G)$. Then there exists a pair of maximal cliques, C_i and C_j, adjacent in $C(G)$ and not in T such that S_{ij} strictly contains a minimal separator of the unique path P_{ij} between C_i and C_j in T. Let S_{kl} be such a minimal separator on P_{ij}. Let $x \in (S_{ij} \backslash S_{kl})$. So x does not belong to at least one of the two nodes C_k and C_l. Hence, the set of nodes containing the vertex x is not a subtree of T and so T is not a clique-tree of G.

$\Leftarrow$ Assume that T is a maximum weighted spanning-tree of $C(G)$. Let C_i and C_j be two non-adjacent maximal cliques of T containing the vertex $x \in V$. Assume that x does not belong to any maximal cliques on the unique path in T between C_i and C_j, P_{ij}.

Consider the subgraph G' of G induced by the maximal cliques of P_{ij}. The subgraph G' is a chordal graph. In [16], it is proved that each vertex of a chordal graph is either a simplicial vertex, or belongs to a minimal separator. Simplicial vertices belong to only one maximal clique. Therefore there exists a minimal separator S of G' such that $x \in S$. Since no maximal clique apart from C_i and C_j contains x, S must be equal to $C_i \cap C_j$. Hence the edge

$(C_i, C_j) \in E_c$, and the maximal cliques of P_{ij} induces a cycle. Let $y \in C_i \setminus S$ and $z \in C_j \setminus S$ be two vertices, then S is a minimal z, y-separator. If no minimal separator on P_{ij} is strictly included in S, we can easily extract a path in G' between y and z which is not cut by S. This leads to a contradiction. Therefore there exists S' on P_{ij} such that $S' \subset S$. Hence T is not a maximum spanning tree. So x belongs to any maximal clique on P_{ij}, and T is a clique tree.

$\square$

The next theorem yields a kind of optimality for a clique graph.

Theorem 7. *Let $C(G)$ be the clique graph of the chordal graph G. Then each edge of $C(G)$ belongs to at least one clique tree.*

Proof: Let T be a clique tree of G. And let C_i and C_k two adjacent maximal cliques in $C(G)$ but not in T. Let P_{ik} be the path between C_i and C_k in T. By the weak triangulation lemma, either the edge (C_1, C_{k-1}) or (C_2, C_k) exists. Assume that (C_1, C_{k-1}) exists. Then C_1, C_k, C_{k-1} is a 3-cycle, and so the triangle lemma can be applied. Since T is a maximum spanning tree, (C_1, C_k) is not the biggest edge of the 3-cycle. So $S_{1,k} = S_{1,k-1}$. Applying the weak triangulation lemma iteratively, we can find an edge E in P_{ij} with the same label than (C_1, C_k). Therefore changing E by (C_1, C_k) in T yields to a new clique tree. We have proved that every edge of $C(G)$ can belong to a clique tree. $\square$

The previous theorem proves that the edges belonging to the clique intersection graph and not to the clique graph, never belong to a clique tree. Since every edges of a clique tree are minimal separator, clique graphs are the minimal structures containing all clique trees.

2   Greedy aspects of recognition algorithms

Let $G = (V, E)$ be a graph with n vertices. When an elimination ordering is computed by BFS or MCS, another procedure must verify if it is a perfect elimination ordering in order to prove that G is triangulated. BFS or MCS computes the elimination ordering in the reverse order. In this section, we will give an explanation of the greedy aspect of the two linear recognition algorithms of chordal graphs: MCS and BFS. The main result proves that both algorithms compute a maximum spanning-tree of the clique graph.

Let us examine how MCS and BFS visit a chordal graph. We just give the proof for MCS, but this proof can be easily transformed for BFS. When MCS chooses a new vertex x, then the mark level of this vertex, noted $mark(x)$, is maximum over all unnumbered vertices. The set of vertices who has marked x will be denoted by $M(x)$.

Lemma 8. *Let $G = (V, E)$ be a chordal graph. In a execution of MCS or BFS on G, maximal cliques of G are visited consecutively.*

Proof: Let α the PEO computed by MCS. Let $N = \{x_n, ..., x_i\}$ be the set of numbered vertices at some step. Then x_i is simplicial in $G[x_n, ..., x_i]$, and so x_i belongs to a unique maximal clique (see [16]). Let us prove that x_{i-1} belongs to a new maximal clique if and only if $mark(x_{i-1}) \leq mark(x_i)$.

$\Leftarrow$ Assume that x_{i-1} and x_i belong to the same maximal clique. Since x_i is simplicial in $G[x_n, ..., x_i]$, all the vertices of $M(x_i)$ belong to this maximal clique. Hence $mark(x_{i-1}) = mark(x_i) + 1$.

$\Rightarrow$ Assume that x_{i-1} belongs to a new maximal clique. Since x_i was a vertex with the biggest level mark over all unnumbered vertices when it was choosen, $mark(x_i) \geq |M(x_{i-1}) \setminus \{x_i\}|$. But there exist at least one vertex of the maximal clique containing x_i in $G[x_n, ..., x_i]$ which does not mark x_{i-1}, otherwise x_{i-1} does not belong to a new maximal clique. Therefore $mark(x_{i-1}) \leq mark(x_i)$.

If we define the trace of α as the sequence of mark levels of the vertices when they are numbered, each maximal clique is represented by an increasing sequence. We can conclude that the maximal cliques are visited consecutively. A similar argument holds for BFS. $\square$

Now the remaining question is, in which ordering the maximal cliques of the input chordal graph are visited by both algorithms? The edges of the clique graph of a chordal graph $G = (V, E)$ represent all possible transitions from a maximal clique to another one.

In fact a clique graph is a very particular weighted graph. Therefore there are many ways for computing maximum weighted spanning trees in $C(G)$. The following algorithm is one of them. We will prove that BFS and MCS can be seen as special case of this algorithm. In order to find such a general framework, in the following algorithm we assume that the edges of $C(G)$ are labelled with the minimal separators (until now it was only required that the edges of $C(G)$ to be weighted by the cardinality of the minimal separator).

Algorithm 1: Maximum weighted spanning-tree of clique graph

Data: A clique graph $C(G)$
Result: A maximum spanning-tree of $C(G)$
Choose a maximal clique C_1
for $i = 2$ **to** n **do**
> choose a maximally (under inclusion) labelled edge adjacent to $C_1, ..., C_{i-1}$
> to connect the new clique C_i

Theorem 9. *Let G be a chordal graph, then algorithm 1 computes a maximum weighted spanning tree (i.e. a clique tree) of the clique graph $C(G)$.*

Proof: Let T_i be the tree built by algorithm 1 at step i. We now prove by induction the following invariant: T_i *can be completed as a maximum spanning tree of $C(G)$.*

Clearly the property holds for T_1. Let us suppose by induction, it holds for T_{i-1} and that (C_j, C_i), with $j < i$ is the edge chosen by the algorithm at step i. Therefore it exists a maximum spanning tree T containing T_{i-1}.

If (C_j, C_i) belongs to T, we have finished. Else it exists a unique path $\mu = [C_j = D_1, \ldots, D_k = C_i]$ from C_j to C_i in T. Let D_h be the last vextex of $C_1, \ldots, C_{i-1}$ belonging to μ. algorithm 1 insures that $A = label[(D_h, D_{h+1})]$ does not contain $B = label[(C_j, C_i)]$. If $\exists b \in B - A$, then $b \in C_j$ and therefore by the definition of clique tree, b must also belong to all cliques in μ, a contradiction.

Therefore $A = B$, and a maximum spanning tree T' can be obtained from T by exchanging the edges (D_h, D_{h+1}) and (C_j, C_i), which finishes the proof.

□

In [3], it is proven that MCS corresponds to Prim's algorithm. It is easy to see that Prim's algorithm is a special case of the algorithm 1. The next corollary proves the same for BFS.

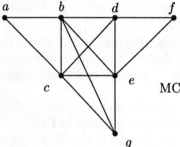

A PEO computed by MCS:
$$b, c, d, e, f, g, a$$

MCS and Prim visit the clique in the ordering:

$$C_1, C_2, C_3, C_4$$

$G = (V, E)$ a triangulated graph

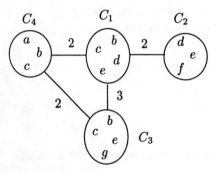

G_C the clique graph of G

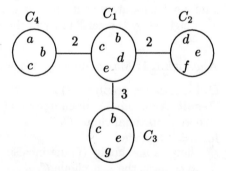

The maximum weighted spanning-tree of G_C generated by Prim's algorithm,

Fig. 1. An execution of MCS on a chordal graph and the corresponding execution of Prim's algorithm on its clique graph

Corollary 10. *MCS and BFS compute maximum spanning trees of $CC(G)$.*

Proof: It suffices to show that MCS and BFS are particular cases of algorithm 1. Let $C_1, ... C_{k-1}$ be the visited cliques and $x_1, ..., x_{i-1}$ be the numbered vertices. When x_i is numbered, a new clique C_k is visited. Since x_i is simplicial in $G[x_1, ..., x_i]$, its neighbourhood is complete and strictly included in a clique C_j, $1 \geq j \geq k - 1$. Therefore the labelof all edges connecting C_k to $C_1, ... C_{k-1}$ must be include in the label of (C_i, C_j). In order to connect, C_k the maximum edge is choosen. This is what algorithm 1 does, since a amximum weighted edge has necessarily a maximal label. Similarly for BFS the lexicographic search insures also a maximal labelled edge to be chosen at each step. □

3 Linear algorithms on the clique trees

In this section, we will first present a linear time and space algorithm which computes a clique tree of G will be presented. This algorithm based on MCS. A similar algorithm was presented in [3] We define the trace of an PEO α computed by MCS as the sequence of mark level of the vertices when they are numbered (see figure 2). In the second part, an extended version of the first algorithm for the recognition of PEO is presented. The reader should note that exactly the same work can be done with BFS.

By the lemma 8, each maximal clique of G corresponds to a strictly increasing sequence of the trace. Let us recall and define some notations. The level mark of a vertex will be denoted by $mark(x)$, and the set of vertices who have marked x is $M(x)$. At each vertex x, we associate $C(x)$, the first maximal clique discovered in α containing x.

Since maximal cliques are visited one after the other, MCS gives an ordering of the maximal cliques: the j-th maximal clique reached by MCS will be noted C_j. At each maximal strictly increasing sequence of marks, L_j in the following (see figure 2), corresponds C_j. Note that the set of L_j is a partition of V. Since α is a PEO, for every x_i, $M(x_i) \subset C(x_i)$. Hence if $x_i \in L_j$, then $C(x_i) = L_j \cup M(x_i) = C_j$.

Let x_i be the first vertex discovered in L_j. Then for every vertex $x_h \in L_j$, $M(x_i) \subseteq M(x_h)$. Henceforth, $mark_j$ will denote the last vertex which has marked all vertices of L_j, i.e $last(x_i)$.

Lemma 11. *Let G be a chordal graph. Let T be a tree whose nodes are the maximal cliques of G, and such that an edge between the maximal cliques C_j and C_k $(k < j)$ exists if and only if $mark_j \in L_k$. Then T is a clique-tree of G.*

Proof: Let x_i be the first vertex discovered in L_j. Let x be a vertex of $M(x_i)$. Let us prove that every maximal clique C on the path between $C(x)$ and C_j in T, contains the vertex x. We have to consider two cases:

1. $C(x)$ and C_j are adjacent (i.e $C(x) = C_k$): trivial.
2. If $C(x)$ and C_j are not adjacent, then $x \neq mark_j$ (i.e. x is not the last vertex which has mark all vertices of C_j). Let us prove that x has marked all vertices of C_k. Since α is a PEO and since $\alpha(x_i) < min(\alpha(x), \alpha(mark_j))$, the edge $(x, mark_j) \in E$. Hence $x \in M(mark_j)$, and so x belongs to C_k.

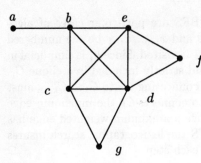
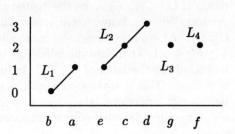

$G = (V, E)$ a triangulated graph

A trace of α, the PEO computed by MCS

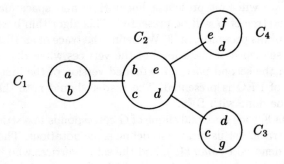

The clique-tree associated to α

Fig. 2. An execution of MCS and the associated clique-tree

This proves that the maximal cliques containing x form a subtree of T. Hence T is a clique-tree of G. □

We are now able to present an extended version of MCS which computes on-line the clique-tree associated to the elimination order.

Theorem 12. *The algorithm 2 computes a PEO and its assciated clique-tree if and only if the input graph $G = (V, E)$ is chordal. The complexity of this algorithm is $O(n + m)$ where $n = |V|$ and $m = |E|$.*

Proof: If the alogrithm 2 computes a PEO and a clique-tree, then the input graph is trivially chordal. The algorithm 2 is an extended version of MCS, hence if G is chordal, the computed elimination ordering is perfect. By lemma 11, the step 1 builds a clique-tree if G is chordal.

Let us have a look at the size of a clique tree. First of all, when a vertex is marked by MCS, this mark corresponds to an edge. Hence the size of all mark sets is $O(m)$, the number of edges in G. Since there is at most n increasing sequences, T contains at most n nodes and so $O(n)$ edges.

Let us examine the size of the set of nodes in T. Since the vertices of minimal separators belong to several maximal cliques, the size of the nodes set is bigger

Algorithm 2: Maximum Cardinality Search and Clique-Tree

Data: A graph $G = (V, E)$
Result: If the input graph is chordal: a PEO and an associated clique-tree $T = (I, F)$ where I is the set of maximal cliques
begin

    each vertex of X is initialized with the empty set
    $previousmark = -1$
    $j = 0$
    for $i = n$ **to** *1* **do**
        choose a vertex x not yet numbered such that $|mark(x)|$ is maximum

1   **if** $mark(x) \leq previousmark$ **then**
            $j = j + 1$
            create the maximal clique $C_j = M(x) \cup \{x\}$
            create the tie between C_j and $C(last(x))$
        else
2            $C_j = C_j \cup \{x\}$

3   **for** *each y neighbour of x* **do**
            $M(y) = M(y) \cup \{x\}$
            $mark(y) = mark(y) + 1$
            $last(y) = x$

        $previousmark = mark(x)$
        x is numbered by i
        $C(x) = j$

end

than n. Let $M(x)$ be the set of marks of x, where x is the first vertex of an increasing sequence L_j. The node associated to x by the algorithm 2 is $M(x) \cup L_j$. But $M(x)$ corresponds to the edges between vertices of $M(x)$ and x. Since every duplicated vertex belongs to the mark set of the first vertex of an increasing sequence, each of them can be associated an edge. Therefore the sum of the cardinality of the nodes of T is smaller than $n + m$, and the space complexity is $O(n + m)$.

All the operations of step 1, 2 and 3 can be done in constant time. Hence this algorithm has the same time complexity as MCS: $O(n + m)$. $\square$

It is well known that MCS computes a PEO if and only if the graph is chordal. Similarly the results of the previous algorithm stand if and only if the algorithm 2 works on a chordal graph, MCS computes a PEO if and only if the graph is chordal. Hence the natural following question is: how the trace of an elimination ordering computed by MCS can be used for the recognition of chordal graph. The following algorithm checks whether the output of the algorithm 2 is a clique tree or not in linear time and space complexity. Since MCS computes a PEO if and only if the input graph is chordal, then the algorithm 2 builds a clique tree if and only if the elimination ordering is perfect. And so the next algorithm can be seen like a recognition procedure for chordal graphs.

Let T be the result of the algorithm 2. The following two lemmas exhibit the properties of simplicial vertices in T. The figure 3 give an illustration of these properties.

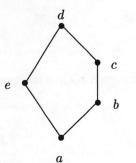

The input graph $G = (V, E)$

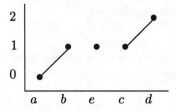

The trace of the elimination ordering

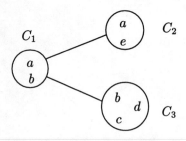

C_3 is not a clique: the mark set
$$M(d) \neq (C_3 \setminus \{d\})$$
Hence, G is not a triangulated graph.

Fig. 3. An execution of the algorithm 3 on a 5-cycle

Lemma 13. *Let x_i be the first vertex of the increasing sequence of marks L_j. Then x_i is simplicial if and only if $M(x_i) \subset C_{father}$.*

Proof: Assume that $M(x_i) \subset C_{father}$ then the neighbourhood of x_i is a clique. If it is not the case, then at least two neighbours of x_i are not in the same clique. Therefore they are not adjacent. Hence its neighbourhood is not a clique. □

Lemma 14. *Let $L_j = \{x_i, ..., x_h\}$ be an increasing sequence of marks such that x_i is simplicial in $G[X_n, ..., x_i]$. Then x_k, with $i < k \leq h$, is simplicial if and only if $M(x_k) = M(x_{k-1} \cup \{x_{k-1}\})$.*

Proof: Let us assume that x_i, the first vertex of the increasing sequence, is simplicial. Then its neighbourhood is a clique. It is clear that x_{i-1} is simplicial if $M(x_{i-1}) = M(x_i) \cup \{x_i\}$. It is the same for each x_k, $i < k \leq h$. If $M(x_k) = M(x_{k-1} \cup \{x_{k-1}\})$, then x_k is simplicial if and only if x_{k-1} is simplicial. Since x_i is simplicial, the property follows. □

These two previous lemmas can lead to an on-line recognition algorithm of chordal graphs. When MCS chooses an unnumbered vertex, one of the two conditions according to the type of this vertex has to be tested. Let us give an idea of such an algorithm. Assume that like in the algorithm 2, the used data structures are sorted lists. Checking whether the first vertex of an increasing sequence is simplicial, can be done just by testing the inclusion of a set in another one. The complexity is in the size of the biggest set. This size can be in $O(n)$, and the test can be done $O(n)$ times. Hence such an algorithm, using linear data structure, has a time complexity in $O(n^2)$. The time complexity can be improved, but then matrix should be used. And so the space complexity is not linear in this case.

We present now a sketch of the recognition algorithm, which works on the whole T. This algorithm can be seen like a parallel version of the previous idea. Since the procedure works on the whole tree T, all the inclusion tests in a given clique can be done at the same time. These inclusion tests are done by the procedures $test - simpliciality - first(C_i, S)$ and $test - simpliciality - next(C_i, S)$. These procedures correspond respectively to the lemmas 13, 14.

Algorithm 3: Recognition of Chordal Graph
Data: The result $T = (C, F)$ of the algorithm 2 as sorted lists
Result: Check whether T is a clique-tree
begin

> **for** *each node C_i of T* **do**
>> $S \leftarrow$ the set of sons of C_i in T
>> **if** *not test-simpliciality-first(C_i, S)* **then**
>>> **return** T *is not a clique-tree, hence G is not chordal*
>>
>> **if** *not test-simpliciality-next(C_i, S)* **then**
>>> **return** T *is not a clique-tree, hence G is not chordal*
>
> **return** T *is a clique-tree, hence G is chordal*

end

Let us study more precisely the complexity of both procedures $test - simpliciality$. The data structures are sorted lists respecting the elimination ordering. Each node of T is defined by two sorted lists: one for its vertex set, and the other for the corresponding increasing sequence. A sorted list of its mark set is associated to every vertex.

The procedure $test - simpliciality - first(C_i, S)$ checks at the same time, if the mark set of the first vertices of the increasing sequences corresponding to the sons of C_i are include in C_i. Since these sets are sorted lists, they are visited just once. Each list has a current element. The idea is to start at the beginning of each list. If all current element of the mark sets are less or equal to the current element of C_i, then the current element of C_i move to the following one. The current elements of the mark sets ore moved to the following if it is equal to the current element of C_i. If one of the current elements of the mark sets is smaller than the current element of C_i, then the test returns false. The end of the test is reached successfully if all the mark sets are visited before the end of C_i.

The procedure $test - simpliciality - next(C_i, S)$ checks whether the mark set of vertices, which are not the first one of an increasing sequence, are prefixes of the node containing it.

Theorem 15. *The recognition algorithm 3 checks whether G is chordal in linear time and space complexity $O(n + m)$.*

Proof: Since both procedures *test-simpliciality* correspond to the lemmas 13, 14, it is clear that the algorithm returns true if and only if the input T is a clique-tree.

The procedure 3 visits the data structure just once and always uses elementary operations. But in the previous section, we have shown that the size of a clique tree is $O(n + m)$. We can conclude that the space complexity of the algorithm 3 is $O(n + m)$. Therefore the time complexity is also $O(m + n)$. □

4 Conclusion

The clique graph of a chordal graph G introduced and studied in the first part, is shown to be the minimal structure containing all clique trees of G. The properties of the clique graph imply a simple algorithm for maximum spanning-trees which cannot be applied for general graphs. As in [3], MCS is compared to Prim's algorithm. Many versions of greedy algorithm for computing a maximal weighted independant set in a matroid are known [18, 11]. It could be worthwhile in this context to consider Kruskal's or Boruvka's alogrithm [4]. Hence, this matroidal approach can justify many versions of recognition algorithms of chordal graphs. It can be very helpful for various generalizations.

Hence this paper presents a new and unified regard on the MCS and BFS algorithms. The trace of these algorithms executions translates the behavior of the algorithm in the input graph. The study of this trace induces a simple way to compute the associated clique tree and the recognition of chordal graphs. This approach, studying the trace of algorithm like MCS and the underlying structure, can be very helpful for various generalizations of elimination orderings, see [13, 7], or generalizations of chordal graphs, see [10, 5].

Acknowledgement

We would like to thank J.X. Rampon for drawing our attention to references [3, 12] and R.H. Möhring for discussions on the BFS algorithm.

References

1. C. Beeri, R. Fagin, D. Maier, and M. Yannakakis. On the desirability of acyclic database schemes. *J. Assoc. Comput.*, 30:479–513, 1983.

2. C. Berge. *Hypergraphs*. North Hollands, 1989.

3. J.R.S. Blair and B. Peyton. An introduction to chordal graphs and clique trees. preprint.

4. O. Boruvka. On a minimal problem. *Prac Moravske Predovedecke Spolecrosti*, 3, 1926.

5. A. Brandstädt, F.F. Dragan, V.D. Chepoi, and V.I Voloshin. Dually chordal graphs. In *Proceedings of the 19th Inter. Workshop on Graph-Theoretic Concept in Computer Science*, 1993. WG93.

6. P. Buneman. A characterization of rigid circuit graphs. *Discrete Math.*, 9:205–212, 1974.

7. E. Dahlhaus, P.L. Hammer, F. Maffray, and S. Olariu. On domination elimination orderings and domination graphs. Technical Report 27-94, Rutgers University Center of Operations Research, P.O. Box 5062, New Brunswick, New Jersey, USA, August 1994.

8. G.A. Dirac. On rigid circuit graphs. *Abh. Math. Sem. Uni. Hamburg 25*, 1961.

9. F. Gavril. The intersection graphs of a path in a tree are exactly the chordal graphs. *Journ. Comb. Theory*, 16:47–56, 1974.

10. Ryan B. Hayward. Weakly triangulated graphs. *Journal of Combinatorial theory*, 39:200–209, 1985. Serie B.

11. B. Korte, L. Lovász, and R. Schrader. *Greedoids*. Number 4 in Algorithms and Combinatorics. Springer Verlag, 1991.

12. J.G Lewis, B.W. Peyton, and A. Pothen. A fast algorithm for reordering sparse matrices for parallel factorization. *SIAM J. Sci. Stat. Comput.*, 10(6):1146–1173, November 1989.

13. S. Olariu. Some aspects of the semi-perfect elimination. *Discrete Applied Mathematics*, 31:291–298, 1991.

14. B. Peyton. *Some applications of clique trees to the solutions of sparse linear systems*. PhD thesis, Clemson University, 1986.

15. R.C. Prim. Shortest connection networks and some generalizations. *Bell System Technical Journal*, 1957.

16. Donald J. Rose. Triangulated graphs and the elimination process. *Journal of Mathematical Analysus and Applications*, 32:597–609, 1970.

17. Donald J. Rose, R. Endre Tarjan, and George S. Lueker. Algorithmic aspects of vertex elimination on graphs. *SIAM Journal of Computing*, 5(2):266–283, June 1976.

18. P. Rosenstielh. L'arbre minimum d'un graphe. *Theory of Graphs*, 1967. P. Rosenstielh, editor, Gordon and Breach, New York.

19. Y. Shibata. On the tree representation of chordal graphs. *Journal of Graph Theory*, 12:421–428, 1988.

20. R.E. Tarjan and M. Yannakakis. Simple linear algorithms to test chordality of graphs, test acyclicity of hypergraphs, and selectively reduce acyclic hypergaphs. *SIAM Journal of Computing*, 13:566–579, 1984.

A Compact Data Structure and Parallel Algorithms for Permutation Graphs*

Jens Gustedt[1]**, Michel Morvan[2]***, and Laurent Viennot[2]†

[1] Technische Universität Berlin,
Sekr. MA 6-1, D-10623 Berlin - Germany.
[2] LITP/IBP Université Paris 7 Denis Diderot,
Case 7014, 2, place Jussieu F-75251, Paris Cedex 05.

Abstract. Starting from a permutation of $\{0, \ldots, n-1\}$ we compute in parallel with a workload of $O(n \log n)$ a compact data structure of size $O(n \log n)$. This data structure allows to obtain the associated permutation graph and the transitive closure and reduction of the associated order of dimension 2 efficiently. The parallel algorithms obtained have a workload of $O(m + n \log n)$ where m is the number of edges of the permutation graph. They run in time $O(\log^2 n)$ on a CREW PRAM.

1 Introduction

Permutation graphs are combinatorial objects that found a lot of attention in recent years. This interest led to many results under a structural point of view as well as algorithmically, see [1, 2, 6, 7, 8].

By definition permutation graphs have a compact encoding of size n, where n is the number of vertices. In sequential computing model, it is possible to pass from the graph to the permutation and vice versa with a workload of $O(n^2)$. For parallel computing this has been open up to now. In this paper we show how to pass from the permutation to the graph on a CREW PRAM with a workload of $O(m + n \log n)$ where m is the number of inversions of the permutation. This also leads to a new representation of permutation graphs in size $O(n \log n)$.

Many graph algorithms require a classical representation of the graph. In order to run these algorithms on a permutation graph, it is necessary to compute the graph from the permutation.

The whole article is written in the directed context. We call permutation digraphs the directed version of permutation graphs. They are also called the orders of dimension 2. The results we give in the last section are specific to the directed context. All the others can obviously be obtained for permutation graphs.

The paper is organized as follows.

---

* The authors are supported by PROCOPE.
** Email: gustedt@math.tu-berlin.de
*** Email: morvan@litp.ibp.fr
† Email: lavie@litp.ibp.fr

After giving the necessary definitions and notations in Sect. 2, we start by computing the number of arcs of the permutation graph. The workload for this computation is $O(n \log n)$ and thus a factor of $\log \log n$ away from the lower bound [4, 5].

In the following section we describe a compact data structure derived from that algorithm that represents the successors sets of all elements in size $O(n \log n)$. In contrast to the one given directly by the permutation this one has the advantage of supporting all classical algorithms on permutation graphs where they are supposed to be represented by adjacency lists.

In Sect. 5 we show how to compute the adjacency lists from our compact representation with a workload of $O(m + n \log n)$.

In the last section we consider the permutation graph as a two-dimensional order and we show how to compute the transitive reduction of the given order from our algorithm.

The problem of computing efficiently representations of an order of dimension d was introduced in these terms by Spinrad in [9]. These problems can be solved with geometric range queries in a d-dimensional space. But this method uses a strong computational model with reals. Moreover it hides the new data structure for representing digraphs that we introduce in Sect. 4. We give here the benefits of a parallel approach to the sequentially well know case of $d = 2$.

2 Definitions and Notations

We always consider simple finite digraphs with no loops.

A *digraph* $G = (V, \mathcal{A})$ is given by a set V of *vertices*, and a set $\mathcal{A}$ of couples of vertices called *arcs*.

If $(u, v) \in \mathcal{A}$, we say that v is a *successor* of u and with respect to undirected graphs terminology that u and v are *adjacent*. Usually, a digraph is represented as *adjacency lists*. For each vertex v, the list of its successors is given. We call this a *classical representation*.

The *degree* $Deg^+(v)$ of a vertex v is the number of its successors.

A permutation π is a bijection from $\{0, \ldots, n-1\}$ into itself, or equivalently a simple word with all the letters $0, \ldots, n-1$. Let $\pi(i)$ denote the image of i, or equivalently the ith letter of π. We write π^{-1} for the inverse of the bijection π, and $\tilde{\pi}$ for the reverse of the word π.

We will always consider digraphs with set of vertices $\{0, \ldots, n-1\}$.

A digraph $G = (V, \mathcal{A})$ is a *permutation digraph* if $\mathcal{A}$ is given by a permutation π of $\{0, \ldots, n-1\}$ such that $(i, j) \in \mathcal{A}$ iff $i < j$ and $\pi(i) < \pi(j)$.

A permutation digraph $G = (V, \mathcal{A})$ is acyclic and transitively closed. It can be seen as an order of dimension two. Its *transitive reduction* is the minimal subgraph whose transitive closure is G. It is equivalently given by the covering relation on V. We say that u is *covered* by v iff $(u, v) \in \mathcal{A}$ and there is no k such that $(u, k) \in \mathcal{A}$ and $(k, v) \in \mathcal{A}$.

3 Computing the Number of Arcs

The number of arcs of the permutation digraph is the number of inversions of $\widetilde{\pi^{-1}}$ since an arc (i, j) with $i < j$ exists iff i appears before j in π^{-1}.

We are going to compute this number by sorting $\widetilde{\pi^{-1}}$ in a "quicksort way". In each dividing phase of quicksort we update a counter c such that the sum of c and the number of inversions of the permutation being sorted remains invariant.

W.l.o.g. we assume that $n = 2^{q+1}$. Since we are sorting exactly all the numbers $0, \ldots, n-1$, we can always find a good pivot element that equally divides the sequences in question. Thus the algorithm requires only a logarithmic number of phases $\varphi = 0, \ldots, q - 1$.

3.1 Partitionning the Permutation

Now we describe how to partition a part of size $2^{q-\varphi}$ of the permutation at phase φ into a sequence of two blocks, detecting some of its inversions without creating any new one.

Like in quicksort, the elements greater (resp. smaller) than the pivot element are put in the right (resp. left) block. Thereby the order of the permutation in each block has to be preserved. Doing this we cancel some inversions. For each vertex v going to the left, we have to count how many vertices previously appearing to its left go to the right and add this number to the counter c, see Fig. 1.

Before the initial phase, we set $W := \widetilde{\pi^{-1}}$ and $c := 0$. In phase φ, we partion the 2^φ consecutive blocks of size $2^{q-\varphi}$ composing W. For the sake of clarity, we just write the algorithm for the first block.

Algorithm 1 Divide and Conquer Phase.

Input: A block $W(0), \ldots, W(2^{q-\varphi} - 1)$ of a permutation.
Output: A number $\Delta(W(x))$ of newly detected inversions relatively to each vertex $W(x)$ for $x \in \{0, \ldots, 2^{q-\varphi} - 1\}$.

Step 1. **Comparisons:** Every vertex $W(x)$ is compared to the pivot element $2^{q-\varphi-1}$. $B(x)$ is set to 0 if $W(x)$ is smaller and 1 otherwise.

Step 2. **Sum up:** Compute the prefix sums $\sum_{i=0}^{x} B(i)$.

Step 3. **Inversions:** If $B(x) = 0$, we have newly detected $\Delta(W(x)) := \sum_{i=0}^{x} B(i)$ inversions relatively to $W(x)$
Otherwise, we set $\Delta(W(x)) := 0$.

Step 4. **Partition:** Divide the block in a quicksort step.

For the other blocks of phase φ we proceed analogously. We just have to compute an offset for each block.

At the end of phase φ, we update the number c of inversions that were detected by adding the sum of all $\Delta(v)$. This can be computed with a prefix sum.

$\pi =$	2 4 7 0 5 6 1 3	
$\pi^{-1} =$	3 6 0 7 1 4 5 2	
$\pi^{-1} =$	2 5 4 1 7 0 6 3	$c = 0$
$\varphi = 1$	2 1 0 3|5 4 7 6	$c = 0 + 0+2+3+4 = 9$
$\varphi = 2$	1 0|2 3|5 4|7 6	$c = 9 + 1+1 + 0+0 = 11$
$\varphi = 3$	0|1|2|3|4|5|6|7	$c = 11 + 1 + 0 + 1 + 1 = 14$
		$m = c = 14$

Fig. 1. A permutation and the different phases of the algorithm.

3.2 Analysis

At the end, W is sorted and has no inversions, so c is the number of inversions of π^{-1}, and we have computed the number of arcs $m = c$ of the permutation digraph.

At each of the $\log n$ phases, we compute two prefix sums, thus requiring a time $O(\log n)$ and a workload of $O(n)$. Hence we have:

Theorem 1. *Our algorithm computes the number of arcs of a permutation digraph given by a permutation on n elements in time $O(\log^2 n)$, and with a workload of $O(n \log n)$.*

This is not far from the best known sequential algorithm which needs time $O\left(\frac{n \log n}{\log \log n}\right)$ [4, 5].

4 A Compact Representation of Adjacency Lists

In fact, the previous algorithm implicitly computes a special representation of the digraph that we introduce now. Therefore we keep copies W_φ and Δ_φ of the vectors obtained in each phase φ, see Fig. 2.

Let us consider the previous algorithm in terms of the permutation digraph. The number $\Delta_\varphi(v)$ of newly detected inversions relative to v corresponds to $\Delta_\varphi(v)$ successors of v. If v goes to a right block, none of its successors is detected. Otherwise v is smaller than the pivot element. Thus the vertices going to the right block are greater than v. And those which in addition appeared on its left before phase φ are among its successors. They are exactly the $\Delta_\varphi(v)$ first elements of the right block and form an interval $I_\varphi(v) = [l_\varphi(v), r_\varphi(v)]$ of W_φ for some indices $l_\varphi(v)$ and $r_\varphi(v)$.

Notice that the other successors of v go to the left block together with v, and will be detected later.

We can compute $I_\varphi(v)$ in phase φ as follows. $l_\varphi(x)$ is the offset calculated for block $W_{\varphi-1}$. We have $r_\varphi(v) = l_\varphi(v) + \Delta_\varphi(v)$. Let x be such that $v = W_{\varphi-1}(x)$ and $x = b_1 \ldots b_q$ be its binary representation. We have $l_\varphi(x) = \underbrace{b_1 \ldots b_q 10 \ldots 0}_{q}$.

We formalize this concept of representing a digraph by intervals over total orders on V in the following way.

Definition 2. Let $G = (V, \mathcal{A})$ be a digraph, and for some integer k let $W_1, \ldots, W_k$ be arrays representing totally ordered subsets of V. For each vertex $v \in V$, let $I_1(v), \ldots, I_k(v)$ be intervals of $W_1, \ldots, W_k$ respectively. $I_\varphi(v) = [l_\varphi(v), r_\varphi(v)]$ is the set $\{W_\varphi(l_\varphi(v)), W_\varphi(l_\varphi(v) + 1), \ldots, W_\varphi(r_\varphi(v))\}$.

We say that $W_1, \ldots, W_k$ and $I_1, \ldots, I_k$ form a k-*compact interval representation* of G if the successor set $Succ(v)$ of each vertex $v \in V$ is given by $Succ(v) = \bigcup_{i=1}^{k} I_i(v)$.

Every graph with n vertices has a trivial n-compact interval representation by setting W_v to the list of successors of v, and by setting $I_v(v) = [1, |W_v|]$ and $I_\varphi(v) = \emptyset$ if $\varphi \neq v$.

We have seen how to compute a k-compact representation of any permutation digraph for $k = \log n$, see Fig. 2. Clearly it also runs in time $O(\log^2 n)$ and with a workload of $O(n \log n)$.

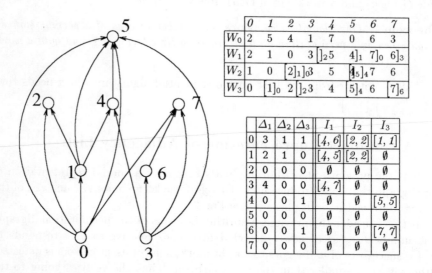

Fig. 2. The permutation digraph of the permutation of Fig. 1 and the intervals of its compact representation.

5 A Parallel Algorithm Which Finds a Classical Representation from a Compact Representation

In order to compute the classical representation of a permutation digraph, we will give in this section an algorithm for finding a classical representation of any digraph from a compact representation. It will lead us to the following results:

Theorem 3. *Let G be a digraph on n vertices and m arcs in a k-compact representation. There is an algorithm that computes a classical adjacency list representation in time $O(\log n)$ and with a workload of $O(m + k.n)$ on a CREW PRAM.*

Combining this with the previous sections, we obtain:

Theorem 4. *It is possible to compute the classical adjacency lists representation of a permutation digraph in time $O(\log^2 n)$ and with a workload of $O(m+n\log n)$ on a CREW PRAM.*

Arc adjacency list representations have size m where m is the number of arcs of the graph. We first have to compute this number which will also be the number of processors our algorithm will require.

Algorithm 2 Outline of the Algorithm.

Input: k arrays $W_1, \ldots, W_k$ and an array of size nk of intervals
$$I_\varphi(v) = [l_\varphi(v), r_\varphi(v)], \text{ for } 0 \leq x < n \text{ and } 1 \leq \varphi \leq k.$$
Output: An adjacency lists representation.
Step 1.   Compute the number m of arcs.
Step 2.   Allocate an array A of size m; each element of A will contain
                 an arc of the digraph.
Step 3.   Compute the tail of each of these arcs.
Step 4.   Compute the head of each of these arcs.

For the following detailed algorithms, recall that for all graphs the set of vertices is $\{0, \ldots, n-1\}$.

Algorithm 3 Computing the Number of Arcs.

Step 1.   For each interval $I_\varphi(v)$, set $\Delta_\varphi(v)$ to its length.
Step 2.   Let D be an array of size nk.
Step 3.   For all $1 \leq \varphi \leq k$ and $0 \leq v \leq n-1$, set $D(kv + \varphi) := \Delta_\varphi(v)$.
Step 4.   Compute the prefix sums $S_\varphi(v) := \sum_{a=0}^{kv+\varphi-1} D(a)$
Step 5.   Set $m := S_{k+1}(n-1)$.

Observe that the values $S_\varphi(v)$ computed in step 4 are equal to $\sum_{u=0}^{v-1} Deg^+(u) + \sum_{\psi=1}^{\varphi-1} \Delta_\psi(v)$.

Algorithm 4 Computing the Tails of the Arcs.

Output: The sorted array A.tail of the tails of the graph's arcs.

Step 1.   Initialize an array T of size m to 0.

Step 2.   **Mark T:**   **for all** $0 \leq v < n$ **do** $T(S_1(v)) := 1$

Step 3.   **tails:**   Compute the prefix sums $A.\text{tail}(a) := \sum_{b=0}^{a} T(b)$.

Now the degree of a vertex v can be found as $S_{k+1}(v) - S_1(v)$. $A.\text{tail}$ now looks as follows:

$$A.\text{tail} = \underbrace{0 \cdots 0}_{Deg^+(0)} \underbrace{1 \cdots 1}_{Deg^+(1)} \cdots\cdots \underbrace{n-1 \cdots n-1}_{Deg^+(n-1)}$$

Algorithm 5 Computing the Heads of the Arcs.

Output: The sorted array A.head of the heads of the graph's arcs.

Step 1.   Initialize an array A.phase of size m to 0.

Step 2.   **Mark $A.$phase:**   **for all** $0 \leq v < n$ **and** $1 \leq \varphi \leq k$ **do**
                                 $A.\text{phase}(S_\varphi(v)) := 1$

Step 3.   **Interval:**   Compute the prefix sums $A.\text{phase}(a) := \bigoplus_{b=0}^{a} A(b)$

Step 4.   **heads:**   **for all** $0 \leq a < m$ **do**
                                 $v := A.\text{tail}(a)$
                                 $\varphi := A.\text{phase}(a)$
                                 $A.\text{head}(a) := W_\varphi \left(l_\varphi(v) + a - S_\varphi(v)\right).$

The operation $\oplus$ is used to compute prefix sums in each adjacency list block. With $A(a) = (A.\text{tail}(a), A.\text{phase}(a)) = (u, \varphi)$ and $A(b) = (v, \psi)$, it is defined by $A(a) \oplus A(b) = (v, \psi)$ if $u < v$ and $A(a) \oplus A(b) = (u, \varphi + \psi)$ if $u = v$. Thus after step 3 A.phase looks as follows:

$$A.\text{phase} = \underbrace{\underbrace{1 \cdots 1}_{\Delta_1(0)} \cdots \underbrace{k \cdots k}_{\Delta_k(0)}}_{Deg^+(0)} \; \underbrace{\underbrace{1 \cdots 1}_{\Delta_1(1)} \cdots \underbrace{k \cdots k}_{\Delta_k(1)}}_{Deg^+(1)} \; \cdots\cdots \; \underbrace{\underbrace{1 \cdots 1}_{\Delta_1(n-1)} \cdots \underbrace{k \cdots k}_{\Delta_k(n-1)}}_{Deg^+(n-1)}$$

The complexity of the algorithm comes from the prefix sum's one. Note that only the last step requires a concurrent read.

6 Computing the Transitive Reduction of a Permutation Digraph

Let us finish by a similar algorithm for computing the transitive reduction. In contrast to the previous one, among the successors of a vertex, we only want to select those corresponding to arcs of the transitive reduction, see Fig. 3.

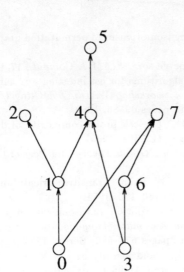

Fig. 3. The transitive reduction of the permutation digraph of Fig. 2.

An arc (i, j) of a permutation digraph is in the transitive reduction if in addition there is no k such that $i < k < j$ and $\pi(i) < \pi(k) < \pi(j)$, or equivalently if there is no k such that $i < k < j$ appearing between i and j in π^{-1}.

Let S_i be the list of the successors of i in the same order as in π^{-1}. The arcs (i, j) of the transitive reduction are those verifying $j \in S_i(p)$ and $j \geq S_i(q), \forall q < p$, or equivalently:

$$j \in S_i \qquad \text{and} \qquad j = \min S_i(1), S_i(2), \ldots, j \ .$$

If the adjacency lists of the transitive closure are given and ordered according to π^{-1}, a prefix sum over all these lists allows us to compute these minimums. Then a test of equality between a successor and this minimum determines whether it corresponds to an arc of the transitive reduction. This requires time $O(\log n)$ and a workload of $O(m)$ where m is the number of arcs of the transitive closure.

Since we can sort the adjacency lists in time $O(\log n)$ with a workload of $O(m \log n)$ [3], we obtain:

Theorem 5. *It is possible to compute the transitive reduction of a permutation digraph in time $O(\log^2 n)$ and a workload of $O((n+m) \log n)$ on a CREW PRAM.*

References

1. K.A. Baker, P.C. Fishburn, and F.S. Roberts. Partial orders of dimension 2. *Networks*, 2:11–28, 1971.

2. C.J. Colbourn. On testing isomorphism of permutation graphs. *Networks*, 11:13–21, 1981.

3. Richard Cole. Parallel merge sort. *SIAM J. Comput.*, 17(4), august 1988.

4. Paul F. Dietz. Optimal algorithms for list indexing and subset rank. In *Algorithms and data structures, Proc. workshop WADS '89, Ottawa/Canada*, number 382 in Lect. Notes Comput. Sci., pages 39–86, 1989.

5. M. Fredman and M. Saks. The cell probe complexity of dynamic data structures. In *21st ACM STOC*, pages 345–354, 1989.

6. M. C. Golumbic. *Algorithmic Graph Theory and Perfect Graphs*. Academic Press, New York, 1985.

7. A. Pnueli, A. Lempel, and W. Even. Transitive orientation of graphs and identification. *Canad. J. Math.*, 23:160–175, 1971.

8. J. Spinrad and J. Valdes. Recognition and isomorphism of two dimensional partial oreders. In *10th Coll. on Automata, Language and Programming*, number 154 in Lect. Notes Comput. Sci., pages 676–686, Berlin, 1983. Springer.

9. Jeremy Spinrad. Dimension and algorithms. In V. Bouchité and M. Morvan, editors, *Orders, Algorithms, and Applications*, number 831 in Lect. Notes Comput. Sci., pages 33–52. Springer-Verlag, 1994.

Homogeneously Orderable Graphs and the Steiner Tree Problem

Andreas Brandstädt[1], Feodor F. Dragan[2], Falk Nicolai[3]

[1] Universität Rostock, Fachbereich Informatik, Lehrstuhl für Theoretische
Informatik,
D 18051 Rostock, Germany
ab@informatik.uni-rostock.de

[2] Department of Mathematics and Cybernetics Moldova State University
A. Mateevici str. 60 Chişinău 277009 Moldova[†]
dragan@university.moldova.su

[3] Gerhard-Mercator-Universität –GH– Duisburg FB Mathematik FG Informatik I
D 47048 Duisburg, Germany[‡]
nicolai@marvin.informatik.uni-duisburg.de

Abstract. In this paper we introduce homogeneously orderable graphs
which are a common generalization of distance–hereditary graphs, dually
chordal graphs and homogeneous graphs. We present a characterization
of the new class in terms of a tree structure of the closed neighbourhoods
of homogeneous sets in 2–graphs which is closely related to the defining
elimination ordering.

The local structure of homogeneously orderable graphs implies a simple
polynomial time recognition algorithm for these graphs. Finally we give a
polynomial time solution for the Steiner tree problem on homogeneously
orderable graphs which extends the efficient solutions of that problem
on distance–hereditary graphs, dually chordal graphs and homogeneous
graphs.

1 Introduction

Some basic intractable algorithmic graph problems such as eg. the Steiner tree
problem on undirected graphs are efficiently solvable on restricted types of graphs
especially using tree structure properties. Several important graph classes have a
kind of tree structure which can be formulated in terms of hypergraph (namely
hypertree) properties. Among them are the well–known chordal graphs (dual
hypertrees of maximal cliques), the dually chordal graphs (hypertrees of maximal cliques, dual hypertrees of closed neighbourhoods [2]) and the distance–
hereditary graphs (dual hypertrees of maximal cographs [14]).

The tree structure of the last two classes turned out to be useful especially
for some distance and domination–like problems (cf. e.g. [8, 3, 4, 1]). The characterization of distance–hereditary graphs as dual hypertrees of cographs in [14] is

[†] Second author supported by the VW–Stiftung Project No. I/69041 and by DAAD
[‡] Third author supported by DFG

used in [15] for designing efficient algorithms solving various Hamiltonian problems.

In [7] homogeneous graphs as a generalization of distance–hereditary graphs are introduced which lead to a polynomial time algorithm for the Steiner tree problem on these graphs.

In this paper we define a new class of graphs which is a common generalization of distance–hereditary graphs, dually chordal graphs and homogeneous graphs. We present a characterization of the new class in terms of a tree structure of the closed neighbourhoods of homogeneous sets in 2–graphs which is closely related to the defining elimination ordering. Moreover we give a polynomial time solution for the Steiner tree problem on homogeneously orderable graphs which extends the efficient solutions of that problem on distance–hereditary graphs, dually chordal graphs and homogeneous graphs. In the full version of this paper we characterize the hereditary homogeneously orderable graphs by forbidden induced subgraphs as the house–hole–domino–sun–free graphs. Moreover, in [9] we present efficient algorithms for several domination–like problems on homogeneously orderable graphs.

2   Preliminaries

Throughout this paper all graphs $G = (V, E)$ are finite, undirected, simple and connected. The *(open) neighbourhood* of a vertex v of G is $N(v) := \{u \in V : uv \in E\}$. For a vertex set $U \subseteq V$ let $N(U) := \bigcup_{u \in U} N(u) \setminus U$.

A nonempty set $U \subseteq V$ is *homogeneous* in $G = (V, E)$ iff all vertices of U have the same neighbourhood in $V \setminus U$:

$$N(u) \cap (V \setminus U) = N(v) \cap (V \setminus U) \text{ for all } u, v \in U,$$

i.e. any vertex $w \in V \setminus U$ is adjacent to either all or none of the vertices from U. A homogeneous set H is *proper* iff $|H| < |V|$. Trivially for each $v \in V$ the singleton $\{v\}$ is a proper homogeneous set.

The *distance* $d_G(u, v)$ of vertices u, v is the minimal length of any path connecting these vertices. The *k–th neighbourhood* $N^k(v)$ of a vertex v of G is the set of all vertices of distance k to v. For convenience we denote by $N_F^k(v)$ the intersection $N^k(v) \cap F$ where $F \subseteq V$.

The *disk* of radius k centered at v is the set of all vertices of distance at most k to v : $D(v, k) := \{u \in V : d_G(u, v) \leq k\}$. Analogously to neighbourhoods of sets we define for $U \subseteq V$ $D(U, k) := \bigcup_{u \in U} D(u, k)$. For convenience we denote by $D_F(U, k)$ the intersection $D(U, k) \cap F$ where $F \subseteq V$.

The *k–th power* G^k of a graph $G = (V, E)$ is the graph with vertex set V and edges between vertices u, v with distance $d_G(u, v) \leq k$.

Let $e(v)$ denote the *eccentricity* of vertex $v \in V$: $e(v) := \max\{d(v, u) : u \in V\}$. The *diameter* $diam(G)$ of a graph G is the maximum over all eccentricities $e(v)$, $v \in V$.

In the sequel a subset U of V is a *k–set* iff U induces a clique in the power G^k, i.e. for any pair x, y of vertices of U we have $d_G(x, y) \leq k$. A graph G is a k-graph iff $diam(G) \leq k$. If U is a subset of $V(G)$ then U is a k-graph of G iff the induced subgraph G_U is a k-graph (i.e. $diam(G_U) \leq k$) and for any set $U' \supset U$ holds $diam(G_{U'}) \geq k+1$, i.e. G_U is a maximal induced subgraph of diameter $\leq k$ in G. Thus each k-graph is a k-set but the converse is in general not true.

Let U_1, U_2 be disjoint subsets of V. U_1 and U_2 form a *join*, denoted by $U_1 \bowtie U_2$, iff every vertex of U_1 is adjacent to every vertex of U_2.

Let $H = (V, \mathcal{E})$ be a hypergraph, i.e. $\mathcal{E}$ is a set of subsets of V. Throughout this paper all hypergraphs are assumed to be reduced, i.e. no hyperedge is properly contained in another one.

For every vertex $v \in V$ let $\mathcal{E}(v) := \{e \in \mathcal{E} : v \in e\}$ be the set of hyperedges incident to vertex v. Then the *dual hypergraph* H^* of H is the hypergraph with vertex set $\mathcal{E}$ and hyperedges $\mathcal{E}(v)$, $v \in V$. It is well-known that $(H^*)^*$ is isomorphic to H.

The *2-section graph* $2SEC(H)$ is the graph with vertex set V, where two vertices are adjacent iff there is a hyperedge in H containing both.

Tree structure of a hypergraph can be defined as follows : A hypergraph $H = (V, \mathcal{E})$ is a *hypertree* iff there is a tree T with vertex set V such that any hyperedge e of H induces a subtree in T. A hypergraph $H = (V, \mathcal{E})$ is a *dual hypertree* iff H^* is a hypertree.

Hypertrees and dual hypertrees are closely related to chordal graphs – the graphs which do not contain any chordless cycle of length ≥ 4. Indeed, it is well-known that a graph G is chordal iff its clique hypergraph is a dual hypertree.

A hypergraph H is *conformal* iff any clique of $2SEC(H)$ is contained in some hyperedge.

Theorem 2.1 ([10], [11]) *A hypergraph H is a dual hypertree iff H is conformal and $2SEC(H)$ is chordal.*

Next we recall the definition and some characterizations of dually chordal graphs. A vertex u is a *maximum neighbour* of a vertex v iff $D(u, 1) = D(v, 2)$. A *maximum neighbourhood ordering* of a graph G is a sequence $(v_1, \ldots, v_n)$ such that for all $i = 1, \ldots, n$ the vertex v_i has a maximum neighbour in $G_i := G(\{v_i, \ldots, v_n\})$. Let $\mathcal{N}(G) = \{D(v, 1) : v \in V\}$ be the *neighbourhood hypergraph* of G and let $\mathcal{C}(G) = \{C : C$ is a maximal clique in $G\}$ be the *clique hypergraph* of G.

In [2] it is shown that a graph G has a maximum neighbourhood ordering iff the clique hypergraph $\mathcal{C}(G)$ is a hypertree iff the neighbourhood hypergraph $\mathcal{N}(G)$ is a hypertree.

Due to the first equivalence graphs with maximum neighbourhood ordering are called *dually chordal*.

Finally we recall the definition and some characterizations of distance–hereditary graphs. An induced subgraph H of G is *isometric* iff the distances $d_H(u, v)$ of any vertices u, v in H are the same as in G. A graph G is *distance–hereditary* iff each connected induced subgraph H is isometric [13].

The following characterizations are due to [14] : A vertex v is called 2-simplicial iff the disk $D(v, 2)$ induces a cograph in G. Hereby a *cograph* is a P_4-free graph, i.e. a connected cograph is a hereditary 2-graph. An ordering $\tau = (v_1, \ldots, v_n)$ of the vertices of G is a *2-simplicial ordering* iff for every index $i = 1, \ldots, n$ the vertex v_i is 2-simplicial in $G_i := G(\{v_i, \ldots, v_n\})$. A 2-simplicial vertex v is *d-extremal* iff $e(v) = diam(G)$. Analogously we can define a d-extremal ordering. Let $CC(G)$ denote the set of maximal connected cographs of G.

In [14] it is proved that a graph G *is distance-hereditary iff the cograph-hypergraph* $CC(G)$ *of* G *is a dual hypertree iff* G *has a 2-simplicial ordering iff* G *has a d-extremal ordering.*

Moreover, d-extremal vertices have nice local properties as shown in [14] : If G is a distance-hereditary graph and v is a d-extremal vertex in G then there is a set $S \subseteq N(v)$ which is homogeneous (in G) and dominates $D(v, 2)$.

Note, that in dually chordal graphs a maximum neighbour u of a vertex v dominates $D(v, 2)$ too. Thus we can generalize these properties in the following way :

A vertex v of $G = (V, E)$ with $|V| > 1$ is *h-extremal* iff there is a proper subset $H \subset D(v, 2)$ which is homogeneous in G and for which $D(v, 2) \subseteq D(H, 1)$ holds. A sequence $\sigma = (v_1, \ldots, v_n)$ is a *h-extremal ordering* iff for any $i = 1, \ldots, n-1$ the vertex v_i is h-extremal in $G_i := G(\{v_i, \ldots, v_n\})$. A graph G is *homogeneously orderable* iff G has a h-extremal ordering. Thus we immediately obtain

Corollary 2.2 *Dually chordal and distance-hereditary graphs are homogeneously orderable graphs.*

Sometimes we will write $\sigma = ((v_1, H_1), \ldots, (v_n, H_n))$ to emphasize the homogeneous dominating sets H_j for v_j in G_j.

Lemma 2.3 *If H is a proper homogeneous set in G and $x, y \in D(H, 1)$ then $d(x, y) \leq 2$.*

We immediately conclude

Corollary 2.4 *If v is h-extremal in G then $D(v, 2)$ is a 2-graph.*

Lemma 2.5 *Let v be a vertex in a graph $G = (V, E)$.*

(i)   *If $e(v) \geq 2$ and v is h-extremal then there is a proper homogeneous set $H \subseteq N(v)$ which dominates $D(v, 2)$.*

(ii) *If $e(v) = 1$ then G is homogeneously orderable with h-extremal ordering $\sigma = ((v_1, \{v\}), \ldots, (v_{n-1}, \{v\}))$ where $V = \{v_1, \ldots, v_{n-1}, v\}$.*

Proof : For point (ii) there is nothing to show. So let $e(v) \geq 2$ and let H be a proper homogeneous dominating set in $D(v, 2)$. If $v \notin H$ then v must be dominated by some $h \in H$. Since H is homogeneous we immediately conclude $H \subseteq N(v)$.

Now consider the case $v \in H$. Assume first that $N^3(v) \neq \emptyset$ and let $u \in N^3(v)$. Since H is homogeneous, $v \in H$ and $vx \notin E$ for any neighbour x of u in

$N^2(v)$ no one of these neighbours x is in $N(H)$. But H dominates $D(v, 2)$, hence $N(u) \cap N^2(v) \subset H$. But now H is not homogeneous.

Finally assume that $N^3(v) = \emptyset$, i.e. G is a 2–graph. Since v is in the homogeneous set H and v is not adjacent to any vertex of $N^2(v)$, but H dominates $D(v, 2)$, the second neighbourhood $N^2(v)$ must be completely contained in H. But now, $N(v) \setminus H$ is homogeneous in G and dominates $D(v, 2)$, so we have the desired set. $\square$

Therefore, for a homogeneously orderable graph G with h–extremal ordering $\sigma = ((v_1, H_1), \ldots, (v_n, H_n))$ we will assume $H_j \subseteq N_{G_j}(v_j)$ for $j = 1, \ldots, n-1$ in the sequel.

3 Homogeneously orderable graphs and corresponding hypergraphs

Let $\mathcal{DH}(G)$ denote the hypergraph defined by the disks $D_F(H, 1)$ where F is a 2–graph of G and H is a proper homogeneous set in F. We will show that a graph G is homogeneously orderable iff the hypergraph $\mathcal{DH}(G)$ is a dual hypertree.

To prove that $\mathcal{DH}(G)$ of a homogeneously orderable graph G is a dual hypertree we use Theorem 2.1, i.e. we show that $2SEC(\mathcal{DH}(G))$ is chordal and $\mathcal{DH}(G)$ is conformal.

Lemma 3.1 *For any graph G we have $2SEC(\mathcal{DH}(G)) = G^2$.*

Proof : Let xy be an edge in $2SEC(\mathcal{DH}(G))$. Then by definition there must be an hyperedge $D_F(H, 1)$ containing both vertices. From Lemma 2.3 we obtain $d(x, y) \leq 2$, hence these vertices are adjacent in G^2.

Let xy be an edge in G^2, that is $d_G(x, y) \leq 2$. Consider a 2–graph F of G containing both vertices, and if $d_G(x, y) = 2$, a vertex w which is adjacent to both. Obviously there is a proper homogeneous set H containing x for $d_G(x, y) = 1$ and containing w for $d_G(x, y) = 2$, respectively. In both cases $\{x, y\}$ is a subset of $D_F(H, 1)$. $\square$

The following straightforward lemma will be used frequently in the sequel.

Lemma 3.2 *Let G be a graph and let v be a h–extremal vertex of G with $e(v) \geq 2$. Then $G \setminus \{v\}$ is an isometric subgraph of G. In particular we have $G^2 \setminus \{v\} = (G \setminus \{v\})^2$.*

Proof : Since $e(v) \geq 2$ we can choose a homogeneous set $H \subseteq N(v)$ dominating $D(v, 2)$ due to Lemma 2.5. Thus the distances in $G \setminus \{v\}$ are the same as in G. $\square$

Lemma 3.3 *For any homogeneously orderable graph G with h–extremal ordering σ the square G^2 is chordal and σ is a perfect elimination ordering of G^2.*

Proof : Let G be a homogeneously orderable graph with h–extremal ordering $\sigma = ((v_1, H_1), \ldots, (v_n, H_n))$. We prove that v_1 is simplicial in G^2. Let x, y be neighbours of v_1 in G^2. Hence, $d_G(x, v_1) \leq 2$ and $d_G(y, v_1) \leq 2$, i.e. both x and y are contained in $D_G(v_1, 2)$ which is dominated by H_1. Thus, Lemma 2.3 implies $d_G(x, y) \leq 2$. Therefore, x and y are adjacent in G^2 and $D_{G^2}(v_1, 1)$ is complete, that is v_1 is simplicial in G^2.

If $e(v_1) \geq 2$ we can proceed by induction on the position in σ due to the preceding lemma. Otherwise, $G = D(v_1, 1)$ and G^2 is complete. $\qquad \square$

Now we prove the conformality of $\mathcal{DH}(G)$.

Lemma 3.4 *If G is homogeneously orderable then $\mathcal{DH}(G)$ is conformal.*

Proof : Let $\sigma = ((v_1, H_1), \ldots, (v_n, H_n))$ be a h–extremal ordering of G and let $C = \{c_1, \ldots, c_k\}$ be a maximal clique in $2SEC(\mathcal{DH}(G)) = G^2$ such that $c_1 = v_l$ is the leftmost vertex of C with respect to σ. Since C is maximal in G^2 it cannot be completely contained in $D_{G_i}(v_i, 2)$ for $i = 1, \ldots, l-1$ implying $e_{G_i}(v_i) \geq 2$ for $i = 1, \ldots, l-1$. Thus, by Lemma 3.2 C is a clique in $(G_l)^2$, i.e. for all $i, j = 1, \ldots, k$ we have $d_{G_l}(c_i, c_j) \leq 2$. Since $v_l = c_1$ is h–extremal in G_l we immediately conclude $C \subseteq D(H_l, 1) = D_{G_l}(v_l, 2)$. Suppose $D_{G_l}(v_l, 2)$ is not a 2–graph of G. Then it must be properly contained in a 2–graph F of G. But F induces a clique in G^2 contradicting the maximality of C. Thus, $D_{G_l}(v_l, 2)$ is a 2–graph of G and C is contained in a hyperedge. $\qquad \square$

In order to prove the converse, i.e. if $\mathcal{DH}(G)$ is a dual hypertree then G is homogeneously orderable, we introduce another hypergraph. Consider a 2–graph F which is dominated by some homogeneous (in F) set H, i.e. $F = D_F(H, 1)$. Then F is splitted into two joined sets, namely H and $N_F(H) : F = H \bowtie N_F(H)$. In general, a set $U \subseteq V$ is *join–splitted* iff U is the join of two nonempty sets, i.e. $U = U_1 \bowtie U_2$. Since any edge of a graph is a join–splitted set each connected graph can be covered by join–splitted sets. Thus we can define the hypergraph $\mathcal{SH}(G)$ of the maximal join–splitted sets of G, and immediately obtain the following

Lemma 3.5 *For any graph G we have $2SEC(\mathcal{SH}(G)) = G^2$.*

So we get

Theorem 3.6 *The following conditions are equivalent for a graph G :*

(i)   *G is homogeneously orderable.*
(ii) *$\mathcal{DH}(G)$ is a dual hypertree.*
(iii) *G^2 is chordal and every maximal 2–set of G is join–splitted.*
(iv) *$\mathcal{SH}(G)$ is a dual hypertree.*

Proof :

(i) $\Longrightarrow$ (ii) Follows from the preceding results.
(ii) $\Longrightarrow$ (iii) By Lemma 3.1 the square G^2 is chordal. Let S be a maximal 2–set

in G. Thus S is a maximal clique in G^2, and the conformality of $\mathcal{DH}(G)$ implies $S \subseteq D_F(H,1)$ for some 2–graph F of G and some homogeneous set H of F. From the maximality of S we conlude $S = D_F(H,1)$ and hence $S = F$. But now $S = H \bowtie N_S(H)$, so we are done.

$(iii) \Longrightarrow (i)$ Let v be a simplicial vertex of the chordal graph G^2. If $e(v) = 1$ we are done by Lemma 2.5. So let $e(v) \geq 2$. We show that v is h–extremal in G. Since v is simplicial in G^2 the disk $D(v,2)$ is complete in G^2. Thus that disk is a maximal 2–set in G and hence join–splitted, say $D(v,2) = X \bowtie Y$. W.l.o.g. assume $v \in X$ implying $Y \subseteq N(v)$. Therefore, Y is the desired homogeneous set dominating $D(v,2)$, and v is h–extremal. By Lemma 3.2 $(G \setminus \{v\})^2$ is chordal, and obviously each maximal 2–set of $G \setminus \{v\}$ is join–splitted.

$(iv) \Longleftrightarrow (iii)$ By Lemma 3.5 and Theorem 2.1 point (iv) is a reformulation of (iii). $\qquad \square$

The following two corollaries are important for algorithmic purposes.

Corollary 3.7 *If G is homogeneously orderable then for each perfect elimination ordering $\sigma = (v_1, \ldots, v_n)$ of G^2 and $k(\sigma) := \min\{i : G^2(\{v_i, \ldots, v_n\})$ is complete\} there exists a h–extremal ordering τ of G such that $\tau(i) = \sigma(i)$ for $i = 1, \ldots, k(\sigma) - 1$.*

Corollary 3.8 *If G is a homogeneously orderable graph and v is an arbitrary vertex of G then there is a h–extremal ordering σ of G with v at the end.*

Proof : Recall that for chordal graphs each vertex can be placed at the end of a perfect elimination ordering. Let τ be such a perfect elimination ordering of G^2. Thus the index of v in τ is at least $k(\tau)$. By the above Corollary and by Lemma 5.1 we are done. $\qquad \square$

4 Homogeneous reductions and extensions

In [7] the authors generalize distance–hereditary graphs. Recall that any distance–hereditary graph can be generated from a single vertex by a sequence of the following three one–vertex extensions. Let $G' = (V', E')$ be a graph, $x' \in V'$ and $x \notin V'$. Define $G := (V' \cup \{x\}), E' \cup E_x)$ where E_x is defined as follows :

PV $E_x := \{xx'\}$ — x is a pendant vertex (leaf) to x,
FT $E_x := \{xy : y \in N_{G'}(x')\}$ — x and x' are false twins,
TT $E_x := \{xy : y \in D_{G'}(x', 1)\}$ — x and x' are true twins.

It is obvious that in the case of twin operations $\{x, x'\}$ forms a homogeneous set in G. Now, in [7] instead of twins (as a special kind of homogeneous sets) arbitrary homogeneous sets are used.

Let H be a proper homogeneous set of G containing at least two vertices and let $v_H \in H$. Then the graph $H\,Red(G, H, v_H)$ obtained from G by deleting $H \setminus \{v_H\}$, i.e. contracting H to a representing vertex v_H, will be called the *homogeneous reduction* of G (via H).

Conversely, the *homogeneous extension* $HExt(G, v, H)$ of G via a graph H in v with $V(H) \cap V(G) = \emptyset$ is the graph obtained by substituting v by H such that the vertices of H have the same neighbours outside of H as v had in G.

Thus, in distance–hereditary graphs the FT operation is the homogeneous extension of G' in x' via the non–edge $\{x, x'\}$. For a TT operation G' is homogeneously extended in x' via the edge $\{xx'\}$.

In what follows we want to clarify the relations between some graph classes.

In the sense of [7] a graph G is a *homogeneous graph* iff the iterated reduction via proper homogeneous sets of 2–connected components leads to a tree.

Let $\mathcal{H}(G)$ denote the set of all maximal proper homogeneous sets of G.

Lemma 4.1 *If $H \in \mathcal{H}(G)$ and S is a maximal 2–set of G such that $H \cap S \neq \emptyset$ then $H \subset S$.*

Proof : First suppose $S \subseteq H$. Since G is connected and H is proper there must be a vertex w in $V \setminus H$ such that $H \subseteq N(w)$ implying $S \subseteq N(w)$, a contradiction. Thus, $S \setminus H \neq \emptyset$. Next suppose $H \setminus S \neq \emptyset$. Define $S' := H \cup S$. We prove that S' is a 2–set which contradicts the maximality of S. Let $w \in H \cap S$ and consider two vertices x, y of S'. If both vertices are in S then $d(x, y) \leq 2$. Now let $x \in S \setminus H$ and $y \in H \setminus S$. If x is adjacent to w then x must be adjacent to y, since H is homogeneous in G. Otherwise there must be a vertex v adjacent to both x and w. Note that v cannot be in H. Thus $yv \in E$ and $d(x, y) = 2$. Finally let both x and y be in H and choose a vertex $v \in S \setminus H$. If $wv \in E$ then both x and y must be adjacent to v. Otherwise there is a vertex u in $V \setminus H$ adjacent to w and v. Hence, x and y are adjacent to u too. So we are done. $\quad\square$

Lemma 4.2 *Let G be a homogeneously orderable graph, H be a nontrivial proper homogeneous set of G and v any vertex of H. Then the graph $G \setminus \{v\}$ is homogeneously orderable too.*

Proof : It is easy to see that $G \setminus \{v\}$ is an isometric subgraph of G. Thus $(G \setminus \{v\})^2 = G^2 \setminus \{v\}$ which is chordal. Consider an arbitrary maximal 2–set S in $G \setminus \{v\}$. If S is not maximal in G then $S' := S \cup \{v\}$ is a maximal 2–set in G. Therefore S' is join–splitted in G, say $S' = X \bowtie Y$ with $v \in Y$. If Y contains at least two vertices we are done. So let $Y = \{v\}$. By the maximality of S' the preceding lemma implies $H \subset S'$. Since H is homogeneous and $X \subseteq N(v)$ we can split S' by $X \setminus \{H\} \bowtie H$. Consequently, S is join–splitted in $G \setminus \{v\}$ because H is nontrivial. The assumption follows from Theorem 3.6. $\quad\square$

Lemma 4.3 *If G is chordal then so is the graph obtained from G by adding a true twin v to some vertex x of G.* $\quad\square$

Lemma 4.4 *Let G be a homogeneously orderable graph and H be a proper homogeneous set of G. Then any graph $G + v := (V \cup \{v\}, E \cup \{vx : x \in N_G(H)\} \cup E')$, where $E' \subseteq \{vx : x \in H\}$, is homogeneously orderable too.*

Proof: Since $N_{V \setminus H}(v) = N(H)$ it is easy to see that G is an isometric subgraph of $G + v$, and $G^2 = (G+v)^2 \setminus \{v\}$. So $(G+v)^2$ can be obtained from the chordal graph G^2 by adding the true twin v to some vertex x of H since $D_G(v, 2) = D_G(x, 2)$. Hence $(G + v)^2$ is chordal by Lemma 4.3.

Consider an arbitrary maximal 2-set S in $G + v$. If S does not contain v then S is a maximal 2-set in G and hence is join–splitted. Otherwise $S' := S \setminus \{v\}$ is a maximal 2-set in G, and Lemma 4.1 implies $H \subset S$. Therefore S' is join–splitted, i.e. $S' = X \bowtie Y$. If H is completely contained in one of the splitting sets we can add v to this one to obtain a splitting for S in $G + v$. So assume $H \cap X \neq \emptyset$ and $H \cap Y \neq \emptyset$. But in this case we can split S into $X \setminus H$ and $Y \cup H$, thus we are again in the preceding case. By Theorem 3.6 we are done. □

Corollary 4.5 *Let G be a homogeneously orderable graph.*

1. *If $H \in \mathcal{H}(G)$ and $G' := H\,Red(G, H, v_H)$ then G' is homogeneously orderable too.*
2. *If $v \in V(G)$, H an arbitrary graph and $G' := H\,Ext(G, v, H)$ then G' is homogeneously orderable too.*

Proof : Follows immediately from the preceding two lemmata. □

Thus we can summarize our results to

Corollary 4.6 *Homogeneously orderable graphs are closed under homogeneous extensions, homogeneous reductions and under deleting and adding of a vertex with maximum neighbour.*

Proof : The first two points follow directly from the above Corollary. To show the third let G be a homogeneously orderable graph and $\sigma = (v_1, \ldots, v_n)$ be a h–extremal ordering of G. Furthermore, let $y \notin G$ be a vertex with maximum neighbour $x \in G$. Obviously $\{x\}$ is a homogeneous set dominating $D(y, 2)$. Thus, $\tau = (y, v_1, \ldots, v_n)$ is a h–extremal ordering of the new graph. □

Theorem 4.7 *The homogeneously orderable graphs are exactly those graphs which can be generated from a single vertex by adding a vertex with maximum neighbour and by homogeneous extensions, i.e. $\Gamma_{\{MN, HExt\}}(K_1)$ is the class of homogeneously orderable graphs.*

Proof : By Corollary 4.6 any graph from $\Gamma_{\{MN, HExt\}}(K_1)$ is homogeneously orderable. To prove the converse note that every h–extremal vertex v either has a maximum neighbour or contains a proper nontrivial homogeneous set H in its neighbourhood. After homogeneously reducing H to v_H vertex v has a maximum neighbour v_H. □

Lemma 4.8 *A graph G is homogeneously orderable iff each 2-connected component of G is homogeneously orderable.*

Proof : Let G be a homogeneously orderable graph with h–extremal ordering $\sigma = (v_1, \ldots, v_n)$. Denote by $\sigma|_K$ the ordering of vertices of a 2–connected component K of G induced by σ. We show that $\sigma|_K = (v_{i_1}, \ldots, v_{i_l})$ is a h–extremal ordering for K. Suppose the contrary and let i_j, $j < l$, be the smallest index such that v_{i_j} is not h–extremal in $K_{i_j} := K(\{v_{i_j}, \ldots, v_{i_l}\})$. By Lemma 3.2 the graph K_{i_j} is an isometric subgraph of K and hence connected. Since v_{i_j} is h–extremal in G_{i_j} there is a proper homogeneous set H_{i_j} dominating $D_{G_{i_j}}(v_{i_j}, 2)$. Obviously, if $H' := H_{i_j} \cap K_{i_j}$ is nonempty then H' dominates $D_{K_{i_j}}(v_{i_j}, 2)$. Thus H' is empty, i.e. $H_{i_j} \cap K = \emptyset$. This implies, together with $H_{i_j} \subseteq N(v_{i_j})$ and $D_{G_{i_j}}(v_{i_j}, 2) = D_{G_{i_j}}(H_{i_j}, 1)$, that $G_{i_j} \cap K = \{v_{i_j}\}$ and hence $j = l$, a contradiction.

In order to prove the converse consider the tree $T(G)$ defined by the 2–connected components of G. Let K be a leaf of $T(G)$ and v be the only cutvertex of G in K. Then by Corollary 3.8 we have a h–extremal ordering $\sigma_K = (v_{i_1}, \ldots, v_{i_r}, v)$ of K with vertex v at the end. By induction hypothesis $G \setminus (K \setminus \{v\})$ possesses a h–extremal ordering $\tau = (v_1, \ldots, v_l)$. Obviously, $\sigma := (v_{i_1}, \ldots, v_{i_r}, v_1, \ldots, v_l)$ is a h–extremal ordering of G. $\qquad\square$

Remark that a graph G is homogeneous iff each 2–connected component of G is a homogeneous graph (cf. [7]). Thus we can prove

Corollary 4.9 *Homogeneous graphs are homogeneously orderable.*

Proof : By Lemma 4.8 and the remark it is sufficient to show that every 2–connected homogeneous graph is homogeneously orderable. If there is no nontrivial homogeneous set then G is a tree and we are done. Otherwise we proceed by induction. Let H be a nontrivial proper homogeneous set of a 2–connected homogeneous graph G. By the definition of homogeneous graphs $G' := HRed(G, H, v_H)$ is homogeneous too. Thus, by induction hypothesis G' is homogeneously orderable. Since $G = HExt(G', v_H, H)$ the assertion follows from Corollary 4.6. $\qquad\square$

5   A recognition algorithm

Our polynomial time recognition algorithm of homogeneously orderable graphs bases on Lemma 2.5, Corollary 3.7 and

Lemma 5.1 *Let $\mathcal{DH}(G)$ be a dual hypertree with a hyperedge $D(H, 1) = G$. Then G is homogeneously orderable and, with $H := \{u_1, \ldots, u_k\}$, $V \setminus H := \{v_1, \ldots, v_l\}$, both*

$$\sigma = ((u_1, V \setminus H), \ldots, (u_{k-1}, V \setminus H), (v_1, \{u_k\}), \ldots, (v_l, \{u_k\})),$$
$$\sigma' = ((v_1, H), \ldots, (v_{l-1}, H), (u_1, \{v_l\}), \ldots, (u_k, \{v_l\}))$$

are h–extremal orderings of G.

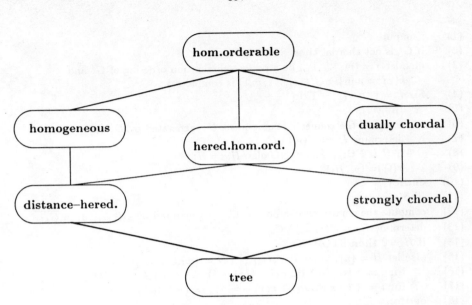

Fig. 1. Inclusion hierarchy of the considered graph classes.

Proof : Follows immediately from $D(H,1) = G$. □

Assume that v is h–extremal with $e(v) \geq 2$ and let $\overline{G} = (V, \overline{E})$ be the complement of G. By Lemma 2.5 there is a proper homogeneous dominating set $H \subseteq N(v)$ dominating $D(v,2)$. Thus, in $\overline{G}$ no vertex of H is adjacent to some vertex of $D_G(v,2) \setminus H$. It suffices to consider the complement of the graph induced by the disk $D(v,2)$. Let C_v denote the connected component of this graph which contains v (and hence $N^2(v)$).

Lemma 5.2 *A vertex v such that $e(v) \geq 2$ is h–extremal iff $H := N(v) \setminus C_v$ is a homogeneous set.*

Proof : Let C_v be the connected component of the graph induced by $D(v,2)$ such that $v \in C_v$. If v is h–extremal then by Lemma 2.5 there is a homogeneous set $H' \subseteq N(v)$ dominating $D(v,2)$. Obviously $H' \cap C_v = \emptyset$. Since $H' \subseteq H$ the set H is nonempty. So it remains to show that H is homogeneous, but this is trivial. The other direction is obvious. □

In the following algorithm all neighbourhoods are restricted to the rest graph $G_i := G(\{v_i, \ldots, v_n\})$.

Algorithm RecHom :
Input: A connected graph $G = (V, E)$.
Output: A h–extremal ordering $\sigma = ((v_1, H_1), \ldots, (v_n, H_n))$ or answer 'NO'.

(1)   Compute G^2.

(2)   **if** G^2 is not chordal **then** STOP.NO

(3)   **else let** $\tau = (v_1, \ldots, v_n)$ be a perfect elimination ordering of G^2 and
          $k(\tau) := \min\{i : G^2(\{v_i, \ldots, v_n\})$ is complete $\}$;

(4)   **for** $i := 1$ **to** $k(\tau) - 1$ **do**

(5)   BFS(v_i);

(6)   compute the connected component C_{v_i} generated by $D(v_i, 2)$ in $\overline{G_i}$;

(7)   determine $H := N(v_i) \setminus C_{v_i}$;

(8)   **if** $H = \emptyset$ **then** STOP.NO **else** $H_i := H$;

(9) $\sigma(i) := (v_i, H_i)$;

(10)   **endfor**;

(11)   BFS($v_{k(\tau)}$);

(12)   compute the connected component $C_{v_{k(\tau)}}$ generated by $D(v_{k(\tau)}, 2)$ in $\overline{G_{k(\tau)}}$;

(13)   determine $H := N(v_{k(\tau)}) \setminus C_{v_{k(\tau)}}$;

(14)   **if** $H = \emptyset$ **then** STOP.NO

(15)   **else let** $H = \{u_1, \ldots, u_r\}$; $V(G_{k(\tau)}) \setminus H = \{w_1, \ldots, w_s\}$;

(16)   **for** $i := 1$ **to** $r - 1$ **do** $\sigma(i + k(\tau) - 1) := (u_i, V(G_{k(\tau)}) \setminus H)$;

(17)   **for** $i := 1$ **to** s **do** $\sigma(i + k(\tau) + r - 2) := (w_i, \{u_r\})$;

(18)   **endfor**

Theorem 5.3 *The algorithm RecHom is correct and works within $O(n^3)$ steps.*

Proof : The correctness of the algorithm immediately follows from Theorem 3.6, Lemma 2.5, Corollary 3.7, Lemma 5.1 and Lemma 5.2.

Time bound : The square G^2 of a graph G can be computed in at most $O(n^3)$ steps. The chordality can be tested in linear time in the size of G^2 and if the graph is chordal one gets a perfect elimination ordering τ in the same time. Thus (1)–(3) are done in $O(n^3)$ steps. Lines (5)–(10) are iterated $k(\tau)$ times. BFS takes $O(|E|)$ steps and finding the connected components in the complement graph $\overline{G}$ takes $O(|\overline{E}|)$ steps. Thus the total amount of time is bounded by $O(n^3)$.   □

6   The Steiner tree problem

In this section we give an efficient algorithm for solving the Steiner tree problem on homogeneously orderable graphs. Recall that given a Steiner set $T \subset V$ we have to compute a minimal set $S \subseteq V$ such that $T \subseteq S$ and $G(S)$ is connected.

Lemma 6.1 *Let v be a h-extremal vertex of a graph $G = (V, E)$ with homogeneous dominating set $H \subset N(v)$ and $u, w \in N(v)$. Define $G' := (V, E \cup \{uw\})$. Then only the distance between u and w is changed in G', i.e. for any vertex x of V and any vertex y of $V \setminus \{u, w\}$ we have $d_G(x, y) = d_{G'}(x, y)$.*   □

Lemma 6.2 *Let G be a homogeneously orderable graph, $\sigma = ((v_1, H_1), \ldots, (v_n, H_n))$ be a h-extremal ordering of G, $u, w \in N(v_1) \setminus H_1$, G' defined as above with $v = v_1$. Then G' is homogeneously orderable and σ is a h-extremal ordering of G'.*

Proof : Since $G^2 = (G')^2$ by Corollary 3.7 it suffices to show that H_i remains homogeneous in G_i'. Suppose the contrary and let i ($i \geq 2$) be the smallest index such that the set H_i is not homogeneous in G_i'. Then, u, w are in G_i and, say, $u \in H_i$, $w \notin H_i$. Since H_i dominates $D_{G_i}(v_i, 2)$ we conclude $d_{G_i}(v_i, w) \geq 3$. Lemma 3.2 implies $d_G(v_i, w) \geq 3$. But then $d_{G_i'}(v_i, w) = 2$ contradicts Lemma 6.1. $\qquad \square$

In order to present the algorithm we consider a homogeneously orderable graph G with a h–extremal vertex v and a homogeneous set $H \subseteq N(v)$ dominating $D(v, 2)$. Furthermore, let $T \subset V$ be a given disconnected set of target vertices. For the sequel define $G' := G \setminus \{v\}$ and let S' be an optimal solution of the Steiner tree problem in G' with respect to a Steiner set T' defined in the different cases :

Case 1. $T \subseteq D(v, 1)$.
This is a trivial case. With $S := T \cup \{v\}$ we are done.

Case 2. $v \in T$ and $T \cap N(v) = \emptyset$ but $T \setminus D(v, 1) \neq \emptyset$.
Define $T' := (T \setminus \{v\}) \cup \{h\}$ for some vertex $h \in H$ and $S := S' \cup \{v\}$. We claim that S is optimal for G with respect to T.

Suppose to the contrary that there is a set F containing T such that $G(F)$ is connected and $|F| < |S|$. Since $v \in T$ we have $v \in F$. Define $F' := F \setminus \{v\}$. For $|F'| = |F| - 1 < |S| - 1 = |S'|$ and by the optimality of S' the set F' cannot be a solution in G' for T'. First assume that F' is not connected. Then v is a cutvertex in $G(F)$. Let $x_1, \ldots, x_k$, $k \geq 2$, be the neighbours of v in F. Since H is homogeneous and a subset of $N(v)$ dominating the whole disk $D(v, 2)$ no x_i, $i = 1, \ldots, k$, can belong to H. But $T \cap N(v) = \emptyset$. So we can replace the vertices $x_1, \ldots, x_k$ by some vertex $h \in H$ obtaining a smaller set, a contradiction. Hence F' is connected and it cannot contain h. But $v \in F$ and $T \setminus D(v, 1) \neq \emptyset$. Thus, F' must contain some vertex from $N(v)$ which can be replaced by h, again a contradiction. Consequently, S is optimal.

Case 3. $v \in T$ and $T \cap N(v) \neq \emptyset$ and $T \setminus D(v, 1) \neq \emptyset$.

Case 3.1. $T \cap H = \emptyset$.
Here we make $T \cap N(v)$ complete and define $G' := (V \setminus \{v\}, (E \setminus \{vx : x \in N(v)\}) \cup \{xy : x, y \in T \cap N(v)\})$ and $S := S' \cup \{v\}$. By Lemma 6.2 adding these edges has no consequences for $\sigma = ((v_1, H_1), \ldots, (v_n, H_n))$.

To verify the correctness note that for each solution F in G for T the set $F \setminus \{v\}$ is a solution in G' for T' accept the case $F \cap (N(v) \setminus T) \neq \emptyset$. But now we consider $F \setminus (D(v, 1) \setminus T) \cup \{h\}$ for some $h \in H$ which is a solution in G' for T'.

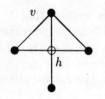 $H = \{h\}$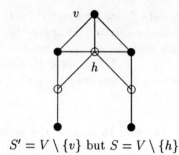

$$S = S' \cup \{v\} = V \qquad\qquad S' = V \setminus \{v\} \text{ but } S = V \setminus \{h\}$$

Case 3.2. $T \cap H \neq \emptyset$.

We define $T' := T \setminus \{v\}$ and $S := S' \cup \{v\}$. The correctness is trivial.

Case 4. $v \notin T$ and $T \setminus D(v,1) \neq \emptyset$.

With $T' := T$ we define $S := S'$. Assume there is a set $F \subseteq V$ containing T such that $G(F)$ is connected and F has a smaller number of vertices than S. Since S' is optimal in G' and v is not in T we conclude $v \in F$. Thus, $F' := F \setminus \{v\}$ includes T and cannot be connected. We proceed as in Case 2.

Theorem 6.3 *The Steiner tree problem on homogeneously orderable graphs can be solved in time $O(n^2)$ provided a h-extremal ordering is given.*

Proof : The algorithm steps through the given h–extremal ordering and computes an optimal solution S recursively as described above. The time bound follows from the fact that all added edges are edges of the square of G, see Lemma 6.1. □

7 Summary

In this paper we defined a new class of graphs which is a common generalization of distance–hereditary graphs, dually chordal graphs and homogeneous graphs. We presented a characterization of the new class in terms of a tree structure of the closed neighbourhoods of homogeneous sets in 2–graphs which is closely related to the defining h–extremal ordering.

Moreover, we gave a polynomial time solution for the recognition and the Steiner tree problem on homogeneously orderable graphs. Thus we obtain :

Class	Recognition		Steiner Tree	
trees	$O(n)$	folk	$O(n)$	folk
distance–hereditary graphs	$O(m)$	[12]	$O(m)$	[1]
dually chordal graphs	$O(m)$	[2, 8]	$O(n^2)$	[3, 8]
homogeneous graphs	polynomial	[7]	polynomial	[7]
homogeneously orderable graphs	$O(n^3)$	here	$O(n^2)$	here

We write $O(m)$ instead of $O(n+m)$ since any graph is connected in our paper.

References

1. A. BRANDSTÄDT and F.F. DRAGAN, A Linear–Time Algorithm for Connected r–Domination and Steiner Tree on Distance–Hereditary Graphs, *Technical Report* Gerhard–Mercator–Universität – Gesamthochschule Duisburg SM–DU–261, 1994.

2. A. BRANDSTÄDT, F.F. DRAGAN, V.D. CHEPOI and V.I. VOLOSHIN, Dually chordal graphs, 19th *International Workshop "Graph-Theoretic Concepts in Computer Science"* 1993, *Springer, Lecture Notes in Computer Science* 790 (Jan van Leeuwen, ed.) 237–251.

3. A. BRANDSTÄDT, V.D. CHEPOI and F.F. DRAGAN, The algorithmic use of hypertree structure and maximum neighbourhood orderings, *Technical Report* Gerhard–Mercator–Universität – Gesamthochschule Duisburg SM–DU–244, 1994, 20th *International Workshop "Graph-Theoretic Concepts in Computer Science"* 1994, to appear.

4. A. BRANDSTÄDT, V.D. CHEPOI and F.F. DRAGAN, Clique r–domination and clique r–packing problems on dually chordal graphs, *Technical Report* Gerhard–Mercator–Universität – Gesamthochschule Duisburg SM–DU–251, 1994.

5. A. BRANDSTÄDT, F.F. DRAGAN and F. NICOLAI, Homogeneously orderable graphs, *Technical Report* Gerhard–Mercator–Universität – Gesamthochschule Duisburg SM–DU–271, 1994.

6. A. D'ATRI and M. MOSCARINI, Distance–hereditary graphs, Steiner trees and connected domination, *SIAM J. Computing* 17 (1988) 521–538.

7. A. D'ATRI, M. MOSCARINI and A. SASSANO, The Steiner tree problem and homogeneous sets, MFCS '88, *Springer, Lecture Notes in Computer Science* 324, 249–261.

8. F.F. DRAGAN, HT–graphs: centers, connected r—domination and Steiner trees, *Computer Science Journal of Moldova*, 1993, Vol. 1, No. 2, 64–83.

9. F.F. DRAGAN and F. NICOLAI, r–domination problems on homogeneously orderable graphs, *Technical Report* Gerhard–Mercator–Universität – Gesamthochschule Duisburg SM–DU–275, 1995 (to appear in *Proceedings of FCT'95*).

10. P. DUCHET, Propriete de Helly et problemes de representation, *Colloq. Intern. CNRS 260*, Problemes Combin. et Theorie du Graphes, Orsay, France 1976, 117–118.

11. C. FLAMENT Hypergraphes arbores, *Discr. Math.* 21 (1978), 223–227.

12. P.L. HAMMER and F. MAFFRAY, Completely separable graphs, *Discr. Appl. Math.* 27 (1990), 85–99.

13. E. HOWORKA, A characterization of distance–hereditary graphs, *Quart. J. Math. Oxford* Ser. 2, 28 (1977) 417-420.

14. F. NICOLAI, A hypertree characterization of distance–hereditary graphs, *Technical Report* Gerhard–Mercator–Universität – Gesamthochschule Duisburg SM–DU–255, 1994.

15. F. NICOLAI, Hamiltonian problems on distance–hereditary graphs, *Technical Report* Gerhard–Mercator–Universität – Gesamthochschule Duisburg SM–DU–264, 1994.

List of Participants

Alimonti, Paola
Università di Roma "La Sapienza"
Dipt. di Informatica e Sistemistica
Via Salaria 113
I–00198 Roma
Tel. ++39 6 49918341
alimon@dis.uniroma1.it

Dr. d'Amore, Fabrizio
Università di Roma "La Sapienza"
Dipt. di Informatica e Sistemistica
Via Salaria 113
I–00198 Roma
Tel. ++39 6 49918333
damore@dis.uniroma1.it

Dr. Babel, Luitpold
TU München, Mathematisches Institut
D–80290 München
Tel. ++49 89 21058203
babel@statistik.tu–muenchen.de

Bauderon, Michel
Université Bordeaux I, LaBRI
F–33405 Talence
Tel. ++33 56845824
bauderon@labri.u–bordeaux.fr

Birchler, Barbara
Michigan State University
Comp. Science Dept.
A714 Wells Hall,
East Lansing, MI, 48824–1027, USA
Tel. ++1 517 353 8666
birchler@cps.msu.edu

Dr. Bodlaender, Hans
Dept. of Computer Science
University of Utrecht
P.O.Box 80.089
NL–3508 TB Utrecht
Tel: ++31 30 534409
hansb@cs.ruu.nl

Brandes, Ulrik
Universität Konstanz
Fakultät f. Mathematik/Inform.
Postfach 5560
D–78434 Konstanz
Tel. ++49 7531 884263
Ulrik.Brandes@uni–konstanz.de

Prof. Dr. Brandstädt, Andreas
Universität Rostock
Fachbereich Informatik
D–18051 Rostock
Tel. ++49 381 4983363
ab@informatik.uni–rostock.de

Dr. Chen, Zhi–Zhong
Tokyo Denki University
Dept. of Math. Sciences
Hatoyama–cho, Saitama 350–03, Japan
Tel. ++81 492 96 2911 ex. 2105
chen@r.dendai.ac.jp

Prof. Dr. Dahlhaus, Elias
University of Sydney
Basser Dept. of Computer Science
Sydney, NSW 2006, Australia
Tel. ++61 2 3513756
dahlhaus@cs.su.oz.au

PD Dr. Damaschke, Peter
FernUniversität Hagen
Theoretische Informatik II
D–58084 Hagen
Tel. ++49 2331 9872185
peter.damaschke@fernuni–hagen.de

Degiorgi, Daniele
Via Maraini 4
CH–6900 Massagno
Tel. ++41 1 6327371
degiorgi@inf.ethz.ch

Della Vedova, Gianluca
Università degli Studi Milano
Dipt. Scienze dell' Informazione
Via 4 Novembre 29a
I–23100 Sondrio SO
Tel. ++39 342 215151
dellavg@pippo.sm.dsi.unimi.it

Dr. Djidjev, Hristo
Rice University
Dept. of Computer Science
P.O.Box 1892
Houston, Texas 77251, USA
Tel. ++1 713 5278750 ext. 3246
hristo@cs.rice.edu

Dr. Dragan, Feodor
Moldova State University
Dept. of Mathematics and Cybern.
A. Mateevichi str. 60
Chisinau 277009, Moldova
Tel. ++ 373 2 240229
dragan@university.moldova.su or
@marvin.informatik.uni–duisburg.de

Fischer, Sophie
University of Amsterdam
Plantage Muidergracht 24
NL–1018 TV Amsterdam
Tel. ++31 20 5256508
sophie@fwi.uva.nl

Dr. Fracchia, David
Simon Fraser University
School of Computing Science
Burnaby, BC, V5A 1S6, Canada
Tel. ++1 604 2915692
fracchia@cs.sfu.ca

Franzke, Angelika
Universität Koblenz
Lehrstuhl für Informatik
Rheinau 1
D–56075 Koblenz
Tel. ++49 261 9119 440
franzke@informatik.uni–koblenz.de

Prof. Dr. Gambosi, Giorgio
Università di Roma "Tor Vergata"
Dipartimento di Matematica
Via della Ricerca Scientifica
I–00133 Roma
Tel. ++39 6 72594724
gambosi@mat.utovrm.it
gambosi@iasi.rm.cnr.it

Giaccio, Roberto
Università di Roma "La Sapienza"
Dipt. di Informatica e Sistemistica
Via Salaria 113
I–00198 Roma
Tel. ++39 6 49918341
giaccio@dis.uniroma1.it

Prof. Dr. Habib, Michel
LIRMM
161 rue Ada
F–34392 Montpellier
Tel. ++33 67418541
habib@lirmm.fr

Hartmann, Sven
Universität Rostock
Fachbereich Mathematik
D–18051 Rostock
Tel. ++49 381 4981533
sven@zeus.math.uni–rostock.de

Prof. Dr. Kaufmann, Michael
Universität Tübingen
W.–Schickard–Institut für Informatik
Sand 13
D–72076 Tübingen
Tel. ++49 7071 297404
mk@informatik.uni–tuebingen.de

Klein, Peter
RWTH Aachen
Lehrstuhl für Informatik III
D–52056 Aachen
Tel. ++49 241 8021316
pk@i3.informatik.rwth–aachen.de

Dr. Kloks, Ton
Eindhoven University of Technology
Dept. Math. and Computing Science
P.O.Box 513
NL–5600 MB Eindhoven
Tel. ++31 40 475152
ton@win.tue.nl

Dr. Kratsch, Dieter
F.–Schiller–Universität
Fak. für Mathematik und Informatik
D–07740 Jena
Tel. ++49 3641 630728
kratsch@minet.uni–jena.de

Krizanc, Danny
Carleton University
School of Computer Science
Ottawa, Ontario, K1S 5B6, Canada
Tel. ++1 613 7882600 ext. 4359
krizanc@scs.carleton.ca

Krumke, Sven Oliver
Universität Würzburg
Lehrstuhl für Informatik I
D–97074 Würzburg
Tel. ++49 931 8885057
krumke@informatik.uni–wuerzburg.de

Prof. Dr. Kucera, Ludek
Charles University
Dept. of Appl. Mathematics
Malostranske namesti 25
118 00 Prague, Czech Republic
Tel. ++422 24510286
ludek@kam.ms.mff.cuni.cz

Dr. Lauther, Ulrich
Siemens AG, ZFE T SE 1
Otto–Hahn–Ring 6
D–81730 München
Tel. ++49 89 63648834
lauther@zfe.siemens.de

Prof. Dr. Marchetti–Spaccamela,
Alberto
Università di Roma "La Sapienza"
Dipt. di Informatica e Sistemistica
Via Salaria 113
I–00198 Roma
Tel. ++39 6 49918327
alberto@dis.uniroma1.it

Prof. Dr. Mayr, Ernst W.
TU München, Institut für Informatik
D–80290 München
Tel. ++49 89 21052680
mayr@informatik.tu–muenchen.de

Prof. Dr. Möhring, Rolf
TU Berlin, Fachbereich Mathematik
Straße des 17. Juni 136
D–10623 Berlin
Tel. ++49 30 31424594
moehring@math.tu–berlin.de

Dr. Müller, Haiko
Friedrich–Schiller–Universität Jena
Universitätshochhaus 17. OG.
D–07740 Jena
Tel. ++49 3641 630728
hm@minet.uni–jena.de

Prof. Dr. Nagl, Manfred
RWTH Aachen
Lehrstuhl für Informatik III
D–52056 Aachen
Tel. ++49 241 8021300
nagl@i3.informatik.rwth–aachen.de

Dr. Nicolai, Falk
Universität Duisburg
Fachbereich 11/Informatik
D–47048 Duisburg
Tel. ++49 203 3792194
nicolai@marvin.informatik.
    uni–duisburg.de

Prof. Dr. Noltemeier, Hartmut
Universität Würzburg
Institut für Informatik I
Am Hubland
D–97074 Würzburg
Tel. ++49 931 8885055
noltemei@informatik.
    uni–wuerzburg.de

Dr. Olariu, Stephan
Old Dominion University
Dept. of Computer Science
Norfolk, Virginia 23529–0162, USA
Tel. ++1 804 6834417
olariu@cs.odu.edu,
ol_s@niniveh.ee.uwa.edu.au

Paul, Christophe
L.I.R.M.M.
161, rue Ada
F–34392 Montpellier cedex 5
Tel. ++33 67418578
paul@lirmm.fr

Prof. Dr. Pfaltz, John
University of Virginia
Dept. of Computer Science
Charlottesville, VA 22903, USA
Tel. ++1 804 9822222
pfaltz@cs.virginia.edu

Dr. Plump, Detlef
Universität Bremen
Fachbereich Math. und Informatik
Postfach 33 04 40
D–28334 Bremen
Tel. ++49 421 2184854
det@informatik.uni–bremen.de

Ruf, Berthold
TU Graz
Institute for Theor. Comp. Science
Klosterwiesgasse 32/2
A–8010 Graz
Tel. ++43 316 810063 24
bruf@igi.tu–graz.ac.at

Sass, Ron
Michigan State University
Comp. Science Dept.
A714 Wells Hall,
East Lansing, MI 48824, USA
Tel. ++1 517 355–8094
sass@cps.msu.edu

Dr. Sykora, Ondrej
Slovak Academy of Science
Instit. for Informatics
P.O.Box 56, Dúbravská 9
SK–840 00 Bratislava, Slovakia
Tel. ++42 7 811282
sykorao@savba.sk

Prof. Dr. Schiermeyer, Ingo
TU Cottbus
Lehrstuhl für Diskrete Math. und
Grundlagen der Informatik
D–3013 Cottbus
Tel. ++49 355 693195
schierme@math.tu–cottbus.de

Dr. Schürr, Andy
RWTH Aachen
Lehrstuhl für Informatik III
D–52056 Aachen
Tel. ++49 241 8021312
andy@i3.informatik.rwth–aachen.de

Prof. Dr. Takaoka, Tadao
Ibaraki University
Dept. of Computer Science
Hitachi 316, Japan
Tel. ++81 294 385130
takaoka@cis.ibaraki.ac.jp

Thilikos, Dimitris
University of Patras
Dept. of Comp. Science & Informatics
GR–26500 Patras, Greece
Tel. ++30 61 997704
SEDTHILK@CTI.GR

Prof. Dr. Tinhofer, Gottfried
TU München, Mathematisches Institut
D–80290 München
Tel. ++49 89 21058209
gottin@statistik.tu–muenchen.de

Viennot, Laurent
Universitè Paris, LITP/IBP
2, place Jussieu
F–75251 Paris Cedex 5
Tel. ++33 1 44272848
lavie@litp.ibp.fr

Dr. Voloshin, Vitaly
Moldova State University
Mateevici str. 60
Kishinev, 277099, Moldova
Tel. ++373 2 243904
voloshin@university.moldova.su

Dr. Vrt'o, Imrich
Slovak Academy of Science
Instit. for Informatics
P.O.Box 56, Dúbravská 9
SK–840 00 Bratislava, Slovakia
Tel. ++42 7 3783201
vrto@savba.sk

Dr. Wada, Koichi
Nagoya Institute of Technology
Dept. of Electrical and Computer Eng.
Gokiso–cho, Showa–ku
Nagoya, Aichi 464, Japan
Tel. ++81 52 7355438
wada@elcom.nitech.ac.jp

Westermann, Matthias
Universität–GH Paderborn
Eringerfelder Str. 30
D–59590 Geseke
Tel. ++49 2942 1635
marsu@uni–paderborn.de

Prof. Dr. Widmayer, Peter
ETH Zürich
CH–8092 Zürich
Tel. ++41 1 6327400
widmayer@inf.ethz.ch

Winter, Andreas
RWTH Aachen
Lehrstuhl für Informatik III
D–52056 Aachen
Tel. ++49 241 8021314
winter@i3.informatik.rwth–aachen.de

Dr. Zündorf, Albert
Universität GH Paderborn
FB 17 – Praktische Informatik
AG Softwaretechnik
D–33095 Paderborn
Tel. ++49 5251 602430
zuendorf@uni–paderborn.de

Authors' Index

List of WG Proceedings

Pape, U. (Ed.): *Graphen–Sprachen und Algorithmen auf Graphen*. 1. Fachtagung Graphentheoret. Konzepte der Informatik, Hanser, Munich, 1976, 236 pages, ISBN 3–446–12215–X.

Noltemeier, H. (Ed.): *Graphen, Algorithmen, Datenstrukturen*. Proc. WG 76, Graphtheoretic Concepts in Computer Science, Hanser, Munich, 1977, 336 pages, ISBN 3–446–12330–4.

Mühlbacher, J. (Ed.): *Datenstrukturen, Graphen, Algorithmen*. Proc. WG 77, Hanser, Munich, 1978, 368 pages, ISBN 3–446–12526–3.

Nagl, M./ Schneider H.–J. (Eds.): *Graphs, Data Structures, Algorithms*. Proc. WG 78, Hanser, Munich, 1979, 320 pages, ISBN 3–446–12748–3.

Pape, U. (Ed.): *Discrete Structures and Algorithms*. Proc. WG 79, Hanser, Munich, 1980, 270 pages, ISBN 3–446–13135–3.

Noltemeier, H. (Ed.): *Graphtheoretic Concepts in Computer Science* Proc. WG 80, Lect. Notes Comput. Sci. 100, Springer, Berlin, 1981, 403 pages, ISBN 0–387–10291–4.

Mühlbacher, J. (Ed.): *Proc. of the 7th Conf. Graphtheoretic Concepts in Computer Science (WG 81)*. Hanser, Munich, 1982, 355 pages, ISBN 3–446–13538–3.

Schneider, H.–J./ Göttler, H. (Eds.): *Proc. of the 8th Conf. Graphtheoretic Concepts in Computer Science (WG 82)*. Hanser, Munich, 1983, 280 pages, ISBN 3–446–13778–5.

Nagl, M./ Perl, J. (Eds.): *Proc. WG 83, Workshop on Graphtheoretic Concepts in Computer Science*. Trauner, Linz, 1984, 397 pages, ISBN 3–853–20311–6.

Pape, U. (Ed.): *Proc. WG 84, Workshop on Graphtheoretic Concepts in Computer Science*, Trauner, Linz, 1985, 381 pages, ISBN 3–853–20334–5.

Noltemeier, H. (Ed.): *Graphtheoretic Concepts in Computer Science*. Proc. WG 85, Trauner, Linz, 1986, 443 pages, ISBN 3–853–20357–4.

Tinhofer, G./ Schmidt, G. (Eds.): *Graph–Theoretic Concepts in Computer Science*. Proc. WG 86, Lect. Notes Comput. Sci. 246, Springer, Berlin, 1987, 305 pages, ISBN 0–387–17218–1.

Göttler, H./ Schneider H.–J. (Eds.): *Graph–Theoretic Concepts in Computer Science*. Proc. WG 87, Lect. Notes Comput. Sci. 314, Springer, Berlin, 1988, 254 pages, ISBN 0–387–19422–3.

van Leeuwen, J. (Ed.): *Graph–Theoretic Concepts in Computer Science*. Proc. WG 88, Lect. Notes Comput. Sci. 344, Springer, Berlin, 1989, 457 pages, ISBN 0–387–50728–0.

Nagl, M. (Ed.): *Graph–Theoretic Concepts in Computer Science*. Proc. WG 89, Lect. Notes Comput. Sci. 411, Springer, Berlin, 1990, 374 pages, ISBN 0–387–52292–1.

Möhring, M. (Ed.): *Graph–Theoretic Concepts in Computer Science*. Proc. WG 90, Lect. Notes Comput. Sci 484, Springer, Berlin, 1991, 360 pages, ISBN 0–387–53832–1.

Schmidt, G./ Berghammer, R. (Eds.): *Graph–Theoretic Concepts in Computer Science*. Proc. WG 91, Lect. Notes Comput. Sci. 570, Springer, Berlin, 1992, 253 pages, ISBN 0–387–55121–2.

Mayr, E. W. (Ed.) : *Graph–Theoretic Concepts in Computer Science*. Proc. WG 92, Lect. Notes Comput. Sci 657, Springer, Berlin, 1993, 350 pages, ISBN 0–387–56402–0.

van Leeuwen, J. (Ed.): *Graph–Theoretic Concepts in Computer Science*. Proc. WG 93, Lect. Notes Comput. Sci. 790, Springer, Berlin, 1994, 431 pages, ISBN 0–387–57889–4.

Mayr, E. W./ Schmidt, G./ Tinhofer G. (Eds.): *Graph–Theoretic Concepts in Computer Science*, Proc. WG 94, Lect. Notes Comput. Sci. 903, Springer, Berlin, 1995, 414 pages, ISBN 3–540–59071–4.

Nagl, M. (Ed.): Graph–Theoretic Concepts in Computer Science, Proc. WG 95, Springer, Berlin, this volume.

Lecture Notes in Computer Science

For information about Vols. 1–935

please contact your bookseller or Springer-Verlag